# Actinomycetes in Marine and Extreme Environments
## Unexhausted Sources for Microbial Biotechnology

*Editor*

**D. İpek Kurtböke**

School of Science, Technology and Engineering
University of the Sunshine Coast
Maroochydore BC, QLD
Australia

CRC Press
Taylor & Francis Group
Boca Raton London New York

CRC Press is an imprint of the
Taylor & Francis Group, an **informa** business

A SCIENCE PUBLISHERS BOOK

Cover illustrations reproduced by kind courtesy of the editor, D. İpek Kurtböke.

First edition published 2024
by CRC Press
2385 NW Executive Center Drive, Suite 320, Boca Raton FL 33431

and by CRC Press
4 Park Square, Milton Park, Abingdon, Oxon, OX14 4RN

*Library of Congress Cataloging-in-Publication Data (applied for)*

ISBN: 978-0-367-25280-9 (hbk)
ISBN: 978-1-032-52057-5 (pbk)
ISBN: 978-0-429-29394-8 (ebk)

DOI: 10.1201/9780429293948

Typeset in Times New Roman
by Radiant Productions

# Preface

The discovery, development and therapeutic use of antibiotics was one of the most significant advances in medicine in the 20th century that set the scene for development of effective drugs. Subsequent advances in fermentation technologies resulted in delivery of numerous potent antibiotics. The first golden era of antibiotics discovery lasted from the 1940s to the late 1960s and in this era actinomycetes most notably the members of genus *Streptomyces* proved to be a particularly rich source of structurally diverse and effective bioactive compounds for treatment of microbial infections. Global search for new and potent compounds from streptomycetes resulted in the establishment of the "*International Streptomyces Project* (ISP)", with collaboration of over 40 laboratories from 18 different countries resulting in the description of over 400 streptomycete species. In the mid-era of antibiotic discovery (1975–2000) focus shifted toward discovery of bioactive compounds from rare actinomycetes. They proved to be the sources of most diverse, unique, unprecedented and occasionally complicated compounds with excellent antibacterial potency and usually low toxicity. Several chemical types such as simple terpenoids or benzenoids were reported to be almost completely absent from these compounds and vancomycin-ristocetin type complicated glycopeptides were reported to be produced exclusively by various rare actinomycete species. Antibiotics like spinosyn and teicoplanin are the products of this era. In the new age of antibiotics (2000–2010) molecular advances and advance sampling techniques played a significant role and marine environments associated actinomycetes were purposefully targeted, examples include detection of *Salinispora* and subsequent discoveries of wide range of bioactive compounds like arenamides, arenicolides, arenimycins or salinisporamycin from the members of this genus. In the second stage of the new age of antibiotics (2010-ongoing) genomics, metagenomics and proteomics facilitating discoveries, and the examples include the use of genome sequencing to reveal complex secondary metabolome of *Salinispora* species.

Mathematical modelling and evolution studies indicate that actinomyetes will undoubtedly continue to play a significant role in antibiotic discovery. Rare actinomycetes inhabiting in those environments will undoubtedly be contributors to such discoveries. The book "Actinomycetes in Marine and Extreme Environments: Unexhausted Sources for Microbial Biotechnology" targets marine and extremophilic bioactive actinomycetes and contributed by experts in the field from four corners of the world. I thank all authors for their contribution and Emeritus Prof. Michael Goodfellow for his feedback and review.

# Contents

# Chapter 1

# Sponge Symbiotic Actinomycetes as Sources of Novel Bioactive Compounds

## Atlantic and Pacific Ocean Examples

Asmaa Boufridi,[1#,2] Candice M. Brinkmann,[1] Chandra Risdian,[3,4]
Joachim Wink[3] and D. İpek Kurtböke[1,*]

## 1. Introduction

### 1.1 Antibiotic Discovery

The discovery of penicillin from *Penicillium notatum* by Alexander Fleming in 1928 paved the way for a new era of treatment of bacterial infections (Drews 2000, Demain and Sanchez 2009). Chain, Florey and co-workers furthered Fleming's findings and introduced the world to a compound with life-saving abilities including protection against fatal streptococcal infections (Chain et al. 1940). This was followed with commercial penicillin production and the discovery of subsequent antibiotics such as streptomycin in 1943, followed by chloramphenicol and tetracycline (Davies 2006a, Kurtböke 2012). Natural products have also been one of the most successful chemotherapeutic tools in the treatment of infections associated with cancers such as acute myeloid leukaemia (treatment with high-dose cytarabine) (Flowers et al. 2013) and have contributed significantly to the control of many infectious diseases (Aminov 2010, Moloney 2016). However, although there have been many advances in the discovery of terrestrial origin microbial natural products over the last

[1] School of Science, Technology and Engineering, University of the Sunshine Coast, Maroochydore BC, QLD 4558, Australia.
[2] Griffith Institute for Drug Discovery, 46 Don Young Road, Nathan, QLD 4111, Australia.
[3] Microbial Strain Collection (MISG), Helmholtz Centre for Infection Research (HZI), 38124 Braunschweig, Germany.
[4] Research Unit for Clean Technology, Indonesian Institute of Sciences (LIPI), Bandung 40135, Indonesia.
* Corresponding author: ikurtbok@usc.edu.au
# Current address

60 years (Baltz 2006, Davies 2006b, Fenical and Jensen 2006) above listed antibiotics since the end of the golden era of the antibiotics (1940–1974) (Kurtböke 2012) have increasingly been rediscovered. Over 90% of all biologically active cultures yielded known compounds thus rendering the process inefficient and costly (Fenical 1993, Baltz 2006, Fenical and Jensen 2006). Furthermore, pathogenic microorganisms developed mechanisms to resist the effects of previously potent antibiotics. Such resistance mechanisms by pathogenic microorganisms developed either due to widespread use of antibiotics within communities (e.g., agricultural practices and human and veterinary therapeutics) and/or in hospitals, where constant selective pressures created "super bugs" (Lowy 2003, Chambers and DeLeo 2009, Sosa et al. 2010, Frieri et al. 2017).

## *1.2  Emergence of Antibiotic Resistance*

The appearance of microbial resistance mechanisms to key therapeutic agents is a result of either chromosomal changes or due to the exchange of genetic material through plasmids or transposons within the microbial communities (Neu 1992, Van Hoek et al. 2011, Zhuang et al. 2021). *Staphylococcus aureus*, which causes skin and soft tissue infections and life-threatening systemic infections, develops resistance mechanisms rapidly with the introduction of each new antibiotic (from penicillin and methicillin leading to more recent antibiotics, daptomycin, linezolid and vancomycin). These resistance mechanisms include: (i) antibiotic inactivation using enzymes, or (ii) the use of enzymes altering target receptors thus resulting in lower drug susceptibility, (iii) entrapment of antibiotics, (iv) development of efflux pumps (active protein transporters) (Walsh and Wright 2005) and (v) the development of spontaneous mutations (Lowy 2003, Pantosti et al. 2007, Zhang and Yew 2009, Peterson and Kaur 2018).

The first penicillin resistant *Streptococcus pneumoniae* strain was seen in Australia in 1967 and in the United States in 1974, by the 1980's 3–5% of such strains were penicillin resistant and by 1998 34% (Criswell 2004). The widespread misuse of antibiotics within communities and hospitals globally, is the main cause of resistance. Strains such as *Streptococcus pneumonia*, *S. pyogenes* and *Staphylococcus aureus* for instance that cause respiratory and cutaneous infections, are now resistant to all of the antibiotics that were previously used to treat these infections. Consequently, drug resistant pathogens such as *S. aureus* and MDR *Mycobacterium tuberculosis* (Chambers and DeLeo 2009, Zhang and Yew 2009, Frieri et al. 2017) flourish within the community and in nosocomial settings and opportunistic pathogens that include enterococci and staphylococci can also be detected. In addition, viral diseases remain a major health threat while severe complications result from antimicrobial resistance in immunocompromised patients (DeNegre et al. 2019). Clearly, new classes of antibiotics with different modes of actions are needed to control such drug resistant pathogens (Bush et al. 2011, Jackson et al. 2018).

## *1.3  Global Need for New Therapeutic Agents*

The renewed quest to find new novel biologically active natural compounds (bio-discovery) from a range of biological resources including the marine ones, is again gaining importance due to the emergence and re-emergence of multi-drug resistant (MDR) microorganisms (Fischbach and Walsh 2009, Dashti et al. 2014, Brinkmann et al. 2017a, Chokshi et al. 2019, Larsson and Flach 2022). Antibiotics have generally come from a small set of microbial scaffold molecules (Wright 2012) that have had their use extended by synthetic modifications. However, given the risk in the numbers of drug resistant pathogens the discovery of new molecular scaffolds has prime importance (Fischbach and Walsh 2009, Jackson et al. 2018). Molecular biological approaches, e.g., use gene expressed proteins/recombinant proteins to synthesise drugs (Fischbach and Walsh 2009) and gene therapies to counteract resistant pathogens in MDR cancers, to inhibit the growth of cancer cells and to prevent or control chronic diseases, such as acquired immunodeficiency syndrome (AIDS), asthma, diabetes and degenerative diseases, such as Alzheimer's (Walsh and Wright 2005, Senthilkumar and Kim 2013) have also been used.

Rediscory of known compounds from terrestrial organisms led to a change in drug discovery approaches, most notably in the introduction of combinatorial chemistry and computer-assisted drug design as alternatives to traditional drug discovery (Verdine 1996, Davies 2006b). However, nature has still been a superior supply of natural products (NPs) and drug discovery attempts utilizing NPs has re-emerged due to an increased understanding of the extent of biodiversity and the need for previously unknown chemical structures as new drug scaffolds (Newman and Cragg 2016). The emergence and re-emergence of multi-drug resistant (MDR) microorganisms have also played a role in this shift (Fischbach and Walsh 2009, Dashti et al. 2014, Brinkmann et al. 2017a, Chokshi et al. 2019, Larsson and Flach 2022).

## 2. Marine Environments: A Source of Biotechnologically Important Natural Products

The marine environment accounts for 70% of the earth's surface has been attracting more attention due to the vast biodiversity, extreme environmental conditions and the recent discoveries of novel bioactive small molecules from marine plants, invertebrates, and their associated microbial communities (Fenical and Jensen 2006, Kurtböke et al. 2015, Lyu et al. 2021). Furthermore, many algal and invertebrate phyla are found only in the sea and can be expected to have different biosynthetic pathways than terrestrial ones (Fenical and Jensen 2006, Leal et al. 2014).

In recent years, investigations of marine life and evidence of biochemical diversity and biotechnological potential has revealed how marine resources can be used to develop new novel compounds for pharmaceutical, nutraceutical and industrial applications (Bhatnagar and Kim 2010, Kiuru et al. 2014, Blockley et al. 2017, Lyu et al. 2021). Researchers observed that fouling by marine microorganisms on marine inanimate structures, as well as on marine animals and plants was not extensive as inanimate objects suggesting other defence mechanisms were involved (Pawlik 1993, Armstrong et al. 2000, Tian et al. 2020). Indeed, several antifouling compounds have been found to have antimicrobial activity responsible for the inhibition of causative microorganisms (Armstrong et al. 2000, Xu et al. 2012,

Satheesh et al. 2016). Given these observations it became clear that such marine natural products may be significant in the search for new antimicrobial compounds (Zotchev et al. 2016, Buedenbender et al. 2018, 2019, Kurtböke et al. 2020).

More than 15,000 natural compounds have been isolated from invertebrates (Datta et al. 2015) determining the identity of natural compounds has contributed significantly to the progression of these products into pre-clinical trials and beyond (Senthilkumar and Kim 2013). Many highly active anti-tumour compounds have been isolated from marine invertebrates, including eleutherobin (*Eleutherobia* corals), sarcodictyin (Stolonigeran coral) and bryostatins (Bryozoan-*Bugula neritine*) (Datta et al. 2015). An earlier comparative study based on statistical data from the United States (US) National Cancer Institute indicated that marine invertebrates are a preferred source of bioactive compounds due to a much higher level of cytotoxicity exhibited by such compounds (Garson 1994, Munro et al. 1999).

The isolation of the uncommon nucleosides spongothymidine and spongouridine from the sponge *Tethya crypta* in the 1950's which led to the production of cytarabine (anticancer agent) and vidarabine (antiviral agent) (Martins et al. 2014, Mehbub et al. 2014). Biologically active metabolites have also been isolated from algae, bryozoans, molluscs, corals, tunicates, ascidians, and microorganisms (Jensen et al. 2007, Tsukimoto et al. 2011, Senthilkumar and Kim 2013, Mehbub et al. 2014, Buedenbender et al. 2017, 2018) and shown to exhibit a range of medicinal properties including anti-cancer, anti-fungal, anti-malarial and antimicrobial activity (Abou-Elela et al. 2009, Bhatnagar and Kim 2010). Marine invertebrates (*Porifera*, *Bryozoa* and *Chordata*) have been the largest contributors to marine natural compound production with some compounds showing potent activity in a range of *in vitro* and *in vivo* assays. The phyla *Porifera* (sponges) and *Cnidaria* (corals, jellyfish) have been main sources of novel marine natural compounds (Mehbub et al. 2014). Sponges (*Porifera*) have provided significant number of novel bioactive metabolites than any other marine taxon (Taylor et al. 2007, Krishnan et al. 2014).

## 2.1 Sponges: Novel Sources of Biologically Active Compounds

Over 5,000 natural compounds have been discovered from sponges with over 200 new discoveries per year for the last ten years (Hentschel et al. 2006, Bell 2008, Sagar et al. 2010, Brinkmann et al. 2017a, b), thereby, contributing nearly 30% of all marine-derived compounds discovered to date (Mehbub et al. 2014). These potent natural compounds have demonstrated a range of pharmacological activities, including antimicrobial, anti-malarial, anti-inflammatory, anti-viral, anti-tumor, and antifouling properties (Atikana et al. 2013, Aguila-Ramirez et al. 2014) and have also been directly linked to secondary metabolites produced by sponge-symbiotic bacteria (Taylor et al. 2007, Atikana et al. 2013).

Sponge secondary metabolites account for 75% of anti-tumour products patent registrations and represent the main source for natural products undergoing clinical trials (Mehbub et al. 2014, Thomas et al. 2010). Examples of antibacterial substances from sponges and antimicrobial agents produced by sponge associated bacteria are shown in Tables 1 and 2.

**Table 1.** Examples of antibacterial substances extracted from marine sponges.

| Substance | Chemical class | Species | Activity spectrum |
|---|---|---|---|
| (S)-(+)-Curcuphenol | Sesquiterpene | *Myrmekioderma styx* | *Mycrobacterium tuberculosis* |
| 6-Hydroxymanzamine E | Alkaloid | *Acanthostrongylophora* sp. | *Mycrobacterium tuberculosis* |
| 8-Hydroxy-2-*N*-methylmanzamine D | Alkaloid | *Petrosia contignata Cribrochalina* sp. | *Mycrobacterium tuberculosis* |
| (+)-Aeroplysinin-1 | Brominated alkaloid | *Aplysina aerophoba* | *Bacillus cereus Bacillus subtilis Staphylococcus aureus Staphylococcus albus Vibrio anguillarum Flexibacter* sp. *Moraxella* sp. |
| (+)-Agelasidine C | Monocyclic sesquiterpenoid | *Agelas nakamurai Agelas dispar* | Antibacterial activity |
| Agelasine D | Purine derivative | *Agelas* sp. | Gram-positive bacteria, including *Mycobacterium tuberculosis* Gram-negative bacteria |
| Arenosclerins A-C | Alkaloids | *Arenosclera brasiliensis* | *Staphylococcus aureus* (MRSA strain) *Pseudomonas aeruginosa* (antibiotic-resistant strain) |
| Axinellamines B-D | Alkaloids | *Axinella* sp./ *Halichondrida* | *Helicobacter pylori* |
| (+)-Axisonitrile 3 | Nitrogenous sesquiterpenes | *Acanthella Axinella Axinyssa Ciocalypta Phyllidia* spp. *Topsentia* | Antimycobacterial activity |
| C14 acetylenic acid | Fatty acid | *Oceanapia* sp. | *Bacillus subtilis Staphylococcus aureus Escherichia coli Pseudomonas aeruginosa* |
| Caminosides A-D | Glycolipids | *Caminus sphaeroconia* | *Escherichia coli* |
| Corallidictyals A-D | Hydroquinones | *Aka coralliphaga* | *Staphylococcus aureus* |
| Crambescidin 800 | Alkaloid | *Clathria cervicornis* | *Acinetobacter baumannii Klebsiella pneumonia Pseudomonas aeruginosa* |
| Cribrostatin 3 | Alkaloid | *Cribrochalina* sp. | *Neisseria gonorrheae* (antibiotic-resistant strain) |
| Cribrostatin 6 | Alkaloid | *Cribrochalina* sp. | *Streptococcus pneumoniae* (antibiotic-resistant strain) |

*Table 1 contd. ...*

*...Table 1 contd.*

| Substance | Chemical class | Species | Activity spectrum |
|---|---|---|---|
| CvL | Lectine | *Cliona varians* | *Bacillus subtilis* *Staphylococcus aureus* |
| Cyclostellettamines A-I, K-L | Nitrogenous | *Pachychalina* sp. | *Mycrobacterium tuberculosis Pseudomonas aeruginosa* (antibiotic resistant strain) *Staphylococcus aureus* (MRSA strain) |
| Deoxytopsentin | Alkaloids | *Spongosorites* sp. | *Staphylococcus aureus* (MRSA strain) |
| 3,5-Dibromo-4-hydroxyphenylacetamide | Amide | *Aplysina fistularis* *Verongia archeri* | Antimicrobial agent |
| Dichloroverongiaquinol | Chlorinated *p*-quinol | *Aplysina cavernicola* *Aplysina fistularis* | Inhibits Gram-positive and Gram-negative bacteria |
| (+)-Discorhabdin D | Alkaloid | *Latrunculia brevis* *Prianos melanos* *Sceptrella* sp. | Antimicrobial agent |
| 5-Epiisospongiaquinone | Quinone | *Hippospongia* sp. *Spongia hispida* | Antibiotic activity |
| Fucosterol | Sterol | *Haliclona* spp. *Lissodendoryx noxiosa* *Stelletta clarella* *Tethya aurantia* | Moderate antibacterial activity |
| Haliclonacyclamine E | Alkylpiperidine alkaloids | *Arenosclera brasiliensis* | *Pseudomonas aeruginosa* (antibiotic-resistant strain) *Staphylococcus aureus* (MRSA strain) |
| Hamacanthin A | Alkaloids | *Spongosorites* sp. | *Pseudomonas aureus* (MRSA strain) |
| Heptaprenylhydroquinone | p-quinol | *Hippospongia communis* *Ircinia* sp. | Antibacterial activity |
| Ingenamine G | Nitrogenous | *Pachychalina* sp. | *Escherichia coli* *Mycrobacterium tuberculosis* *Pseudomonas aureus* (MRSA strain) |
| Isojaspic acid cacospongin D jaspaquinol | Meroditerpenes | *Cacospongia* sp. | *Staphylococcus epidermidis* |
| Kalihinene | Diterpene isonitrile | *Acanthella cavernosa* *Acanthella klethra* *Phakellia pulcherrima* | Inhibitor of bacterial folate biosynthesis |
| Latrunculins | Macrolides | *Negombata magnifica* | *Bacillus cereus* *Staphylococcus aureus* |

*Table 1 contd. ...*

*...Table 1 contd.*

| Substance | Chemical class | Species | Activity spectrum |
|---|---|---|---|
| (+)-Manzamine A | Alkaloid | *Haliclona* sp. *Pellina* sp. *Xestospongia* sp. | Gram-positive bacteria, including *Mycobacterium tuberculosis* Gram-negative bacteria |
| Manzamine B | Alkaloid | *Amphimedon* spp. *Haliclona* *Ircina* | *Mycrobacterium tuberculosis* |
| Manzamine D | Alkaloid | *Amphimedon* spp. *Haliclona* *Ircina* | *Mycrobacterium tuberculosis* |
| Melophlin C | Nitrogen heterocycles | *Melophlus sarassinorum* | *Bacillus subtilis* *Staphylococcus aureus* |
| Mimosamycin | Benzoquinone | *Halichondria* *Oceanapia* sp. *Petrosia* spp. *Reniera* *Xestospongia* | Mainly active against mycobacteria |
| Oroidin | Alkaloid | *Acantella* spp. *Agelas* spp. *Axinella* spp. *Hymeniacidon* sp. *Pseudaxinyssa cantharella* | Active against Gram-positive bacteria |
| Petrosamine B | Alkaloids | *Oceanapia* sp. | *Helicobacter pylori* |
| Polydiscamide A | Peptide | *Discodermia* sp. | *Bacillus subtilis* |
| Polygodial | Sesquiterpene dialdehyde | *Dysidea* spp. | Antibacterial activity |
| Psammaplin A | Bromotyrosine-derived | *Psammaplysilla* | *Staphylococcus aureus* (MRSA strain) |
| Smenospongine | Quinoterpenoid | *Smenospongia* sp. *Dactylospongia elegans* *Petrosaspongia metachromia* | Antimicrobial agent |

Data adapted from Laport et al. 2009 and Dictionary of Natural Products 2020.

Although a large number of sponge compounds show promising activity as potential therapeutic agents (Perdicaris et al. 2013) the ability to obtain continuous large supplies of novel compounds from sponges has been challenging (Belarbi et al. 2003, Gandhimathi et al. 2008, Perdicaris et al. 2013), as only a few milligrams of the compounds can usually be isolated from the sponge samples (Mehbub et al. 2014). In the last decade some of these issues have been overcome using techniques such as aquacultural (increase sponge biomass), isolation and fermentation of metabolite producers, and genomics (Banik and Brady 2010). When sponge symbiotic

**Table 2.** Examples of antimicrobial substances produced by sponge-associated bacteria.

| Class | Bacteria | Sponge | Activity spectrum |
|---|---|---|---|
| Alkaloids | *Pseudomonas aeruginosa* | *Isodictya setifera* | *Bacillus cereus* *Micrococcus luteus* *Staphylococcus aureus* |
| Alkaloids | *Pseudomonas* sp. | *Hymeniacidon perleve* | Antimicrobial activities |
| Alkaloids | unknown | *Ircinia variabilis* | Cell-cell signaling mechanism |
| Amicoumacins | *Bacillus pumillus* | *Dendrilla* sp. | Antibacterial activities |
| Ansamycin family | *Salinispora* sp. | *Pseudoceratina clavata* | Gram-positive bacteria |
| Benz[α]anthracene | *Streptomyces* sp. HB202 | *Halichondria panicea* | Methicillin-resistant *Staphylococcus aureus* (MRSA) *Pseudomonas aeruginosa* |
| Benzo[α]naphthacene quinones | *Streptomyces* sp. strain RV15 | *Dysidea tupha* | Gram-positive bacteria Mycoplasmas Gram-negative bacteria |
| Bromo-phenolic | *Vibrio* sp. | *Dysidea* sp. | *Staphylococcus aureus* |
| Cyclic lipopeptides | *Bacillus subtilis* | *Aplysina aerophoba* | *Agrobacterium tumefaciens* *Bacillus megaterium* *Candida albicans* *Clavibacter michiganensis* *Escherichia coli* *Paecilomyces variotii* *Proteus vulgaris* *Saccharomyces cerevisiae* *Staphylococcus* *Vibrio* sp. |
| Cyclic peptides | *Ruegeria* strain | *Suberites domuncula* | *Bacillus subtilis* |
| Cyclic thiopeptides | *Bacillus cereus* | *Halichondria japonica* | *Staphylococcus* *Enterococcus*-antibiotic-resistant strains |
| Glycoglycerolipids | *Micrococcus luteus* | *Xestospongia* sp. | *Staphylococcus aureus* |
| Lactones | *Streptomyces* sp. | *Tethya* sp. | Antimicrobial activities |
| Macrolactams | *Saccharopolyspora cebuensis* SPE 10-1 | *Haliclona* sp. | Antibacterial activities |
| Manzamines | *Micromonospora* sp. | *Petrosiidae* | Antimicrobial activities |
| Phenazines | *Brevibacterium* sp. KMD 003 | *Callyspongia* sp. | *Enterococcus hirae* *Micrococcus luteus* |
| Polyketide-Peptide | *Vibrio* sp. | *Hyatella* sp. | *Bacillus* sp. |

*Table 2 contd. ...*

*...Table 2 contd.*

| Class | Bacteria | Sponge | Activity spectrum |
|---|---|---|---|
| Pyrones | *Pseudomonas* sp. | Marine sponge | *Bacillus subtilis* *Candida albicans* *Enterococcus faecium* *Escherichia coli* *Staphylococcus aureus* *Streptococcus pneumoniae* *Moraxella catarrhalis* |
| Quinolones | *Pseudomonas* sp. | *Homophymia* sp. | *Plasmodium falciparum* HIV-1 *Staphylococcus aureus* |
| Quinolones | *Pseudomonas* sp. | *Suberea creba* | *Staphylococcus aureus* |
| Tetromycins | *Streptomyces axinellae* Pol001T | *Axinella polypoides* | Gram-positive bacteria, including methicillin-resistant *Staphylococcus aureus* (MRSA) |

Data adapted from Laport et al. 2009, Abdelmohsen et al. 2014a, and Dictionary of Natural Products 2020.

microorganisms were found to be responsible for the production of an important compound, scaled up process have been easier as microorganisms can flourish independently of the sponge (Mehbub et al. 2014). Currently the screening of sponge associated microorganisms in pure culture is reported to be the most effective way of discovering novel bioactive compounds (Graça et al. 2013).

## 2.2 Supply of Natural Products Produced by Sponge Symbionts

Bacterial numbers within sponges are usually two to four folds higher than that of the surrounding seawater (Laport et al. 2009, Hentschel et al. 2012, Mehbub et al. 2014) and due to their abundant microbial community, sponges are seen as 'fermenters' (Hentschel et al. 2003) that hold a variety of potentially novel natural products produced by symbiotic microorganisms (Hentschel et al. 2012, Müller 2012). When natural compounds are produced only under conditions specific to the marine environment (e.g., certain salt concentration, hydrostatic pressure and marine nutrients) standard laboratory fermentation conditions may not be conducive to the expression of microbial biosynthetic pathways and the subsequent production of potentially novel compounds (Fenical 1993, Jensen and Fenical 1994). The mass fermentation of sponge-symbiotic microorganisms however, is a highly promising and stable option for discovering novel compounds from sponges (Jimenez et al. 2009, Brinkmann et al. 2017b) while conserving natural sponge populations (Kumar et al. 2012). Therefore, examination of ideal submerged fermentation parameters is important as they help to improve methods for large-scale compound production from microorganisms (Jimenez et al. 2009) and offer an optimised medium for the identification of secondary metabolite producers (Thomas et al. 2010, Buedenbender et al. 2019). The production of secondary metabolites by the members of microbial community is highly dependent on the fermentation conditions and different parameters are needed to be tested to determine optimal fermentation conditions for sponge symbiotic bacterial cultures as bioactive compound production varies from strain to strain (Graça et al. 2013).

Sponges provide a highly suitable environment for housing microorganisms in terms of nutritional requirements; they produce ammonia as a waste product available to bacteria as a nitrogen source and digested microorganisms provide carbon sources (Hentschel et al. 2006). However, only a small fraction of sponge symbiotic bacteria have been cultivated in the laboratory as most sponge symbioants may require a range of nutrients at varying concentrations in the growth media in order to be cultivated (Webster et al. 2001). Once successfully cultivated the correct parameters for fermentation optimisation needs to be conducted to be able to get novel compounds produced by these microorganisms. Limiting nutritional components of the media used and introducing stressful parameters may enhance the production of microbial natural compounds (Finore et al. 2014) as well as using ingredients similar to natural sponge environment. Examples include, the use of varying concentrations of sodium chloride (Jensen and Fenical 1994), glucose as a carbon source, chitin, yeast extract, sulphated polysaccharides, marine proteins, seawater (Fenical 1993, Tasharrofi et al. 2011) and other nutrients, such as *Laminaria* (also known as Kobu-cha; powdered Japanese seaweed containing iodine) (Fenical 1993, Bernan et al. 1997). Conditions such as pH, incubation period, inoculum size and temperature should also be monitored and adjusted accordingly (Kiruthika et al. 2011, Abdelwahed et al. 2014). Designed growth media, such as, Zobell Marine Broth (ZMB) (Devi et al. 2010) and SYP-SW broth (Kennedy et al. 2009), have been trialled for metabolite production and have allowed observations of some compounds that may be produced under routine laboratory conditions. There is however, little understanding of how inorganic elements such as lithium and silicon (abundant in marine sediments) might effect compound production (Fenical 1993). Moreover, investigations into triggering bacterial cryptic biosynthetic pathways for the enzymatic expression of potentially novel compounds would be of interest Further research into fermentation media is required, e.g., trialling a range of organic and inorganic elements at varying concentrations for the expression of biosynthetic genes that are involved in the synthesis of novel compounds (Pan et al. 2019).

## 3. Genes and Enzymes Involved in Metabolite Production

Microbial polyketide synthase (PKS) and non-ribosomal peptide synthetase (NRPS) genes are of considerable interest due to their apparent involvement in the synthesis of a wide range of natural marine products isolated from marine invertebrates (e.g., sponges) (Hochmuth and Piel 2009). Marine natural product synthesis involve multi-functional enzymes that are also known to be present in the synthesis of terrestrial products. However, marine natural products also possess a range of functional groups that have not previously been described for terrestrial metabolites. The difference in structures of these compounds are reported to be a result of biosynthetic pathways that use completely new enzymes (Jimenez et al. 2009).

Genome sequencing technologies have advanced greatly in the last decade and multiple genomes can now be efficiently sequenced. Sequencing of *Streptomyces* genome revealed that strains contain genes that encode enzymes predicted to synthesize more than 20 secondary metabolites that are not always seen under standard fermentation conditions (Ochi and Hosaka 2013). Sequencing the genome of marine-derived *Salinispora tropica*, has also revealed a complex secondary metabolome and the most diverse assemblage of polyketide biosynthetic mechanisms observed in a single organism (Udwary et al. 2007). The diversity of PKS genes is also important even though they may not contribute to the production of a complete bioactive compound, such diverse genes and gene fragments can provide the building blocks for combinatorial polyketide synthesis through gene recombination (Kim and Fuerst 2006). Cloning and heterologous expression of gene clusters encoding novel secondary metabolites may thus assist in creating an efficient supply route and may thus aid in the molecular engineering of new therapeutic agents (Jimenez et al. 2009).

The analysis of secondary metabolism in microbial genomes can be done using *in silico* methods, which are based on identification of biosynthetic gene clusters (BGCs) (Gaudêncio and Pereira 2015). A BGC is a physically clustered group of several genes in a genomes responsible in the biosynthesis of a metabolite. The analysis of BCGs for secondary metabolites in microbial genome data is primarily based on homology with previously known BCGs. This homology can be determined from the classification of biosynthetic pathway types, structure of domains, conserved motifs, hidden Markov models and chemical structural classes of the compound (Medema et al. 2015). The two best known programs for conducting such analysis, are antiSMASH (Medema et al. 2011) and PRISM (Skinnider et al. 2015).

## 4. Quorum Sensing: Communication within the Microbial Community

Quorum sensing is a small compound signalling between microorganisms, including bacteria and some fungi which allow regulation of gene expression in a population density-dependent manner. Communication is carried out by a bacterial signalling system, the acyl homoserine lactone (AHL) regulatory system (Fuqua and Greenberg 2002). Bacteria capable of producing AHLs have been reported in sponge species. One study found that 24 out of 31 sponge species had traces of the small signalling molecule (SSM) N-[AHL] present in extracts of sponge tissue samples (Taylor et al. 2004). This system is used to mediate the colonising traits of the bacterial community, such as virulence, congregation and biofilm formation (Marketon et al. 2003, Quiñones et al. 2005, González and Keshavan 2016) which are essential for establishment of symbiotic or pathogenic relationships with their host organism (Sperandio 2004). This is of particular interest for research, as symbiont mediated production of novel secondary metabolites may be influenced by SSM concentrations. Trialling the addition of quorum sensing SSMs (or co-cultivation for bacterial communication) within fermentation media could be of benefit by triggering biosynthetic pathways and subsequent secondary metabolite production (Reen et al. 2017).

## 5. Sponge-Specific Microbial Communities

Microorganisms representing the three domains of life- *Bacteria, Archaea* and *Eukarya-* are known to be associated with sponges with bacteria being the most diverse of the three (Simister et al. 2012, Brinkmann et al. 2017a, b). Sponge-symbiotic microbial communities make up 35–60% (Webster and Taylor 2012, Aguila-Ramirez et al. 2014) of the total sponge mass (Anand et al. 2006, Taylor et al. 2007) with densities ranging from $\times 10^9$ microbial cells per mL of sponge tissue (Hentschel et al. 2012). Based on both cultivation-dependent and -independent techniques Twenty-eight bacterial phyla have been reported to be associated with sponges (Hentschel et al. 2012, Trindade-Silva et al. 2012, Hentschel et al. 2012, Abdelmohsen et al. 2014a). Phylogenetic analysis of biologically active compound producing taxa include *Actinobacteria, Acidobacteria Alphaproteobacteria, Bacteriodetes, Betaproteobacteria, Chloroflexi, Cyanobacteria, Fumicates, Gammaproteobacteria, Proteobacteria, 'Poribacteria'*, and *Verrucomicrobia*, some being sponge-specific according to the phylogenetic analyses performed on sponge derived bacterial sequences (Müller 2012, Graça et al. 2013).

Several 16S rDNA libraries provide useful information about sponge-symbiotic microbial communites such as the detection of phylogenetically diverse microbial communities from *Aplysina aerophoba, Rhopaloeides ordorabile*, and *Theonella swinhoei* that were sourced from the Mediterranean Sea, the Great Barrier Reef, and Red Sea (Taylor et al. 2007, Abdelmohsen et al. 2014b). Sponge specimens collected from marine environments in different countries (e.g., France, Japan and Egypt) were also found to contain a diverse range of microorganisms that were different from previously described 16S rDNA signatures from 600 species living in typical seawater thereby indicating sponge specifity (Hentschel et al. 2012). Furthermore, ribosomal tag pyrosequencing on bacterial colonies from three different sea sponges from the Red Sea as well as surrounding sea water indicated extensive diversity between the bacterial communities within sponges from different locations. These results support the view that many bacterial communites are sponge specific (Lee et al. 2011). Sponge symbionts are transferred by vertical transmission (passed on through sponge larvae) (Webster and Blackall 2009, Hentschel et al. 2012, Müller 2012) as well as by horizontal transmission (release of symbionts into the surrounding seawater due to host tissue damage then taken up by neighbouring sponges) (Schmitt et al. 2008, Hentschel et al. 2012, Sipkema et al. 2015).

A 454-tag sequencing study of sponges collected from the Great Barrier Reef showed that the main bacterial taxa were *Acidobacteria, Actinobacteria, Chlorofelxi*, and *Alpha-, Delta-* and *Gammaproteobacteria* (Webster and Taylor 2012). The total cultivable community of the sponge *Candidaspongia flabellata* consisted of 228 bacteria, 25 fungi, 9 cyanobacteria and 3 actinomycetes. The 9 cyanobacteria and 8 eubacteria were consistently associated with *C. flabellata* but were absent from the surrounding seawater (Burja and Hill 2001).

## 6. Fermentation Optimization of Sponge-Symbiotic Microorganisms

The Plackett-Burman (PB) design is a highly efficient method of identifying suitable variables needed for optimal fermentation (Treichel et al. 2010, Tasharrofi et al. 2011). It uses a two-level fractional factorial design and identifies critical nutrient and variables (Tasharrofi et al. 2011, Sathiyanarayanan et al. 2014) needed for elevated growth and optimal production of bioactive compounds. However, it does not consider interactions between variables, therefore, this design may be further supported by using Response Surface Methodology (statistical and mathematical optimisation tool) (Tasharrofi et al. 2011).

The mechanism of co-fermentation of microorganisms may also trigger the expression of biosynthetic pathways encoding gene clusters involved in the production of important metabolites (Betrand et al. 2014). Many biosynthetic pathways are silent and are not expressed under routine laboratory conditions. This greatly limits the chemical diversity of microbial compounds produced through laboratory fermentation. Inoculating fermentation media with two or more microorganisms may mimic the sponges' natural environment where microorganisms co-exist within residing microbial communities and may induce a competitive environment (Marmann et al. 2014) ultimately leading to secondary metabolite production (Dashti et al. 2014, Bertrand et al. 2014). Competition experienced by the co-cultivation of microorganisms has been shown to induce significantly greater production of compounds and the accumulation of cryptic compounds (compounds usually unexpressed under fermentation conditions with individual strains) (Marmann et al. 2014).

It is increasingly becoming evident that sponge associated bacteria are of immense potential for biodiscovery and industrial applications. To recover these industrially important products, the physiological and metabolic capabilities of the producing organisms need to be understood to be able to trigger previously unknown bioactive compound secretion. Therefore, there is a need to obtain a better understanding of the interactions in real sponge-microbial environment to be able to mimic these conditions within the laboratory. From these optimised fermentation conditions, it is possible to scale up the supply of compounds of interest by large-scale fermentations of the producer microorganisms. Furthermore, sequencing of the entire microbial genome may lead to the detection of cryptic genes that may be used to produce intact bioactive compounds through recombination methods.

## 7. Sponge-Symbiotic Actinomycetes

Members of the proposed Class *Actinomycetia* (Salam et al. 2020) of the Phylum *Actinomycetota* (Oren and Garrity 2021) (formerly Actinobacteria, Nouioui et al.

2018), commonly known as actinomycetes and have been frequently reported to be present in sponge microbiome, including rare and bioactive genera such as *Actinokineospora* (Dashti et al. 2014), *Nocardiopsis* (Selvin et al. 2009) and *Salinispora* (Bose et al. 2017), although most actinomycete genera isolated from marine sponges generally classified in the suborder Micrococcineae like *Arthrobacter*, *Microbacterium* and *Micrococcus*. However, because of the random distribution of actinomycetes in host sponges, it is difficult to establish a clear pattern that explains the specific host-symbiont relationship between marine sponges and actinomycetes (Abdelmohsen et al. 2014a).

Actinomycete species, especially members of the genus *Streptomyces*, are known for their ability to yield diverse biologically active secondary metabolites (Watve et al. 2001, Kurtböke 2012). Bioactive compounds produced by actinomycetes include antimicrobial agents (e.g., activity against bacteria, fungi, viruses, and protozoans), as well as producing cytotoxic compounds, and antioxidants. During sponge symbiosis these compounds may also have a protective role for their hosts such as protecting sponge from pathogenic bacteria and eukaryotic antagonists (Schneemann et al. 2010). Actinomycetes may also gain benefits from their sponge hosts (Pita et al. 2018) such as nutrition and protection.

In the light of the above presented information the chapter provides examples from Pacific and Atlantic Oceans highlighting bioactive potential of sponge symbiotic actinomycetes.

### 7.1 Pacific Ocean Example: Actinomycetes Associated with Candidaspongia flabellate *and* Rhopaloeides odorabile

In contrast to *Micrococcus* species bioactive compounds have rarely been isolated from *Kytococcus* and *Sanguibacter* species associated with sponges. However, bioactive compound search from sponge symbiotic *Kocuria* species are on the increase (Brinkmann et al. 2017b). Examples include isolation of *Kocuria* species from sponge samples in Australia (*Candidaspongia flabellate*) (Burja and Hill 2001), the South China Sea (*Hymeniacidon perleve*) (Xi et al. 2012) and the United States (*Halichondria panicea*) (Christensen and Martin 2017). Detection of *Kocuria* spp. From marine environments is of interest as *K. rhizophila* strain DC2201 has been reported to carry the smallest actinomycete genomes (2.7 Mb) (Takarada et al. 2008). The genome of this strain contains a limited number of genes that expressed for secondary metabolite pathways, in addition PKS-III and NRPS genes are detected. The genome of this strain encodes only smaller numbers of proteins necessary for secondary metabolism, transcriptional regulation, and lateral gene transfer (Takarada et al. 2008). The fact that *Kocuria* genome is relatively small compared with other actinomycetes might suggest that *Kocuria* species are high adaptable to their ecological niches which might include the sponge hosts. During such adaptation they might produce bioactive compounds as they alter their metabolic activities.

Thiazolyl peptide antibiotics are a class of natural products that have been shown to produce *in vitro* activity against Gram-positive bacteria (Palomo et al.

2013, Linares-Otoya et al. 2017). Kocurin a thiazolyl peptide that was first isolated from sponge associated *Kocuria palustris* (Palomo et al. 2013) and has an MIC value of 0.24 µg/mL against methicillin-resistant *Staphylococcus aureus* MB5393 (Martin et al. 2013). The different subunit structures of kocurin were identified by NMR and the sequence of the amino acids established by HMBC and MS/MS. Kocurin subsequently, was isolated from a strain of *Micrococcus* (Palomo et al. 2013). Both *Kocuria* and *Micrococcus* strains produced kocurin under one out of 12 fermentation conditions tested (Palomo et al. 2013). Apart from the kocurin (Fig. 1) other compounds known to be synthesized by *K. rhizophila* include carbapenem, monobactam, streptomycin and novobiocin. In addition, although not sponge derived a new *Kocuria* strain isolated from a brown macroalgae yielded a new compound, kocumarin (2), which showed an anti-MRSA activity at an MIC value of 10 µg/disk (Uzair et al. 2018).

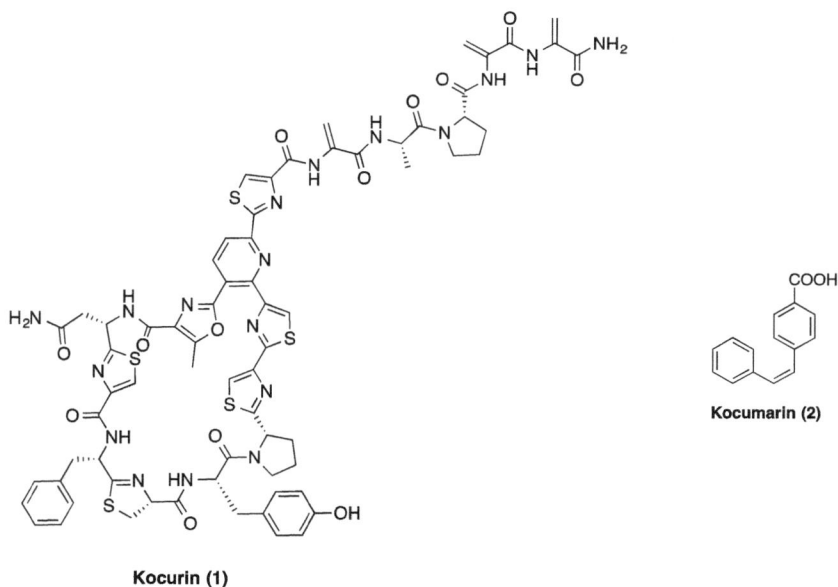

Kocurin (1)

Kocumarin (2)

**Figure 1.** Chemical structures of kocurin (1) and kocumarin (2) (from Dictionary of Natural Products 2020).

Chemical investigations on a marine *Micrococcus* resulted in the isolation of different classes of natural products. Diketopiperazines (3–5) (Fig. 2) have been isolated from the sponge *Tedania ignis* (Schmitz et al. 1983), subsequent studies

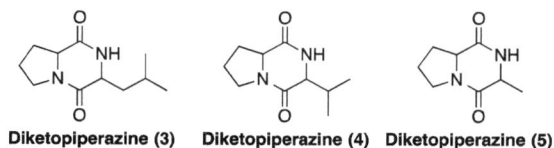

Diketopiperazine (3)   Diketopiperazine (4)   Diketopiperazine (5)

**Figure 2.** Chemical structures of diketopiperazines (3–5) (from Dictionary of Natural Products 2020).

on a *Micrococcus* strain associated with this sponge also yielded these natural products, thereby confirming the true origin of the bioactive compound (Stierle et al. 1988).

Sponge symbiotic bacteria from the microbial collection of the *Australian Institute of Marine Science* (AIMS) (https://www.aims.gov.au/) originating from two different Australian sponge species of the Great Barrier Reef (*Candidaspongia flabellate* (105 isolates) and *Rhopaloeides odorabile* (18 isolates)) were investigated for their bioactivity at the University of the Sunshine Coast (Brinkmann et al. 2017b) and the structure elucidation of the chemical compounds were determined at the Griffith Institute for Drug Discovery.

Phylogenetic analysis based on 16S rRNA gene sequencing of these isolates revealed actinomycete species were present among the bacterial isolates (Brinkmann et al. 2017b). Members of the genera, *Kocuria, Kytococcus Micrococcus* and *Sanguibacter* were isolated from *C. flabellate*, whereas only the members of the genus *Micrococcus* were isolated from *R. odorabile* (Brinkmann et al. 2017b). *Sanguibacter* (AIMS-53634) and *Micrococcus* (AIMS-50899) strains when co-cultured in M11 medium (Wael 2006) yielded previously described compounds ¹H-indole-3-carboxaldehyde (6) and pyriculamide (7) (Fig. 3) (Gómez-Betancur et al. 2019, Vaca and Chàvez 2019) using ¹H-NMR guided isolation (Camp et al. 2012). The results provided further evidence for the bioactive potential of rare actinomycetes associated with sponges when different fermentation parameters are used.

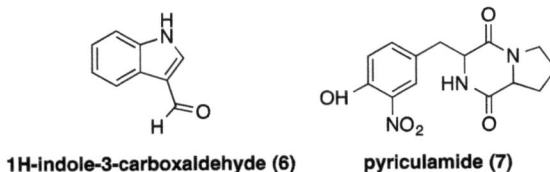

**1H-indole-3-carboxaldehyde (6)**          **pyriculamide (7)**

**Figure 3.**  Chemical structures of ¹H-indole-3-carboxaldehyde (6) and pyriculamide (7) (from Dictionary of Natural Products 2020).

*Kocuria* sp., previously has been shown to have antimicrobial activity (Palomo et al. 2013). Similarly, the *Kocuria* strain (AIMS-53209) showed high antimicrobial activity, as well as the presence of genes encoding for non-ribosomal peptide synthetases (NRPSs) were detected (Brikmann et al. 2017b). To activate the silent genes in *Kocuria* strain (AIMS-53209) and obtain antimicrobial compounds, *One Strain Many Compounds* (OSMAC) method (Bode et al. 2002) was used. Changes in the cultivation conditions and/or co-cultivation are known to effectively shift the metabolic profile of bacterial isolates (Liu et al. 2017, Newman et al. 2017). *Kocuria* strain (AIMS-53209) fermented on a nutritionally poor R2A agar (Reasoner and Geldreich 1985) was found to exhibit a broad range of antimicrobial activity.

HPLC fractionation, ¹H-NMR fingerprint and antimicrobial assays on the fractions lead to the isolation and purification of 2-hydroxyphenazine (8) (Fig. 4).

Figure 5 shows a comparison of ¹H-NMR fingerprints of the active fraction of the *Kocuria* strain (AIMS-53209) fermented on R2A agar The structure of the

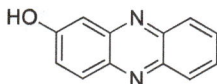

**2-hydroxyphenazine (8)**

**Figure 4.** Chemical structure of of 2-hydroxyphenazine (8) (from Dictionary of Natural Products 2020).

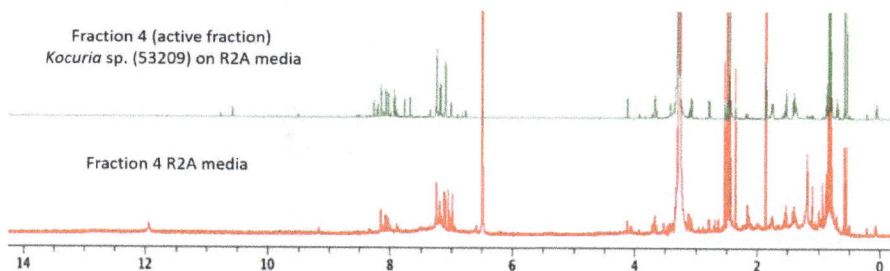

Fraction 4 (active fraction)
*Kocuria* sp. (53209) on R2A media

Fraction 4 R2A media

**Figure 5.** ¹H-NMR fingerprint of fraction 4, from *Kocuria* strain AIMS-53209 on R2A agar (green), R2A agar (red).

active fraction was confirmed by NMR and MS analysis as well as comparison with the published data (Pierson and Thomashow 1992). This is the first record of 2-hydroxyphenazine (8) detection from a *Kocuria* strain when solid state fermentation was used.

When tested against a panel of pathogenic bacterial reference strains compound **8** was found to be active against *Aspergillus niger* (ATCC® 16888™), *Bacillus subtilis* (ATCC® 19659™), *Candida albicans* (ATCC® 10231™) *Enterococcus faecalis* (ATCC® 51575™), *Escherichia coli* (ATCC® 25922™), *Klebsiella pneumoniae* (ATCC® BAA-1705™) and *Staphylococcus aureus* (ATCC® 29247™) with MIC between 20–100 µg/mL.

## 7.2 Atlantic Ocean Example: Streptomyces bathyalis

*Streptomyces* strain (ASO4wetᵀ) (Fig. 6), which has been proposed as the novel species *Streptomyces bathyalis*, was isolated from an unidentified sponge (SO4) collected from the deep-sea of the North Atlantic Ocean near Madeira Island (Risdian et al. 2021). This halotolerant actinomycete shares 16S rRNA gene sequence similarities within the range 97.53%–98.87% with the type-strains *Streptomyces karpasiensis*, *Streptomyces glycovorans*, and *Streptomyces abyssalis*. The *S. karpasiensis* was isolated from a soil sample taken from Cyprus (Veysioğlu et al. 2014) and the others from marine sediments collected from the deep-sea of the South China Sea (Xu et al. 2012).

*S. bathyalis* (Fig. 6) has a moderately high genome size (7,377,472 bp) a digital G+C content of 70.24 mol and contains 6,332 coding sequences, 59 tRNA genes, and six rRNA operons. Four out of the 23 secondary metabolite biosynthesis gene clustersdetected in the genome showed 100% identity with bioclusters coding for planosporicin (9), geosmin (10), ectoine (11), and desferrioxamine E (12) (Risdian et al., 2021).

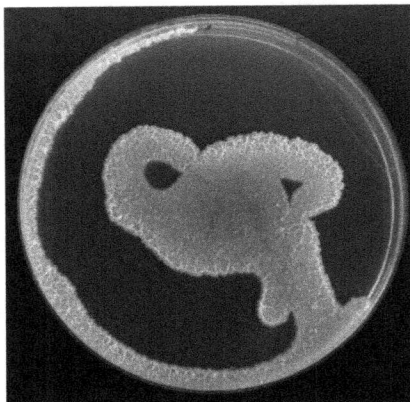

**Figure 6.** Growth morphology of *Streptomyces bathyalis* ASO4wet[T] growing on ISP2 agar after 10 days of incubation at 30°C.

Planosporicin (9) (Fig. 7) is a lantibiotic isolated from a *Planospora* strain is a polypeptide (size 2194 Da) which contains 24 proteinogenic amino acid residues and inhibits peptidoglycan biosynthesis (Maffioli et al. 2009, Castiglione et al. 2007), geosmin (10) is a sesquiterpene alcohol a volatile terpene produced by *Nocardia* and *Streptomyces* strains (Jüttner and Watson 2007) and ectoine (11) amino acid derivatives first isolated from *Ectothiorhodospira halochloris* (Fig. 8) which helps overcome osmotic stress in halophilic and halotolerant microorganisms, including streptomycetes (Galinski and Trüper 1994, Sadeghi et al. 2014) and desferrioxamine E (12), a siderophore compound, which promotes the growth by acquiring environmental ferric ion which enhances secondary metabolite production (Yamanaka et al. 2005).

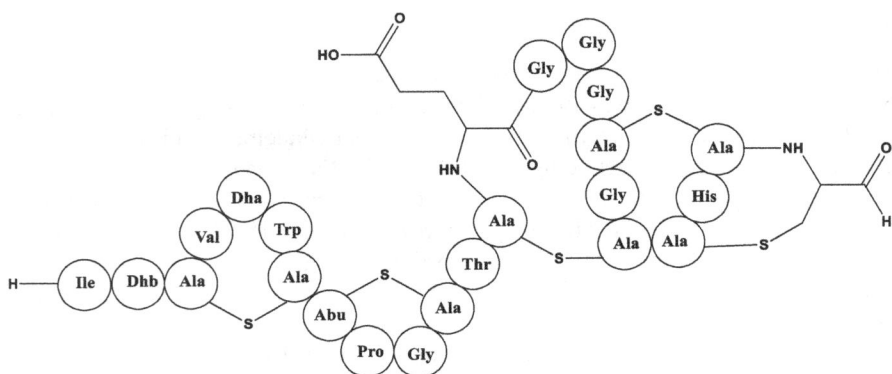

**Planosporicin (9)**

**Figure 7.** Chemical structure of of planosporicin (9) (from Dictionary of Natural Products 2020).

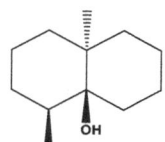

**(-)-Geosmin (10)**

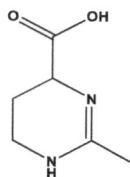

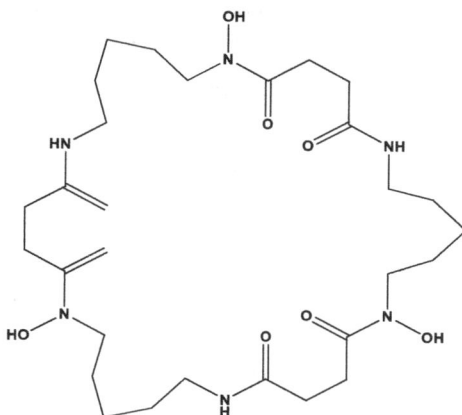

**Ectoine (11)**          **Desferrioxamine E (12)**

**Figure 8.** Chemical structures of geosmin (10), ectoine (11) and desferrioxamine E (12) (from Dictionary of Natural Products 2020).

## 8. Conclusions

It can be concluded that sponge-associated actinomycetes like the ones from the Atlantic and Pacific Oceans covered in this chapter are an attractive source for bioactive compounds with biotechnological importance. Exploring new cultivation processes (Kurtböke 2017) and fermentation protocols (Pan et al. 2019) together with advances in natural product chemistry and genome sequencing will undoubtedly facilitate new discoveries of natural products from marine and sponge symbiotic actinomycetes.

## Acknowledgements

The authors from Australia gratefully acknowledge Ms Elisabeth Evans-Illidge and Dr. Phil Kearns at the Australian Institute of Marine Science who provided the sponge symbiotic bacterial samples and Emeritus Prof. Ronald Quinn at the Griffith Institute for Drug Discovery for expert advice on natural product chemistry. The authors from Germany thank Dr. Wiebke Landwehr, Prof. Dr. Manfred Rohde, Dr. Peter Schumann, Dr. Richard L. Hahnke, Dr. Cathrin Spröer, Dr. Boyke Bunk, Prof. Dr. Peter Kämpfer, and Prof. Dr. Peter J. Schupp for their contribution to the novel species description of *Streptomyces bathyalis*.

## References

Abdelmohsen, U.R., Bayer, K. and Hentschel, U. (2014a). Diversity, abundance, and natural products of marine sponge-associated actinomycetes. *Natural Products Reports*, 31: 381–399.

Abdelmohsen, U.R., Yang, C., Horn, H., Hajjar, D., Ravasi, T. and Hentschel, U. (2014b). Actinomycetes from Red Sea sponges: Sources for chemical and phylogenetic diversity. *Marine Drugs*, 12: 2771–2789.

Abdelwahed, N.M., Ahmed, E., El-Gammal, E. and Hawas, U. (2014). Application of statistical design for the optimization of dextranase production by a novel fungus isolated from Red Sea sponge. *3 Biotech*, 4: 533–544.

Abou-Elela, G., Abd-Elnaby, H., Ibrahim, H. and Okbah, M. (2009). Marine natural products and their potential applications as anti-infective agents. *World Applied Sciences Journal*, 7: 872–880.

Aguila-Ramírez, R.N., Hernández-Guerrero, C.J., González-Acosta, B., Id-Daoud, G., Hewitt, S., Pope, J. et al. (2014). Antifouling activity of symbiotic bacteria from sponge *Aplysina gerardogreeni*. *International Biodeterioration and Biodegradation*, 90: 64–70.

Aminov, R.I. (2010). A brief history of the antibiotc era: Lessons learned and challenges for the future. *Frontiers in Microbiology*, 1: 1–7.

Anand, T.P., Bhat, A.W., Shouche, Y.S., Roy, U., Siddharth, J. and Sarma, S.P. (2006). Antimicrobial activity of marine bacteria associated with sponges from the waters off the coast of southeast India. *Microbiological Research*, 161: 252–262.

Armstrong, E., Boyd, K.G. and Burgess, J.G. (2000). Prevention of marine biofouling using natural compounds from marine organisms. *Biotechnology Annual Review*, 6: 221–241.

Atikana, A., Naim, M.A. and Sipkema, D. (2013). Detection of keto synthase (KS) gene domain in sponges and bacterial sponges. *Annales Bogoriensis*, 17: 27–33.

Baltz, R.H. (2006). Marcel Faber Roundtable: Is our antibiotic pipeline unproductive because of starvation, constipation or lack of inspiration? *Journal of Industrial Microbiology and Biotechnology*, 33: 507–513.

Banik, J.J. and Brady, S.F. (2010). Recent application of metagenomic approaches towards the discovery of antimicrobials and other bioactive small molecules. *Current Opinion in Microbiology*, 13: 603–609.

Belarbi, E.H., Gomez, A.C., Chisti, Y., Camacho, F.G. and Grima, E.M. (2003). Producing drugs from marine sponges. *Biotechnology Advances*, 21: 585–598.

Bell, J.J. (2008). The functional roles of marine sponges. *Estuarine, Coastal Shelf Science*, 79: 341–353.

Bernan, V.S., Greenstein, M. and Maiese, W.M. (1997). Marine microorganisms as a source of new natural products. *Advances in Applied Microbiology*, 43: 57–90.

Bertrand, S., Bohni, N., Schnee, S., Schumpp, O., Gindro, K. and Wolfender, J.L. (2014). Metabolite induction via microorganism co-culture: A potential way to enhance chemical diversity for drug discovery. *Biotechnology Advances*, 32: 1180–1204.

Bhatnagar, I. and Kim, S.K. (2010). Immense essence of excellence: Marine microbial bioactive compounds. *Marine Drugs*, 8: 2673–2701.

Blockley, A., Elliott, D.R., Roberts, A.P. and Sweet, M. (2017). Symbiotic microbes from marine invertebrates: Driving a new era of natural product drug discovery. *Diversity*, 9(4): 49.

Bode, H.B., Bethe, B., Höfs, R. and Zeeck, A. (2002). Big effects from small changes: Possible ways to explore nature's chemical diversity. *ChemBioChem*, 3: 619–627.

Bose, U., Ortori, C.A., Sarmad, S., Barrett, D.A., Hewavitharana, A.K., Hodson, M.P., Fuerst, J.A. and Shaw, P.N. (2017). Production of N-acyl homoserine lactones by the sponge-associated marine actinobacteria *Salinispora arenicola* and *Salinispora pacifica*. *FEMS Microbiology Letters*, 364(2): 1–7.

Brinkmann, C.M., Kearns, P.S., Evans-Illidge, E. and Kurtböke, D.İ. (2017b). Diversity and bioactivity of marine bacteria associated with the sponges *Candidaspongia flabellata* and *Rhopaloeides odorabile* from the Great Barrier Reef in Australia. *Diversity*, 9: 39.

Brinkmann, C.M., Marker, A. and Kurtböke, D.İ. (2017a). An overview on marine sponge-symbiotic bacteria as unexhausted sources for natural product discovery. *Diversity*, 9: 40.

Buedenbender, L., Carroll, A.R., Ekins, M. and Kurtböke, D.İ. (2017). Taxonomic and metabolite diversity of actinomycetes associated with three Australian ascidians. *Diversity*, 9(4): 53.

Buedenbender, L., Carroll, A.R. and Kurtböke, D.İ. (2019). *Integrated Approaches for Marine Actinomycete Biodiscovery*. E-book chapter, Vol. 5, p. 1. Bentham Science Publishers.

Buedenbender, L., Robertson, L.P., Lucantoni, L., Avery, V.M., Kurtböke, D.İ. and Carroll, A.R. (2018). HSQC-TOCSY fingerprinting-directed discovery of antiplasmodial polyketides from the marine ascidian-derived *Streptomyces* sp. (USC-16018). *Marine Drugs*, 16(6): 189.

Burja, A.M. and Hill, R.T. (2001). Microbial symbionts of the Australian Great Barrier Reef sponge, *Candidaspongia flabellata*. *Hydrobiologia*, 461: 41–47.

Bush, K., Courvalin, P., Dantas, G., Davies, J., Eisenstein, B., Huovinen, P. et al. (2011). Tackling antibiotic resistance. *Nature Reviews Microbiology*, 9: 894–896.

Camp, D., Davis, R.A., Campitelli, M., Ebdon, J. and Quinn, R.J. (2012). Drug-like properties: Guiding principles for the design of natural product libraries. *Journal of Natural Products*, 75: 72–81.

Castiglione, F., Cavaletti, L., Losi, D., Lazzarini, A., Carrano, L., Feroggio, M. et al. (2007). A novel lantibiotic acting on bacterial cell wall synthesis produced by the uncommon actinomycete *Planomonospora* sp. *Biochemistry*, 46: 5884–5895.

Chain, E., Florey, H.W., Gardner, A.D., Heatley, N.G., Jennings, M.A., Orr-Ewing J. et al. (1940). Penicillin as a chemotherapeutic agent. *Lancet*, 236: 226–228.

Chambers, H.F. and DeLeo, F.R. (2009). Waves of resistance: *Staphylococcus aureus* in the antibiotic era. *Nature Reviews Microbiology*, 7: 629–641.

Christensen, A. and Martin, G.D.A. (2017). Identification and bioactive potential of marine microorganisms from selected Florida coastal areas. *Microbiology Open*, 6: e00448.

Chokshi, A., Sifri, Z., Cennimo, D. and Horng, H. (2019). Global contributors to antibiotic resistance. *Journal of Global Infectious Diseases*, 11(1): 36–42.

Criswell, D. (2004). The "evolution" of antibiotic resistance. Vital Articles on Science Creation, Institute for Creation Research, 378: 1–4.

Dashti, Y., Grkovic, T., Abdelmohsen, U.R., Hentschel, U. and Quinn, R.J. (2014). Production of induced secondary metabolites by a co-culture of sponge-associated actinomycetes, *Actinokineospora* sp. EG49 and *Nocardiopsis* sp. RV163. *Marine Drugs*, 12: 3046–3059.

Datta, D., Talapatra, S. and Swarnakar, S. (2015). Bioactive compounds from marine invertebrates for potential medicines—An overview. *International Letters of National Sciences*, 34: 42–61.

Davies, J. (2006a). Are antibiotics naturally antibiotics? *Journal of Industrial Microbiology and Biotechnology*, 33: 496–499.

Davies, J. (2006b). Where have all the antibiotics gone? *Canadian Journal of Infectious Diseases and Medical Microbiology*, 17(5): 287–290.

Demain, A.L. and Sanchez, S. (2009). Microbial drug discovery: 80 years of progress. *Journal of Antibiotics*, 62: 5–16.

DeNegre, A.A., Ndeffo Mbah, M.L., Myers, K. and Fefferman, N.H. (2019). Emergence of antibiotic resistance in immunocompromised host populations: A case study of emerging antibiotic resistant tuberculosis in AIDS patients. *PLoS One*, 14(2): e0212969.

Devi, P., Wahidullah, S., Rodrigues, C. and Souza, L.D. (2010). The sponge-associated bacterium *Bacillus licheniformis* SAB1: A source of antimicrobial compounds. *Marine Drugs*, 8(4): 1203–1212.

Dictionary of Natural Products on DVD. June (2020). Version 29:1. CRC Press: Boca Raton, Florida, USA.

Drews, J. (2000). Drug discovery: A historical perspective. *Science*, 287: 1960–1964.

Fenical, W. (1993). Chemical studies of marine bacteria: Developing a new resource. *Chemical Reviews*, 93: 1673–1683.

Fenical, W. and Jensen, P.R. (2006). Developing a new resource for drug discovery: Marine actinomycete bacteria. *Nature Chemical Biology*, 2: 666–673.

Finore, I., Di Donato, P., Mastascusa, V., Nicolaus, B. and Poli, A. (2014). Fermentation technologies for the optimization of marine microbial exopolysaccharide production. *Marine Drugs*, 12: 3005–3024.

Fischbach, M.A. and Walsh, C.T. (2009). Antibiotics for emerging pathogens. *Science*, 325: 1089–1093.

Flowers, C.R., Seidenfeld, J., Bow, E.J., Karten, C., Gleason, C., Hawley, D.K. et al. (2013). Antimicrobial prophylaxis and outpatient management of fever and neutropenia in adults treated for malignancy: American Society of Clinical Oncology clinical practice guideline. *Journal of Clinical Oncology*, 31: 794–810.

Frieri, M., Kumar, K. and Boutin, A. (2017). Antibiotic resistance. *Journal of Infection and Public Health*, 10(4): 369–378.

Fuqua, C. and Greenberg, E.P. (2002). Listening in on bacteria: Acyl-homoserine lactone signalling. *Nature Reviews Molecular Cell Biology*, 3(9): 685–695.

Galinski, E.A. and Trüper, H.G. (1994). Microbial behaviour in salt-stressed ecosystems. *FEMS Microbiology Reviews*, 15: 95–108.

Gandhimathi, R., Arunkumar, M., Selvin, J., Thangavelu, T., Sivaramakrishnan, S., Kiran, G.S. et al. (2008). Antimicrobial potential of sponge associated marine actinomycetes. *Journal of Medical Mycology*, 18: 16–22.

Garson, M.J. (1994). The biosynthesis of sponge secondary metabolites: Why it is important. pp. 427–440. *In*: van Soest, R.W.M., van Kempen, T.M.G. and Braekman, J.-C. (eds.), *Sponges in Time and Space*. Balkema, Rotterdam.

Gaudêncio, S.P. and Pereira, F. (2015). Dereplication: Racing to speed up the natural products discovery process. *Natural Products Reports*, 32: 779–810.

Gómez-Betancur, I., Zhao, J., Tan, L., Chen, C., Yu, G., Rey-Suárez, P. and Preciado, L. (2019). Bioactive compounds isolated from marine bacterium *Vibrio neocaledonicus* and their enzyme inhibitory activities. *Marine Drugs*, 17(7): 401.

Graça, A.P., Bondoso, J., Gaspar, H., Xavier, J.R., Monteiro, M.C., de la Cruz, M. et al. (2013). Antimicrobial activity of heterotrophic bacterial communities from the marine sponge *Erylus discophorus* (Astrophorida, Geodiidae). *PLoS One*, 8: e78992.

Hentschel, U., Fieseler, L., Wehrl, M., Gernert, C., Steinert, M., Hacker, J. and Horn, M. (2003). Microbial diversity of marine sponges. pp. 59–88. *In*: Müller, W.E.G. (ed.). *Sponges (Porifera). Progress in Molecular and Subcellular Biology*, vol 37. Springer: Berlin, Heidelberg.

Hentschel, U., Piel, J., Degnan, S.M. and Taylor, M.W. (2012). Genomic insights into the marine sponge microbiome. *Nature Reviews Microbiology*, 10: 641–654.

Hentschel, U., Usher, K.M. and Taylor, M.W. (2006). Marine sponges as microbial fermenters. *FEMS Microbiology Ecology*, 55: 167–177.

Hochmuth, T. and Piel, J. (2009). Polyketide synthases of bacterial symbionts in sponges–evolution-based applications in natural products research. *Phytochemistry*, 70(15-16): 1841–1849.

Jackson, N., Czaplewski, L. and Piddock, L.J. (2018). Discovery and development of new antibacterial drugs: learning from experience? *Journal of Antimicrobial Chemotherapy*, 73(6): 1452–1459.

Jensen, P.R. and Fenical, W. (1994). Strategies for the discovery of secondary metabolites from marine bacteria: ecological perspectives. *Annual Reviews in Microbiology*, 48: 559–584.

Jensen, P.R., Williams, P.G., Oh, D.C., Zeigler L. and Fenical, W. (2007). Species-specific secondary metabolite production in marine actinomycetes of the genus *Salinispora*. *Applied and Environmental Microbiology*, 73: 1146–1152.

Jimenez, J.T., Šturdíková, M. and Šturdík, E. (2009). Natural products of marine origin and their perspectives in the discovery of new anticancer drugs. *Acta Chimica Slovenica*, 2: 63–74.

Jüttner, F. and Watson, S.B. (2007). Biochemical and ecological control of geosmin and 2-methylisoborneol in source waters. *Applied and Environmental Microbiology*, 73: 4395–4406.

Kennedy, J., Baker, P., Piper, C., Cotter, P., Walsh, M., Mooij, M. et al. (2009). Isolation and analysis of bacteria with antimicrobial activities from the marine sponge *Haliclona simulans* collected from Irish waters. *Marine Biotechnology*, 11: 384–396.

Kim, T.K. and Fuerst, J.A. (2006). Diversity of polyketide synthase genes from bacteria associated with the marine sponge *Pseudoceratina clavata*: Culture-dependent and culture-independent approaches. *Environmental Microbiology*, 8: 1460–1470.

Kiruthika, P., Nisshanthini, S.D., Saraswathi, A., Angayarkanni, J. and Rajendiran, R. (2011). Application of statistical design to the optimization of culture medium for biomass production by *Exiguobacterium* sp. HM 119395. *International Journal of Advanced Biotechnological Research*, 2: 422–430.

Kiuru, P., D'Auria, M.V., Muller, C.D., Tammela, P., Vuorela, H. and Yli-Kauhaluoma, J. (2014). Exploring marine resources for bioactive compounds. *Planta Medica*, 80: 1234–1246.

Krishnan, P., Balasubramaniam, M., Dam Roy, S., Sarma, K., Hairun, R. and Sunder, J. (2014). Characterization of the antibacterial activity of bacteria associated with *Stylissa* sp., a marine sponge. *Advances in Animal and Veterinary Sciences*, 2: 20–25.

Kumar, P.S., Krishna, E.R., Sujatha, P. and Kumar, B.V. (2012). Screening and isolation of associated bioactive microorganisms from *Fasciospongia cavernosa* from of Visakhapatnam Coast, Bay of Bengal. *E-Journal of Chemistry*, 9(4): 2166–2176.

Kurtböke, D.İ. (2012). From Actinomycin onwards: Actinomycete success stories. *Microbiology Australia*, 33: 108–110.

Kurtböke, D.İ. (ed.). (2017). *Microbial Resources: From Functional Existence in Nature to Applications*. Academic Press, Elsevier Inc.

Kurtböke, D.İ., Grkovic, T. and Quinn, R.J. (2015). Marine actinomycetes in biodiscovery. pp. 663–676. *In*: Kim, S.-K. (ed.). *Springer Handbook of Marine Biotechnology*. Chapter 27. Springer: Berlin. Heidelberg.

Kurtböke, D.İ., Okazaki, T. and Vobis, G. (2020). Actinobacteria in marine environments: From terrigenous origin to adapted functional diversity. pp. 1951–1978. *In*: Kim, S.-K. (ed.). *Encyclopedia of Marine Biotechnology*. Wiley-Blackwell.

Laport, M.S., Santos, O.C.S. and Muricy, G. (2009). Marine sponges: Potential sources of new antimicrobial drugs. *Current Pharmaceutical Biotechnology*, 10(1): 86–105.

Larsson, D.G.J. and Flach, C.F. (2022). Antibiotic resistance in the environment. *Nature Reviews Microbiology*, 20(5): 257–269.

Leal, M.C., Sheridan, C., Osinga, R., Dionísio, G., Rocha, R.J., Silva, B., Rosa, R. and Calado, R. (2014). Marine microorganism-invertebrate assemblages: Perspectives to solve the "supply problem" in the initial steps of drug discovery. *Marine Drugs*, 2014 Jul; 12(7): 3929–52.

Lee, O.O., Wang, Y., Yang, J., Lafi, F.F., Al-Suwailem, A. and Qian, P.Y. (2011). Pyrosequencing reveals highly diverse and species-specific microbial communities in sponges from the Red Sea. *ISME Journal*, 5: 650–664.

Linares-Otoya, L., Linares-Otaya, V., Armas-Mantilla, L., Blanco-Olano, C., Crusemann, M., Ganoza-Yupanqui, M.L. et al. (2017). Identification and heterologous expression of the kocurin biosynthetic gene cluster. *Microbiology*, 163: 1409–1414.

Liu, M., Grkovic, T., Liu, X., Han, J., Zhang, L. and Quinn, R.J. (2017). A systems approach using OSMAC, Log P and NMR fingerprinting: An approach to novelty. *Synthetic and Systems Biotechnology*, 2: 276–286.

Lowy, F.D. (2003). Antimicrobial resistance: The example of *Staphylococcus aureus*. *Journal of Clinical Investigation*, 111: 1265.

Lyu, C., Chen, T., Qiang, B., Liu, N., Wang, H., Zhang, L. and Liu, Z. (2021). CMNPD: A comprehensive marine natural products database towards facilitating drug discovery from the ocean. *Nucleic Acids Research*, 49(D1): D509–D515.

Maffioli, S.I., Potenza, D., Vasile, F., De Matteo, M., Sosio, M., Marsiglia, B. et al. (2009). Structure revision of the lantibiotic 97518. *Journal of Natural Products*, 72: 605–607.

Marketon, M.M., Glenn, S.A., Eberhard, A. and González, J.E. (2003). Quorum sensing controls exopolysaccharide production in *Sinorhizobium meliloti*. *Journal of Bacteriology*, 185: 325–331.

Marmann, A., Aly, A.H., Lin, W., Wang, B. and Proksch, P. (2014). Co-Cultivation-A powerful emerging tool for enhancing the chemical diversity of microorganisms. *Marine Drugs*, 12: 1043–1065.

Martins, A., Vieira, H., Gaspar, H. and Santos, S. (2014). Marketed marine natural products in the pharmaceutical and cosmeceutical industries: Tips for success. *Marine Drugs*, 12: 1066–1101.

Martín, J., Sousa, D.S., Crespo, G., Palomo, S., González, I., Tormo, J.R., De la Cruz, M., Anderson, M., Hill, R.T., Vicente, F. and Genilloud, O. (2013). Kocurin, the true structure of PM181104, an anti-methicillin-resistant *Staphylococcus aureus* (MRSA) thiazolyl peptide from the marine-derived bacterium *Kocuria palustris*. *Marine Drugs*, 11: 387–398.

Medema, M.H., Blin, K., Cimermancic, P., De Jager, V., Zakrzewski, P., Fischbach, M.A. et al. (2011). AntiSMASH: Rapid identification, annotation and analysis of secondary metabolite biosynthesis gene clusters in bacterial and fungal genome sequences. *Nucleic Acids Research*, 39: 339–346.

Medema, M.H., Kottmann, R., Yılmaz, P., Cummings, M., Kiggins, J.B., Blin, K. et al. (2015). Minimum information about a biosynthetic gene cluster. *Nature Chemical Biology*, 11: 625–631.

Mehbub, M.F., Lei, J., Franco, C. and Zhang, W. (2014). Marine sponge derived natural products between 2001 and 2010: Trends and opportunities for discovery of bioactives. *Marine Drugs*, 12: 4539–4577.

Moloney, M.G. (2016). Natural products as a source for novel antiobiotics. *Trends in Pharmacological Sciences*, 37: 689–701.

Müller, W.E.G. (ed.). (2012). *Sponges (Porifera)*. Springer: Berlin, Heidelberg.

Munro, M.H., Blunt, J.W., Dumdei, E.J., Hickford, S.J., Lill, R.E., Li, S. et al. (1999). The discovery and development of marine compounds with pharmaceutical potential. *Journal of Biotechnology*, 70: 15–25.

Neu, H.C. (1992). The crisis in antibiotic resistance. *Science*, 257: 1064–1073.

Newman, D.J. and Cragg, G.M. (2016). Natural products as sources of new drugs from 1981 to 2014. *Journal of Natural Products*, 79(3): 629–661.

Newman, D.J., Cragg, G.M. and Grothaus, P. (eds.). (2017). *Chemical Biology of Natural Products*. CRC Press.

Nouioui, I., Carro, L., García-López, M., Meier-Kolthoff, J.P., Woyke, T., Kyrpides, N.C., Pukall, R., Klenk, H.-P. Goodfellow, M. and Göker, M. (2018). Genome-based taxonomic classification of the phylum *Actinobacteria*. *Frontiers in Microbiology*, 9: 2007.

Ochi, K. and Hosaka, T. (2013). New strategies for drug discovery: Activation of silent or weakly expressed microbial gene clusters. *Applied Microbiology and Biotechnology*, 97: 87–98.

Oren, A. and Garrity, G.M. (2021). Valid publication of the names of forty-two phyla of prokaryotes. *International Journal of Systematic and Evolutionary Microbiology*, 71(10).

Palomo, S., González, I., de la Cruz, M., Martín, J., Tormo, J.R., Anderson, M. et al. (2013). Sponge-derived *Kocuria* and *Micrococcus* spp. as sources of the new thiazolyl peptide antibiotic kocurin. *Marine Drugs*, 11: 1071–1086.

Pan, R., Bai, X., Chen, J., Zhang, H. and Wang, H. (2019). Exploring structural diversity of microbe secondary metabolites using OSMAC strategy: A literature review. *Frontiers in Microbiology*, 10: 294.

Pantosti, A., Sanchini, A. and Monaco, M. (2007). Mechanisms of antibiotic resistance in *Staphylococcus aureus*. *Future Microbiology*, 2: 323–334.

Pawlik, J.R. (1993). Marine invertebrate chemical defenses. *Chemical Reviews*, 93: 1911–1922.

Perdicaris, S., Vlachogianni, T. and Valavanidis, A. (2013). Bioactive natural substances from marine sponges: New developments and prospects for future pharmaceuticals. *Natural Products Chemistry and Research*, 1: 1–8.

Peterson, E. and Kaur, P. (2018). Antibiotic resistance mechanisms in bacteria: Relationships between resistance determinants of antibiotic producers, environmental bacteria, and clinical pathogens. *Frontiers in Microbiology*, 2928.

Pierson III, L.S. and Thomashow, L.S. (1992). Cloning and heterologous expression of the phenazine biosynthetic locus from *Pseudomonas aureofaciens* 30–84. *Molecular Plant-Microbe Interactions*, 5: 330–339.

Pita, L., Rix, L., Slaby, B.M., Franke, A. and Hentschel, U. (2018). The sponge holobiont in a changing ocean: From microbes to ecosystems. *Microbiome*, 6: 46.

Quiñones, B., Dulla, G. and Lindow, S.E. (2005). Quorum sensing regulates exopolysaccharide production, motility, and virulence in *Pseudomonas syringae*. *Molecular Plant-Microbe Interactions*, 18: 682–693.

Reasoner, D.J. and Geldreich, E.E. (1985). A new medium for the enumeration and subculture of bacteria from potable water. *Applied and Environmental Microbiology*, 49: 1–7.

Reen, F.J., Gutiérrez-Barranquero, J.A. and O'Gara, F. (2017). Mining microbial signals for enhanced biodiscovery of secondary metabolites. pp. 287–300. *In*: Streit, W.R. and Daniel, R. (eds.). *Metagenomics, Methods and Protocols* (2nd ed). Humana Press, New York, NY.

Risdian, C., Landwehr, W., Rohde, M., Schumann, P., Hahnke, R.L., Spröer, C., Bunk, B., Kämpfer, P., Schupp, P.J. and Wink, J. (2021). *Streptomyces bathyalis* sp. nov., an actinobacterium isolated from the sponge in a deep sea. *Antonie van Leeuwenhoek*, 114(4): 425–435.

Sadeghi, A., Soltani, B.M., Nekouei, M.K., Jouzani, G.S., Mirzaei, H.H. and Sadeghizadeh, M. (2014). Diversity of the ectoines biosynthesis genes in the salt tolerant *Streptomyces* and evidence for inductive effect of ectoines on their accumulation. *Microbiological Research*, 169(9-10): 699–708.

Sagar, S., Kaur, M. and Minneman, K.P. (2010). Antiviral lead compounds from marine sponges. *Marine Drugs*, 8: 2619–2638.

Salam, N., Jiao, J.Y., Zhang, X.T. and Li, W.J. (2020). Update on the classification of higher ranks in the phylum Actinobacteria. *International Journal of Systematic and Evolutionary Microbiology*, 70(2): 1331–1355.

Satheesh, S., Ba-akdah, M.A. and Al-Sofyani, A.A. (2016). Natural antifouling compound production by microbes associated with marine macroorganisms—A review. *Electronic Journal of Biotechnology*, 21: 26–35.

Sathiyanarayanan, G., Gandhimathi, R., Sabarathnam, B., Seghal Kiran, G. and Selvin, J. (2014). Optimization and production of pyrrolidone antimicrobial agent from marine sponge-associated *Streptomyces* sp. MAPS15. *Bioprocess and Biosystems Engineering*, 37(3): 561–573.

Schmidtz, F.J., Vanderah, D.J., Hollenbeak, K.H., Enwall, C.E., Gopichand, Y., SenGupta, P.K., Hossain, M.B. and Van der Helm, D. (1983). Metabolites from the marine sponge *Tedania ignis*. A new atisanediol and several known diketopiperazines. *The Journal of Organic Chemistry*, 48: 3941–3945.

Schmitt, S., Angermeier, H., Schiller, R., Lindquist, N. and Hentschel, U. (2008). Molecular microbial diversity survey of sponge reproductive stages and mechanistic insights into vertical transmission of microbial symbionts. *Applied and Environmental Microbiology*, 74(24): 7694–7708.

Schneemann, I., Nagel, K., Kajahn, I., Labes, A., Wiese, J. and Imhoff, J.F. (2010). Comprehensive investigation of marine Actinobacteria associated with the sponge *Halichondria panicea*. *Applied and Environmental Microbiology*, 76(11): 3702–3714.

Selvin, J., Shanmughapriya, S., Gandhimathi, R., Seghal Kiran, G., Rajeetha Ravji, T., Natarajaseenivasan, K. and Hema, T.A. (2009). Optimization and production of novel antimicrobial agents from sponge associated marine actinomycetes *Nocardiopsis dassonvillei* MAD08. *Applied Microbiology and Biotechnology*, 83(3): 435–445.

Senthilkumar, K. and Kim, S.-K. (2013). Marine invertebrate natural products for anti-inflammatory and chronic diseases. *Evidence-Based Complementary* and *Alternative Medicine*, 2013: 572859.

Simister, R.L., Deines, P., Botte, E.S., Webster, N.S. and Taylor, M.W. (2012). Sponge-specific clusters revisited: A comprehensive phylogeny of sponge-associated microorganisms. *Environmental Microbiology*, 14: 517–524.

Sipkema, D., de Caralt, S., Morillo, J.A., Al-Soud, W.A., Sørensen, S.J., Smidt, H. et al. (2015). Similar sponge-associated bacteria can be acquired via both vertical and horizontal transmission. *Environmental Microbiology*, 17: 3807–3821.

Skinnider, M.A., Dejong, C.A., Rees, P.N., Johnston, C.W., Li, H., Webster, A.L.H. et al. (2015). Genomes to natural products Prediction Informatics for Secondary Metabolomes (PRISM). *Nucleic Acids Research*, 43: 9645–9662.

Sosa, A.D.J., Byarugaba, D.K., Amábile-Cuevas, C.F., Hsueh, P.R., Kariuki, S. and Okeke, I.N. (eds.). (2010). *Antimicrobial Resistance in Developing Countries* (p. 554). New York: Springer.

Sperandio, V. (2004). Striking a balance: Inter-kingdom cell-to-cell signaling, friendship or war? *Trends in Immunology*, 25: 505–507.

Stierle, A.C., Cardellina, J.H. and Singleton, F.L. (1988). A marine micrococcus produces metabolites ascribed to the sponge *Tedania ignis*. *Experientia*, 44: 1021–1021.

Takarada, H., Sekine, M., Kosugi, H., Matsuo, Y., Fujisawa, T., Omata, S. et al. (2008). Complete genome sequence of the soil actinomycete *Kocuria rhizophila*. *Journal of Bacteriology*, 190: 4139–4146.

Tasharrofi, N., Adrangi, S., Fazeli, M., Rastegar, H., Khoshayand, M.R. and Faramarzi, M.A. (2011). Optimization of chitinase production by *Bacillus pumilus* using Plackett-Burman design and response surface methodology. *Iranian Journal* of *Pharmaceutical Research*, 10: 759.

Taylor, M.W., Schupp, P.J., Baillie, H.J. Charlton, T.S., De Nys, R., Kjelleberg, S. et al. (2004). Evidence for acyl homoserine lactone signal production in bacteria associated with marine sponges. *Applied and Environmental Microbiology*, 70: 4387–4389.

Taylor, M.W., Radax, R., Steger, D. and Wagner, M. (2007). Sponge-associated microorganisms: Evolution, ecology, and biotechnological potential. *Microbiology and Molecular Biology Reviews*, 71: 295–347.

Thomas, T.R.A., Kavlekar, D.P. and LokaBharathi, P.A. (2010). Marine drugs from sponge-microbe association—A review. *Marine Drugs*, 8: 1417–1468.

Tian, L., Yin, Y., Jin, H., Bing, W., Jin, E., Zhao, J. and Ren, L. (2020). Novel marine antifouling coatings inspired by corals. *Materials Today Chemistry*, 17: 100294.

Treichel, H., de Oliveira, D., Mazutti, M.A., Di Luccio, M. and Oliveira, J.V. (2010). A review on microbial lipases production. *Food and Bioprocess Technology*, 3: 182–196.

Trindade-Silva, A.E., Rua, C., Silva, G.G.Z., Dutilh, B.E., Moreira, A.P.B., Edwards, R.A. et al. (2012). Taxonomic and functional microbial signatures of the endemic marine sponge *Arenosclera brasiliensis*. *PLoS One*, 7: e39905.

Tsukimoto, M., Nagaoka, M., Shishido, Y., Fujimoto, J., Nishisaka, F., Matsumoto, S. et al. (2011). Bacterial production of the tunicate-derived antitumor cyclic depsipeptide didemnin B. *Journal of Natural Products*, 74: 2329–2331.

Udwary, D.W., Zeigler, L., Asolkar, R.N., Singan Lapidus, Fenical, W. et al. (2007). Genome sequencing reveals complex secondary metabolome in the marine actinomycete *Salinispora tropica*. *Proceedings of the National Academy of Sciences*, 104: 10376–10381.

Uzair, B., Menaa, F., Khan, B.A., Mohammad, F.V., Ahmad, V.U., Djeribi, R. and Menaa, B. (2018). Isolation, purification, structural elucidation, and antimicrobial activities of kocumarin, a novel antibiotic isolated from actinobacterium *Kocuria marina* CMG S2 associated with the brown seaweed *Pelvetia canaliculata*. *Microbiological Research*, 206: 186–197.

Vaca, I. and Chávez, R. (2019). Bioactive compounds produced by Antarctic filamentous fungi. pp. 265–283. *In*: Rosa, L.H. (ed.). *Fungi of Antarctica: Diversity, Ecology and Biotechnological Applications*. Springer: Cham.

Van Hoek, A.H., Mevius, D., Guerra, B., Mullany, P., Roberts, A.P. and Aarts, H.J. (2011). Acquired antibiotic resistance genes: An overview. *Frontiers in Microbiology*, 2: 203.

Verdine, G.L. (1996). The combinatorial chemistry of nature. *Nature*, 384(6604): 11–13.

Veysioğlu, A., Tatar, D., Çetin, D., Güven, K. and Şahin, N. (2014). *Streptomyces karpasiensis* sp. nov., isolated from soil. *International Journal of Systematic and Evolutionary Microbiology*, 64: 827–832.

Wael, A.-Z. (2006). PhD Thesis: Natural products from marine bacteria. https://kluedo.ub.rptu.de/frontdoor/index/index/docId/1780, Technische Universität Kaiserslautern, Germany.

Walsh, C. and Wright, G. (2005). Introduction: Antibiotic resistance. *Chemical Reviews*, 105: 391–394.

Watve, M.G., Tickoo, R., Jog, M.M. and Bhole, B.D. (2001). How many antibiotics are produced by the genus *Streptomyces*? *Archives of Microbiology*, 176(5): 386–390.

Webster, N.S. and Blackall, L.L. (2009). What do we really know about sponge-microbial symbioses? *ISME Journal*, 3: 1–3.

Webster, N.S. and Taylor, M.W. (2012). Marine sponges and their microbial symbionts: Love and other relationships. *Environmental Microbiology*, 14: 335–346.

Webster, N.S., Luter, H.M., Soo, R.M., Botté, E.S., Simister, R.L., Abdo, D. and Whalan, S. (2012). Same, same but different: Symbiotic bacterial associations in GBR sponges. *Frontiers in Microbiology*, 3: 444.

Webster, N.S., Wilson, K.J., Blackall, L.L. and Hill, R.T. (2001). Phylogenetic diversity of bacteria associated with the marine sponge *Rhopaloeides odorabile*. *Applied and Environmental Microbiology*, 67(1): 434–444.

Wright, G.D. (2012). Antibiotics: A new hope. *Chemistry and Biology*, 19: 3–10.

Xi, L., Ruan, J. and Huang, Y. (2012). Diversity and biosynthetic potential of culturable actinomycetes associated with marine sponges in the China seas. *International Journal of Molecular Sciences*, 13: 5917–5932.

Xu, Y., He, J., Tian, X.P., Li, J., Yang, L.L., Xie, Q. et al. (2012). *Streptomyces glycovorans* sp. nov., *Streptomyces xishensis* sp. nov. and *Streptomyces abyssalis* sp. nov., isolated from marine sediments. *International Journal of Systematic and Evolutionary Microbiology*, 62: 2371–2377.

Yamanaka, K., Oikawa, H., Ogawa, H.O., Hosono, K., Shinmachi, F., Takano, H. et al. (2005). Desferrioxamine E produced by *Streptomyces griseus* stimulates growth and development of *Streptomyces tanashiensis*. *Microbiology*, 151: 2899–2905.

Zhang, Y. and Yew, W. (2009). Mechanisms of drug resistance in *Mycobacterium tuberculosis* [State of the art series. Drug-resistant tuberculosis. Edited by CY. Chiang. Number 1 in the series]. *International Journal of Tuberculosis and Lung Disease*, 13: 1320–1330.

Zhuang, M., Achmon, Y., Cao, Y., Liang, X., Chen, L., Wang, H., Siame, B.A. and Leung, K.Y. (2021). Distribution of antibiotic resistance genes in the environment. *Environmental Pollution*, 285: 117402.

Zotchev, S.B., Sekurova, O.N. and Kurtböke, D.İ. (2016). Metagenomics of marine actinomycetes: From functional gene diversity to biodiscovery. Chapter 9, pp. 185–206. *In*: Kim, S.-K. (ed.). *Marine OMICS: Principles and Applications*. CRC Press, Taylor and Francis Group.

# Chapter 2

# Actinomycetes from Tropical Marine Environments of Thailand and their Biotechnological Applications

*Wasu Pathom-aree,*[1,*] *Pharada Rangseekaew,*[1]
*Manita Kamjam*[1] and *Kannika Duangmal*[2]

## 1. Introduction

Thailand is centrally located in the Indo-China Peninsula with a total area of 513,000 square kilometres. The coastal area is in the southern part of the country with a 3,148 km long shoreline including the Gulf of Thailand (2,055 km) to the east and the Andaman Sea (1,093 km) to the west (Lange et al. 2019). The Gulf of Thailand is a shallow inlet in the southwestern South China Sea with an average depth of 45 m and a maximum depth of 83 m (Aschariyaphotha and Wongwises 2012). The coast is characterized by wide and long mainland beaches with lagoons and bays. Geographically, it is divided into Upper and Lower Gulfs. The Upper Gulf is the catchment basin of four large rivers namely, the Bangpakong River, the Chaopraya River, the Mae klong River, and the Thachin River, and consists of an inverted U-shaped area of approximately 10,000 square kilometres (Department of Marine and Coastal Resources 2013). The west coast along the Andaman Sea is 1,093 km long and is dominated by pocket beaches, and an extensive preserved tidal flat vegetated with mangrove forests; it also includes many cliffs and includes about 545 islands lying offshore (Lange et al. 2019). Inshore areas within three kilometres have an average depth of about three metres. Marine resources of the area comprise many marine ecosystems including mangroves, seagrass beds, coral reefs, sandy beaches, rocky and muddy shores (Department of Marine and Coastal Resources 2012).

[1] Research Center of Microbial Diversity and Sustainable Utilization, Department of Biology, Faculty of Science, Chiang Mai University, Chiang Mai 50200 Thailand.
[2] Department of Microbiology, Faculty of Science, Kasetsart University, Bangkok 10900 Thailand.
* Corresponding author: wasu.p@cmu.ac.th

Actinomycetes are a large group of Gram-positive organisms with a characteristic high ratio of guanine plus cytosine (G+C) in their genome. They are the most prolific producers of bioactive compounds especially antibiotics (Berdy 2012, Barka et al. 2016). Currently, there are, at least 120 actinomycete antibiotics in the market, including erythromycin, gentamicin, streptomycin and vancomycin, most of which are produced by members of the genus *Streptomyces* of this group of bacteria. Actinomyetes are widely distributed in both aquatic and terrestrial habitats, notably soil (Barka et al. 2016) though they are increasingly being found in extreme biomes including marine habitats. Indeed, the ones from marine environments are proving to be an attractive source of novel bioactive metabolites (Subramani and Albersberg 2012, Manivasagan et al. 2014, Hassan et al. 2017, Kamjam et al. 2017, Subramani and Sipkema 2019). Research on actinomycetes from marine environments in Thailand can be traced back to 2000 though such studies have not been documented in a systematic way. This chapter aims to rectify this situation with an emphasis on the biotechnological potential of actinomycetes isolated from Thai marine habitats.

## 2. Selective Isolation and Cultivation of Marine Actinomycetes

The isolation of microorganisms of interest from environmental samples is of paramount importance in biotechnology. Microbial communities in natural habitats are complex, hence in order to isolate actinomycetes it is necessary to eliminate or reduce the growth of unwanted microorganisms, mostly fast growing bacteria and fungi on isolation plates using selective procedures. Most selective isolation strategies involve the pretreatment of environmental samples, and the use of selective media and appropriate incubation regimes.

### 2.1 Sample Pretreatment

Pretreatment methods are designed to aid the isolation of slow-growing and difficult to culture actinomycetes by inhibiting the growth of unwanted microorganisms, especially fast-growing bacteria and fungi. Pretreatment of environmental samples involve chemical and/or physical procedures (Goodfellow and Fiedler 2010). The pretreatment methods that have been used to isolate these bacteria from Thai marine environments are summarized in Table 1.

Phenol treatment is a chemical method designed to isolate rare, that is, infrequently isolated actinomycetes such as members of the genera *Microbispora* and *Micromonospora*, and *Streptomyces violaceusniger* strains (Hayakawa et al. 2004, Hayakawa 2008). However, pretreatment of Thai coastal marine sediments with 1.5% phenol failed to isolate microbisporae or micromonosporae (Ruttanasutja and Pathom-aree 2015).

Heat treatment is the most common physical procedure used by Thai researchers, for example, heating a coastal soil samples at 55°C for 15 min to reduce the numbers of unwanted microorganisms and at 100°C for 60 minutes to activate dormant actinomycete spores (Srivibool et al. 2004, Srivibool and Sukchotiratana

**Table 1.** Isolation strategy for actinomycetes from tropical marine environments of Thailand.

| Location | Sample | Pretreatment procedure | Isolation medium/Incubation temperature and time | Genus | References |
|---|---|---|---|---|---|
| Samae Sarn, Juang, Raet and Kram islands | Island soils | Pre-treated at 55°C for 15 min and at 100°C for 60 min | Actinomycetes isolation agar, and starch casein agar | *Actinomadura* | Srivibool 2000 |
| Chonburi, Rayong and Trat Provinces | Coastal sediment | Pre-treated at 55°C for 15 min and at 100°C for 60 min | Actinomycetes isolation agar, and starch casein agar | Thermotolerant *Streptomyces* | Srivibool et al. 2004 |
| Chang, Hwai, Lao-yanai (Trat Province) and Pai Islands (Chonburi Province) | Island soils | Pre-treated at 55°C for 15 min and at 100°C for 60 min | Actinomycetes isolation agar, Starch casein agar and glucose asparagine agar incubated at 32°C for 4 weeks | Not specified | Srivibool and Sukchotiratana 2006 |
| Andaman seashore, Trang Province | Marine sediments | Wet-heated at 70°C for 15 min | Starch–casein nitrate agar (SCA) supplemented with novobiocin and mycostatin | *Micromonospora* | Thawai 2010 |
| Mangrove forests along the inner gulf of Thailand (Samut Prakan and Samut Songkarm Provinces) | Mangrove soils | - Samples were dried both at room temperature for 1 week and at 110°C for 1 h.<br>- Soil suspensions heated at 60-65°C (15 min) | Starch–casein nitrate agar (SCA) supplemented with novobiocin and nystatin | *Streptomyces* | Hunadanamra et al. 2013 |
| East and west of the Gulf of Thailand | Mangrove sediments | Heated for 1 h at 100°C and/or 30 min treated with 1.5% phenol before isolation | Starch casein agar, actinomycetes isolation agar, ISP2 medium and SCA, M4 and humic acid vitamin agar supplemented with novobiocin, and nystatin incubated at 30°C for 4 weeks | *Micromonospora, Nocardia, Nocardiopsis, Salinispora, Spirilliplanes, Streptomyces, Streptoalloteichus, Virgisporangium* | Srivibool and Watanadilok 2015 |

*Table 1 contd....*

*...Table 1 contd.*

| Location | Sample | Pretreatment procedure | Isolation medium/Incubation temperature and time | Genus | References |
|---|---|---|---|---|---|
| Coastal area from Klong-Taguan, Rayong Province, Thailand | Marine sediments | Shaken at 125 rpm for 60 min at room temperature | Marine agar (MA), starch casein agar (SCA), Gause No. 2 agar (GNO2A), marine soil extract agar (MSA), starch casein soil extract agar (SCSA), Gause No.2 soil extract agar (GNO2SA), soil extract agar (SEA), 1:100MA, 1:100SCA, 1:100GNO2A, 1:100MSA,1:100SCSA, 1:100GNO2SA, 1:2 SEA, and 1:100SEA | *Curtobacterium, Dermacoccus, Micromonospora, Microbispora, Pseudonocardia, Rhodococcus, Streptomyces, Tsukamurella* | Ruttanasutja and Pathom-aree 2015 |
| Eastern coast of, Chonburi and Chanthaburi Provinces | Mangrove sediments | Not specified | Starch casein agar (SCA) medium with sodium chloride (NaCl) 3% (w/v) supplemented with nalidixic acid and ketoconazole, incubated at 30°C for 7–10 days | Not specified | Tangjitjaroenkun et al. 2017 |
| General Prem Tinsulanonda Historical Park, Songkhla Province | Mangrove sediments | Various pretreatment methods including treatment with liquid nitrogen; 0.05% sodium dodecyl sulfate (SDS) and 5% yeast extract; 1.5% phenol, dry heat, and rehydration and centrifugation | Actinomycetes isolation agar, modified soil extract agar, humic acid vitamin agar and starch nitrate agar supplemented with cycloheximide and nalidixic acid, incubated at room temperature for 4 weeks | *Nocardiaceae* *Micromonosporaceae* *Pseudonocardiaceae* *Streptomycetaceae* *Thermomonosporaceae* | Sangkanu et al. 2017 |

| Andaman Sea and the Gulf of Thailand | Sea sediments (3–30 m) | Heated at 55 °C for 60 min | 3% Artificial seawater agar, starch casein agar and starch yeast extract agar (3% artificial seawater) supplemented with ampicillin and cycloheximide, incubated at 28°C up to six weeks | *Actinopolymorpha, Actinomycetospora, Dietzia, Micromonospora, Mycobacterium, Nocardiopsis, Streptomyces* | Leetanasaksakul and Thamchaipenet 2018 |
|---|---|---|---|---|---|
| Chumphon beach (Chumphon Province), Bangsaen beach (Chonburi Province), Phanwa beach (Phuket Province), Koh Rok Nork, Koh Rok Nai, Koh Mah (Trang Province) and mangrove forest (Krabi Province) | Marine samples, including sand, sediments and marine sponges | Not specified | M1, M2 and seawater-proline media supplemented with cycloheximide and nalidixic acid, incubated at 28°C for 30 days | *Micromonospora, Nocardia, Salinispora, Streptomyces, Verrucosispora* | Phongsopitanun et al. 2019 |

**Starch casein agar medium (SCA)\*:** 0.4 g·L$^{-1}$ casein, 1.0 g·L$^{-1}$, starch, 0.1 g·L$^{-1}$ CaCO$_3$, 0.2 g·L$^{-1}$ KH$_2$PO$_4$, 0.5 g·L$^{-1}$ KNO$_3$, 0.1 g·L$^{-1}$ MgSO$_4$·7H$_2$O, 30.0 g·L$^{-1}$ NaCl, 15.0 g·L$^{-1}$ agar; **Starch yeast extract agar\*:** 10.0 g·L$^{-1}$ soluble starch, 4.0 g·L$^{-1}$ yeast extract, 2.0 g·L$^{-1}$ peptone, 18.0 g·L$^{-1}$ agar, 3% artificial seawater agar; **Actinomycetes isolation agar\*:** 2.0 g·L$^{-1}$ Sodium caseinate, 0.1 g·L$^{-1}$ L-Asparagine, 4.0 g·L$^{-1}$ Sodium propionate, 0.5 g·L$^{-1}$ K$_2$HPO$_4$ 0.1 g·L$^{-1}$ MgSO$_4$, 0.001 g·L$^{-1}$ FeSO$_4$, 15.0 g·L$^{-1}$ agar; **Modified soil extract agar\*:** 5.0 g·L$^{-1}$ soluble starch, 1.0 g·L$^{-1}$ KNO$_3$, 1000 mL soil extracts, 10.0 g·L$^{-1}$ agar; **Humic acid-salts vitamin agar\*:** 2.0 g·L$^{-1}$ humic acid, 1.0 g·L$^{-1}$ Asparagine, 0.5 g·L$^{-1}$ K$_2$HPO$_4$ 0.5 g·L$^{-1}$ FeSO$_4$·7H$_2$O, 20.0 g·L$^{-1}$ agar; **Starch nitrate agar\*:** 20.0 g·L$^{-1}$ starch, 1.0 g·L$^{-1}$ KNO$_3$, 0.5 g·L$^{-1}$ K$_2$HPO$_4$, 0.5 g·L$^{-1}$ MgSO$_4$·7H$_2$O, 0.5 g·L$^{-1}$ NaCl, 0.01 g·L$^{-1}$ FeSO$_4$·7H$_2$O, 20.0 g·L$^{-1}$ agar; **Glucose Asparagine Agar\*:** 10.0 g·L$^{-1}$ glucose, 0.5 g·L$^{-1}$ asparagine, 0.5 g·L$^{-1}$ K$_2$HPO$_4$, 15.0 g·L$^{-1}$ agar; **ISP 2 medium\*:** 4.0 g·L$^{-1}$ yeast extract, 4.0 g·L$^{-1}$ glucose, 10.0 g·L$^{-1}$ malt extract, 20.0 g·L$^{-1}$ agar; **M1\*:** 10.0 g·L$^{-1}$ soluble starch, 4.0 g·L$^{-1}$ yeast extract, 2.0 g·L$^{-1}$ peptone, 18.0 g·L$^{-1}$ agar; **M2\*:** 6 mL 100% glycerol, 1.0 g·L$^{-1}$ arginine, 1.0 g·L$^{-1}$ K$_2$HPO$_4$, 0.5 g·L$^{-1}$ MgSO$_4$, 18.0 g·L$^{-1}$ agar; **M4\*:** not specified; **Seawater-proline media\*:** 1.0% proline, seawater.

**\* Shown in the Family-level taxonomic assignments of 16S rRNA gene.**

2006). Hunadanamra et al. (2013) heated suspensions of sediment at 60–65°C for 15 minutes to reduce the number of non-thermotolerant microorganisms prior to plating onto selective isolation plates. Ruttanasutja and Pathom-aree (2015) found that it was necessary to shake sediment samples at 125 rpm for 60 minutes to isolate actinomycetes from coastal marine sediments.

## 2.2 Selective Media

The successful isolation of marine actinomycetes requires knowledge of their taxonomy and physiology (e.g., carbon and nitrogen sources, growth factors and inhibitors) and their tolerance to environmental factors (e.g., temperature, oxygen and pH) (Jiang et al. 2016, Subramani and Sipkema 2019). Such information can be used to design selective media for the isolation of specific fractions of actinomycete communities present in natural habitats (Goodfellow and Fiedler 2010). Actinomycetes can use both simple and complex substrates as carbon and nitrogen sources for growth. Their ability to use complex carbon (e.g., glycerol, humic acid and starch) and nitrogen (e.g., casein, chitin and nitrate) sources are used to formulate selective media to target actinomycete of interest (Manivasagan et al. 2014). Most fast-growing bacteria, such as *Escherichia* (Welch 2006) and *Pseudomonas* (Moore et al. 2006) do not grow on isolation media based on complex substrates. Common selective media used for the isolation of actinomycetes from marine habitats in Thailand are (1) Actinomycetes Isolation, (2) Gause No. 2, (3) Humic acid-salts vitamin, (4) Humic acid vitamin, (5) M1, (6) Starch casein and (7) Starch-casein nitrate agars (Tables 1 & 2). Starch is the most effective carbon source as more than half of new actinomycete species isolated in Thailand were recovered from starch based selective media. On the other hand, marine actinomycetes have been isolated on AV medium which combines the use of simple (glucose) and complex (glycerol) carbon compounds (Phongsopitanun et al. 2016a, b). This medium contains L-arginine as a nitrogen source which enhances the growth and development of aerobic soil actinomycetes on isolation plates (Porter et al. 1960, El-Nakeeb and Lechevalier 1963).

   In general, marine habitats are characterized by low concentrations of nutrients. The use of low nutrient selective media to mimic nutrient availability in marine ecosystems has been used with considerable success to increase the recovery of actinomycetes (Cannon and Giovannoni 2002, Gontang et al. 2007, Wang et al. 2014, Pulschen et al. 2017). A range of diluted selective media, namely 1:100 marine agar, 1:100 marine soil extract agar, 1:100 starch casein soil extract agar and 1:100 Gause No. 2 agar were used successfully to increase the numbers of actinomycetes from coastal marine sediments by at least 17% compared to corresponding media rich in nutrients (Ruttanasutja and Pathom-aree 2015). Similarly, media prepared using concentrations of seawater or synthetic sea water that mimic the osmotic pressure of natural marine environments provide an effective way of recovering actinomycetes from environments (Subramani and Sipkema 2019). All novel species of actinomycetes isolated from marine habitats in Thailand involved the use of either seawater or synthetic seawater (Table 2).

**Table 2.** Novel actinobacterial species isolated from tropical marine environments in Thailand from 2008–2020.

| Family | Species | Location | Sample | Selective medium | Incubation temperature and time | References |
|---|---|---|---|---|---|---|
| *Jiangellaceae* | *Jiangella mangrovi* | Laemson National Park, Ranong Province, Thailand | Mangrove soil | 10-fold-diluted marine agar 2216 (MA; Difco) supplemented with nalidixic acid and nystatin | 28°C for 1 month | Suksaard et al. 2015 |
| *Kineosporiaceae* | *Kineococcus mangrovi* | Phetchaburi Province | Mangrove sediment | Starch casein agar supplemented with nalidixic acid and ketoconazole | 28°C for 14 days | Duangmal et al. 2016 |
| *Micromonosporaceae* | *Micromonospora krabiensis* | Krabi Province | Marine soil | Starch-casein nitrate seawater agar | 30°C for 21 days | Jongrungruangchok et al. 2008 |
| | *Micromonospora pattaloongensis* | Pattaloong Province | Mangrove forest soil | Starch-casein nitrate agar | 30°C for 21 days | Thawai et al. 2008 |
| | *Micromonospora marina* | Hua-Hin, Prajuabkirikhun Province | Sea sand | Starch-casein nitrate agar | 30°C for 21 days | Tanasupawat et al. 2010 |
| | *Micromonospora maritima* | Samut Sakhon Province | Mangrove soil | Starch-casein nitrate agar supplemented with nalidixic acid and nystatin | 30°C for 14 days | Songsumanus et al. 2013 |
| | *Micromonospora spongicola* | Gulf of Thailand | Marine sponge | Modified starch-casein nitrate seawater agar | 30°C for 21 days | Supong et al. 2013a |
| | *Verrucosispora andamanensis* | Andaman Sea, Phuket Province | Marine sponge *Xestospongia* sp. | Modified starch-casein nitrate seawater agar | 30°C for 21 days | Supong et al. 2013b |
| | *Micromonospora sediminicola* | Andaman Sea, Phuket Province | Marine sediment | Modified starch-casein nitrate seawater agar | 30°C for 21 days | Supong et al. 2013c |

*Table 2 contd. ...*

...*Table 2 contd.*

| Family | Species | Location | Sample | Selective medium | Incubation temperature and time | References |
|---|---|---|---|---|---|---|
| | *Micromonospora fluostatini* | Panwa Cape, Phuket Province | Nearshore marine sediments | M1 medium supplemented with cycloheximide and nalidixic acid | 28°C for 21 days | Phongsopitanun et al. 2015 |
| | *Micromonospora sediminis* | Chonburi Province | Mangrove sediment | AV medium | 30°C for 1 month | Phongsopitanun et al. 2016a |
| *Nocardiaceae* | *Nocardia xestospongiae* | Andaman Sea, Phuket Province | Marine sponge *Xestospongia* sp. | Modified starch-casein nitrate seawater agar supplemented with nalidixic acid and nystatin | 30°C for 21 days | Thawai et al. 2017 |
| | *Gordonia sediminis* | Songkhla Province | Mangrove sediment | Humic acid-vitamin agar supplemented with nalidixic acid and cycloheximide | 28°C for 21 days | Sangkanu et al. 2019 |
| *Nocardiopsaceae* | *Nocardiopsis sediminis* | Laem Son National Park, Ranong Province | Mangrove sediment | Starch casein agar supplemented with nalidixic acid and nystatin | 28°C for 2 weeks | Muangham et al. 2016 |
| *Promicromonosporaceae* | *Promicromonospora thailandica* | Andaman Sea, Trang Province | Marine sediment | Humic acid-salts vitamin agar supplemented with cycloheximide and nalidixic acid | Not specified | Thawai and Kudo 2012 |
| *Pseudonocardiaceae* | *Pseudonocardia mangrovi* | Mangrove forest, Samut Prakan Province | Mangrove soil | Humic acid-salts vitamin agar supplemented with cycloheximide and nystatin | Not specified | Chanama et al. 2018 |
| | *Saccharopolyspora maritima* | Ranong Province | Mangrove sediment | Starch casein agar medium supplemented with nalidixic acid and nystatin | 28°C for 1 month | Suksaard et al. 2018 |
| *Streptomycetaceae* | *Streptomyces similanensis* | Similan Island National Park, Phanga Province | Marine soil | Starch-casein nitrate agar supplemented with nystatin and tetracycline | Not specified | Sripreechasak et al. 2013 |

| | Species | Location | Source | Medium | Conditions | Reference |
|---|---|---|---|---|---|---|
| | *Streptomyces chumphonensis* | Chumphon beach Chumphon Province | Marine sediment | Seawater–proline medium supplemented with cycloheximide and nalidixic acid | 28°C for 2–3 weeks | Phongsopitanun et al. 2014 |
| | *Streptomyces ferrugineus* | Not specified | Mangrove soil | Humic acid-vitamin agar (HV) medium supplemented with $K_2Cr_2O_7$ | 28°C for 5–7 days | Ruan et al. 2015 |
| | *Streptomyces andamanensis* | Similan Islands, Phang-Nga Province | Marine soil | Starch-casein nitrate agar supplemented with nystatin | 28°C for 14 days | Sripreechasak et al. 2016 |
| | *Streptomyces verrucosisporus* | Chumphon beach Chumphon Province | Marine sediment | Seawater–proline medium supplemented with cycloheximide and nalidixic acid | 28°C for 2–3 weeks | Phongsopitanun et al. 2016b |
| *Streptosporangiaceae* | *Sphaerisporangium krabiense* | Krabi Province | mangrove forest soil | Humic acid-salts vitamin agar supplemented with cycloheximide, terbinafine and nalidixic acid | Not specified | Suriyachadkun et al. 2011 |
| | *Nonomuraea purpurea* | Ranong Province | Mangrove sediment | 10-fold-diluted marine agar 2216 (MA; Difco) supplemented with nalidixic acid and nystatin | 28°C for 1 month | Suksaard et al. 2016 |
| | *Nonomuraea suaedae* | Phetchaburi Province | Rhizosphere soil of *Suaeda maritima* | Humic acid-vitamin agar supplemented with nalidixic acid and nystatin | 30°C for 21 days | Lipun et al. 2019 |

**Modified starch-casein nitrate seawater agar\***: 10.0 g soluble starch, 1.0 g sodium caseinate, 0.5 g $KH_2PO_4$, 0.5 g $MgSO_4$ and 18.0 g agar in 1 l seawater, pH 8.3; **Seawater–proline medium\***: 10.0 g proline, 15.0 g agar, 1000 ml artificial seawater; **M1 medium\***: 10.0 g soluble starch,4.0 g yeast extract, 2.0 g peptone,18.0 g agar, 1l seawater; **Humic acid-salts vitamin agar\***: 2.0 g Humic acid, 1.0 g Asparagine, 0.5 g $K_2HPO_4$ 0.5 g $FeSO_4 \cdot 7H_2O$, 20.0 g agar, 1000 ml sea-water, pH 7.0–7.4; **Starch-casein nitrate seawater agar\***: 10.0 g starch, 0.3 g sodium caseinate (Difco), 2.0 g $KNO_3$ and 15.0 g agar, pH 7.0–7.4; **Starch casein agar medium\***: 0.4 g casein, 1.0 g. starch, 0.1 g $CaCO_3$, 0.2 g $KH_2PO_4$ 0.5g $KNO_3$, 0.1 g $MgSO_4 \cdot 7H_2O$, 15.0 g agar, 1000 ml marinium synthetic sea salt, pH 7.0–7.4; **AV medium\***: 1.0 g glucose, 1.0 g glycerol, 0.3 g L-arginine, 0.3 g $K_2HPO_4$, 0.2 g $MgSO_4 \cdot 7H_2O$, 0.3 g NaCl, 18 g agar, artificial seawater added up to 1 litre, pH 7.4.

The addition of anti-bacterial and anti-fungal antibiotics to selective media is used to inhibit the growth of un-wanted microorganisms thereby facilitating the growth of actinomycetes on isolation plates (Jiang et al. 2016, Kamjam et al. 2017). Widely used fungal inhibitors include cycloheximide (50 mg/l), ketoconazole (100 mg/ml) and nystatin (25–50 mg/l). Similarly, nalidixic acid (0.01–25 mg/ml) is used to control the growth of Gram-negative bacteria on isolation plates. A combination of nalidixic acid and anti-fungal antibiotics has been used to facilitate the isolation of actinomycetes from marine sources (Jiang et al. 2016, Kamjam et al. 2017).

Actinomycetes from marine environments tend to grow slowly compared to their terrestrial counterparts, hence long incubation times can be crucial factor in the growth on agar plates. This simple practice has been used to isolate a range of novel actinomycetes from diverse environments (Busarakam et al. 2014, Pulschen et al. 2017, Rangseekaew and Pathom-aree 2019). Extending incubation times to one month led to the isolation and characterization of *Jiangella mangrovi* (Suksaard et al. 2015), *Micromonospora sediminis* (Phongsopitanun et al. 2016a), *Nonomuraea purpurea* (Suksaard et al. 2016) and *Saccharopolyspora maritima* (Suksaard et al. 2018) from mangrove sediments. Similarly, actinomycetes were isolated from Thai habitats by incubating selective isolation plates for 2–4 weeks (Tables 1 & 2). The isolation of slow-growing actinomycetes following prolonged incubation is enhanced by the use of deep set media (Jiang et al. 2016).

## 3. Diversity and Distribution of Marine Actinomycetes

Actinomycetes are widely distributed in tropical marine habitats in Thailand, ranging from sea sediments (Leetanasaksakul and Thamchaipenet 2018), mangrove sediments (Sangkanu et al. 2017, Tangjitjaroenkun et al. 2017), rhizosphere soil (Lipun et al. 2019), coastal soils (Srivibool and Sukchotiratana 2006), island soils (Srivibool and Sukchotiratana 2006) to marine sponges (Phongsopitanun et al. 2019) (Table 1). They are isolated from sea sediments of the Andaman and the Gulf of Thailand coasts were dominated by members of the genus *Streptomyces* though strains assigned to the genera *Actinomycetospora*, *Actinopolymorpha*, *Dietzia*, *Nocardiopsis*, *Micromonospora* and *Mycobacterium* were also isolated (Leetanasaksakul and Thamchaipenet 2018).

Mangrove ecosystems account for approximately 60–75% of the world's transition zones (Holguin et al. 2001). Mangroves are saline, highly rich in organic matter with high nitrogen and sulfur contents (Kizhekkedathu and Parukuttyamma 2005). Sediment microorganisms are responsible for energy flow and major nutrient transformations within mangrove ecosystems. Mangrove regions in Thailand occur predominantly in the southern and eastern region of the country. Thai mangrove habitats are a rich source of actinomycetes. Sangkanu et al. (2017) isolated actinomycetes from a mangrove forest in Songkhla Province, Southern Thailand and discovered that the dominant family was the *Streptomycetaceae*, followed by the *Micromonosporaceae*, *Nocardiaceae*, *Pseudonocardiaceae* and *Thermomonosporaceae*. Diverse actinomycetes were isolated from the Nakorn Si-Thammarat mangrove though the major components were *Streptomyces*,

*Micromonospora, Nocardiopsis, Streptoalloteichus, Nocardia,* and a number of unidentified genera. In contrast, actinomycetes from Rayong and Chumporn mangroves were mainly members of the family *Micromonosporaceae,* notably the genera *Micromonospora, Salinispora, Spirilliplanes* and *Virgisporangium* (Srivibool and Watanadilok 2015).

## 4. Novel Marine Actinomycetes

Actinomycetes isolated from Thai tropical marine environments have been assigned to an impressive number of novel species with the highest number obtained from mangrove sediments/soils (48%), followed by marine sediments/soils (32%), marine sponges (12%), rhizosphere soil (4%) and sea sand (4%) (Fig. 1).

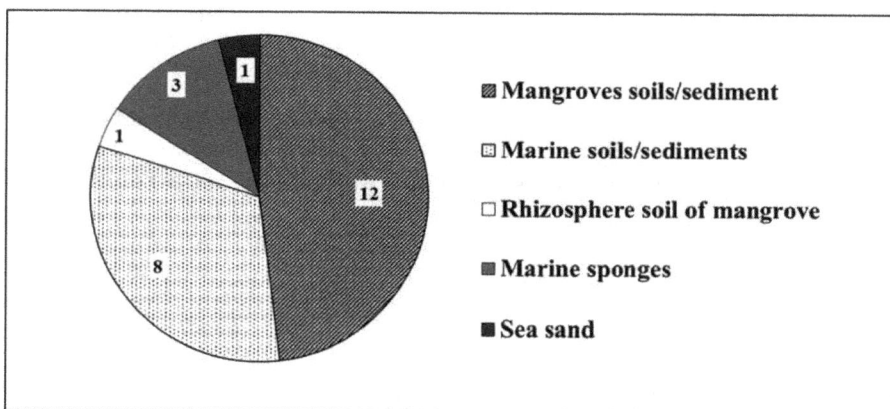

**Figure 1.** Number of novel actinomycetes species isolated from tropical marine environments of Thailand.

These novel species were obtained from around 60% of the seaside provinces and from the central (Phetchaburi, Prajuabkirikhun, Samut Prakan), east (Ranong, Samut Sakhon, Chonburi) and south (Krabi, Phuket, Pattaloong, Songkhla, Ranong, Trang, Phang-Nga, Chumphon) of the country. Twenty-five of the novel species were classified in 9 families and 13 genera. Most of the species were assigned to the genus *Micromonospora* (n = 8), followed by *Streptomyces* (n = 5), *Nonomuraea* (n = 2) with single species shown to belong to the genera *Gordonia, Jiangella, Kineococcus, Nocardia, Nocardiopsis, Promicromonospora, Pseudonocardia, Saccharopolyspora, Sphaerisporangium* and *Verrucosispora* (now subsumed into the genus *Micromonospora* (Nouioui et al. 2018) (Table 2).

Novel actinomycetes were recovered mainly from three types of samples (1) mangroves samples, (2) marine samples and (3) marine sponges. It is not surprising that the highest number and the most diverse novel actinomycete species are from mangrove sites with *Micromonospora* (23%) as the dominant genus (Fig. 2a). Mangroves are the most unique marine habitats in tropical regions, including Thailand. It is the only known ecosystem between land and sea with blackish water. Marine samples (marine sediments/soils and sea sand) yielded the second highest

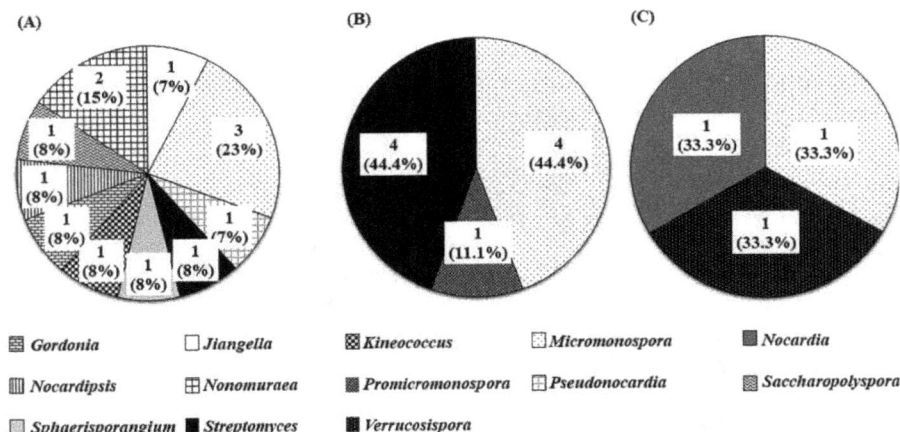

**Figure 2.** Number of novel marine actinomycetes in (A) Mangrove samples (B) Marine samples and (C) Marine sponges.

number of novel species with *Micromonospora* (44%) and *Streptomyces* (44%) as the predominant genera (Fig. 2b). Three novel species of *Nocardia* and *Micromonospora* (including *Verrucosispora*) were recovered from *Xestospongia* sp. and un-identified marine sponges (Fig. 2c).

## 5.  Biotechnological Applications

### *5.1  Bioactive Compounds*

#### *5.1.1  Bioactivities of Marine Actinomycetes from Tropical Marine Environments of Thailand*

Marine actinomycetes assigned to 12 genera were reported to show beneficial activities (Table 3), notably streptomycetes which displayed the widest range of bioactivities, including antimicrobial activities against several Gram-positive and Gram-negative bacterial pathogens, they also inhibited the growth of fungal pathogens such as *Candida albicans*. Members assigned to rare genera also exhibited bioactivities, as exemplified by antimicrobial properties shown by *Actinomadura*, *Salinispora* and *Micromonospora*, *Microbispora*, *Nocardia*, *Pseudonocardia*, *Streptoalloteichus*, *Saccharomonospora* and *Streptoverticillium* strains. Anti-oxidant activity was detected in cultures of *Nocardia*, *Nocardiopsis*, *Streptoalloteichus* and *Virgisporangium* strains. *Nocardiopsis* strains were also found to have anti-biofilm activity.

Emerging and re-emerging infectious diseases caused by antibiotic resistant bacteria are threatening human health on a global scale. Pathogenic bacteria such as *Acinetobacter baumannii*, *Pseudomonas aeruginosa* and methicillin resistant *Staphylococcus aureus* (MRSA) strains (Ventola 2015) have developed resistance to antibiotics that were once commonly used to treat them (Sangkanu et al. 2017). Consequently, new drugs are urgently needed. Actinomycetes from marine sources are a promising source of new bioactive compounds (Subramani and Albersberg

**Table 3.** Bioactivity of actinobacteria isolated from tropical marine environments in Thailand.

| Location | Sample | Genus | Bioactivity | References |
|---|---|---|---|---|
| Samae Sarn, Juang, Raet and Kram islands | Soil samples | *Actinomadura* | - Antimicrobial activity against *B. subtilis, Micrococcus luteus, S. aureus, C. albicans* | Srivibool 2000 |
| Chonburi, Rayong and Trat Provinces | Coastal sediment | Thermotolerant *Streptomyces* | *N*-acylamino acid racemase activity | Srivibool et al. 2004 |
| Chang, Hwai, Lao-yanai (Trat Province) and Pai Islands (Chonburi Province) | Island soils | *Actinomadura, Micromonospora, Microbispora, Nocardia, Pseudonocardia, Streptoalloteichus, Saccharomonospora Streptomyces, Streptoverticillium* | - Antimicrobial activity against at least one of the following pathogens: *B. subtilis, S. aureus, M. luteus, P. aeruginosa, E. coli* | Srivibool and Sukchotiratana 2006 |
| East coast of the Gulf of Thailand | Marine sediment | *Streptomyces* | - Antimicrobial activity against Gram-positive (*B. subtilis*, methicillin-resistant *S. aureus* (MRSA), *M. luteus* Gram-negative bacteria (*P. aeruginosa* and fungi (*Aspergillus niger, Candida albicans, Debaryomyces hansenii, Mucor racemosus, Schizosaccharomyces pombe, Penicillium chrysogenum*<br>- Anti-breast cancer cells | Srivibool et al. 2010 |
| Andaman seashore, Trang Province | Marine sediments | *Micromonospora* | - Antimicrobial activities against *B. subtilis, S. aureus, M. luteus, P. aeruginosa, E. coli, C. albicans* | Thawai 2010 |
| Mangrove forests along the inner Gulf of Thailand (Samut Prakan and Samut Songkarm Provinces) | Mangrove soils | *Streptomyces* | - Antimicrobial activities against *S. aureus, B. subtilis, K. rhizophila, E. coli* and *P. aeruginosa* | Hunadanamra et al. 2013 |

*Table 3 contd. ...*

*...Table 3 contd.*

| Location | Sample | Genus | Bioactivity | References |
|---|---|---|---|---|
| East and west of the Gulf of Thailand | Mangrove sediments | *Nocardia, Nocardiopsis, Streptomyces, Streptoalloteichus* | Antimicrobial activities against at least one of *B. subtilis*, MRSA and *C. albicans* | Srivibool and Watanadilok 2015 |
| | | *Nocardiopsis, Streptomyces, Streptoalloteichus, Virgisporangium, Micromonospora* | Antioxidant activity | |
| Panwa Cape, Phuket Province | Nearshore marine sediment | *Micromonospora fluostatini* | Fluostatins B and C antibiotics | Phongsopitanun et al. 2015 |
| Not specified | Mangrove sediment | *Streptomyces* sp. J6.2 | Actinomycin-D | Kengnipat et al. 2016 |
| Eastern coast of, Chonburi and Chanthaburi Provinces | Mangrove sediments | *Streptomyces* | - Antimicrobial activities against *B. subtilis*, *S. aureus*, *P. aeruginosa*, *C. albicans*<br>- Cytotoxic activities on human fibrosarcoma cell line, HT1080 | Tangjitjaroenkun et al. 2017 |
| General Prem Tinsulanonda Historical Park, Songkhla Province | Mangrove sediments | *Streptomyces* | - Antibacterial activity against *S. aureus, S. epidermidis*, MRSA, *A. baumannii* and *E. coli*<br>- Anti-biofilm activity against *S. epidermidis*<br>- Inhibition of violacein | Sangkanu et al. 2017 |
| Andaman sea and the Gulf of Thailand | Sea sediments | *Nocardiopsis, Streptomyces* | - Anti-biofilm activities against *E. coli* and *S. aureus* | Leetanasaksakul and Thamchaipenet 2018 |

| Chumphon beach (Chumphon Province), Bangsaen beach (Chonburi Province), Panwa beach (Phuket Province), Koh Rok Nork, Koh Rok Rok Nai, Koh Mah (Trang Province) and mangrove forest (Krabi Province) | Marine samples, including sand, sediments and marine sponges | *Micromonospora Salinispora, Streptomyces,* | - Antimicrobial activities against *S. aureus, K. rhizophila, B. subtilis, E. coli, C. albicans,* and *M. racemosus* | Phongsopitanun et al. 2019 |

*A. baumannii\**: *Acinetobacter baumannii*; *B. subtilis\**: *Bacillus subtilis*; *C. albicans\**: *Candida albicans*; *E. coli\**: *Escherichia coli*; *K. rhizophila\**: *Kocuria rhizophila*; *M. luteus\**: *Micrococcus luteus*; *M. racemosus\**: *Mucor racemosus*; **MRSA**\*: Methicillin-resistant *Staphylococcus aureus*; *P. aeruginosa\**: *Pseudomonas aeruginosa*; *S. aureus\**: *Staphylococcus aureus*; *S. epidermidis\**: *Staphylococcus epidermidis*.

2012, Manivasagan et al. 2014, Hassan et al. 2017, Kamjam et al. 2017, Subramani and Sipkema 2019). Sangkanu et al. (2017) isolated actinomycetes from mangrove forests in the south of Thailand (Songkhla province), screened them against various drug-resistant bacteria and found that *Streptomyces* isolates showed antibacterial activities against *A. baumannii*, *Escherichia coli*, *S. aureus* and *Staphylococcus epidermidis*. In addition, ethyl acetate extracts of *Streptomyces* isolate AMA11 cells displayed anti-bacterial activity against methicillin resistant *S. aureus* by destroying MRSA cells as seen in scanning electron micrographs. Moreover, GC-MS analysis of active compounds extracted from this strain revealed the presence of 3-nitro-1,2-benzenedicarboxylic acid and quinoxaline-2-carboxamide which are anti-MRSA and anti-biofilm compounds. Interestingly, 3-nitro-1,2-benzenedicarboxylic acid partially purified from *Nocardia levis* MK-V_113 was reported to have antibacterial and antifungal activities (Kavitha et al. 2010). Quinoxaline-2-carboxamide is a derivative of quinoxaline; quinoxaline compounds such as actinoleukin and levomycin produced by *Streptomyces* strains were reported to be antibacterial antibiotics (Katagiri et al. 1975). In general, streptomycetes from Thai marine habitats display antimicrobial activities against Gram-negative bacteria, such as *E. coli* (Thawai 2010, Hunadanamra et al. 2013, Phongsopitanun et al. 2019), *P. aeruginosa* (Srivibool et al. 2010), anti-Gram positive activity against *B. subtilis* and *Micrococcus luteus* (Srivibool and Sukchotiratana 2006, Thawai 2010) and *Kocuria rhizophila* (Hunadanamra et al. 2013, Phongsopitanun et al. 2019). They also inhibit the growth of fungi, including *Aspergillus niger*, *Candida albicans*, *Debaryomyces hansenii*, *Mucor racemosus*, *Penicillium chrysogenum* and *Schizosaccharomyces pombe* (Srivibool et al. 2010). Similarly, anti-microbial activities are shown by members of rare genera, for example, *Actinomadura* strains isolated from Samae Sarn, Juang, Raet and Kram islands inhibited the growth of *B. subtilis*, *M. luteus*, *S. aureus* and *C. albicans* (Srivibool 2000), while strains from the Gulf of Thailand belonging to the genera *Nocardia*, *Nocardiopsis*, *Streptoalloteichus*, *Streptomyces* and *Virgisporangium* inhibited the growth of *B. subtilis*, MRSA and *C. albicans* using the cross-streak method (Srivibool and Watanadilok 2015). *Salinispora arenicola* strains isolated from un-identified marine sponges were shown to inhibit various microorganisms including, *C. albicans*, *B. subtilis*, *E. coli*, *K. rhizophila*, *M. racemosus* and *S. aureus* (Phongsopitanun et al. 2019).

Biofilm formation is a survival strategy that helps protect bacteria from environmental stress (Leetanasaksakul and Thamchaipenet 2018). Biofilms are a major cause of persistent and recurrent infections by clinically important pathogens such as *E. coli*, *P. aeruginosa* and *S. aureus* (Verderosa et al. 2019), hence the need for anti-biofilm inhibition and anti-biofilm establishment agents. Actinomycetes from marine biomes with anti-biofilm activity against several pathogenic bacteria have been reported (Sangkanu et al. 2017, Leetanasaksakul and Thamchaipenet 2018). Streptomycetes from mangrove habitats have been found to show anti-biofilm activity against *S. epidermidis*, for example, extracts from *Streptomyces* strains AMA11 and AMA12 (0.5 × MIC concentration) reduced biofilm formation by 50% at an MIC concentration of 0.5, though these extracts did not inhibit the growth of *S. epidermidis* (Sangkanu et al. 2017). In addition, cell ethyl acetate extracts of *Streptomyces* isolates AMA11, AMA12 and AMA21CE and ethyl acetate extracts from the culture broth of isolate AMA12 weakly inhibited the growth of *Chromobacterium violaceum*. Quorum sensing has a major impact on biofilm formation. Anti-biofilm agents and quorum sensing inhibitors are needed to prevent biofilm formation. For the quantitative analysis of anti-quorum sensing, the inhibition of violacein synthesis (%) from the extracts of the *Streptomyces* strains was in a concentration dependent manner, though violacein inhibition was not observed at $0.125 \times$ MIC of cell extracts of *Streptomyces* strain AMA11. *Nocardiopsis* and *Streptomyces* strains isolated from Andaman Sea and the Gulf of Thailand sediments reduced biofilm formation of *E. coli* and *Staphylococcus* by 61 to 80% and over 60%, respectively (Leetanasaksakul and Thamchaipenet 2018).

Reactive oxygen species (ROS) cause oxidative stress and cardiovascular, cancer and inflammatory disorders in humans (Karthik et al. 2013) hence the need to find antioxidant compounds of therapeutic value. In general, the 2,2'-diphenyl-1-picrylhydrazyl (DPPH) radical scavenging assay is used to evaluate the ability of antioxidants to neutralize free radicals (Kedare and Singh 2011). Streptomycetes isolated from mangrove areas in the Gulf of Thailand were found to have strong DPPH radical scavenging activity while *Micromonospora, Nocardiopsis, Streptoalloteichus* and *Virgisporangium* strains showed potent antioxidant activity (Srivibool and Watanadilok 2015).

Approximately 60% of cancer in Thailand belong to five types: breast, cervix, colorectal, liver and lung cancer (Virani et al. 2017). In 2020, the total mortality from cancer in the Thai population was over 66% with liver cancer as the main cause (14.4%), followed by lung cancer (12.4%) (The Global Cancer Observatory 2021). The incidence of cancer shows an increasing trend hence the need for effective, new anticancer drugs. Actinomycetes isolated from mangrove sites show anticancer activity, as exemplified by *Streptomyces* strain CH54-4 from mangrove sediment in the Gulf of Thailand showing strong activity against breast cancer cells with an IC50 of 2.91 µg ml$^{-1}$ (Srivibool et al. 2010). Similarly, streptomycetes from a mangrove soil from the east coast of Thailand showed strong cytotoxic effects on human fibrosarcoma cell lines HT1080 (Tangjitjaroenkun et al. 2017).

In the amino acid production industry, N-acylamino acid racemase is an essential enzyme for converting N-acetyl-DL-methionine into L-methionine. However, a thermotolerant or thermostable enzyme is needed to complete the conversion of N-acetyl-DL-methionine into L-methionine. *Streptomyces tendae* Sal 35-16 and *Streptomyces maritimus* Sal 31-15, thermotolerant strains isolated from coastal sediment from the Gulf of Thailand produce N-acylamino acid racemase leading to the production of L-methionine at 0.68 mM and 0.21 mM, respectively (Srivibool et al. 2004).

The examples given above show the potential of actinomycetes from Thai marine habitats to produce bioactive compounds. However, there are only two reports where the chemical structure of such compounds have been determined. *Micromonospora fluostatini* strain PWB-003T was found to produce fuostatins B and C (Phongsopitanun et al. 2015), dipeptidyl peptidase III inhibitors which hydrolyse dipeptides from N-terminal oligopeptides from three to 10 amino acid residues. This enzyme is involved in treating primary ovarian carcinoma, oxidative stress (Nrf2 nuclear localization), pain, inflammation and cataractogenesis (Prajapati and Chauhan 2011). *Streptomyces parvulus* strain J6.2 isolated from mangrove soil produces actinomycin D, an anti-cancer drug which has been used to treat a kidney tumor (Wilm's tumor) in children (Kengpipat et al. 2016). This strain also shows anti-microbial activitiy against *A. niger*, *B. subtilis*, *C. albicans*, *E. coli*, *P. aeruginosa*, *S. aureus*, and *S. cerevisiae*.

## 5.2  Potential to Promote Plant Growth

Actinomycetes are regarded as the most prolific producers of biologically active compounds, including plant growth promoting substances (Palaniyandi et al. 2013, Hamedi and Mohammadipanah 2015, Betancur et al. 2017, Bhatti et al. 2017). The use of plant growth promoting actinomycetes (PGPA) in agriculture is attracting increasing attention (Nouioui et al. 2019). Commercial bioinoculants from *Streptomyces* strains have been developed to improve plant growth and control soil-borne pathogens, for example, *Streptomyces lydicus* WYEC 108 which is commercialized either as Actinovate® or as Actino-iron®, and *Streptomyces griseoviridis*, a fungicide used to control soil and seed-borne pathogens which is sold under the trade name Mycostop® (Sousa and Olivares 2016).

The application of actinomycetes to promote plant growth and to control plant diseases offers an eco-friendly alternative to pesticides for sustainable agriculture. Actinomycetes have been shown to give benefit to plants through direct and indirect plant growth promoting mechanisms (Sousa and Olivares 2016, Nouioui et al. 2019). In Thailand, actinomycetes isolated from marine environments have been shown to promote plant growth. Suksaard et al. (2017) demonstrated that *Nocardiopsis* 1SM5-02, *Pseudonocardia* 3WH5-01, *Streptomyces* 2SH3-07 and *Streptomyces* 3SH5-05, isolated from sea water and mangrove sediments enhanced the growth of rice seedling under saline (100–200 mM NaCl) and non-saline conditions. Further, Pathom-aree et al. (2019) showed that a mangrove isolates, *Streptomyces* strain S2-SC16 promoted the growth of rice seedlings by increasing several growth parameters, including root length, shoot height, leaf width and fresh weight in pot experiments. This organism also showed an ability to control Bakanae disease in Thai jasmine rice variety KDML 105. These studies show that actinomycetes from mangrove sites have the potential to promote plant growth under normal and salt stress conditions.

### 5.2.1  Direct Promotion of Plant Growth

#### 5.2.1.1  Indole-Acetic Acid (IAA) Production

IAA is a key phytohormone which takes part in the regulation of basic cellular processes in plants, including cell division, elongation and differentiation (Matsukawa et al. 2007, Bhatti et al. 2017). It also increases the rate of xylem and root development, stimulates seed and tuber germination, photosynthesis, pigment formation and resistance to stress conditions (Glick 2012, Kour et al. 2019). IAA production was detected in 228 out of 488 actinomycetes (50.9%) isolated from sea water and marine sediments; the amount of IAA produced fell within the range of 0.21 to 165.74 (μg/ml) (Suksaard et al. 2017). These workers also found that

*Nocardiopsis*, *Pseudonocardia* and *Streptomyces* strains were able to produce IAA in the presence of 100, 200 and 300 mM of sodium chloride.

### 5.2.1.2 Phosphate Solubilization

Phosphorus is one of the most essential macronutrients needed for plant growth, it is involved in cell division, development, photosynthesis, sugar breakdown and nutrient transportation (Behera et al. 2014). In nature, most soil phosphorus is found in insoluble forms which cannot be used by plants. Plants only absorb phosphorus in soluble forms as monobasic ($H_2PO_4^-$) and dibasic ($HPO_4^{2-}$) ions. Lack of phosphorus can lead to atrophy of healthy stems and leaves, and in a delay of fruit maturation (Kalayu 2019). Microorganisms, including actinomycetes, produce organic and inorganic acids which solubilize inorganic phosphate and release phosphorus in soil (Rodriguez and Fraga 1999, Behera et al. 2014, Nouioui et al. 2019). Phosphate-solubilizing actinomycetes occur in diverse environments, including marine habitats (Dastager and Damare 2013) and have been recovered from mangrove sites in Thailand (Suksaard et al. 2017). These investigators found that nearly 40% of actinomycetes isolated from sea water and mangrove sediments showed phosphate solubilizing activity while *Pseudonocardia* and *Streptomyces* solubilized tri-calcium phosphate under NaCl concentration up to 300 mM.

### 5.2.1.3 ACC Deaminase

In plants, 1-aminocyclopropane-1-carboxylic acid (ACC) is the immediate precursor of the hormone ethylene. PGP bacteria can reduce stress in plants by lowering the level of ethylene via ACC hydrolysis to ammonia and α-ketobutyrate which is mediated by the enzyme ACC deaminase. It is widely reported that plants inoculated with ACC deaminase-producing bacteria are more resistant to both abiotic (salinity, drought, flood) and biotic (plant pathogens and pests attack) stress (Gupta and Pandey 2019, Karthikeyan et al. 2012, Saikia et al. 2018). Eleven out of 50 actinomycetes isolated from Thai mangrove sediments produced ACC deaminase which enabled them to grow on ACC supplemented media as a sole nitrogen source (Suksaard et al. 2017). Seven of these strains showed ACC deaminase activity when the concentration of NaCl in the test medium was increased to 300 mM.

An alternative strategy to improve salt tolerance of crops may be achieved by introducing salt-tolerant microbes that enhance crop growth (Dodd and Pérez-Alfocea 2012). Endophytic actinomycetes isolated from halophilic plants (e.g., *Salicornia bigelovii*) have been reported to produce ACC deaminase thereby enhancing plant growth under stress conditions (El-Tarabily et al. 2019).

### 5.2.2 *Indirect Plant Growth Promotion*

### 5.2.2.1 Siderophore Production

Siderophores are low molecular weight, high-affinity iron chelating compounds that serve to bind and transport $Fe^{3+}$ into microbial cells (Glick 2012). These compounds can enhance plant growth both directly and indirectly. Microorganisms produce siderophores to bind $Fe^{3+}$ which is then taken up for their metabolism. In addition, siderophores can act as biocontrol agents to prevent phytopathogens from acquiring sufficient iron thereby limiting their growth (Glick 2012, Sathya et al. 2017). Rungin et al. (2012) isolated the endophytic *Streptomyces* strain GMKU 3100 from roots of a Thai jasmine rice plant. and showed that a siderophore-deficient mutant produced siderophores that enhanced the growth of rice and mung beans. Suksaard et al. (2017) found that 397 of 448 actinomycetes (89%) from mangrove sediments and sea water produced siderophores on chrome azurol S (CAS) agar. They also showed that strains assigned to the genera *Nocardiopsis*, *Pseudonocardia* and *Streptomyces* produced siderophores under salt stress up to 300 mM NaCl, and enhanced rice seedling growth under salt stress at 100 and 200 mM NaCl as exemplified from the observed shoot/root length and shoot/root weight.

### 5.2.2.2 Antibiotic Production

Actinomycetes can also produce secondary metabolites, especially antibiotics, which inhibit the growth of plant pathogens. These organisms include strains isolated from marine sources which show antifungal and antimicrobial activity. *Streptomyces chumphonensis* isolate BDK01 recovered from a marine sediment displayed antimicrobial activities against clinical importance strains and produced various bioactive compounds such as salicyl alcohol and N-phenyl benzamide (Manikandan et al. 2019). Similarly, strains isolated from mangroves of the Colombian Caribbean Sea exhibited antifungal activity against plant pathogenic fungi, such as *Colletotrichum gloeosporioides* and *Fusarium oxysporum* (Betancur et al. 2017). Suksaard et al. (2017) tested the antagonistic potential of 50 plant growth promoting actinomycetes from marine habitats against the rice pathogenic bacteria *Xanthomonas oryzae* pv. *oryzae* and *X. oryzae* pv. *oryzicola* and found that almost half of them inhibited at least one of the bacterial pathogens; 14 of the isolates (28%) showed inhibitory activity against both pathogens. Pathom-aree et al. (2019) reported that *Streptomyces* strain S2-SC16, isolated from a Thai mangrove sediment, inhibited the germination of *Fusarium fujikuroi* CMU-F02 conidia and suppressed the ability of this pathogen to infect rice seeds. Furthermore, this strain showed an ability to promote the growth of rice seedlings by increasing several growth parameters, including root length, shoot height, leaf width and weight in pot experiments. These results show that actinomycetes from mangroves have the potential to control Bakanae disease in Thai jasmine rice variety KDML 105.

## 6.  Conclusions and Future Prospects

Marine habitats in Thailand contain taxonomically diverse actinomycetes taxa, as shown by the recognition of at least 25 novel species assigned to 12 genera over the past 13 years. The majority of these taxa belong to the genus *Micromonospora*, which is second only to *Streptomyces* as a source of new antibiotics (Carro et al. 2018). Most of the novel species encompass isolates from coastal sediments collected from either the Andaman Sea or Gulf of Thailand. These initial studies need to be extended to determine the composition of actinomycete communities in deep-sea sediments of the Thai coast. Deep-sea sediments are a rich source of novel bioactive actinomycetes (Kamjam et al. 2017, Sayed et al. 2019, Bull and Goodfellow 2019). Initial studies show that *Amycolatopsis* and *Streptomyces* strains are associated with seaweed (e.g., genus *Sargassum*) from the Gulf of Thailand (W.P., unpublished data). It seems likely that other marine organisms from Thai coastal water, such as corals, sea grasses and marine invertebrates will prove to be a source of novel actinomycetes.

Actinomycetes from Thai marine biomes show anticancer and antimicrobial properties though associated natural product chemistry is in its infancy (Phongsopitanun et al. 2015, Prajapati and Chauhan 2011). It can be expected that a continuation of such studies will yield additional compounds given the ability of marine derived actinomycetes to synthesize new antibiotics (Lu et al. 2019, Subramani and Sipkema 2019, Nweze et al. 2020). It is also encouraging that some of the actinomycetes from marine sources produce plant growth promoting metabolites (Suksaard et al. 2017, Pathom-aree et al. 2019). It is also interesting that seaweed associated *Amycolatopsis* strains have been shown to promote the growth of Chinese kale under salinity stress (W.P., unpublished data). However, further work is needed, to understand the PGP properties of such strains before they can be developed for use in sustainable agriculture.

The importance of integrated multidisciplinary approaches in the search for new natural products from actinomycetes has been stressed (Barka et al. 2016, Goodfellow et al. 2018, Baltz 2019). It is understandable that initial stages in natural product pipelines tend to be overlooked given the focus on outcomes, that is, on the detection of new and effective specialized (secondary) metabolites. However, in a broader context the isolation, recognition of "gifted" actinomycetes from poorly explored habitats are key elements in culture-dependent-natural products campaigns.

## Acknowledgements

Wasu Pathom-aree, Pharada Rangseekaew and Manita Kamjam are grateful for financial support from Chiang Mai University. Pharada Rangseekaew is supported by the Graduate School, Chiang Mai University TA/RA scholarship for 2019–2021. Kannika Duangmal is indebted to support from Kasetsart University. Special thanks, are also extended to Professor Michael Goodfellow for his constructive comments on the chapter.

## References

Aschariyaphotha, N. and Wongwises, S. (2012). Simulations of seasonal current circulations and its variabilities forced by runoff from freshwater in the Gulf of Thailand. *Arabian Journal for Science and Engineering*, 37: 1389–1404.

Baltz, R.H. (2019). Natural product discovery in the genomic era: Realities, conjectures, misconceptions and opportunities. *Journal of Industrial Microbiology and Biotechnology*, 46: 281–299.

Barka, E.A., Vatsa, P., Sanchez, L., Gaveau-Vaillant, N., Jacquard, C., Klenk, H.-P., Clément, C., Ouhdouch, Y. and Wezeld, G.P.v. (2016). Taxonomy, physiology, and natural products of *Actinobacteria*. *Microbiology and Molecular Biology Reviews*, 80: 1–43.

Behera, B.C., Singdevsachan, S.K., Mishra, R.R., Dutta, S.K. and Thatoi, H.N. (2014). Diversity, mechanism and biotechnology of phosphate solubilizing microorganism in mangrove—A review. *Biocatalysis and Agricultural Biotechnology*, 3: 97–110.

Berdy, J. (2012). Thoughts and facts about antibiotics: Where we are now and where we are heading. *Journal of Antibiotics*, 65: 385–395.

Betancur, L.A., Naranjo-Gaybor, S.J., Vinchira-Villarraga, D.M., Moreno-Sarmiento, N.C., Maldonado, L.A., Suarez-Moreno, Z.R. et al. (2017). Marine actinobacteria as a source of compounds for phytopathogen control: An integrative metabolic-profiling/bioactivity and taxonomical approach. *PLoS One*, 12(2): e0170148.

Bhatti, A.A., Haq, S. and Bhat, R.A. (2017). Actinomycetes benefaction role in soil and plant health. *Microbial Pathogenesis*, 111: 458–467.

Bull, A.T. and Goodfellow, M. (2019). Dark, rare and inspirational microbial matter in the extremobiosphere: 16 000 m of bioprospecting campaigns. *Microbiology*, 165: 1252–1264.

Busarakam, K., Bull, A.T., Girard, G., Labeda, D.P., Wezel, G.P. and Goodfellow, M. (2014). *Streptomyces leeuwenhoekii* sp. nov., the producer of chaxalactins and chaxamycins, forms a distinct branch in *Streptomyces* gene trees. *Antonie van Leeuwenhoek*, 105: 849–861.

Cannon, S.A. and Giovannoni, S.J. (2002). High-throughput methods for culturing microorganisms in very-low-nutrient media yield diverse new marine isolates. *Applied and Environmental Microbiology*, 68(8): 3878–3885.

Carro, L., Nouioui, I., Sangal, V., Meier-Kolthoff, J.P., Trujillo, M.E. and del Carmen Montero-Calasanz, M. et al. (2018). Genome-based classification of micromonosporae with a focus on their biotechnological and ecological potential. *Scientific Reports*, 8: 525.

Chanama, S., Janphen, S., Suriyachadkun, C. and Chanama, M. (2018). *Pseudonocardia mangrovi* sp. nov., isolated from soil. *International Journal of Systematic and Evolutionary Microbiology*, 68: 2949–2955.

Dastager, S.G. and Damare, S. (2013). Marine actinobacteria showing phosphate-solubilizing efficiency in Chorao Island, Goa, India. *Current Microbiology*, 66: 421–427.

Department of Marine and Coastal Resources, Thailand. (2012). Marine Protected Area Database of Thailand: Executive Summary. https://www.dmcr.go.th/detailLib/2635. Accessed 9 August 2020.

Department of Marine and Coastal Resources, Thailand. (2013). The Oceanography of the Gulf of Thailand. https://km.dmcr.go.th/en/c_51/d_1132. Accessed 9 August 2020.

Dodd, I.C. and Pérez-Alfocea, F. (2012). Microbial amelioration of crop salinity stress. *The Journal of Experimental Botany*, 63: 3415–3428.

Duangmal, K., Muangham, S., Mingma, R., Yimyai, T., Srisuk, N., Kitpreechavanich, V. et al. (2016). *Kineococcus mangrovi* sp. nov., isolated from mangrove sediment. *International Journal of Systematic and Evolutionary Microbiology*, 66: 1230–1235.

El-Nakeeb, M. and Lechevalier, H. (1963). Selective isolation of aerobic actinomycetes. *Applied Microbiology*, 11: 75–77.

El-Tarabily, K.A., AlKhajeh, A.S., Ayyash, M.M., Alnuaimi, L.H., Sham, A., El-Baghdady, K.Z. et al. (2019). Growth promotion of *Salicornia bigelovii* by *Micromonospora chalcea* UAE1, an endophytic 1-aminocyclopropane-1-carboxylic acid deaminase-producing actinobacterial isolate. *Frontiers in Microbiology*, 10: 1694.

Glick, B.R. (2012). Plant growth-promoting bacteria: Mechanisms and applications. *Scientifica*, 2012: 963401.

Gontang, E.A., Fenical, W. and Jensen, P.R. (2007). Phylogenetic diversity of gram positive bacteria cultured from marine sediments. *Applied and Environmental Microbiology*, 73(10): 3272–3282.

Goodfellow, M. and Fiedler, H.-P. (2010). A guide to successful bioprospecting: Informed by actinobacterial systematics. *Antonie van Leeuwenhoek*, 98: 119–142.

Goodfellow, M., Nouioui, I., Sanderson, R., Xie, F. and Bull, A.T. (2018). Rare taxa and dark microbial matter: Novel bioactive actinobacteria abound in Atacama Desert soils. *Antonie van Leeuwenhoek*, 111(8): 1315–1332.

Gupta, S. and Pandey, S. (2019). ACC deaminase producing bacteria with multifarious plant growth promoting traits alleviates salinity stress in French bean (*Phaseolus vulgaris*) plants. *Frontiers in Microbiology*, 10: 1506.

Hamedi, J. and Mohammadipanah, F. (2015). Biotechnological application and taxonomical distribution of plant growth promoting actinobacteria. *Journal of Industrial Microbiology and Biotechnology*, 42: 157–171.

Hassan, S.S.u., Anjum, K., Abbas, S.Q., Akhter, N., Shagufta, B.I., Shah, S.A.A. and Tasneem, U. (2017). Emerging biopharmaceuticals from marine actinobacteria. *Environmental Toxicology and Pharmacology*, 49: 34–47.

Hayakawa, M. (2008). Studies on the isolation and distribution of rare actinomycetes in soil. *Actinomycetologica*, 22(1): 12–19.

Hayakawa, M., Yoshida, Y. and Iimura, Y. (2004). Selective isolation of bioactive soil actinomycetes belonging to the *Streptomyces violaceusniger* phenotypic cluster. *Journal of Applied Microbiology*, 96: 973–981.

Holguin, G., Vazquez, P. and Basha, Y. (2001). The role of sediment microorganisms in the productivity, conservation, and rehabilitation of mangrove ecosystems: An overview. *Biology and Fertility of Soils*, 33: 265–278.

Hunadanamra, S., Akaracharanya, A. and Tanasupawat, S. (2013). Characterization and antimicrobial activity of *Streptomyces* strains from Thai mangrove soils. *International Journal of Bioassays*, 02(05): 775–779.

Jiang, Y., Li, Q., Chen, X. and Jiang, C. (2016). Isolation and cultivation methods of Actinobacteria. pp. 39–57. *In*: Dhanasekaran, D. and Jiang, Y. (eds.). *Actinobacteria-Basics and Biotechnological Applications*. InTech, BoD – Books on Demand, ISBN 9535122487, 9789535122487.

Jongrungruangchok, S., Tanasupawat, S. and Kudo, T. (2008). *Micromonospora krabiensis* sp. nov., isolated from marine soil in Thailand. *Journal of General and Applied Microbiology*, 54(2): 127–133.

Kalayu, G. (2019). Phosphate solubilizing microorganisms: Promising approach as biofertilizers. *International Journal of Agronomy*, 2019: 4917256.

Kamjam, M., Sivalingam, P., Deng, Z. and Hong, K. (2017). Deep sea actinomycetes and their secondary metabolites. *Frontiers in Microbiology*, 8: 760.

Karthik, L., Kumar, G. and Rao, K.V.B. (2013). Antioxidant activity of newly discovered lineage of marine actinobacteria. *Asian Pacific Journal of Tropical Medicine*, 6(4): 325–332.

Katagiri, K., Yoshida, T. and Sato, K. (1975). Quinoxaline antibiotics. pp. 234–251. *In*: Corcoran, J.W. and Hanh, F.F. (eds.). *Mechanism of Action of Antimicrobial and Antitumor Agent, in Antibiotics* Volume III. Springer-Verlag, Heidelberg, New York, USA.

Karthikeyan, B., Joe, M.M., Islam, M.R. and Sa, T. (2012). ACC deaminase containing diazotrophic endophytic bacteria ameliorate salt stress in *Catharanthus roseus* through reduced ethylene levels and induction of antioxidative defense systems. *Symbiosis*, 56: 77–86.

Kavitha, A., Prabhakar, P., Narasimhulu, M., Vijayalakshmi, M., Venkateswarlu, Y., Rao, K.V. et al. (2010). Isolation, characterization and biological evaluation of bioactive metabolites from *Nocardia levis* MK-VL_113. *Microbiological Research*, 165: 199–210.

Kedare, S.B. and Singh, R.P. (2011). Genesis and development of DPPH method of antioxidant assay. *The Journal of Food Science and Technology*, 48(4): 412–422.

Kengpipat, N., Pornpakakul, S., Piapukiew, J., Palaga, T., Robson, G., Whalley, A.J.S. et al. (2016). Effect of NaCl concentration on production of actinomycin-D by *Streptomyces* sp. strain J6.2 isolated from Thailand. *Chiang Mai Journal of Science*, 43(3): 682–686.

Kizhekkedathu, N.N. and Parukuttyamma, P. (2005). Mangrove actinomycetes as the source of lignolytic enzymes. *Actinomycetologica*, 19: 40–47.

Kour, D.K., Rana, K.L., Yadav, N., Yadav, A.N., Kumar, A. and Meena, V.S. et al. (2019). Rhizospheric microbiomes: Biodiversity, mechanisms of plant growth promotion, and biotechnological applications for sustainable agriculture. pp. 19–65. *In*: Kumar, A. and Meena, V.S. (eds.). *Plant Growth Promoting Rhizobacteria for Agricultural Sustainability*. Springer, Singapore.

Lange, I.D., Schoenig, E. and Khokiattiwong, S. (2019). Chapter 22 – Thailand. pp. 491–513. *In*: Charles S. (ed.). *World Seas: An Environmental Evaluation* (2nd ed). Academic Press, London.

Leetanasaksakul, K. and Thamchaipenet, A. (2018). Potential anti-biofilm producing marine actinomycetes isolated from sea sediments in Thailand. *Agriculture and Natural Resources*, 52(3): 228–233.

Lipun, K., Teo, W.F.A., Tongpan, J., Matsumoto, A. and Duangmal, K. (2019). *Nonomuraea suaedae* sp. nov., isolated from rhizosphere soil of *Suaeda maritima* (L.) Dumort. *Journal of Antibiotics*, 72: 518–523.

Lu, Q.P., Ye, J.J., Huang, Y.M., Liu, D., Liu, L.F., Dong, K., Razumova, E.A., Osterman, I.A., Sergiev, P.V., Dontsova, O.A., Jia, S.H., Huang, D.L. and Sun, C.H. (2019). Exploitation of potentially new antibiotics from mangrove actinobacteria in Maowei Sea by combination of multiple discovery strategies. *Antibiotics*, 8: 236.

Manikandan, M., Gowdaman, V., Duraimurugan, K. and Prabagaran, S.R. (2019). Taxonomic characterization and antimicrobial compound production from *Streptomyces chumphonensis* BDK01 isolated from marine sediment. *3 Biotech*, 9: 167.

Manivasagan, P., Kang, K.-H., Sivakumar, K., Li-Chan, E.C.Y., Oh, H.-M. and Kim, S.-K. (2014). Marine actinobacteria: An important source of bioactive natural products. *Environmental Toxicology and Pharmacology*, 38(1): 172–188.

Matsukawa, E., Nakagawa, Y., Limura, Y. and Hayakawa, M. (2007). Stimulatory effect of indole-3-acetic acid on aerial mycelium formation and antibiotic production in *Streptomyces* spp. *Actinomycetologica*, 21: 32–39.

Moore, E.R.B., Tindall, B.J., Santos, V.A.P.M.D., Pieper D.H., Ramos, J.-L. and Palleroni, N.J. (2006). Chapter 3.3.21 Non-medical *Pseudomonas*. pp. 646–703. *In*: Dworkin, M., Falkow, S., Rosenberg, E., Schleifer, K.-H. and Stackebrandt, E. (eds.). *The Prokaryotes* (3rd ed). Volume 6. Springer Science+Business Media, LLC, New York, USA.

Muangham, S., Suksaard, P., Mingma, R., Matsumoto, A., Takahashi, Y. and Duangmal, K. (2016). *Nocardiopsis sediminis* sp. nov., isolated from mangrove sediment. *International Journal of Systematic and Evolutionary Microbiology*, 66: 3835–3840.

Nouioui, I., Carro, L., García-López, M., Meier-Kolthoff, J.P., Woyke, T., Kyrpides, N.C., Pukall, R., Klenk, H.-P., Goodfellow, M. and Göker, M. (2018). Genome-based taxonomic classification of the phylum *Actinobacteria*. *Frontiers in Microbiology*, 9: 2007.

Nouioui, I., Cortés-albayay1, C., Carro, L., Castro, J.F., Gtari, M., Ghodhbane-Gtar, F., Klenk, H.-P., Tisa, L.S., Sangal, V. and Goodfellow, M. (2019). Genomic insights into plant-growth-promoting potentialities of the genus *Frankia*. *Frontiers in Microbiology*, 10: 1457.

Nweze, J.A., Mbaoji, F.N., Huang, G., Li, Y., Yang, L., Zhang, Y., Huang, S., Pan, L. and Yang, D. (2020). Antibiotics development and the potentials of marine-derived compounds to stem the tide of multidrug-resistant pathogenic bacteria, fungi, and protozoa. *Marine Drugs*, 18: 145.

Palaniyandi, S.A., Yang, S.H., Zhang, L. and Suh, J.-W. (2013). Effects of actinobacteria on plant disease suppression and growth promotion. *Applied Microbiology and Biotechnology*, 97: 9621–9636.

Pathom-aree, W., Kreawsa, S., Kamjam, M., Tokuyama, S., Yoosathaporn, S. and Lumyong, S. (2019). Potential of selected mangrove *Streptomyces* as plant growth promoter and rice bakanae disease control agent. *Chiang Mai Journal of Science*, 46: 261–276.

Phongsopitanun, W., Kudo, T., Mori, M., Shiomi, K., Pittayakhajonwut, P., Suwanborirux, K. et al. (2015). *Micromonospora fluostatini* sp. nov., isolated from marine sediment. *International Journal of Systematic and Evolutionary Microbiology*, 65: 4417–4423.

Phongsopitanun, W., Kudo, T., Ohkuma, M., Pittayakhajonwut, P., Suwanborirux, K. and Tanasupawat, S. (2016a). *Micromonospora sediminis* sp. nov., isolated from mangrove sediment. *International Journal of Systematic and Evolutionary Microbiology*, 66: 3235–3240.

Phongsopitanun, W., Kudo, T., Ohkuma, M., Pittayakhajonwut, P., Suwanborirux, K. and Tanasupawat, S. (2016b). *Streptomyces verrucosisporus* sp. nov., isolated from marine sediments. *International Journal of Systematic and Evolutionary Microbiology*, 66: 3607–3613.

Phongsopitanun, W., Suwanborirux, K. and Tanasupawat, S. (2019). Distribution and antimicrobial activity of Thai marine actinomycetes. *Journal of Applied Pharmaceutical Science*, 9(02): 129–134.

Phongsopitanun, W., Thawai, C., Suwanborirux, K., Kudo, T., Ohkuma, M. and Tanasupawat, S. (2014). *Streptomyces chumphonensis* sp. nov., isolated from marine sediments. *IInternational Journal of Systematic and Evolutionary Microbiology*, 64: 2605–2610.

Porter, J., Wilhelm, J. and Tresner, H. (1960). Method for the preferential isolation of actinomycetes from soils. *Applied Microbiology*, 8: 174–178.

Prajapati, S.C. and Chauhan, S.S. (2011). Dipeptidyl peptidase III: A multifaceted oligopeptide N-end cutter. *The FEBS Journal*, 278(18): 3256–3276.

Pulschen, A.A., Bendia, A.G., Fricker, A.D., Pellizari, V.H., Galante, D. and Rodrigues, F. (2017). Isolation of uncultured bacteria from Antarctica using long incubation periods and low nutritional media. *Frontiers in Microbiology*, 8: 1346.

Rangseekaew, P. and Pathom-aree, W. (2019). Cave actinobacteria as producers of bioactive metabolites. *Frontiers in Microbiology*, 10: 387.

Rodriquez, H. and Fraga, R. (1999). Phosphate solubilizing bacteria and their role in plant growth promotion. *Biotechnology Advances*, 17: 319–339.

Ruan, C.Y., Zhang, L., Ye, W.W., Xie, X.C., Srivibool, R., Duangmal, K. et al. (2015). *Streptomyces ferrugineus* sp. nov., isolated from mangrove soil in Thailand. *Antonie van Leeuwenhoek*, 107(1): 39–45.

Rungin, S., Indananda, C., Suttiviriya, P., Kruasuwan, W., Jaemsaeng, R. and Thamchaipenet, A. (2012). Plant growth enhancing effects by a siderophore-producing endophytic streptomycete isolated from a Thai jasmine rice plant (*Oryza sativa* L. cv. KDML105). *Antonie van Leeuwenhoek*, 102: 463–472.

Ruttanasutja, P. and Pathom-aree, W. (2015). Selective isolation of cultivable actinomycetes from Thai coastal marine sediment. *Chiang Mai Journal of Science*, 42(2): 88–103.

Saikia, J., Sarma, R.K., Dhandia, R., Yadav, A., Bharali, R., Gupta, V.K. et al. (2018). Alleviation of drought stress in pulse crops with ACC deaminase producing rhizobacteria isolated from acidic soil of Northeast India. *Scientific Reports*, 8: 3560.

Sangkanu, S., Rukachaisirikulc, V., Suriyachadkun, C. and Phongpaichit, S. (2017). Evaluation of antibacterial potential of mangrove sediment-derived actinomycetes. *Microbial Pathogenesis*, 112: 303–312.

Sangkanu, S., Suriyachadkun, C. and Phongpaichit, S. (2019). *Gordonia sediminis* sp. nov., an actinomycete isolated from mangrove sediment. *International Journal of Systematic and Evolutionary Microbiology*, 69: 1814–1820.

Sathya, A., Vijayabharathi, R. and Gopalakrishnan, S. (2017). Plant growth-promoting actinobacteria: A new strategy for enhancing sustainable production and protection of grain legumes. *3 Biotech*, 7: 102.

Sayed, A.M., Hassan, M.H.A., Alhadrami, H.A., Hassan, H.M., Goodfellow, M. and Rateb, M.E. (2019). Extreme environments: Microbiology leading to specialized metabolites. *Journal of Applied Microbiology*, 128: 630–657.

Songsumanus, A., Tanasupawat, S., Igarashi, Y. and Kudo, T. (2013). *Micromonospora maritima* sp. nov., isolated from mangrove soil. *International Journal of Systematic and Evolutionary Microbiology*, 63: 554–559.

Sousa, J.A.J. and Olivares, F.L. (2016). Plant growth promotion by streptomycetes: Ecophysiology, mechanisms and applications. *Chemical and Biological Technologies in Agriculture*, 3: 24.

Sripreechasak, P., Matsumoto, A., Suwanborirux, K., Inahashi, Y., Shiomi, K., Tanasupawat, S. et al. (2013). *Streptomyces siamensis* sp. nov., and *Streptomyces similanensis* sp. nov., isolated from Thai soils. *Journal of Antibiotics*, 66: 633–640.

Sripreechasak, P., Tamura, T., Shibata, C., Suwanborirux, K. and Tanasupawat, S. (2016). *Streptomyces andamanensis* sp. nov., isolated from soil. *International Journal of Systematic and Evolutionary Microbiology*, 66: 2030–2034.

Srivibool, R. (2000). Antimicrobial activity of *Actinomadura* isolates from tropical island soils. *Actinomycetes*, 10: 10–12.

Srivibool, R. and Sukchotiratana, M. (2006). Bioperspective of actinomycetes isolates from coastal soils: A new source of antimicrobial producers. *Songklanakarin Journal of Science and Technology*, 28(3): 493–499.

Srivibool, R. and Watanadilok, R. (2015). Distribution of actinomycetes in Thai mangrove sediments. *Proc. Burapha Univ. Int. Conf.*, 944–951.

Srivibool, R., Jaidee, K., Sukchotiratana, M., Tokuyama, S. and Pathom-aree, W. (2010). Taxonomic characterization of *Streptomyces* strain CH54-4 isolated from mangrove sediment. *Annals of Microbiology*, 60: 299–305.

Srivibool, R., Kurakami, K., Sukchotiratana, M. and Tokuyama, S. (2004). Coastal soil actinomycetes: Thermotolerant strains producing N-acylamino acid racemase. *ScienceAsia*, 30: 123–126.

Subramani, R. and Albersberg, W. (2012). Marine actinomycetes: An ongoing source of novel bioactive metabolites. *Microbiological Research*, 167(10): 571–580.

Subramani, R. and Sipkema, D. (2019). Marine rare actinomycetes: A promising source of structurally diverse and unique novel natural products. *Marine Drugs*, 17: 249.

Suksaard, P., Duangmal, K., Srivibool, R., Xie, Q.Y., Hong, K. and Pathom-aree, W. (2015). *Jiangella mangrovi* sp. nov., isolated from mangrove soil. *International Journal of Systematic and Evolutionary Microbiology*, 65: 2569–2573.

Suksaard, P., Mingma, R., Srisuk, N., Matsumoto, A., Takahashi, Y. and Duangmal, K. (2016). *Nonomuraea purpurea* sp. nov., an actinomycete isolated from mangrove sediment. *International Journal of Systematic and Evolutionary Microbiology*, 66: 4987–4992.

Suksaard, P., Pathom-aree, W. and Duangmal, K. (2017). Diversity and plant growth promoting activities of actinomycetes from mangroves. *Chiang Mai Journal of Science*, 44: 1210–1223.

Suksaard, P., Srisuk, N. and Duangmal, K. (2018). *Saccharopolyspora maritima* sp. nov., an actinomycete isolated from mangrove sediment. *International Journal of Systematic and Evolutionary Microbiology*, 68: 3022–3027.

Supong, K., Suriyachadkun, C., Pittayakhajonwut, P., Suwanborirux, K. and Thawai, C. (2013a). *Micromonospora spongicola* sp. nov., an actinomycete isolated from a marine sponge in the Gulf of Thailand. *Journal of Antibiotics*, 66: 505–509.

Supong, K., Suriyachadkun, C., Suwanborirux, K., Pittayakhajonwut, P. and Thawai, C. (2013b). *Verrucosispora andamanensis* sp. nov., isolated from a marine sponge. *International Journal of Systematic and Evolutionary Microbiology*, 63: 3970–3974.

Supong, K., Suriyachadkun, C., Tanasupawat, S., Suwanborirux, K., Pittayakhajonwut, P., Kudo, T. et al. (2013c). *Micromonospora sediminicola* sp. nov., isolated from marine sediment. *International Journal of Systematic and Evolutionary Microbiology*, 63: 570–575.

Suriyachadkun, C., Chunhametha, S., Ngaemthao, W., Tamura, T., Kirtikara, K., Sanglier, J.J. and Kitpreechavanich, V. (2011). *Sphaerisporangium krabiense* sp. nov., isolated from soil. *International Journal of Systematic and Evolutionary Microbiology*, 61: 2890–2894.

Tanasupawat, S., Jongrungruangchok, S. and Kudo, T. (2010). *Micromonospora marina* sp. nov., isolated from sea sand. *International Journal of Systematic and Evolutionary Microbiology*, 60: 648–652.

Tangjitjaroenkun, J., Tangchitcharoenkhul, R., Yahayo, W. and Supabphol, R. (2017). *In vitro* antimicrobial and cytotoxic activities of mangrove actinomycetes from eastern Thailand. *Chiang Mai Journal of Science*, 44(2): 322–337.

Thawai, C. (2010). Isolation, identification, and antimicrobial activity of *Micromonospora* strains from Thai marine sediments. *Proceedings of the 8th International Symposium on Biocontrol and Biotechnology*, 273–282.

Thawai, C. and Kudo, T. (2012). *Promicromonospora thailandica* sp. nov., isolated from marine sediment. *International Journal of Systematic and Evolutionary Microbiology*, 62: 2140–2144.

Thawai, C., Rungjindamai, N., Klanbut, K. and Tanasupawat, S. (2017). *Nocardia xestospongiae* sp. nov., isolated from a marine sponge in the Andaman Sea. *International Journal of Systematic and Evolutionary Microbiology*, 67: 1451–1456.

Thawai, C., Tanasupawat, S. and Kudo, T. (2008). *Micromonospora pattaloongensis* sp. nov., isolated from a Thai mangrove forest. *International Journal of Systematic and Evolutionary Microbiology*, 58: 1516–1521.

The Global Cancer Observatory. (2021). Thailand population fact sheets. https://gco.iarc.fr/today/data/factsheets/populations/764-thailand-fact-sheets.pdf. Accessed 21 April 2021.

Verderosa, A.D., Totsika, M. and Fairfull-Smith, K.E. (2019). Bacterial biofilm eradication agents: A current review. *Frontiers in Chemistry*, 7: 824.

Ventola, C.L. (2015). The antibiotic resistance crisis: Part 1: Causes and threats. *Pharmacology & Therapeutics*, 40(4): 277–283.

Virani, S., Bilheem, S., Chansaard, W., Chitapanarux, I., Daoprasert, K., Khuanchana, S. et al. (2017). National and subnational population-based incidence of cancer in Thailand: Assessing cancers with the highest burdens. *Cancers*, 9: 108.

Wang, D.-S., Xue, Q.-H., Ma, Y.-Y., Wei, X.-L., Chen, J. and He, F. (2014). Oligotrophy is helpful for the isolation of bioactive actinomycetes. *Indian Journal of Microbiology*, 54(2): 178–184.

Welch, R.A. (2006). Chapter 3.3.3 The genus *Escherichia*. pp. 60–71. *In*: Dworkin, M., Falkow, S., Rosenberg, E., Schleifer, K.-H. and Stackebrandt, E. (eds.). *The Prokaryotes* (3rd ed). Volume 6. Springer Science+Business Media, LLC, New York, USA.

# Chapter 3

# Nocardiae Associated with Foaming Coastal Marine Waters of the Sunshine Coast in Australia

*Luke Wright, Kerry E. Aitken, Tara Nielsen, Domenico Mattiucci, Briana Knox, Laura Dionysius, Christina Neuman* and *D. İpek Kurtböke**

## 1. Introduction

*Nocardia* Trevisan 1889, the nomenclatural type-genus of the family *Nocardiaceae* (Castellani and Chalmers 1913) of the order *Mycobacteriales* (Janke 1924) has had a long and complex pedigree due to the uneven weight given to morphological properties (Lechevalier 1976). Most *Nocardia* species have been identified using combinations of genotypic and phenotypic properties in the past (Tamura et al. 2018, Nouioui et al. 2020), however, the genus is now well-classified following the application of polyphasic procedures (Goodfellow and Maldonado 2012, Benndorf et al. 2020, Nouioui et al. 2020). The improved classification of the genus provides a sound framework for the recognition of new *Nocardia* species. Nocardiae encompass aerobic, Gram positive, acid and alcohol fast properties (Nouioui et al. 2022) and found widely distributed in the environment (Hashemi-Shahraki et al. 2015, McTaggart et al. 2015, Camozzota et al. 2017, Wright et al. 2021a). The species of the genus can be located in faecal matter, soil, and water (Torres et al. 2000, Hashemi-Shahraki et al. 2015). The genus currently includes 130 validly published species (htpps://www.bacterio.net) which form a monophyletic clade within the order *Mycobacteriales* (Nouioui et al. 2018).

Nocardiae are common in marine and terrestrial ecosystems where they are mostly successful saprophytes given their ability to degrade complex organic compounds (Luo et al. 2014, Orchard 2020). They are also associated with the formation of stable foams in activated sludge plants (Goodfellow and Maldonado

School of Science, Technology and Engineering, University of the Sunshine Coast, Maroochydore BC, QLD 4558 AUSTRALIA.
* Corresponding author: ikurtbok@usc.edu.au

2012, Bafghi and Yousef 2016, Batinovic et al. 2021). However, nocardiae are known best as casual agents of mycetoma and nocardiosis (Lerner 1996) and granulomatous diseases in animals (Conville et al. 2018, Verma and Jha 2019). Human causes of nocardiosis can be distinguished clinically as skin, pulmonary, and systemic infections of two or more body parts (Schaal and Beaman 1984, Hamid et al. 2001, Fatahi-Bafghi 2018). Common agents of nocardiosis include *Nocardia asteroides*, *N. brasiliensis*, *N. farcinica* and *N. nova*, though other species may be involved such as *N. cyriacigeorgica* and *N. otitiscaviarium* (Goodfellow and Maldonado 2012, Wilson 2012). The tropical disease mycetoma is caused by several nocardial species, notably by *N. brasiliensis* (Goodfellow and Maldonado 2012, Fahal 2017), in contrast, nocardiosis has global reach. Several molecular methods are available for the identification of clinically significant nocardiae (Chun and Goodfellow 1995, Conville et al. 2018, Kosova-Maali et al. 2018, Watson et al. 2022). Nocardiae show moderate variability in drug susceptibility patterns and some effective antibiotic treatment of mycetoma and nocardiosis depend on accurate identification of the casual organisms (Cercenado et al. 2007, Conville et al. 2018, Huang et al. 2019, Duggall and Chung 2020, Watson et al. 2022).

Treatment with amikacin and sulfanamides is effective in the case of immunocompetent patients (Brown-Elliott et al. 2006, Yan et al. 2020) though mortality rates for immunocompromised patients with systemic nocardiosis is high (Agterof et al. 2007, Minero et al. 2009, Ercibengoa et al. 2020). Linezolid and Amikacin were found to be the most effective antibiotics for the treatment of nocardiosis (Duggal and Chugh 2020). Morevoer, amikacin was found to be the highly effective antibiotic against clinical *Nocardia* isolates from a leading tertiary laboratory in Australia (Tan et al. 2020). A recent study by Wright et al. (2021a) also communicated similar findings for Australian isolates. Increased resistance against these medications, however, has raised interest into alternative treatments, including phage therapy for antibiotic resistant *Nocardia* (Lai et al. 2009, Kurtböke 2017a, Taylor et al. 2019, Batinovic et al. 2019, 2021, Wright et al. 2021a).

### 1.1 *Antibiotic-resistant* Nocardia

Due to the variation in drug susceptibility and the number of increasing *Nocardia* species being identified and published in the literature as clinically significant, six different categories are constructed based on the drug susceptibility patterns of nocardiae (Larruskain et al. 2011, Conville et al. 2018, Duggal and Chugh 2020). These six drug susceptibility patterns of *Nocardia* species may serve as reference source for clinicians for accurately determining the target directed antibiotics to prescribe in the event of a nocardiae linked infections once the causative species has been isolated and identified from its host (Wallace Jr. et al. 1991, Cercenado et al. 2007, Uhde et al. 2010, Schlaberg et al. 2014, Valdezate 2017, Duggal and Chugh 2020).

## 2. Ecology of Nocardiae

Actinomycetes, in the cluster of nocardiae, occur in a range of natural and man-made habitats including soil, fresh and salt water, dust, agricultural fields and wastewater treatment plants all over the globe (Zann 1995, Kachuei et al. 2012, Rasouli-Nasab et al. 2017), however, their presence is variable and dependent on environmental factors such as temperature, humidity and vegetation (Kachuei et al. 2012). Although their primary habitat is soil, some strains have formed endophytic associations with plants (Kaewkla and Franco 2013, Xing et al. 2011, Zhao et al. 2011, 2020, Nouioui et al. 2022) while some strains form mutualistic associations with blood-sucking insects (Johnston and Cross 1976, Yamamura et al. 2003). Forming filamentous, branched cells with aerial hyphae and occur in a range of man-made and natural habitats such as activated sludge within sewage treatment plant (STPs), soil, water, and the tissues of plants and animals including humans (Thomas et al. 2002, Zhuang et al. 2021). There is evidence that they have a role in the degradation of organic matter and pollutants such as polycyclic hydrocarbons and rubber (Johnston and Cross 1976, Goodfellow and Williams 1983, Luo et al. 2014, Sarkar et al. 2021). *Nocardia* species also produce a diverse range of natural bioactive metabolites (antimicrobials, immunomodifiers and enzyme inhibitors) (Glupczynski et al. 2006, Lai 2009, Carrasco et al. 2020). Most ecological studies on *Nocardia* species though have been focused on presence/absence, abundance and their role in waste treatment and activated sludge plants (Seviour and Nielsen 2010).

## 2.1 Occurrence of Nocardiae in Coastal and Marine Environments

The world's oceans contain roughly 5,000 trillion kilograms of salt and have a salinity of between 3.3% and 3.7% varying in concentration with evaporation, precipitation and freshwater runoff (Garrison 2012). Nocardiae, have been isolated from marine environments and some isolates reported to possess the features of bacteria indigenous to the sea (Mincer et al. 2002, Mudryk et al. 2013). They have also been grown on seawater-based media at increased hydrostatic pressure and are reported to be well-adapted and functional members in the marine community (Jensen et al. 1991). *Nocardia farcinica* was reported to have a salt tolerance of 5% (w/v) NaCl (Naidoo et al. 2011).

Nocardiae in marine environments, with their ability to adhere to natural or artificial surfaces and biofilms, became adapted as well as becoming functional members of the aquatic microbial community (Ogunmwonyi et al. 2010). Moreover, their ability to grow on low concentrations of carbon containing substances as well as to degrade recalcitrant organic matter ensures nocardiae survival and wide-spread presence in various bodies of water (Ogunmwonyi et al. 2010, Wright et al. 2021a).

The marine environment possesses extreme ranges in temperature, pressure salinity nutrients, oxygen and light, these extremes allow for many different adaptations to occur (Skropeta and Wei 2014). These adaptations can lead to the production of a diverse range of metabolites and bioactive compounds (Skropeta and Wei 2014). Many species of *Nocardia* found in the marine or coastal environment also have the ability to degrade toxic substances such as oil, rubber and fuel (Lepo et al. 2003, Wasmund et al. 2009).

## 2.2 Nocardiae in Sewage Treatment Plants

### 2.2.1 Nocardiae-mediated Foaming

Different foam types and reasons of formation have been reviewed by Schilling and Zessner (2011) in a range of diverse settings. In the sewage treatment facilities a characteristic feature of the members of the genus *Nocardia* (Goodfellow 1998, Goodfellow et al. 1998) is their ability to form a viscous and brown, stable foam (Fig. 1) which results in activated sludge bulking and potential overflow into further systems, as well as blockage of piping and interference with monitoring equipment (Övez and Orhon 2005, Bafghi and Yousefi 2016, Pajdak-Stós et al. 2017).

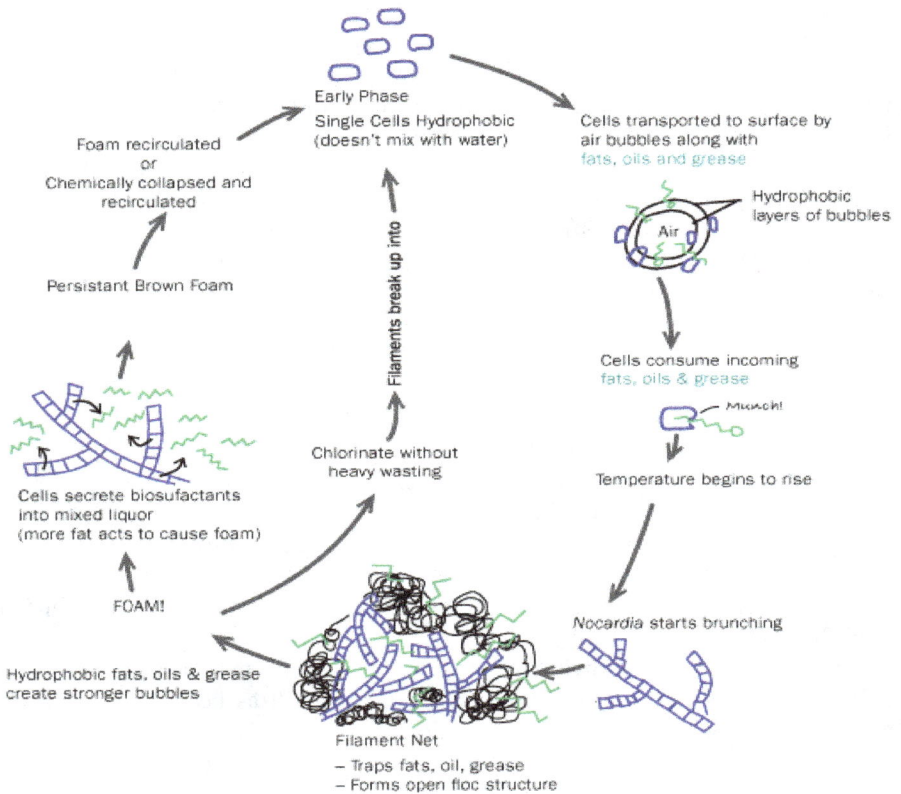

**Figure 1.** Cyclic nature of the foam formation and common stimulating factors.©https://teamaquafix.com/nocardia-in-wastewater/(published with permission from Aquafix).

Typically, foam is formed when gas bubbles rise up from within a liquid, the liquid surface expands to enclose the trapped gas and the bubbles are then retained at the surface without collapsing (Blackall et al. 1991, Iwahori et al. 2001). Biosurfactants produced by *Nocardia* species facilitate the formation of these foams as well as other genera of the order *Mycobacteriales* like *Gordonia*, *Millisia*, *Skermania*, *Tsukamurella* (Iwahori et al. 2001). Droplets of micro-emulsified oil are created by the interaction of oil and water with released biosurfactants; these droplets sit between the bubbles and allow for its formation and stabilisation (Iwahori et al. 2001). Where three or more bubbles are in contact, a "plateau-border" is formed, with a lower liquid pressure than at the surface of the bubble. This liquid pressure gradient encourages the liquid forming the bubble to run from the surface into the plateau border, destroying the bubble (Blackall et al. 1991). The released biosurfactants prevent the thinning of the bubble and aided by emulsified oil droplets situated in the plateau border and bacterial cells containing hydrophobic cells walls, stabilise the foam (Blackall et al. 1991). *Nocardia* species have a hydrophobic surface which can selectively attach to rising air bubbles within the liquid medium; the cells closely pack within the foam at the surface and reduce liquid drainage from the bubble surface stabilizing the foam (Blackall et al. 1991, Iwahori et al. 2001, Kurtböke 2017a).

In sewage treatment facilities, it has been found that the abundance of *Nocardia* cells is much higher in the foam than it is in the activated sludge liquid (Pitt and Jenkins 1990, Iwahori et al. 2001). These results indicate that when *Nocardia* cells initially attach to the air bubbles within the foam and float to the surface, it is this initial accumulation of cells which enhances stabilization of the foam (Iwahori et al. 2001, Lederman and Crum 2004).

## 3. Pollutants Contributing to Foam Formation

Foam forms naturally in the environment from a process like that described in sewage treatment facilities and consist of materials commonly found on the surface micro layer of the natural water body (Napolitano and Richmond 1995, Schilling and Zessner 2011). Natural foams form where fine colloidal particles, humic and fulvic acid substances, and lipids and protein originating from plants are all available (Schilling and Zessner 2011). Anthropogenic compounds such as phosphates from farm fertilisers, and the organic and inorganic compounds discharged into waterways from factories, ships and storm water are also known to contribute to the formation of foam (Fisenko et al. 2004, Ruzicka et al. 2009, Schilling and Zessner 2011, Varjani, 2017).

### *3.1 Composition of Oil and Environmental Hydrocarbon Contamination*

Crude and processed oils are organic compounds comprised of a complex mixture of hydrocarbons. The hundreds of components of oil can be divided into three basic classes based on their structure (Atlas, 1981):

1. Saturated/aliphatic/alkane fraction
2. Aromatic ring fraction
3. Asphaltene/polar fraction

While these components are related and share a chemical precursor, their impact can vary drastically.

Urban runoff is a significant source of the hydrocarbon pollution found in waterways destined for the sea (Ince and Ince 2019). Latimer et al. (1990), found that that in all of the four land use sites they monitored (commercial, residential, interstate highway and industrial areas), some degree of oil pollution, primarily crankcase oil with a small amount of fuel oil (both petroleum and diesel), was present in stormwater runoff. This has been further demonstrated by Göbel et al. (2007) who showed that hydrocarbons are the biggest pollutant in storm water runoff, although drip loss levels from vehicles have reduced in recent decades due to modernisation.

## 3.2 *Marine Hydrocarbon Biodegradation*

Physical means of oil hydrocarbon degradation in marine environments include evaporation, dissolution, photooxidation, sedimentation, emulsification, dispersion, spreading and drifting (Pavitran et al. 2006). While physical degradation is common, the main pathway of hydrocarbon degradation in marine sediments is via microbial biodegradation (Pavitran et al. 2006, Wasmund et al. 2009). Microorganisms can possess the ability to enzymatically degrade oil hydrocarbons with some preferring to degrade alkanes (normal, branched and cyclic), other aromatics, and others both alkanes and aromatics (Atlas 1995).

Alkanes are a major component of oil but are also produced by many living organisms such as plants, green algae, bacteria, and animals (Rojo 2005, 2010). Alkanes within living organisms provide many functions such as chemo-attractants and provide protection against water loss, insects or pathogens (Rojo 2005, 2010). Due to their simplicity alkanes are easily reduced and provide an excellent carbon and energy source for those organisms able to digest them and a low level of alkanes persist naturally within the environment, likely maintained through continual biodegradation and biosynthesis (Rojo 2005, 2010). This organic production of alkanes results in a low background level found naturally within most soils and waters (Rojo 2005, 2010, Ankit et al. 2017).

### 3.2.1 *Biodegradation of Hydrocarbons by Marine-derived* Nocardia *Species*

Increased availability of hydrocarbons such as oil in a marine environment leads to a rapid increase in the concentration of marine microorganisms. In response, some *Nocardia* strains release a low concentration of trehalolipids, a biosurfactant which adsorbs onto the surface of oil droplets (Blackall et al. 1991, Pavitran et al. 2006). This biosurfactant lowers the interfacial tension between the seawater and oil and forms a microemulsion. These surfactants are ampiphilic in nature and allow the molecule to sit between the interfaces of liquids, water and gases and mix or disperse emulsions, or stabilize foams (Banat et al. 2000, Pavitran et al. 2006). Incredibly versatile compounds, these surfactants display excellent dispersing, emulsifying, foaming and detergent characteristics in the field (Iwahori et al. 2001, Pavitran et al. 2006). This versatility translates into potential commercial applications such as oil recovery, crude oil drilling, lubricants, surfactant aided bioremediation of water insoluble pollutants, health care and food processing, as well as cosmetics and soap formulations, food, and both dermal and transdermal drug delivery systems (Banat et al. 2000).

The formation of a stable microemulsion allows *Nocardia* cells to form a foam and situate in close proximity to the oil droplets. *Nocardia amarae* is a non-pathogenic and commonly found species in *Nocardia*-mediated foams where they are considered a foam promoting organism (Hao et al. 1988). *Nocardia otitidiscaviarum* is an ear pathogen with limited pathogenicity that has been isolated from oil contaminated sea water and sediments (Vyas and Dave 2011). This strain of *Nocardia* has been demonstrated to produce biosurfactants necessary for foam formation and oil biodegradation (Vyas and Dave 2011). Foam formation enables internalisation and digestion of the oil droplets by these bacteria (Blackall et al. 1991, Pavitran et al. 2006).

Alkane monooxygenases are a group of enzymes capable degrading medium chain length ($C_5$ to $C_{11}$) hydrocarbons and especially longer chain length ($> C_{12}$) hydrocarbons, specifically straight chain, branched, cycloalkanes, and fatty acids but not alkenes and arenes (Hamamura et al. 1999). They are found in a variety of non-related bacteria and is a rubredoxin dependant [2FeO] protein encoded by the *alkB* gene (Kloos et al. 2006, Cappelletti et al. 2011, Tourova et al. 2016). The *alkB* proteins are located in 6 hydrophobic stretches that likely span the cytoplasmic membrane, and eight to nine histidines, which are essential in alkane hydroxylizing activity (Wang et al. 2010). PCR Primers have been designed to allow amplification of the *alkB* encoding gene based on its functional group structure and are used in the detection of the gene in bacteria (Wang et al. 2010).

Other alkane degrading pathways may also contribute to hydrocarbon degradation such as the cytochrome P450 alkane hydroxylases found in both yeasts and bacteria, or *AlmA* another alkane monoxygenase but of the flavin-binding family which is involved in long chain n-alkane degradation. Several alkane degrading pathways can co-exist within a single bacterial strain with overlapping substrate specificity (Rojo 2005, 2010, Wang et al. 2010). Short chain ($C_2$–$C_4$) alkane degraders possess enzymes related to the soluble and particulate methane monooxygenases, enzymes capable of degrading a broad range of small hydrocarbons and even certain aromatic compounds and cyclic alkanes, albeit slowly (Wang et al. 2010). In addition to alkane preferencing enzymatic pathways, enzymes specific to degradation of toxic aromatic fraction of oil also exist. Bacteria capable of degrading aromatic hydrocarbons use dioxygenase enzymes which activate and cleave aromatic ring structures (Mesarch et al. 2000). One such dioxygenase, the enzyme catechol 2, 3-dioxygenase (C230), is considered capable of degrading alkyl substituted aromatics (Mesarch et al. 2000).

Malik and Ahmed (2012) revealed that alkane degradation rates vary over time with alkanes of C16 length degrading early on in the process, and $C_{28}$ and $C_{32}$ only beginning to degrade after a long (16 to 24 days) delay. The more toxic aromatics are also readily susceptible to biodegradation (Atlas 1995). In general, most components of oil are biodegradable by a variety of organisms under both aerobic and anaerobic conditions, but this process may be very slow (Chaillan et al. 2004, Malik and Ahmed 2012).

## 4. Microplastics

At present, there is no universally accepted definition for microplastics (Van Cauwenberghe et al. 2015). First described in 2004, the term microplastic was adopted to refer to microscopic plastic debris in the size range of 20 μm (Thompson et al. 2004), however, most scientific studies are now adopting the definition put forward by Arthur et al. (2009) where microplastics are particles in a size range of less than 5mm. However, the definition of 1mm is also commonly used in the wider scientific community, as it is more intuitive as 'micro' refers to the micrometre range (Van Cauwenberghe et al. 2015). When plastic debris is exposed to UV irradiation from sunlight it undergoes photo oxidation (Lucas et al. 2008, Sivan 2011). This makes the plastic lose its tensile strength, go brittle and crumble and break up into small fragments and particles called microplastics (Lucas et al. 2008, Sivan 2011, Lambert et al. 2014).

Plastic has changed the way we live and due to a unique set of properties that plastic possesses, it is popular in everyday life; it can be used at a wide variety of temperatures, has low thermal conductivity, a high strength to weight ratio, is bio-inert, durable and is cheap (Andrady and Neal 2009, Andrady 2011). Due to the popularity of plastic, global production has increased significantly since the 1950s, with 322 million metric tonnes of plastic being produced in 2015 (Plastics Europe 2018). Even though the benefits to everyday life that plastic has, there is no denying the impact that it has on the environment. Globally, every minute, 18 tons of plastic enter the ocean, with this figure constantly increasing (Plastics Europe 2018, Ostle et al. 2019, Watt et al. 2021). Once at sea, only 1% of plastic floats on or near the surface, the remaining 99% sink or break down into micro-particles littering the sea floor (Eriksen et al. 2014).

## 4.1 Detrimental Effects of Microplastics on Macrofauna

As plastic pollution increases, microplastics also increase and become more available in marine sediments, thus giving the opportunity for ingestion by benthic macrofaunal communities. However, many factors such as size, density, colour, aggregation, and abundance of the particles affect their potential bioavailability to a wide range of marine organisms (Kach and Ward 2008, Wright et al. 2013, 2020, 2021b). Uptake and ingestion of microplastics by marine biota have been studied in organisms in both natural conditions and in laboratory settings (Van Cauwenberghe et al. 2015). Most studies undertaken in both natural and laboratory conditions studied polychaetes, crustaceans, bivalves, and fish, showing that microplastics were readily ingested and retained in the flesh of the animals and was shown to have a negative effect on energy budgets and feeding activity (Thompson et al. 2004, Wright et al. 2013, Van Cauwenberghe et al. 2015).

## 4.2 Biodegradation of Microplastics

Microorganisms such as bacteria and fungi are involved in the degradation of both natural and synthetic plastics (Gu et al. 2000). Biodegradation of plastic is governed by numerous factors that include polymer characteristics, the type of organism involved, and the nature of the pre-treatment, for example, sunlight exposure or weathering (Bonhomme et al. 2003, Thompson et al. 2004, Lucas et al. 2008, Lambert et al. 2014). The use of marine bacteria for biodegradation of both natural and synthetic substances is increasingly drawing attention due to the immense potential the isolates possess for environmental restoration.

Polyethylene is the most found non-degradable waste that has been recently recognised as a major threat to marine life, representing up to 64% of synthetic plastics that are discarded after short term use (Lee et al. 1991, Bonhomme et al. 2003, Harshvardhan and Jha 2013). However, microorganisms have the potential to degrade the plastic to non-toxic forms. In recent studies it was found that *Rhodococcus ruber* degrades 8% of dry weight in plastic after 30 days in concentrated liquid culture *in vitro* (Andrady 2011). The genus *Rhodococcus* has a powerful battery to degrade such compounds as it belongs to the family nocardiaceae (McKay et al. 1997, Wasmund et al. 2009, see Chapter 8 of this book).

# 5. Sandy Beaches

Sandy beaches harbour a large variety of organisms ranging from microscopic algae and meiofauna living in the interstitial spaces between the sand grains to the larger macrofauna burrowing through the sand (Armonies and Reise 2000, MacLachlan and Brown 2006). Most invertebrate phyla are represented on sandy beaches either as interstitial forms, or macrofauna, or as both. However, the most common representatives are polychaetes mollusks and crustaceans (McLachlan 1977, 2001, Dugan et al. 2003).

## 5.1 Sandy Beach Macrofaunal Communities

Ocean-exposed beaches usually receive large inputs of organic matter from the wind, waves, phytobenthic assimilates, products washed and leached from seaweed, animal faeces, the products from meio and macrofauna, and plant and animal remain (Koop and Griffiths 1982, Brown and McLachlan 2002, Jędrzejczak 1999, Mudryk et al. 2013). Large accumulations of organic matter coupled with organic runoff and anthropogenic pollution in the soil of sandy beaches generate optimal growth conditions for a high population of organisms, such as bacteria, fungi, yeasts and actinomycetes that can adapt to a constantly changing environment, living between the grains of sand; these microorganisms are involved heavily in food chain interactions (Koop et al. 1982, Malm et al. 2004, Mudryk and Podgorska 2007, Zakaria et al. 2011, Mudryk et al. 2013).

The majority of sandy beach macrofaunal communities are crustaceans, with the most dominant being amphipods and isopods, thus producing several million tonnes of chitin annually (Greco et al. 1990, Kirchner 1995). Bacteria and fungi mainly mineralise chitin, cellulose and organic substances, therefore crustaceans might be prone to infection by nocardiae and other bacteria within the marine environment (Vogan et al. 2008). Chitinolytic bacteria produce chitinase and are believed to be the main source of chitin breakdown within the marine environment and largely restricted to moulted shells or dead animals (Vogan et al. 2008). Chitinases are a group of enzymes that break down and decompose chitin, the most abundant natural polymer in the marine environment (Bansode and Bajekal 2006, Das et al. 2016). Marine microbial chitinase producing bacteria include species of actinomycetes (Gooday 1990, Vogan et al. 2002, Costa-Ramos and Rowley 2004, Bhattacharya et al. 2007). However, in some cases exoskeleton degradation can occur on living animals, commonly known as shell disease syndrome (Vogan et al. 2002, 2008). Shell disease syndrome is a worldwide condition affecting a wide range of crustaceans due to the bacterial degradation of the exoskeleton leading to unsightly lesions and even death, if the underlying tissues become infected (Vogan et al. 2002, 2008). Typically, marine crustaceans are reported to have exoskeletons with approximately 70% of the organic compound being chitin; microbial chitinases are believed to be the primary cause of shell disease (Brimacombe and Webber 1964, Fisher et al. 1978, Vogan et al. 2002, 2008).

Apart from the shell diseases encountered nocardiosis has also been reported in marine animals and in general, the primary route of pathogens in fish is through damaged or injured skin (Austin 2005, Austin and Austin 2007). The infection route of *Nocardia* spp. is still unclear, however, it could plausibly occur through the gills, muscles and injury sites (Maekawa et al. 2018). This is thought to be the case due to reports of naturally occurring nocardiosis in yellowtail (*Seriola quinqueradiata*) in the gills, muscles, and dermis (Itano et al. 2006). It has also been reported that crayfish are susceptible to nocardiosis through a damaged exoskeleton (Alderman et al. 1986, Beaman and Beaman 1994, Bower et al. 1994).

While there are very few studies undertaken on nocardiae infections on amphipods and isopods, there has been recorded evidence of bacterial colonisation on the body of copepods by rod-shaped bacteria (Kirchner 1995). Although it has been found that nocardiae have the potential to be potent chitinase producers (Bansode and Bajekal 2006), it is unknown whether or not they are pathogenic to sandy beach macrofauna, in particular amphipods and isopods, therefore, there is a need to further this line of investigation, as there could be a potential mechanism of gaining entry to the macrofauna and subsequently causing alterations in the ecosystem balance in sandy beaches which in turn can have impact on the food chain.

## 5.2  *Foam Introduced Nocardiae on Sandy Beaches*

Foaming events on the coast transport a large amount of marine carbon and nutrients inland which, in turn, forms a major part of the macrofaunal production as they (amphipods, isopods and dipteran larvae) are the major primary consumers of beach wrack, as the sea foam traps algae, seaweed, fungi and insects (Schilling and Zessner 2011). The foam especially affects areas near the shore, but these effects might reach further inland via aerosols (Schilling and Zessner 2011). Furthermore, once the foam is crusted and dried out in the beach areas, it can also travel during heavy winds in an associated form with the sand particles (Schilling and Zessner 2011) (Fig. 2). Recent study by Wright et al. (2021a) confirms the presence of nocardiae on these beaches.

The potential detrimental effects of these sea-foam-associated nocardiae on macrofaunal communities are understudied, due to the intertidal zone being such a dynamic environment. Spilmont et al. (2009) found that macrofaunal primary production rates on sandy beaches increased with foam deposits, therefore suggesting it may be an important food source for animal consumers. Accordingly, unpolluted foam has importance for macrofauna. However, if polluted foam is continuously

**Figure 2.** Example of foam crusting onto a sandy beach, Sunshine Coast, Australia.

present on these beaches, such continual input may substantially alter the natural microfloral composition and primary productivity of sandy beaches and coastal systems (Spilmont et al. 2009). As coastal polluted waters carry nocardiae to the beaches via seafoam, the foam then forms a crust allowing the nocardiae to either sink into the faunal layers, potentially impacting the fauna via enzymatic shell digestion and/or ingested by the macrofauna (Spilmont et al. 2009). The cumulative effect of such foam on the fauna remains unknown including the fate of smeared or ingested nocardiae by the fauna, thus further studies are warranted.

## 6. Coastal Pollution

Williams (1996) noted that there is only 'one pollution' because every pollutant, in the air or on land, tends to end up in the ocean. While agriculture is supportive of the local economy, it endangers the coastal marine environment through land clearing and cropping of fertilizer laden soil which results in elevated nutrients and sediments in catchment areas and subsequent on-flow out to sea (Williams 1996). It is estimated that agriculture contributes about 50% of the total pollution in surface water worldwide through nutrient enrichment (Islam and Tanaka 2004). It is a known fact that during land development and urbanisation, infrastructure such as roads, parking surfaces and footpaths are designed to collect precipitation and shift it out through stormwater channels and into streams and rivers (Göbel et al. 2007). With the movement of urban, rural and industrial stormwater runoff from collection surfaces, pollutants like heavy metals, and hydrocarbons such as polycyclic aromatic hydrocarbons and mineral oil hydrocarbons, are carried into local creeks and waterways (Göbel et al. 2007). Urban stormwater may also contain significant nutrient outflow in the form of garden fertilisers, animal faeces, septic system leachate and sewage system overflows (Brodie 1995, Zann 1995).

Sewage is also a major source of nutrients polluting waterways within urban areas and threatens marine environments with eutrophication (Zann 1995). With sewage in metropolitan and most inland regions only secondarily treated, outflow remains high in nutrients with phosphorus discharge estimated at 10,000 tonnes annually and nitrogen at 100,000 tonnes (Zann 1995). In certain locations effluent is reused for industrial purposes, irrigation and horticulture, thus risking on flow and downstream impacts on local waterways and coastlines (Brodie 1995).

## 7. The Sunshine Coast Region

Situated 100 km north of Brisbane, the Sunshine Coast Region stretches roughly 100 km from Caloundra to Noosa. Covering an area of 3,127 km², the Sunshine Coast is a tourist destination renowned for its pristine beaches and green hinterland (Spearritt 2009) (https://profile.id.com.au/sunshine-coast/about). Originally a farming region, the expansion of the Sunshine Coast Region began when land allotments went on sale in Caloundra in 1900, followed by land in Maroochydore in 1908 (Spearritt 2009). This region is growing at an ever-increasing rate and is currently considered part of the fastest expanding metropolitan region in Australia, along with the Gold Coast and Brisbane (Spearritt 2009). Doubling in size over the last 10 years, the Sunshine Coast also has ever-expanding transient population of tourists who are crucial to the local economy (https://profile.id.com. au/sunshine-coast/about). The tourist industry predominates on the Sunshine Coast, while forestry timber production industry and the processing of resulting materials, as well as the farming of sugar cane, pineapples and fruit trees such as macadamias are also present (https://profile.id.com.au/sunshine-coast/about).

The Sunshine Coast is also actively promoting an increase in light industrial business (https://profile.id.com.au/sunshine-coast/about). The region now supports three large industrial suburbs, the existing Kunda Park industrial suburb, the new 215-hectare Sunshine Coast Industrial Park located in Caloundra, the largest land subdivision ever carried out by the Queensland Government and the Coolum Eco-Industrial Park. As the Sunshine Coast Region expands road use, industrial and urban runoff increase thus naturally results in microbially mediated biodegradation.

### 7.1 Sunshine Coast Foaming Events

In Australia, *Nocardia*-mediated foams have proven more likely to occur in STPs, due to the warmer climate which encourages a higher growth rate and supports the mechanisms involved in foaming (Soddell and Seviour 1995, Thomas et al. 2002, Naidoo et al. 2011). Varying levels of *N. amarae* and *N. pinensis* have been discovered in many Queensland sewage treatment facilities where sewage foams have been problematic (Seviour et al. 1990). *Nocardia* have also been identified in the Sunshine Coast Region, with high levels of *Nocardia amarae* discovered at the now closed Caloundra sewage treatment facility (Seviour et al. 1990). While *Nocardia* species are ubiquitous at low numbers within the environment, the greater concentrations present in sewage treatment facilities can be released into the ocean and waterways through overspills of treated and untreated sewage. In recent past, it was not unusual for untreated sewage on the Sunshine Coast to overflow into local waterways after heavy rainfall (*Courier Mail* 2011, Hoffman 2012, Hoffman and Furler 2012). Recent upgrades to the sewage network however, reduced household sewage overflows during the wet season from 40 incidents in 2012 to just 10 in 2013.

*Nocardia* foaming is mostly promoted by warmer growth temperatures no lower than 18°C, this is because of a correlation between the ambient growth temperature and the functioning of the enzymes involved in foaming (Naidoo et al. 2011). The optimum growth temperatures for nocardiae range from 28 to 37°C though they have been demonstrated to grow at a slower growth rate at temperatures as low as 10°C. However, nocardiae associated foaming not frequently seen at lower temperatures (Naidoo et al. 2011). Nocardiae and associated foams have also been identified on beaches and in local waterways of the Sunshine Coast Region (Kurtböke 2008, 2016, 2017a, b).

A high fatty acid and lipid content associated with any substrate encourages the rapid growth of *Nocardia* populations which has been directly linked to foaming intensity (Naidoo et al. 2011). Recognised for an ability to degrade a wide variety of substrates, nocardiae possess a competitive advantage over other microorganisms allowing them to colonize and proliferate (Seviour et al. 1990). It is likely that the presence of nocardiae in orangy-browny stable sea foam occurring on Sunshine Coast beaches and waterways, is the result of hydrocarbon and nitrogen pollution, as well as the existence of favourable growth conditions such as in marinas (Fig. 3). The presence of foaming in the unpolluted areas, e.g., waterfalls, national parks (Fig.

**Figure 3.** Example of foaming in an ocean merging river where man made pollution exists.

**Figure 4.** Example of foaming in at the entrance of a national park where no man made pollution exists.

4)might also suggest that a natural carbon source such as tree oil, and nutrients in the form of organic leachate from surrounding trees, may also be sufficient in promoting foaming events (Eggeling and Sahm 1980).

The Sunshine Coast offers a sheltered holding region for tankers and cargo ships awaiting entry into the Port of Brisbane. At night ships can often be seen lined up offshore from Sunshine Coast beaches at anchor waiting. Ship-originated pollution poses a significant risk to both the local environment and the global environment in the form of discharged ballast water containing foreign biological materials and contaminants, bilge and wastewater containing human waste, oil, detergents, solvents, chemicals, pitch, and other materials, as well as the constant potential risk of spills (Doğan and Burak 2007). In March 2009, a cargo ship named the MV Pacific Adventurer, on its way from Newcastle to Brisbane and anchored 7 miles east of Cape Moreton, was overcome during gale force winds. The ship then spilled 21 containers of ammonium nitrate overboard, piercing two of the ships fuel oil bunker tanks and releasing 270 tonnes of heavy fuel oil into the sea as they sank (Australian Maritime Safety Authority 2009). During the following 2 days oil washed ashore on to Moreton Island, Bribie Island and the Sunshine Coast foreshores (Australian Maritime Safety Authority 2009, Schlacher et al. 2011). The impact of this oil spill on the beach environment is not completely understood, but studies have shown that on the lower shores of beaches impacted by heavy fuel oil, the communities of organisms living close to the ocean floor (or microbenthic assemblages) have a significantly lower abundance, species diversity and a distinct shift in community structure (Schlacher et al. 2011). Beaches are generally considered resilient to oil spills and in the past, it was believed that a combination of mechanically removing oil contaminated sand and allowing natural processes to re-profile the beach was adequate in rehabilitation, though this has since been disproven (Schlacher et al. 2011). Experimental trials after the Pacific Adventurer oil spill found that the ecological recovery of lower shores were incomplete after 3 months despite no observable oil presence on the beach surface or in the beach sediment (Schlacher et al. 2011). While numerous studies have investigated the immediate impact of oil spills, documentation of long-term biological recovery has been limited (Bejarano and Michel 2016). Accordingly, Sunshine Coast Region needs continuous monitoring to determine the long-term effect of oil spills on the macrofauna.

## 8. Diversity, Pathogenicity, and Foam Forming Abilities of Local Nocardiae Isolates

Numerous studies to date have predominantly focused on examining the occurrence and diversity of nocardiae from foam generated during the activated sludge process of wastewater recycling (Thomas et al. 2002, Övez and Orhon 2005, Pajdak-Stós et al. 2017). However, little is known outside the STPs about the occurrence, diversity and potential disease-causing capabilities of nocardiae on humans and marine animals and macrofauna, as well as about their pollutant degrading activities of nocardiae in particular the ones associated with the foaming coastal marine waters. Precursor studies performed at the University of the Sunshine Coast (UniSC) indicated that nocardiae were capable of growth in sterile seawater as well as producing foams in the presence of different oils and sewage (Kurtböke 2008, 2016, 2017a). Foam formation was more prominent in temperatures 24°C and above. On average the annual temperature range for the Sunshine Coast Region heavily overlaps with the optimal temperatures reported for nocardiae growth and the nocardiae numbers was found to be the highest during the hotter summer months of the year at the UniSC studies (Wright et al. 2021a). Example species of nocardiae and species from closely related genera found to be associated with the foaming coastal marine waters at the UniSC since 2003 is illustrated in Figs. 5 and 6a, b (Chun and Goodfellow 1995, Goodfellow 1998). The details of the ones that belong only to the genus *Nocardia* can be seen in the recent publication by Wright et al. (2021a), including the ones closely related to validly described pathogenic *Nocardia* species. Preliminary studies also confirmed the adhesion of nocardiae isolates to renal cells (A-498) (Fig. 7), which was also the case when Calu-3 cells (representing human lung cells) were tested. Nocardiae not only adhered to the Calu-3 cells but also translocated between these cells (Wright et al. 2021a) clearly indicating their pathogenic potentials.

## 9. Hydrocarbon Degradation Abilities of Local Nocardiae Isolates

Alkanes in the range of $C_{10}$ to $C_{26}$ are considered the most readily degraded, though biodegradation of alkanes up to $C_{44}$ has also been demonstrated (Hamamura and Arp 2000). Experiments conducted at the UniSC also found that *Nocardia* isolated from Sunshine Coast beaches were capable of degrading hydrocarbon components of diesel fuel (derived from petroleum oil). Ability to biodegrade hydrocarbons varied between the 25 strains however, when compared to a sterile control, on average they degraded $C_{12}$–$C_{25}$ length alkanes, $C_{14}$–$C_{15}$ length branched alkanes, $C_{16}H_{32}$ cycloalkane as well as the aromatic hydrocarbons trimethylbenzene and diethybenzene. As an example, isolate USC-10104 (see Fig. 8) degraded straight chain alkanes (85.9%), branched alkanes (45.6%), the cyclic alkane $C_{16}H_{32}$ (47%) and aromatics (90.0%), when compared to the control. In addition, 6 of the 25 isolates carried the *alkB*-monooxygenase suggesting a possible pathway for previously identified for hydrocarbon biodegradation. Comparative values for the isolates with and without the possession of *alkB* genes are given in Table 1. It is immediately apparent that there was an increase in straight chain n-alkane degradation ($C_{13}$–$C_{17}$) in those *Nocardia* strains carrying the *alkB* gene. There was also an increase, albeit small, in degradation of $C_{12}$ straight chain alkanes, $C_{18}$–$C_{21}$ straight chain alkanes, $C_{15}$ branched alkanes, and the aromatic compound trimethylbenzene in those strains carrying the *alkB* gene. While alkane monooxygenase likely contributed to these

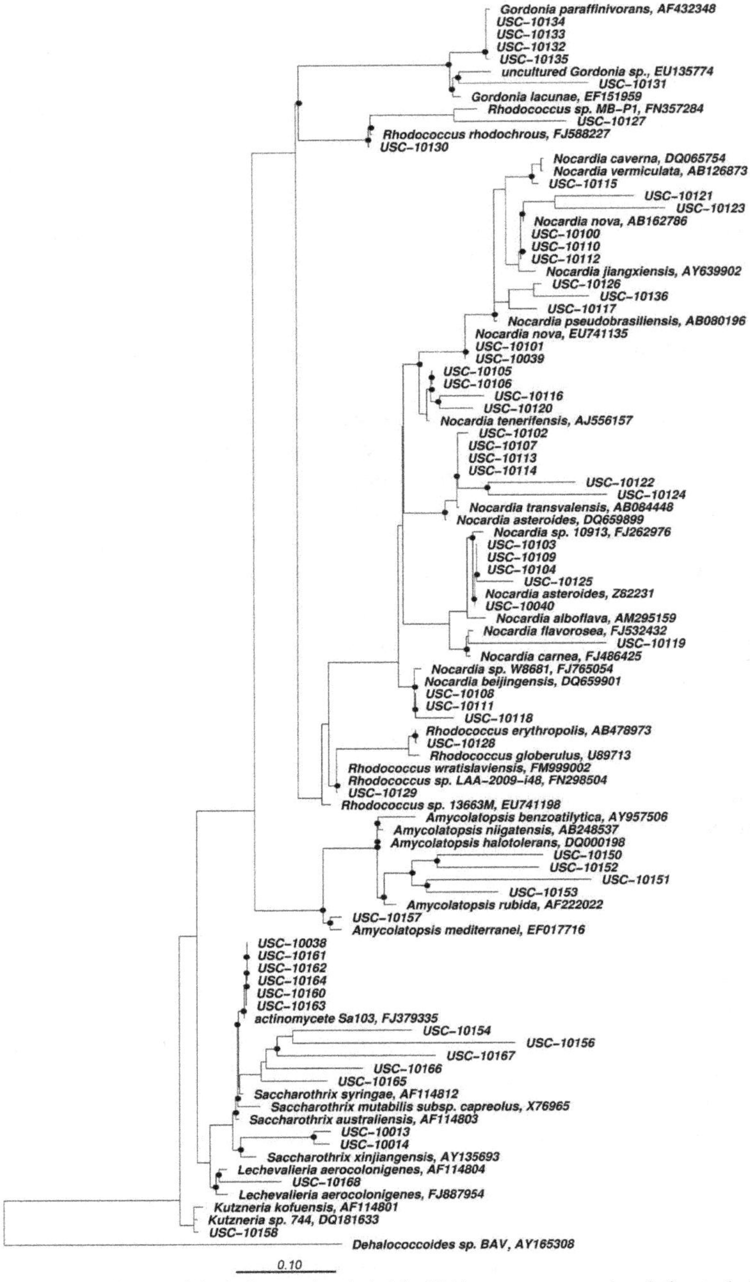

**Figure 5.** Phylogenetic tree of the isolates using their 16s rRNA gene sequences in relation to their closest relatives.

**Footnote:** Bootstrap values (≥ 50%) are indicated at nodes. The scale bar represents percentage (%) divergence. The accession number of the closest relatives is included. *Dehalococcoides* sp. BAV, AYI65308 was chosen as an out-group sequence to root the tree.

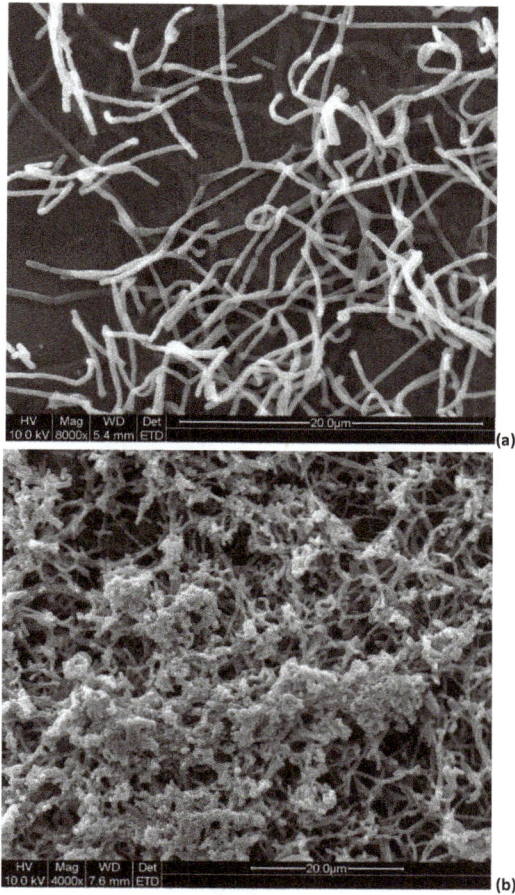

**Figure 6.** (a) USC-10157 (*Amycolatopsis* sp.) and (b) USC-10039 (*Nocardia* sp.) (Kurtböke personal collection).

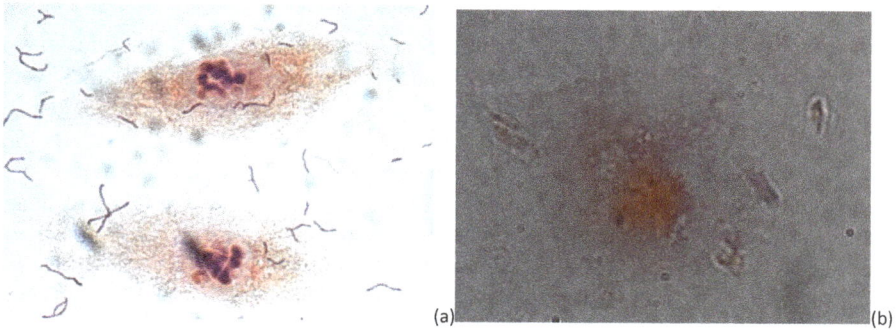

**Figure 7.** (a) Adhesion of a foam associated *Nocardia* isolate onto renal cells (A-498) and (b) subsequent lysis of the cells.

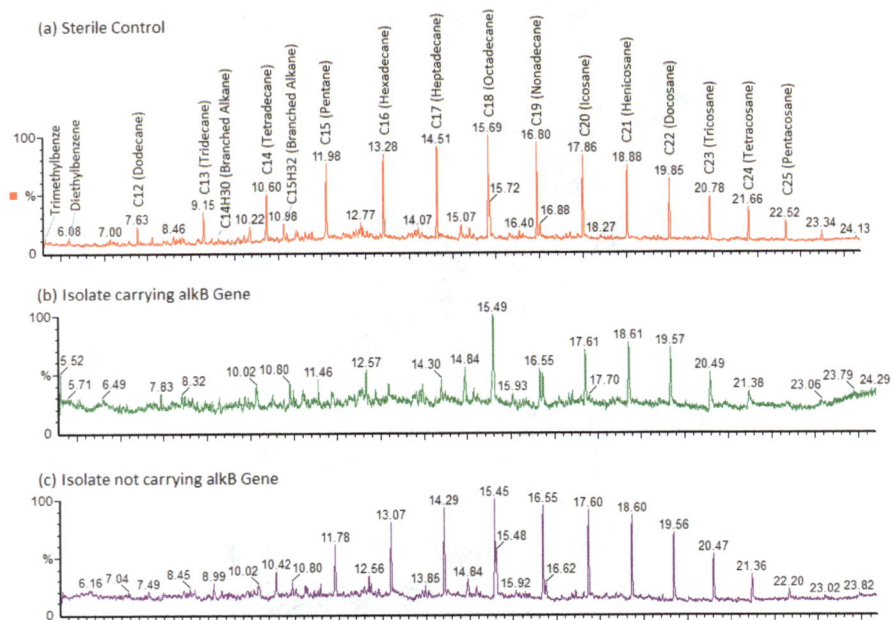

**Figure 8.** (a) GCMS analysis of aseptic negative control (seawater and commercially available diesel fuel) incubated at 28°C with shaking for 28 days.

**Footnote:** Peaks for straight chain alkanes $C_{12}$–$C_{25}$ and other identifiable hydrocarbons have been indicated, (b) Comparative GCMS analysis of sample inoculated with USC-10040 (*Nocardia* sp.) carrying the *alkB* gene, (c) Comparative GCMS analysis of sample inoculated with USC-10120 (*Nocardia* sp.) not carrying the *alkB* gene.

increases in degradation of hydrocarbons, it must be noted that there are also other possible biochemical pathways for microbial biodegradation (Koma et al., 2001, Varjani 2017).

## 10.  Bacteriophage Mediated Control of Foam

Tailed phages are grouped into the order *Caudovirales*, which includes three families: the *Siphorviridae*, *Myoviridae*, and *Podoviridae*. Nocardiae phages were reported to be present in the STP (Thomas et al. 2002, Petrovski et al. 2012) with a host range covering genera *Dietzia*, *Gordonia*, *Nocardia*, *Rhodoccous*, *Tsukamurella* and *Mycobacterium*. Similar polyvalent nocardiae phages were also located in the Sunshine Coast Region (Kurtböke 2016, 2017a) (Fig. 9). Given the fact that these phages can reduce the numbers of foam causing nocardiae (Liu et al. 2015), they naturally constitute a significant biological control potential and can  effectively be used in clinical and environmental settings. Tests conducted at the UniSC proved foam collapse immediately after addition of phages into the foam generating host containing flasks (Fig. 10).

**Table 1.** Percentage degradation of measured hydrocarbon compounds when comparing samples inoculated with *alkB* gene carrying *Nocardia* spp. (n = 6) and isolates inoculated with *Nocardia* spp. not carrying *alkB* gene (n = 19).

| | % Degradation (Mean, n=6) | % Degradation (Mean, n=19) | Δ Degradation |
|---|---|---|---|
| ***alkB* gene** | Positive | Negative | |
| **Hydrocarbon Compound** | | | |
| **Straight Chain** | | | |
| $C_{12}H_{26}$ | 29.2 | 24.7 | 4.5 |
| $C_{13}H_{28}$ | 51.7 | 31.2 | 20.5 |
| $C_{14}H_{30}$ | 47.9 | 31.9 | 16.0 |
| $C_{15}H_{32}$ | 38.6 | 18.1 | 20.5 |
| $C_{16}H_{34}$ | 26.8 | 10.3 | 16.5 |
| $C_{17}H_{36}$ | 36.5 | 15.3 | 21.2 |
| $C_{18}H_{38}$ | 21.5 | 18.1 | 3.4 |
| $C_{19}H_{40}$ | 20.2 | 13.3 | 6.9 |
| $C_{20}H_{42}$ | 21.1 | 12.6 | 8.6 |
| $C_{21}H_{44}$ | 11.4 | 6.9 | 4.4 |
| $C_{22}H_{46}$ | 12.1 | 16.8 | -4.8 |
| $C_{23}H_{48}$ | 7.7 | 7.9 | -0.2 |
| $C_{24}H_{50}$ | 26.4 | 24.4 | 1.9 |
| $C_{25}H_{52}$ | 42.1 | 43.3 | -1.2 |
| **Branched** | | | |
| $C_{14}H_{30}$ | 29.4 | 30.6 | -1.2 |
| $C_{15}H_{32}$ | 37.6 | 30.5 | 7.1 |
| **Cyclic** | | | |
| $C_{16}H_{32}$ | 40.0 | 38.5 | 1.5 |
| **Aromatics** | | | |
| Trimethylbenzene | 85.2 | 70.3 | 14.9 |
| Diethylbenzene | 55.3 | 66.6 | -11.4 |

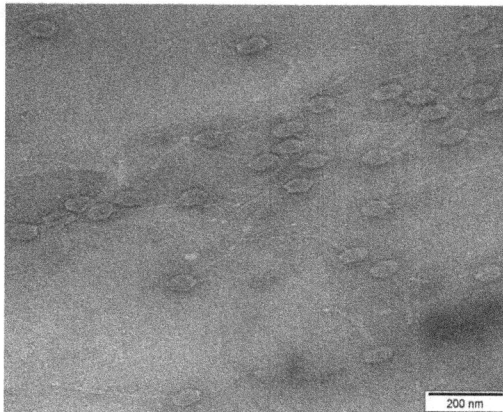

**Figure 9.** Transmission electron micrograph of *Nocardia* phage (ϕ5) displaying typical siphoviridae morphology (Kurtböke personal collection).

**Footnote:** Bar indicate: 200 nm.

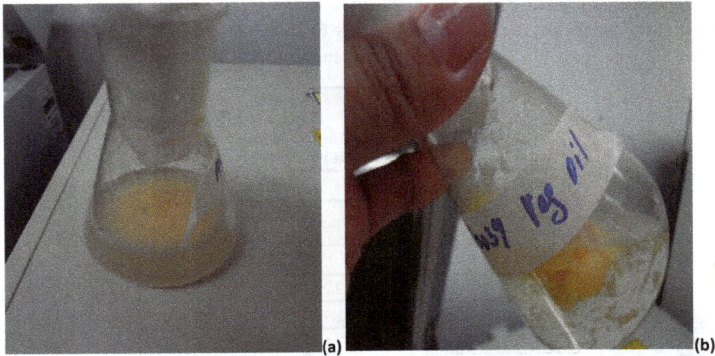

**Figure 10.** (a) *Nocardia* foam caused by the isolate USC-10039 on vegetable oil in sterile sea water and (b) its collapse after *Nocardia*-specific phage application.

## 11. Conclusions

Information presented in the chapter on nocardiae ranging from their antibiotic and phage susceptibility, from hydrocarbon degradation to pathogenicity, clearly indicate that a holistic understanding on nocardiae is required to be able to manage this cluster of bacteria in clinical and natural settings. Comprehensive and still ongoing study (2003-onward) further increases our understanding of the occurrence, diversity and human pathogenic potential of *Nocardia* isolates obtained from foaming coastal marine waters in the Sunshine Coast Region of Australia. As recently indicated by Mehta and Shamoo (2020) advancements in rapid molecular diagnostic technology will soon place nocardiae in the "extended pantheon of medically important pathogens" (Beaman et al. 1995). From the human health point of view the presence of such isolates in the coastal foams may present a risk, especially for swimmers and beach goers with underlying immune conditions. Identification of such nocardiae might thus lead to the development of preventative measures by the local authorities. In addition, as indicated by Porri et al. (2021) sea foam is important for "retaining larvae of polychaetes, mussels and barnacles near to the shore" thus foams occurring under unpolluted pristine conditions are of significance for survival of near shore marine life as well. Information like the one presented in this chapter on the occurrence and diversity of nocardiae in different environments (Kurtböke 2003, 2006, Wright et al. 2023) is of significance in terms of linking the clinical data to the presence of disease-causative species in the environment.

## Acknowledgements

Authors thank Dr. Ken Wasmund, University of Portsmouth, UK for the construction of the phylogenetic tree. The authors also thank Ms. Rachel Hancock at the Central Analytical Research Facility (CARF) of the Queensland University of Technology (QUT), a *Microscopy Australia* linked laboratory, Brisbane, Australia for the support with the SEM analysis. Authors also gratefully acknowledge the technical support by Daniel Powell and Daniel Shelley at the UniSC over the years of study.

# References

Agterof, M.J., Van der Bruggen, T., Tersmette, M., Ter Borg, E.J., Van den Bosch, J.M. and Biesma, D.H. (2007). Nocardiosis: A case series and a mini review of clinical and microbiological features. *Netherland Journal of Medicine*, 65(6): 199–202.

Alderman, D., Feist, S. and Polglase J.L. (1986). Possible nocardiosis of crayfish, *Austropotamobius pallipes*. *Journal of Fish Diseases*, 9(4): 345–347.

Andrady, A.L. (2011). Microplastics in the marine environment. *Marine Pollution Bulletin*, 62(8): 1596–1605.

Andrady, A.L. and Neal, M.A. (2009). Applications and societal benefits of plastics. *Philosophical Transactions of the Royal Society B: Biological Sciences*, 364(1526): 1977–1984.

Ankit, Y., Mishra, P.K., Kumar, P., Jha, D.K., Kumar, V.V., Ambili, B. and Ambili, A. (2017). Molecular distribution and carbon isotope of n-alkanes from Ashtamudi Estuary, South India: Assessment of organic matter sources and paleoclimatic implications. *Marine Chemistry*, 196: 62–70.

Arthur, C., Baker, J.E. and Bamford, H.A. (2009). *Proceedings of the International Research Workshop on the Occurrence, Effects, and Fate of Microplastic Marine Debris*, September 9–11, 2008, University of Washington Tacoma, Tacoma, WA, USA.

Armonies, W. and Reise, K. (2000). Faunal diversity across a sandy shore. *Marine Ecology Progress Series*, 49–57.

Atlas, R.M. (1981). Microbial degradation of petroleum hydrocarbons: An environmental perspective. *Microbiological Reviews*, 45(1): 180–209.

Atlas, R.M. (1995). Petroleum biodegradation and oil spill bioremediation. *Marine Pollution Bulletin*, 31(4): 178–182.

Austin, B. (2005). Bacterial pathogens of marine fish. *Oceans and Health: Pathogens in the Marine Environment*, Springer New York, 391–413 (DOI: https://doi.org/10.1007/b102184).

Austin, B. and Austin, D. (eds.). (2007). Pathogenicity. *Bacterial Fish Pathogens: Diseases of Farmed and Wild Fish*, pp. 283–335. Springer Dordrecht (Springer Praxis Books), DOI: https://doi.org/10.1007/978-1-4020-6069-4.

Australian Maritime Safety Authority (2009). https://www.amsa.gov.au/marine-environment/incidents-and-exercises/pacific-adventurer-11-march-2009.

Bafghi, M.F. and Yousefi, N. (2016). Role of *Nocardia* in activated sludge. *Malaysian Journal of Medical Sciences*, 23(3): 86–88.

Banat, I., Makkar, R. and Cameotra, S. (2000). Potential commercial applications of microbial surfactants. *Applied Microbial Biotechnology*, 53: 495–508.

Bansode, V.B. and Bajekal, S.S. (2006). Characterization of chitinases from microorganisms isolated from Lonar lake. *Indian Journal of Biotechnology*, 5: 357–363.

Batinovic, S., Rose, J.J., Ratcliffe, J., Seviour, R.J. and Petrovski, S. (2021). Cocultivation of an ultrasmall environmental parasitic bacterium with lytic ability against bacteria associated with wastewater foams. *Nature Microbiology*, 6(6): 703–711.

Batinovic, S., Wassef, F., Knowler, S.A., Rice, D.T., Stanton, C.R., Rose, J., Tucci, J., Nittami, T., Vinh, A., Drummond, G.R. and Sobey, C.G. (2019). Bacteriophages in natural and artificial environments. *Pathogens*, 8(3): 100.

Bhattacharya, D., Nagpure, A. and Gupta, R.K. (2007). Bacterial chitinases: Properties and potential. *Critical Reviews in Biotechnology*, 27(1): 21–28.

Beaman, B.L. and Beaman, L. (1994). *Nocardia* species: host-parasite relationships. *Clinical Microbiology Reviews*, 7(2): 213–264.

Beaman, B.L., Saubolle, M.A. and Wallace, R.J. (1995). *Nocardia*, *Rhodococcus*, *Streptomyces*, *Oerskovia*, and other aerobic actinomycetes of medical importance. pp. 379–399. *In*: Murray, P.R., Baron, E.J., Pfaller, M.A., Tenover, F.C. and Yolken, R.H. (eds.). *Manual of Clinical Microbiology*, 6th ed. American Society for Microbiology, Washington, DC.

Bejarano, A.C. and Michel, J. (2016). Oil spills and their impacts on sand beach invertebrate communities: A literature review. *Environmental Pollution*, 2018: 709–722.

Benndorf, R., Schwitalla, J.W., Martin, K., de Beer, Z.W., Vollmers, J., Kaster, A.-K., Poulsen, M. and Beemelmanns, C. (2020). *Nocardia macrotermitis* sp. nov. and *Nocardia aurantia* sp. nov., isolated from the gut of the fungus-growing termite *Macrotermes natalensis*. *International Journal of Systematic and Evolutionary Microbiology*, 70(10): 5226–5234.

Blackall, L., Harbers, A., Greenfield, P. and Hayward, A. (1991). Activated sludge foams: Effects of environmental variables on organism growth and foam formation. *Environmental Technology*, 12: 241–248.

Brimacombe, J.S. and Webber, J.M. (1964). *Mucopolysaccharides: Chemical Structure, Distribution and Isolation* (Vol. 6). Elsevier Publishing Company, Amsterdam.

Bonhomme, S., Cuer, A., Delort, A.M., Lemaire, J., Sancelme, M. and Scott, G. (2003). Environmental biodegradation of polyethylene. *Polymer Degradation and Stability*, 81(3): 441–452.

Bower, S.M., McGladdery, S.E. and Price, I.M. (1994). Synopsis of infectious diseases and parasites of commercially exploited shellfish. *Annual Review of Fish Diseases*, 4: 1–199.

Brodie, J. (1995). The problems of nutrients and eutrophication in the Australian marine environment. *State of the Marine Environment Report for Australia: Pollution-Technical Annex, 2*. Townsville: Great Barrier Reef Marine Park Authority.

Brown, A. and McLachlan, A. (2002). Sandy shore ecosystems and the threats facing them: Some predictions for the year 2025. *Environmental Conservation*, 29(1): 62–77.

Brown-Elliott, B.A., Brown, J.M., Conville, P.S. and Wallace, R.J. (2006). Clinical and laboratory features of the *Nocardia* spp. based on current molecular taxonomy. *Clinical Microbiology Reviews*, 19(2): 259–82.

Camozzota, C., Goldman, A., Tchernev, G., Lotti, T. and Wollina, U. (2017). A primary cutaneous nocardiosis of the hand. *Open access Macedonian Journal of Medical Sciences*, 5(4): 470.

Cappelletti, M., Fedi, S., Frascari, D., Ohtake, H., Turner, R.J. and Zannoni, D. (2011). Analyses of both the alkB gene transcriptional start site and alkB promoter-inducing properties of *Rhodococcus* sp. strain BCP1 grown on n-alkanes. *Applied and Environmental Microbiology*, 77(5): 1619–1627.

Carrasco, G., Monzón, S., San Segundo, M., García, E., Garrido, N., Medina-Pascual, M.J., Villalón, P., Ramírez, A., Jiménez, P., Cuesta, I. and Valdezate, S. (2020). Molecular characterization and antimicrobial susceptibilities of *Nocardia* species isolated from the soil; a comparison with species isolated from humans. *Microorganisms*, 8(6): 900.

Castellani, A. and Chalmers, A. (1913). *Manual of Tropical Medicine*, 2nd Ed., p. 816, William Wood & Co, New York.

Cercenado, E., Marin, M., Sánchez-Martínez, M., Cuevas, O., Martínez-Alarcón, J. and Bouza, E. (2007). *In vitro* activities of tigecycline and eight other antimicrobials against different *Nocardia* species identified by molecular methods. *Antimicrobial Agents and Chemotherapy*, 51(3): 1102–1104.

Chaillan, F., Le Flèche, A., Bury, E., Phantavong, Y.H., Grimont, P., Saliot, A. and Oudot, J. (2004). Identification and biodegradation potential of tropical aerobic hydrocarbon-degrading microorganisms. *Research in Microbiology*, 155(7): 587–595.

Chun, J. and Goodfellow, M. (1995). A phylogenetic analysis of the genus *Nocardia* with 16S rRNA gene sequences. *International Journal of Systematic and Evolutionary Microbiology*, 45(2): 240–245.

Conville, P.S., Brown-Elliott, B.A., Smith, T. and Zelazny, A.M. (2018). The complexities of *Nocardia* taxonomy and identification. *Journal of Clinical Microbiology*, 56(1): e01419–17.

*Courier Mail* (2011). Sewage spill closes popular beaches on Sunshine Coast. Brisbane, 5 December 2011 by Sam Smallbone and Sherine Conyers and Maroochy Journal.

Costa-Ramos, C. and Rowley, A.F. (2004). Effect of extracellular products of *Pseudoalteromonas atlantica* on the edible crab *Cancer pagurus*. *Applied and Environmental Microbiology*, 70(2): 729–735.

Das, S., Pitts, N.L., Mudron, M.R., Durica, D.S. and Mykles, D.L. (2016). Transcriptome analysis of the molting gland (Y-organ) from the blackback land crab, Gecarcinus lateralis. *Comparative Biochemistry and Physiology Part D: Genomics and Proteomics*, 17: 26–40.

Doğan, E. and Burak, S. (2007). Ship-originated pollution in the Istanbul Strait (Bosphorus) and Marmara Sea. *Journal of Coastal Research*, 23(2): 388–394.

Dugan, J.E., Hubbard, D.M., McCrary, M.D. and Pierson, M.O. (2003). The response of macrofauna communities and shorebirds to macrophyte wrack subsidies on exposed sandy beaches of southern California. *Estuarine, Coastal and Shelf Science*, 58: 25–40.

Duggal, S.D. and Chugh, T.D. (2020). Nocardiosis: A neglected disease. *Medical Principles and Practice*, 29(6): 514–23.

Eggeling, L. and Sahm, H. (1980). Degradation of coniferyl alcohol and other lignin-related aromatic compounds by *Nocardia* sp. DSM 1069. *Archives of Microbiology*, 126(2): 141–148.

Ercibengoa, M., Càmara, J., Tubau, F., García-Somoza, D., Galar, A., Martín-Rabadán, P., Marin, M., Mateu, L., García-Olivé, I., Prat, C. and Cilloniz, C. (2020). A multicentre analysis of *Nocardia pneumonia* in Spain: 2010–2016. *International Journal of Infectious Diseases*, 90: 161–166.

Eriksen, M., Lebreton, L.C., Carson, H.S., Thiel, M., Moore, C.J., Borerro, J.C., Galgani, F., Ryan, P.G. and Reisser, J. (2014). Plastic pollution in the world's oceans: More than 5 trillion plastic pieces weighing over 250,000 tons afloat at sea. *PloS One*, 9(12): e111913.

Fahal, A.H. (2017). Mycetoma.pp. 355–380. *In*: Mora-Montes, M.H. and Lopes-Bezerra, L.M. (eds.). *Current Progress in Medical Mycology*. Springer, Cham. DOI: https://doi.org/10.1007/978-3-319-64113-3.

Fatahi-Bafghi, M. (2018). Nocardiosis from 1888 to 2017. *Microbial Pathogenesis*, 114: 369–384.

Fisenko, A.I., Hromnysky, R. and Fisenko, R. (2004). The future for our small streams and rivers: Froth formation and natural purification (Practical Policy Proposals), 1st ed, *Protea Pubulishing Company*, Toronto.

Fisher, W.S., Nilson, E.H., Steenbergen, J.F. and Lightner, D.V. (1978). Microbial diseases of cultured lobsters: A review. *Aquaculture*, 14(2): 115–140.

Garrison, T. (2012). *Essentials of Oceanography*. 6th ed. Belmont: Brooks/Cole.

Glupczynski, Y., Berhin, C., Janssens, M. and Wauters, G. (2006). Determination of antimicrobial susceptibility patterns of *Nocardia* spp. from clinical specimens by Etest. *Clinical Microbiology & Infection*, 12(9): 905–912.

Göbel, P., Dierkes, C. and Coldewey, W. (2007). Storm water runoff concentration matrix for urban area. *Journal of Contaminant Hydrology*, 91(1-2): 26–42.

Gooday, G.W. (1990). The ecology of chitin degradation. *Advances in Microbial Ecology*, Springer, 387–430.

Goodfellow, M. (1998). *Nocardia* and related genera. pp. 463–489. *In*: Balows, A. and Duerden, B.I. (eds.). *Topley and Wilson's Microbiology and Microbial Infections*, 9th Ed., vol. 2: Systematic Bacteriology. London: Arnold: 1998.

Goodfellow, M. and Maldonado, L.A. (2012). Genus *Nocardia* Trevisan 1889, pp. 376–419. *In*: Goodfellow, M., Kämpfer, P., Busse, H.J., Trujillo, M.E., Suzuki, K., Ludwig, W. and Whitman, W. (eds.). *Bergey's Manual of Systematic Bacteriology*, 2nd ed., vol 5. Springer, New York, NY.

Goodfellow, M. and Williams, S.T. (1983). Ecology of actinomycetes. *Annual Review of Microbiology*, 37(1): 189–216.

Goodfellow, M., Stainsby, F.M., Davenport, R., Chun, J. and Curtis, T. (1998). Activated sludge foaming: The true extent of actinomycete diversity. *Water Science and Technology*, 37(4-5): 511–519.

Greco, N., Bussers, J.-C., Van Daele, Y. and Goffinet, G. (1990). Ultrastructural localization of chitin in the cystic wall of *Euplotes muscicola* Kahl (Ciliata, Hypotrichia). *European Journal of Protistology*, 26: 75–80.

Gu, Y.-Z., Hogenesch, J.B. and Bradfield, C.A. (2000). The PAS superfamily: Sensors of environmental and developmental signals. *Annual Review of Pharmacology and Toxicology*, 40(1): 519–561.

Hamamura, N. and Arp, D. (2000). Isolation and characterization of alkane-utilizing *Nocardioides* sp. strain CF8. *FEMS Microbiology Letters*, 186(1): 21–26.

Hamamura, N., Storfa, R.T., Semprini, L. and Arp. D.J. (1999). Diversity in Butane Monooxygenases among Butane-grown bacteria. *Applied and Environmental Microbiology*, 65(10): 4586–93.

Hamid, M.E., Maldonado, L., Sharaf Eldin, G.S., Mohamed, M.F., Saeed, N.S. and Goodfellow, M. (2001). *Nocardia africana* sp. nov., a new pathogen isolated from patients with pulmonary infections. *Journal of Clinical Microbiology*, 39(2): 625–30.

Hao, O., Strom, P. and Wu, Y. (1988). A review of the role of *Nocardia*-like filaments in activated sludge foaming. *Water SA*, 14(2): 105–110.

Harshvardhan, K. and Jha, B. (2013). Biodegradation of low-density polyethylene by marine bacteria from pelagic waters, Arabian Sea, India. *Marine Pollution Bulletin*, 77(1): 100–106.

Hashemi-Shahraki, A., Heidarieh, P., Bostanabad, S.Z., Hashemzadeh, M., Feizabadi, M.M., Schraufnagel, D. and Mirsaeidi, M. (2015). Genetic diversity and antimicrobial susceptibility of *Nocardia* species among patients with nocardiosis. *Scientific Reports*, 5(1): 1–9.

Hoffman, B. (2012). Sewage flows again. *Sunshine Coast Daily*, 26 January, Maroochydore, Australia.

Hoffman, B. and Furler, M. (2012). Sewage stations overflow in deluge. *Sunshine Coast Daily*, 25 January, Maroochydore, Australia.

Huang, L., Chen, X., Xu, H., Sun, L., Li, C., Guo, W., Xiang, L., Luo, G., Cui, Y. and Lu, B. (2019). Clinical features, identification, antimicrobial resistance patterns of *Nocardia* species in China: 2009–2017. *Diagnostic Microbiology and Infectious Disease*, 94(2): 165–172.

Ince, M. and Ince, O.K. (eds.). (2019). *Hydrocarbon Pollution and its Effect on the Environment*. BoD-Books on Demand, Intech Open, Croatia.

Islam, M. and Tanaka, M. (2004). Impacts of pollution on coastal and marine ecosystems including coastal and marine fisheries and approach for management: A review and synthesis. *Marine Pollution Bulletin*, 48(7-8): 624–649.

Itano, T., Kawakami, H., Kono, T. and Sakai, M. (2006). Experimental induction of nocardiosis in yellowtail, Seriola quinqueradiata Temminck & Schlegel by artificial challenge. *Journal of Fish Diseases*, 29(9): 529–534.

Iwahori, K., Tokutomi, T., Miyata, N. and Fujita, M. (2001). Formation of stable foam by the cells and culture supernatant of *Gordonia (Nocardia) amarae*. *Journal of Bioscience and Bioengineering*, 92(1): 77–79.

Janke, A. (1924). Allgemeine Technische Mikrobiologie. I. Teil. Die Mikroorganismen. Dresden; Leipzig: T.Steinkopf.

Jędrzejczak, M. (1999). The degradation of stranded carrion on a Baltic Sea sandy beach. *Oceanological Studies*, 28(3-4): 109–141.

Jensen, P., Dwight, R. and Fenical, W. (1991). Distribution of Actinomycetes in near-shore tropical marine sediments. *Applied and Environmental Microbiology*, 57(4): 1102–1108.

Johnston, D.W. and Cross, T. (1976). The occurrence and distribution of actinomycetes in lakes of the English Lake District. *Freshwater Biology*, 6(5): 457–463.

Kach, D.J. and Ward, J.E. (2008). The role of marine aggregates in the ingestion of picoplankton-size particles by suspension-feeding molluscs. *Marine Biology*, 153(5): 797–805.

Kachuei, R., Emami, M., Mirnejad, R. and Khoobdel, M. (2012). Diversity and frequency of *Nocardia* spp. in the soil of Isfahan province, Iran. *Asian Pacific Journal of Tropical Biomedicine*, 2(6): 474–478.

Kaewkla, O. and Franco, C.M. (2013). Rational approaches to improving the isolation of endophytic Actinobacteria from Australian native trees. *Microbial Ecology*, 65: 384–93.

Kirchner, M. (1995). Microbial colonization of copepod body surfaces and chitin degradation in the sea. *Helgoländer Meeresuntersuchungen*, 49(1): 201–212.

Kloos, K., Munch, J.C. and Schloter, M. (2006). A new method for the detection of alkane-monooxygenase homologous genes (alkB) in soils based on PCR-hybridization. *Journal of Microbiological Methods*, 66: 486–496.

Koma, D., Hasumi, F., Yamamoto, E., Ohta, T., Chung, S.Y. and Kubo, M. (2001). Biodegradation of long-chain n-paraffins from waste oil of car engine by *Acinetobacter* sp. *Journal of Bioscience and Bioengineering*, 91(1): 94–96.

Koop, K. and Griffiths, C. (1982). The relative significance of bacteria, meio-and macrofauna on an exposed sandy beach. *Marine Biology*, 66(3): 295–300.

Kosova-Maali, D., Bergeron, E., Maali, Y., Durand, T., Gonzalez, J., Mouniée, D., Sandoval Trujillo, H., Boiron, P., Salinas-Carmona, M.-C. and Rodriguez-Nava, V. (2018). High intraspecific genetic diversity of *Nocardia brasiliensis*, a pathogen responsible for cutaneous nocardiosis found in France: phylogenetic relationships by using sod and hsp65 Genes. *BioMed Research International*, 2018: e7314054.

Kurtböke, D.İ. (2003) (ed.). *Selective Isolation of Rare Actinomycetes*, Queensland Complete Printing Services, Nambour, Australia.

Kurtböke, D.İ. (2006). From culture collections to biological resource centres. *Microbiology Australia*, 27(1): 4–5.

Kurtböke, D.İ. (2008). "Chocolate mousse" on Sunshine Coast beaches. *Microbiology Australia*, 29(2): 104–5.

Kurtböke, D.İ. (2016). Actinomycetes in biodiscovery: Genomic advances and new horizons. Chapter 35, pp. 567–590. *In*: Gupta, V.K., Sharma, G.D., Tuohy, M.G. and Gaur, R. (eds.). *The Handbook of Microbial Resources*. CAB International Publications, Oxfordshire, UK.

Kurtböke, D.İ. (2017a). Ecology and habitat distribution of actinobacteria. pp. 123–49. *In*: Wink, J., Mohammadipanah, F. and Hamedi, J. (eds.). *Biology and Biotechnology of Actinobacteria*, Springer, Berlin, Germany.

Kurtböke, D.İ. (2017b). Bioactive actinomycetes: Reaching rarity through sound understanding of selective culture and molecular diversity. pp. 45–76. *In*: Kurtböke, D.İ. (ed.). *Microbial Resources: From Functional Existence in Nature to Applications*, Elsevier, London, UK.

Lai, C.C., Tan, C.K., Lin, S.H., Liao, C.H., Chou, C.H., Hsu, H.L., Huang, Y.T. and Hseuh, P.R. (2009). Comparative *in vitro* activities of nemonoxacin, doripenem, tigecycline and 16 other antimicrobials against *Nocardia brasiliensis*, *Nocardia asteroides* and unusual *Nocardia* species. *Journal of Antimicrobial Chemotherapy*, 64(1): 73–78.

Lambert, S., Sinclair, C. and Boxall, A. (2014). Occurrence, degradation, and effect of polymer-based materials in the environment. *In*: *Reviews of Environmental Contamination and Toxicology*, Volume 227 (pp. 1–53). Springer, Cham.

Latimer, J., Hoffman, E. and Hoffman, G. (1990). Sources of petroleum hydrocarbons in urban runoff. *Water, Air and Soil Pollution*, 52(1-2): 1–21.

Larruskain, J., Idigoras, P., Marimón, J.M. and Pérez-Trallero, E. (2011). Susceptibility of 186 *Nocardia* sp. isolates to 20 antimicrobial agents. *Antimicrobial Agents and Chemotherapy*, 55(6): 2995–2998.

Lederman, E.R. and Crum, N.F. (2004). A case series and focused review of nocardiosis: Clinical and microbiologic aspects. *Medicine*, 83(5): 300-313.

Lee, B., Pometto, A.L., Fratzke, A. and Bailey, T.B. (1991). Biodegradation of degradable plastic polyethylene by *Phanerochaete* and *Streptomyces* species. *Applied and Environmental Microbiology*, 57(3): 678–685.

Lepo, J.E., Cripe, C.R., Kavanaugh, J.L., Zhang, S. and Norton, G.P. (2003). The effect of amount of crude oil on extent of its biodegradation in open water-and sandy beach-laboratory simulations. *Environmental Technology*, 24(10): 1291–1302.

Lerner, P.I. (1996). Nocardiosis. *Clinical Infectious Diseases*, 22(6): 891–903.

Liu, M., Gill, J.J., Young, R. and Summer, E.J. (2015). Bacteriophages of wastewater foaming associated filamentous *Gordonia* reduce host levels in raw activated sludge. *Scientific Reports*, 5(1): 1–13.

Lucas, N., Bienaime, C., Belloy, C., Queneudec, M., Silvestre, F. and Nava-Saucedo, J.-E. (2008). Polymer biodegradation: Mechanisms and estimation techniques—A review. *Chemosphere*, 73(4): 429–442.

Luo, Q., Hiessl, S., Poehlein, A., Daniel, R. and Steinbüchel, A. (2014). Insights into the microbial degradation of rubber and gutta-percha by analysis of the complete genome of *Nocardia nova* SH22a. *Applied and Environmental Microbiology*, 80(13): 3895–3907.

Maekawa, S., Yoshida, T., Wang, P.C. and Chen, S.C. (2018). Current knowledge of nocardiosis in teleost fish. *Journal of Fish Diseases*, 41(3): 413–419.

Malik, Z.A. and Ahmed, S. (2012). Degradation of petroleum hydrocarbons by oil field isolated bacterial consortium. *African Journal of Biotechnology*, 11(3): 650–658.

Malm, T., Råberg, S., Fell, S. and Carlsson, P. (2004). Effects of beach cast cleaning on beach quality, microbial food web, and littoral macrofaunal biodiversity. *Estuarine, Coastal and Shelf Science*, 60(2): 339–347.

McLachlan, A. (1977). Composition, distribution, abundance and biomass of the macrofauna, and meiofauna of four sandy beaches. *African Zoology*, 12(2): 279–306.

McLachlan, A. (2001). Coastal beach ecosystems. *Encyclopedia of Biodiversity*, 1: 741–751. Academic Press.

McKay, D.B., Seeger, M., Zielinski, M., Hofer, B. and Timmis, K.N. (1997). Heterologous expression of biphenyl dioxygenase-encoding genes from a gram-positive broad-spectrum polychlorinated biphenyl degrader and characterization of chlorobiphenyl oxidation by the gene products. *Journal of Bacteriology*, 179(6): 1924–1930.

McTaggart, L.R., Doucet, J., Witkowska, M. and Richardson, S.E. (2015). Antimicrobial susceptibility among clinical *Nocardia* species identified by multilocus sequence analysis. *Antimicrobial Agents and Chemotherapy*, 59(1): 269–275.

Mehta, H.H. and Shamoo, Y. (2020). Pathogenic *Nocardia*: A diverse genus of emerging pathogens or just poorly recognized? *PLoS Pathog.*, 16: e1008280.

Mesarch, M.B., Nakatsu, C.H. and Nies, L. (2000). Development of Catechol 2,3-Dioxygenase-specific primers for monitoring bioremediation by competitive quantitive PCR. *Applied and Environmental Microbiology*, 66(2): 678–683.

Mincer, T.J., Jensen, P.R., Kauffman, C.A. and Fenical, W. (2002). Widespread and persistent populations of a major new marine actinomycete taxon in ocean sediments. *Applied and Environmental Microbiology*, 68(10): 5005–5011.

Minero, M.V., Marín, M., Cercenado, E., Rabadán, P.M., Bouza, E. and Muñoz, P. (2009). Nocardiosis at the turn of the century. *Medicine*, 88(4): 250–261.

Mudryk, Z.J. and Podgorska, B. (2007). Culturable microorganisms in sandy beaches in south Baltic Sea. *Polish Journal of Ecology*, 55(2): 221–231.

Mudryk, Z., Perliński, P., Skórczewski, P., Wielgat, M. and Zdanowicz, M. (2013). Distribution and abundance of microflora in sandy beaches on the southern coast of the Baltic Sea. *Oceanological and Hydrobiological Studies*, 42(3): 324–331.

Naidoo, D., Kumari, S. and Bux, F. (2011). Characterization of *Nocardia farcinica*, a filamentous bacterium isolated from foaming activated sludge samples. *Water Environment Research*, 83(6): 527–531.

Napolitano, G.E. and Richmond, J.E. (1995). Enrichment of biogenic lipids, hydrocarbons and PCBs in stream-surface foams. *Environmental Toxicology and Chemistry*, 14(2): 197–201.

Nouioui, I., Carro, L., García-López, M., Meier-Kolthoff, J.P., Woyke, T., Kyrpides, N.C., Pukall, R., Klenk, H.P., Goodfellow, M. and Göker, M. (2018). Genome-based taxonomic classification of the phylum Actinobacteria. *Frontiers in Microbiology*, 9: 2007.

Nouioui, I., Cortés-Albayay, C., Neumann-Schaal, M., Vicente, D., Cilla, G., Klenk, H.-P., Marimón, J.M. and Ercibengoa., M. (2020). Genomic virulence features of two novel species *Nocardia barduliensis* sp. nov. and *Nocardia gipuzkoensis* sp. nov., isolated from patients with chronic pulmonary diseases. *Microorganisms*, 8(10): 1517.

Nouioui, I., Ha, S.M., Baek, I., Chun, J. and Goodfellow, M. (2022). Genome insights into the pharmaceutical and plant growth promoting features of the novel species *Nocardia alni* sp. nov. *BMC Genomics*, 23(1): 1–13.

Ogunmwonyi, I.H., Mazomba, N., Mabinya, L., Ngwenya, E., Green, E., Akinpelu, D.A., Olaniran, A.O., Bernard, K. and Okoh, A.I. (2010). Studies on the culturable marine actinomycetes isolated from the Nahoon beach in the Eastern Cape Province of South Africa. *African Journal of Microbiology Research*, 4(21): 2223–2230.

Orchard, D.J. (2020). *The Role of Helper Bacteria in Facilitating Mycorrhization of Biserrula pelecinus L., a Pasture Legume New to Australia* [Doctoral dissertation, Charles Sturt University]. CSU Research Output. https://researchoutput.csu.edu.au/ws/portalfiles/portal/79631096/DOrchard_Thesis.pdf.

Ostle, C., Thompson, R.C., Broughton, D., Gregory, L., Wootton, M. and Johns, D.G. (2019). The rise in ocean plastics evidenced from a 60-year time series. *Nature Communications*, 10(1): 1–6.

Övez, S. and Orhon, D. (2005). Microbial ecology of bulking and foaming activated sludge treating tannery wastewater. *Journal of Environmental Science and Health. Part A Toxic Hazardous Substances and Environmental Engineering*, 40(2): 409–422.

Pajdak-Stós, A., Kocerba-Soroka, W., Fyda, J., Sobczyk, M. and Fiałkowska, E. (2017). Foam-forming bacteria in activated sludge effectively reduced by rotifers in laboratory-and real-scale wastewater treatment plant experiments. *Environmental Science and Pollution Research*, 24(14): 13004–13011.

Pavitran, S., Jagtap, C.B., Subramanian, S.B., Titus, S., Kumar, P. and Deb, P.C. (2006). Microbial bioremediation of fuel oil hydrocarbons in marine environment. *Defence Science Journal*, 56(2): 209–224.

Petrovski, S., Tillett, D. and Seviour, R.J. (2012). Genome sequences and characterization of the related *Gordonia* phages GTE5 and GRU1 and their use as potential biocontrol agents. *Applied and Environmental Microbiology*, 78(1): 42–47.

Pitt, P. and Jenkins, D. (1990). Causes and control of *Nocardia* in activated sludge. *Research Journal (Water Pollution Control Federation)*, 62(2): 143–150.

Plastics Europe (2018). Plastics – The facts 2017. *An Analysis of European Plastics Production, Demand and Waste Data.* Brussels, Belgium. Available at: http://www.plasticseurope.org/application/files/5715/1717/4180/Plastics_the_facts_2017_FINAL_for_website_one_page.pdf.

Porri, F., Puccinelli, E., Weidberg, N. and Pattrick, P. (2021). Lack of match between nutrient-enriched marine seafoam and intertidal abundance of long-lived invertebrate larvae. *Journal of Sea Research,* 170: 102009.

Rasouli-Nasab, M., Fatahi-Bafghi, M., Habibnia, S., Heidarieh, P. and Eshraghi, S.S. (2017). Comparison of various methods for isolation of *Nocardia* from soil. *Zahedan Journal of Research in Medical Sciences,* 19(2): e6107.

Rojo, F. (2005). Specificity at the End of the Tunnel: Understanding substrate length discrimination by the AlkB alkane hydroxylase. *Journal of Bacteriology,* 187(1): 19–22.

Rojo, F. (2010). Enzymes for aerobic degradation of alkanes. pp. 781–797. *In*: Timmis, K.N. (ed.). *Handbook of Hydrocarbon and Lipid Microbiology.* Berlin, Heidelberg: Springer-Verlag.

Ruzicka, K., Gabriel, O., Bletterie, U., Winkler, S. and Zessner, M. (2009). Cause and effect relationship between foam formation and treated wastewater effluents in a transboundary river. *Physics and Chemistry of the Earth, Parts A/B/C,* 34(8-9): 565–573.

Sarkar, B., Gupta, A.M. and Mandal, S. (2021). Insights from the comparative genome analysis of natural rubber degrading *Nocardia* species. *Bioinformation,* 17(10): 880–890.

Schilling, K. and Zessner, M. (2011). Foam in the aquatic environment. *Water Research,* 45(15): 4355–4366.

Schlaberg, R., Fisher, M.A. and Hanson, K.E. (2014). Susceptibility profiles of *Nocardia* isolates based on current taxonomy. *Antimicrobial Agents and Chemotherapy,* 58(2): 795–800.

Schlacher, T.A., Holzheimer, A., Stevens, T. and Rissik, D. (2011). Impacts of the Pacific Adventurer oil spill on the macrobenthos of subtropical sandy beaches. *Estuaries and Coasts,* 34: 937–949.

Seviour, R. and Nielsen, P.H. (eds.) (2010). *Microbial Ecology of Activated Sludge.* IWA publishing.

Seviour, E.M., Williams, C.J., Seviour, R.J., Soddell, J.A. and Lindrea, K.C. (1990). A survey of filamentous bacterial populations from foaming activated sludge plants in eastern states of Australia. *Water Research,* 24(4): 493–498.

Sivan, A. (2011). New perspectives in plastic biodegradation. *Current Opinion in Biotechnology,* 22(3): 422–426.

Skropeta, D. and Wei, L. (2014). Recent advances in deep-sea natural products. *Natural Product Reports,* 31(8): 999–1025.

Spearritt, P. (2009). The 200 km city: Brisbane, the gold coast, and the sunshine coast. *Australian Economic History Review,* 49(1): 87–106.

Spilmont, N., Denis, L., Artigas, L.F., Caloin, F., Courcot, L., Créach, A., Desroy, N., Gevaert, F., Hacquebart, P., Hubas, C., Janquin, M.-A., Lemoine, Y., Luczak, C., Migné, A., Rauch, M. and Davoult, D. (2009). Impact of the *Phaeocystis globosa* spring bloom on the intertidal benthic compartment in the eastern English Channel: A synthesis. *Marine Pollution Bulletin,* 58(1): 55–63.

Soddell, J. and Seviour, R. (1995). Relationship between temperature and growth of organisms causing *Nocardia* foams in activated sludge plants. *Water Research,* 29(6): 1555–1558.

Tan, Y.E., Chen, S.C.-A. and Halliday, C.L. (2020). Antimicrobial susceptibility profiles and species distribution of medically relevant *Nocardia* species: Results from a large tertiary laboratory in Australia, *Journal of Global Antimicrobial Resistance,* 20: 110–7.

Taylor, S., Brown, T.L., Tucci, J., Lock, P., Seviour, R.J. and Petrovski, S. (2019). Isolation and characterization of bacteriophage NTR1 infectious for *Nocardia transvalensis* and other *Nocardia* species. *Virus Genes,* 55(2): 257–265.

Thomas, J.A., Soddell, J.A. and Kurtböke, D.I. (2002). Fighting foam with phages? *Water Science and Technology,* 46(1-2): 511–8.

Thompson, R.C., Olsen, Y., Mitchell, R.P., Davis, A., Rowland, S.J., John, A.W.G., McGonigle, D. and Russell, A.E. (2004). Lost at Sea: Where Is All The Plastic? *Science,* 304(5672): 838.

Torres, O.H., Domingo, P., Pericas, R., Boiron, P., Montiel, J.A. and Vazquez, G. (2000). Infection caused by *Nocardia farcinica*: Case report and review. *European Journal of Clinical Microbiology and Infectious Diseases,* 19(3): 205–212.

Tourova, T.P., Sokolova, D.S., Semenova, E.M., Shumkova, E.S., Korshunova, A.V., Babich, T.L., Poltaraus, A.B. and Nazina, T.N. (2016). Detection of n-alkane biodegradation genes *alkB* and *ladA* in thermophilic hydrocarbon-oxidising bacteria of the genera *Aeribacillus* and *Geobacillus*. *Microbiology*, 85(6): 693–707.

Uhde, K.B., Pathak, S., McCullum Jr, I., Jannat-Khah, D.P., Shadomy, S.V., Dykewicz, C.A., Clark, T.A., Smith, T.L. and Brown, J.M. (2010). Antimicrobial-resistant *Nocardia* isolates. United States, 1995–2004. *Clinical Infectious Diseases*, 51(12): 1445–1448.

Valdezate, S., Garrido, N., Carrasco, G., Medina-Pascual, M.J., Villalón, P., Navarro, A.M. and Saéz-Nieto, J.A. (2017). Epidemiology and susceptibility to antimicrobial agents of the main *Nocardia* species in Spain. *Journal of Antimicrobial Chemotherapy*, 72(3): 754–761.

Van Cauwenberghe, L., Devriese, L., Galgani, F., Robbens, J. and Janssen, C.R. (2015). Microplastics in sediments: A review of techniques, occurrence and effects. *Marine Environmental Research*, 111: 5–17.

Varjani, S.J. (2017). Microbial degradation of petroleum hydrocarbons. *Bioresource Technology*, 223: 277–286.

Verma, P. and Jha, A. (2019). Mycetoma: Reviewing a neglected disease. *Clinical and Experimental Dermatology*, 44(2): 123–129.

Vogan, C.L., Costa-Ramos, C. and Rowley, A.F. (2002). Shell disease syndrome in the edible crab, *Cancer pagurus* – isolation, characterization and pathogenicity of chitinolytic bacteria. *Microbiology*, 148(3): 743–754.

Vogan, C.L., Powell, A. and Rowley, A.F. (2008). Shell disease in crustaceans–just chitin recycling gone wrong? *Environmental Microbiology*, 10(4): 826–835.

Vyas, T. and Dave, B. (2011). Production of biosurfactant by *Nocardia otitidiscaviarum* and its role in biodegradation of crude oil. *International Journal of Science Technology*, 8(2): 425–432.

Wallace Jr., R.J., Brown, B.A., Tsukamura, M., Brown, J.M. and Onyi, G.O. (1991). Clinical and laboratory features of *Nocardia nova*. *Journal of Clinical Microbiology*, 29(11): 2407–2411.

Wang, W., Wang, L. and Shao, Z. (2010). Diversity and abundance of oil-degrading bacteria and alkane hydroxylase (alkB) genes in the subtropical seawater of Xiamen Island. *Microbial Ecology*, 60: 429–439.

Wasmund, K., Burns, K.A., Kurtböke, D.İ. and Bourne, D.G. (2009). Novel alkane hydroxylase gene (alkB) diversity in sediments associated with hydrocarbon seeps in the Timor Sea, Australia. *Applied and Environmental Microbiology*, 75(23): 7391–7398.

Watt, E., Picard, M., Maldonado, B., Abdelwahab, M.A., Mielewski, D.F., Drzal, L.T., ... and Mohanty, A.K. (2021). Ocean plastics: environmental implications and potential routes for mitigation—A perspective. *RSC Advances*, 11(35): 21447–21462.

Williams, C. (1996). Combating marine pollution from land-based activities: Australian initiatives. *Ocean & Coastal Management*, 33(1-3): 87–112.

Wilson, J.W. (2012). Nocardiosis: Updates and clinical overview. *Mayo Clinic Proceedings*, 87(4): 403–407.

Wright, L., Katouli, M. and Kurtböke, D.İ. (2021a). Isolation and characterization of Nocardiae associated with foaming coastal marine waters. *Pathogens*, 10(5): 579.

Wright, L., Nouioui, I., Mast, Y., Bunk, B., Spröer, C., Neumann-Schaal, M., Wolf, J., Katouli1, M. and Kurtböke, D.İ. (2023). *Nocardia australiensis* sp. nov. and *Nocardia spumae* sp. nov., isolated from sea foam in Queensland, Australia. *International Journal of Systematic and Evolutionary Microbiology*, 73(8): 005952.

Wright, R.J., Langille, M.G. and Walker, T.R. (2021b). Food or just a free ride? A meta-analysis reveals the global diversity of the Plastisphere. *The ISME Journal*, 15(3): 789–806.

Wright, R.J., Erni-Cassola, G., Zadjelovic, V., Latva, M. and Christie-Oleza, J.A. (2020). Marine plastic debris: A new surface for microbial colonization. *Environmental Science & Technology*, 54(19): 11657–1672.

Wright, S.L., Thompson, R.C. and Galloway, T.S. (2013). The physical impacts of microplastics on marine organisms: A review. *Environmental Pollution*, 178: 483–492.

Xing, K., Qin, S., Fei, S.M., Lin, Q., Bian, G.K., Miao, Q., Wang, Y., Cao, C.L., Tang, S.K., Jiang, J.H. and Li, W.J. (2011). *Nocardia endophytica* sp. nov., an endophytic actinomycete isolated from the oil-seed plant *Jatropha curcas* L. *International Journal of Systematic and Evolutionary Microbiology*, 61(8): 1854–1858.

Yamamura, H., Hayakawa, M. and Iimura, Y. (2003). Application of sucrose-gradient centrifugation for selective isolation of *Nocardia* spp. from soil. *Journal of Applied Microbiology*, 95(4): 677–685.

Yan, Y., Dai, Q., Hu, G., Jiao, Q., Mei, L. and Fu, W. (2020). Effects of vegetation type on the microbial characteristics of the fissure soil-plant systems in karst rocky desertification regions of SW China. *Science of the Total Environment*, 712: 136543.

Zakaria, L., Yee, T.L., Zakaria, M. and Salleh, B. (2011). Diversity of microfungi in sandy beach soil of Teluk Aling, Pulau Pinang. *Tropical Life Sciences Research*, 22(1): 71.

Zhao, G.Z., Li, J., Zhu, W.Y., Klenk, H.P., Xu, L.H. and Li, W.J. (2011). *Nocardia artemisiae* sp. nov., an endophytic actinobacterium isolated from a surface-sterilized stem of *Artemisia annua* L. *International Journal of Systematic and Evolutionary Microbiology*, 61(12): 2933–2937.

Zhao, J., Han, X., Hu, H., Ling, L., Zhang, X., Guo, X., Wang, X. and Xiang, W. (2020). *Nocardia stercoris* sp. nov., a novel actinomycete isolated from the cow dung. *International Journal of Systematic and Evolutionary Microbiology*, 70(1): 493–498.

Zann, L. (1995). Our Sea, Our Future: Major findings of the State of the Marine Environment Report for Australia. [Online] Available at: http://www.environment.gov.au/archive/coasts/publications/somer/index.htm.

Zhuang, K., Liu, Y., Dai, Y., Xu, J., Li, W., Ming, H., Pradhan, S., Ran, X., Zhang, C., Feng, Y. and Ran, Y. (2021). *Nocardia huaxiensis* sp. nov., an actinomycete isolated from human skin. *International Journal of Systematic and Evolutionary Microbiology*, 71(8): p.004970.

# Chapter 4

# Actinomycetes in Thermal Ecosystems

*Manik Prabhu Narsing Rao* and *Wen-Jun Li**

## 1. Introduction

Thermal environments are often thought of as pinpoint anomalies against a background of ambient life, yet they carry the deepest insights into the earliest life on Earth and perhaps even to the origin of life (Walter 1996). Indeed, a thermophilic lifestyle has been proposed for the common ancestor of extant life (Woese 1987, Stetter 1994). Microbial life left a poor fossil record hindering our interpretation about the origin of life (Schuler et al. 2017); however, the evidence for the presence of life in the thermal ecosystem dated back 3.5 Ga (Djokic et al. 2017). Thermal ecosystems are common in the oceans and on land (Walter 1996), including in China (Fig. 1).

Microbial communities inhabiting thermal habitats have unique characteristics compared with microbiomes of non-geothermal environments (Li and Ma 2019). In the past few years, bioresources in thermal environments have gained interest due to their scientific and biotechnological importance. Studying thermophilic microorganisms, for instance, is necessary for a better understanding of the origin of life (Hussein et al. 2017). The best example for biotechnological importance of thermophiles is *Thermus aquaticus*, the source of Taq polymerase (Chien et al. 1976).

Actinomycetes are ubiquitous, diverse and form one of the largest lineages in the domain *Bacteria* involved in the turnover of organic matter (Valverde et al. 2012). Stackebrandt et al. (1997) reported the first hierarchal phylogenetic clustering of what is now known as the phylum *Actinomycetota* (Oren and Garrity 2021), formerly *Actinobacteria* (Goodfellow 2012) with the introduction of class *Actinobacteria*. The higher ranks in the phylum were recently updated to include six classes, 46 orders, and 79 families, with 16 new orders and 10 new families (Salam et al. 2020).

State Key Laboratory of Biocontrol and Guangdong Provincial Key Laboratory of Plant Resources, School of Life Sciences, Sun Yat-Sen University, Guangzhou PR China, 510275.
Email: deene.manik@gmail.com
* Corresponding author: liwenjun3@mail.sysu.edu.cn

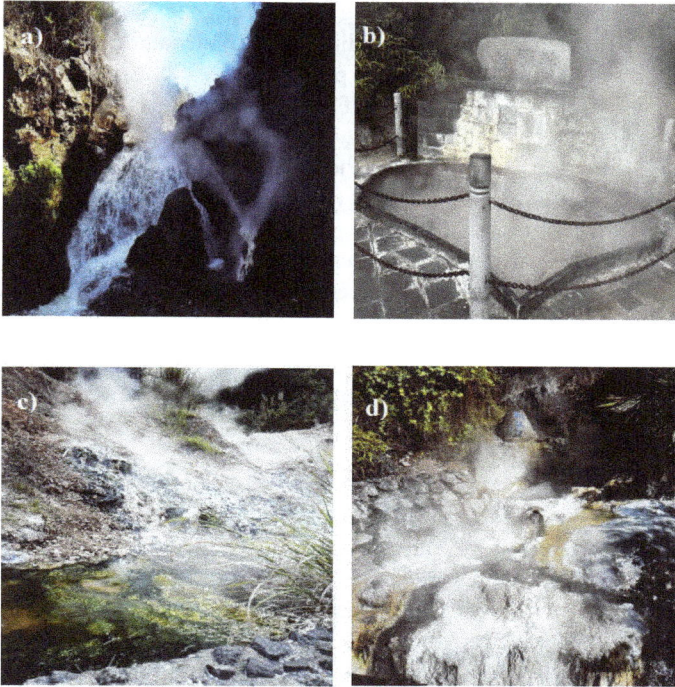

**Figure 1.** Examples of hot springs in (Yunnan) China. (a) Frog Mouth spring, (b) Pearl spring, (c) Hydrothermal explosion, (d) Drum beating spring.

Actinomycetes are an important source of industrial enzymes and pharmaceutical compounds and account for nearly two-thirds of the antibiotics produced by microorganisms (Liu et al. 2016). Apart from antibiotic production, they play a vital role in several biological processes, such as biogeochemical cycles, bioremediation (Alvarez et al. 2017), bio-weathering (Abdulla 2009), plant growth promotion in general (Palaniyandi et al. 2013) and under abiotic stress (Dong et al. 2019). Such actinomycetes are not only present in common environments, but also in extreme ecosystems where they exhibit several adaptive strategies, such as antibiosis, switching between different metabolic modes and production of specific enzymes which allow them to cope with extreme environmental conditions (Shivlata and Satyanarayana 2015).

The prevalence of thermotolerant and thermophilic actinomycetes has been reported in various environments, such as arid soils (Zucchi et al. 2012a, b), composts (Yan et al. 2011), desert soils (Busarakam et al. 2016), geothermally heated soils (Jiao et al. 2015), hydrothermal plumes (Zhang et al. 2017) and sediments of hot spring (Duan et al. 2014). Heat adaptive species have been assigned to diverse genera, including, *Acidimicrobium*, *Aciditerrimonas*, *Acidothermus*, *Actinomadura*, *Streptomyces*, *Thermobispora*, *Thermocatellispora*, *Thermoleophilum*, *Thermomonospora*, *Thermopolyspora*, and *Thermotunica* (Itoh et al. 2011, Jiao et al. 2015, Wu et al. 2018a, b, Shivlata and Satyanarayana 2015).

Actinomycetes have acquired strategies to overcome thermal stress including the presence of chaperones (aid in refolding partially denatured proteins), high GC content, substitution of amino acids in proteins, specific components in the cell wall and a comparatively high number of charged amino acids (Asp, Glu, Arg, and Lys) compared with polar amino acids (Asn, Gln, Ser, and Thr) in their proteins (Shivlata and Satyanarayana 2015). In the past few years, both culture-dependent (Liu et al. 2016) and culture-independent (Song et al. 2009) actinomycetes diversity analyses of thermal environments have been carried out. In the present chapter, we provide insights into the diversity of actinomyetes in thermal environments and their scientific and biotechnological importance using culture-dependent and independent methods.

## 2. Culture-independent Actinomycetes Diversity in Thermal Environments

Next-generation sequencing allows culture-free microbial diversity detection (Sabat et al. 2017). Compared with culture-based procedures, 16S rRNA gene-based culture-independent studies reveal a much broader microbial diversity in natural habitats (López-López et al. 2013). Several culture-independent microbial diversity analyses of thermal environments have been reported. Walker et al. (2005) detected many sequences belonging to *Mycobacterium* species from samples collected from the Norris Geyser Basin in the Yellowstone National Park, USA. Miller et al. (2009) analyzed bacterial communities of two alkaline hot springs along temperature gradients (38.5–72.5°C) and found three OTUs (operational taxonomic units) corresponding to actinomycetes. A microbial diversity analysis of Tuwa hot springs (temperature profile 54–65°C) in India found 22 bacterial phyla though only 0.4% were attributed to actinomycetes (Mangrola et al. 2015). In contrast, an investigation of the microbial flora of the Yumthang and Reshi hot springs (temperatures 41 and 47.4°C respectively) in Sikkim, India showed that actinomycetes accounted for 23–25% of the bacterial diversity (Najar et al. 2020). Actinomycetes in the 'Hammam Essalihine' hot spring in Khenchela Province (temperature around 70°C) in Algeria accounted for only 2% (Benammar et al. 2020). Similarly, Chaudhuri et al. (2017) reported low actinomycetes numbers in two hot springs in Bakreshwar, West Bengal (temperature 54 and 65°C) in India. These reports suggest that actinomycetes abundance and diversity were noticed at low temperature. In contrast, Song et al. (2009) were the first to report, taxonomically diverse actinomycetes in hot springs with temperatures up to 81°C. Valverde et al. (2012) evaluated actinomycetes diversity of hot springs located in China, Kenya, New Zealand and Zambia where temperatures ranged from 44.5 and 86.5°C, 28 significant actinomycetes OTUs were discovered in the resultant 16S rRNA gene libraries. These reports suggest that temperature plays an important role in the structure and functioning of actinomycetes communities in thermal environments.

The diversity and community composition of actinomycetes in microbial mats of five Tibetan hot springs (temperatures 26°C to 81°C) highlighted the presence of *Actinobacteridae, Acidimicrobidae*, and unclassified taxa (Jiang et al. 2012). Unlike some of the reports mentioned above, this study found that actinomycetes diversity was not significantly correlated with temperature suggesting that in this instances temperature was not a key factor. It is known that actinomycetes diversity in thermal environments is influenced by various physicochemicals parameters (Ahmed et al. 2020).

Recent advancements in metagenomics and meta-transcriptomics of microbial communities have provided insight into the diversity of active genes (Hua et al. 2018, Tripathy et al. 2016). Meta-transcriptome analysis of the microbial communities of an alkaline hot sulfur spring revelaed genes related to nitrate reductase and large subunits of carbon-monoxide dehydrogenase (*cutL* and *coxL*) in *Saccharopolyspora erythraea* and *Nocardioides* strains, respectively (Tripathy et al. 2016). The genome of *Acidothermus cellulolyticus* 11B, a thermo-cellulolytic strain which grows optimally at 55°C, has been evaluated for clues into its eco-physiological and evolutionary properties (Barabote et al. 2009). Associated genome analyses showed that it contains at least 43 genes encoding for 35 glycoside hydrolases and eight carbohydrate esterases while neither the genome nor the proteome segregates with those of other thermophiles. It has been suggested that *A. cellulolyticus* 11B has had a short history in thermal pools as its genome and proteome show meso-thermophilic features. Similarly, the mechanism underlying the thermal stability of *Corynebacterium efficiens* has been evaluated (Nishio et al. 2003). Analysis of the amino acid substitutions suggested that three substitutions, lysine to arginine; serine to alanine and serine to threonine have an important role in the stability of thermostable proteins in this organism while an increase in GC content were linked to the accumulation of these amino acid changes. Genome sequencing on *Saccharopolyspora hirsuta* subsp. *hirsuta* VKM Ac-666[T], a moderately thermophilic actinomycetes, showed the presence of genes related to steroid catabolism, rings A/B and C/D degradation, and steroid uptake (Lobastova et al. 2020).

## 3. Culture-dependent Actinomycetes Diversity in Thermal Environments

The isolation and description of microorganisms are essential for understanding their relationships with other organisms within ecosystems (Abdallah et al. 2017). The emergence of culture-independent microbial diversity analyses marginalized the application of culture-based methods which not only were seen to be slow but grossly underestimated the extent of microbial communities in natural ecosystems (López-López et al. 2013, Suyal et al. 2019). However, such analyses have several drawbacks, including overlooking minor populations (present at concentrations lower than $\times 10^5$ CFU/ml) and unreliable taxonomic characterization of species (Abdallah et al. 2017, Lagier et al. 2012). The application of culture-based techniques can overcome such drawbacks.

Isolation of thermotolerant and thermophilic actinomycetes depends on various factors such as medium composition, pH, temperature and incubation time. Waithaka et al. (2017) evaluated the taxonomic diversity of actinomycetes in geothermal vents and noticed that the number of isolates varied significantly between isolation media. Similarly, Prieto-Barajas et al. (2017) evaluated the effect of seasonality and physicochemical parameters on bacterial communities in two hot spring microbial mats and found that fluctuations in bacterial composition were correlated mainly with salt content, temperature, pH, and arsenic content. A range of media have been used to isolate thermophilic actinomycetes, as exemplified in Table 1.

Liu et al. (2016) analyzed culturable actinomycetes in hot spring (40–99°C) samples collected from Tengchong County, Yunnan Province, southwestern China. A total of 58 thermophilic isolates were affiliated to several genera (*Actinomadura, Microbispora, Micromonospora, Micrococcus, Nonomuraea, Nocardiopsis, Promicromonospora, Pseudonocardia, Streptomyces, Thermoactinospora, Thermocatellispora* and *Verrucosispora* (now is a synonym of *Micromonospora*, Nouioui et al. 2018). *Streptomyces* was found to be predominant in all sampling sites however more diverse actinomycetes were noticed at low temperature. Similarly, Kumar et al. (2013) assigned culturable thermotolerant actinomycetes from Manikaran hot springs (89 to 95°C) to the genera *Kocuria, Microbacterium* (including *Microbacterium oxydans*), *Micrococcus* and *Rhodococcus* (including *Rhodococcus baikunurensis*). In turn, Kumar et al. (2014) found thermotolerant *Cellulosimicrobium cellulans, Kocuria palustri* and *Rhodococcus* strains in four hot springs in India.

In a phylogenetic study, Song et al. (2001) assigned thermophilic actinomycetes from 21 mushroom composts to the genera *Pseudonocardia, Saccharomonospora, Saccharopolyspora, Streptomyces* and *Thermobifida*. Further, Kurapova et al. (2012) evaluated the diversity of actinomycetes isolated from soils from the Mongolian desert and steppes and found that periodic heating to high temperatures favored the growth of thermotolerant and moderately thermophilic strains over their mesophilic counterparts. Furthermore, the mycelial length of the srains was greater than that of the mesophiles. The most common thermophilic strains of the desert soils belonged to the genera *Actinomadura, Micromonospora, Streptomyces*, and *Streptosporangium*. Fink et al. (1971) isolated taxonomically diverse thermophilic actinomycetes from residential heating systems.

Recently, many novel actinomycetes have been isolated from thermal environments, as exemplified by *Gandjariella thermophila* a novel species of a new genus classified in the family *Pseudonocardiaceae*, which was recovered from a forest soil collected from a geothermal area (Ningsih et al. 2019). Similarly, Zhou et al. (2012b) reported a novel species, *Thermocatellispora tengchongensis* which belongs to the family *Streptosporangiaceae*. In turn, novel species from compost samples have been classified as *Thermomonospora catenispora* (Wu et al. 2019) and *Thermasporomyces composti* (Yabe et al. 2011) while others from arid soils were characterized as *Amycolatopsis granulosa, A. thermalba, A. thermophila, A. ruanii* and *A. viridis* (Zucchi et al. 2012a, b). In addition, some novel strains have been found to grow well under extreme and varied environmental conditions, examples include the moderately thermophilic and acidophilic species; *Aciditerrimonas ferrireducens*, which was isolated from solfataric field and found to grow optimally at 50°C and at pH 3.0. Further, *Rubrobacter taiwanensis*, a radiation-resistant thermophile, was

**Table 1.** List of some medium and their composition for the isolation of thermophilic actinomycetes.

| Medium | Composition | References |
|---|---|---|
| Reasoner's 2A | Yeast extract 0.05%, peptone 0.05%, casamino acids 0.05%, glucose 0.05%, starch 0.05%, sodium pyruvate 0.03%, $K_2HPO_4$ 0.03%, $MgSO_4$ 0.005%, agar 1.5%, pH 7. | Liu et al. (2016), Shirling and Gottlieb (1966) |
| ISP 2 | Yeast extract 0.4%, glucose 0.4%, malt extract 0.6%, agar 1.5%, pH 7. | |
| Modified T5 | Glucose 0.1%, yeast extract 0.2%, tryptone 0.05%, $CaCl_2$ 0.1%, starch 0.1%, lotus powder 0.1%, agar 1.5%, pH 7. | |
| ISP 1 | Casein enzymic hydrolysate 5.0 g/L, yeast extract 3.0 g/L, pH 7. | Ningsih et al. (2019), Shirling and Gottlieb (1966) |
| ISP 4 | Soluble starch 10 g/L, dipotassium phosphate 1 g/L, magnesium sulfate 1 g/L, sodium chloride 1 g/L, ammonium sulfate 2 g/L, calcium carbonate 2 g/L, ferrous sulfate 1 mg/L, manganous chloride 1 mg/L, zinc sulfate 1 mg/L, agar 20 g/L, pH 7.2 ± 0.2 | Wu et al. (2019), Shirling and Gottlieb (1966) |
| Gause synthetic agar | Soluble starch 20 g/L, sodium chloride 0.5 g/L, ferrous sulfate 0.01 g/L, potassium nitrate 1 g/L, dipotassium hydrogen phosphate 0.5 g/L, magnesium sulfate 0.5 g/L, agar 15 g/L, pH 7.3 ± 0.2 | Yan et al. (2011), Borgmeyer and Crawford (1985) |
| Humic acid-vitamin agar | Humic acid 1.0 g/L*, $Na_2HPO_4$ 0.5 g/L, KCl 1.71 g/L, $MgSO_4.7H_2O$ 0.05 g/L, $FeSO4.7H_2O$ 0.01 g/L, $CaCO_3$ 0.02 g/L, B-vitamins **, cyelohexinaide 50 mg/L, agar 18 g/L, pH 7.2. Note: *Dissolved in 10 ml of 0.2 N NaOH. ** 0.5 mg each of thiamine-HCl, riboflavin, niacin, pyridoxin-HCl, inositol, Ca-pantothenate, p-aminobenzoic acid, and 0.25 mg of biotin. B-vitamins and cycloheximide were filter-sterilized. | Zhou et al. (2012a), Hayakawa and Nonomura (1987) |
| SM1, SM2, SM3 | SM1: yeast nitrogen base 67.0 g and casamino acids 100 mg were added to a liter of distilled water and the solution was sterilized using cellulose filters (0.20 mm) prior to the addition of sterilized dipotassium hydrogen phosphate (200 ml; 10%, w/v); 100 ml of this basal medium was added to 900 ml of sterilized molten agar (1.5%, w/v) followed by filter sterilized solutions of D (–) sorbitol (final concentration 1%, w/v), cycloheximide (50 µg ml⁻¹), neomycin sulphate (4 µg ml⁻¹) and nystatin (50 µg ml⁻¹). SM2 was prepared as described above but with D (+) melezitose (1%, w/v) replacing the D (–) sorbitol. SM3 consisted of Gauze's medium 2 (glucose, 10 g; peptone, 5 g; tryptone, 3 g; NaCl, 5 g; agar, 15 g; distilled water, 1 l; pH 7.0) supplemented with filter sterilised solutions of cycloheximide (50 µg ml⁻¹), nalidixic acid (10 µg ml⁻¹), novobiocin (10 µg ml⁻¹) and nystatin (50 µg ml⁻¹) | Tan et al. (2006) |
| Czapek's agar | Sucrose 30 g/L, Sodium nitrate 2.0 g/L, Dipotassium phosphate 1.0 g/L, Magnesium sulphate 0.5 g/L, Potassium chloride 0.5 g/L, Ferrous sulphate 0.01 g/L, Agar 15.0 g/L, pH 7.2 | Zhou et al. (2012b) |

isolated from a hot spring (Chen et al. 2004). Similarly, a thermophilic and alkali-tolerant isolate from a tropical garden soil from Yogyakarta, Indonesia was classifed as *Streptomyces thermoalcalitolerans* (Kim et al. 1999) and a thermophilic isolate from poultry feaces as *S. thermocoprophilus* (Kim et al. 2000). Additional novel thermophilic actinomycetes are shown in Table 2.

**Table 2.** List of some thermotolerant and thermophilic actinomycetes isolated from various environments.

| Novel strains | Isolation source | Optimum temperature | References |
|---|---|---|---|
| *Microbispora soli* | Hot spring soil sample | 40°C | Kittisrisopit et al. (2018) |
| *Amycolatopsis deserti* | Arid desert soils | 45°C | Busarakam et al. (2016) |
| *Thermoactinomyces guangxiensis* | Mushroom compost | 45–50°C | Wu et al. (2015) |
| *Thermostaphylospora grisealba* | Mushroom compost | 45–55°C | Wu et al. (2018b) |
| *Actinomadura amylolytica* and *Actinomadura cellulosilytica* | Geothermally heated soil | 45°C | Jiao et al. (2015) |
| *Georgenia sediminis* | Sea sediment | 50°C | You et al. (2013) |
| *Streptomyces thermoalkaliphilus* | Tropical rainforest soil | 37–50°C | Wu et al. (2018a) |
| *Rubrobacter spartanus* | Volcanic soil | 50°C | Norman et al. (2017) |
| *Rubrobacter calidifluminis* and *Rubrobacter naiadicus* | Hot stream | 60°C | Albuquerque et al. (2014) |
| *Rubrobacter xylanophilus* | Thermally polluted effluent | 60°C | Carreto et al. (1996) |
| *Marmoricola caldifontis* | Hot spring sediment | 37–45°C | Habib et al. (2020) |
| *Planosporangium thailandense* | Soil of hot spring | 40°C | Thawai et al. (2013) |

# 4. Applications of Thermotolerant and Thermophilic Actinomycetes

Filamentous actinomycetes, especially *Streptomyces*, are a prolific source of novel bioactive compounds (Manikprabhu and Li 2015a,b). Indeed *Streptomyces* account for nearly two third of known antibiotics (Asolkar et al. 2010, Barka et al. 2016, Maiti and Mandal 2021, Li et al. 2019).

Thermotolerant and thermophilic actinomycetes as well as being sources of antibiotics they also play a vital role in several biological processes such as biogeochemical cycles, bioremediation (Alvarez et al. 2017, Manikprabhu and Li 2015a) and the production of enzymes (Barabote et al. 2010) that are active under extreme conditions. A fibrinolytic enzyme has been reported from a thermophilic *Streptomyces* sp. (Chitte and Dey 2002). *Marinactinospora thermotolerans* was reported to produce antimalarial β-carboline and indolactam alkaloids (Huang et al. 2011).

Actinomycetes strains isolated from geothermal springs have also been reported to have antimicrobial, anti-biofilm and anticancer activities (Mehetre et al. 2019). Demir et al. (2013) found that an alkali and thermotolerant ribonuclease was produced by an alkaliphilic *Streptomyces* strain (M49-1) while Barabote et al. (2010) found that *Acidothermus cellulolyticus* (11B) synthesized a thermostable endo-xylanase. A highly thermostable alkaline protease has been reported from a salt-tolerant alkaliphilic strain of *Nocardiopsis alba* (Gohel and Singh 2012). Cellulolytic activity has been reported from a thermo-acidophilic *Acidothermus cellulolyticus* strain (Mohagheghi et al. 1986) and a cellulase-free xylanase from thermotolerant *Streptomyces* strain (Ab106) (Techapun et al. 2002). A thermo-alkalophilic dextranase produced by *Streptomyces* strain (NK458) has been used to remove dextran during sugar manufacture (Purushe et al. 2012). A detergent-stable keratinase has been reported from thermophile *Actinomadura keratinilytica* (Cpt29) (Habbeche et al. 2014). The ability of actinomycetes enzymes to withstand harsh conditions can be translated into industrial applications such as in detergent, brewing, baking, leather, dairy industries as well as in medicine (Edwards 1993).

Gadkari et al. (1990) isolated a novel thermophilic species, *Streptomyces thermoautotrophicus*, which oxidizes CO and $H_2$. Kim et al. (1998) found that two moderately thermophilic carboxydotrophic *Streptomyces* strains classified as *Streptomyces thermocarboxydus* and *S. thermocarboxydovorans* oxidized hydrogen and used carbon monoxide as a sole source of carbon for energy and growth (Kim et al. 1998). A novel heterotrophic, thermophilic and extremely acidophilic species *Ferrithrix thermotolerans* was shown to oxidize iron (Johnson et al. 2009). These reports suggest that thermophilic actinomycetes members play an important role in biogeochemical cycles.

Thermophilic actinomycetes also have potential as agents of bioremediation applications. A thermophilic strain of *Thermobifida fusca*, for instance, produces an extracellular thermo-alkali-stable laccase which oxidizes dye intermediates (Chen et al. 2013). Some thermophilic actinomycetes have been reported to synthesize nanoparticles, such as a novel extremophilic *Thermomonospora* strain which synthesize extracellular monodisperse gold nanoparticles which can replace protocols in which toxic chemicals are used in the process (Ahmad et al. 2003).

## 5. Conclusion and Future prospects

It is apparent from the studies outlined above that actinomycetes are integral part of high temperature habitats and that cultivable representatives of diverse taxonomic groups are a potential source of new chemical compounds, notably antibioitcs and enzymes. Further investigations into the occurence, diversity and bioactivity of actinomycetes in thermal environments using culture-dependent and culture-independent methods can be expected to increase our understanding on their physiology thereby creating opportunities for industrial applications.

## Acknowledgments

This study was supported by the National Natural Science Foundation of China (Grants 91951205 and 32061143043).

## References

Abdallah, R.A., Beye, M., Diop, A., Bakour, S., Raoult, D. and Fournier, P.E. (2017). The impact of culturomics on taxonomy in clinical microbiology. *Antonie van Leeuwenhoek*, 110: 1327–1337.

Abdulla, H. (2009). Bioweathering and biotransformation of granitic rock minerals by actinomycetes. *Microbial Ecology*, 58: 753–761.

Ahmad, A., Senapati, S., Khan, M.I., Kumar, R. and Sastry, M. (2003). Extracellular biosynthesis of monodisperse gold nanoparticles by a novel extremophilic actinomycetes *Thermomonospora* spp. *Langmuir, ACS*, 19: 3550–3553.

Ahmed, R.N., Daniel, F., Gbala, I.D. and Sanni, A. (2020). Potentials of Actinomycetes from reserved environments as antibacterial agents against drug-resistant clinical bacterial strains. *Ethiopian Journal of Health Sciences*, 30(2): 251.

Albuquerque, L., Johnson, M.M., Schumann, P., Rainey, F.A. and da Costa, M.S. (2014). Description of two new thermophilic species of the genus *Rubrobacter*, *Rubrobacter calidifluminis* sp. nov. and *Rubrobacter naiadicus* sp. nov., and emended description of the genus *Rubrobacter* and the species *Rubrobacter bracarensis*. *Systematics and Applied Microbiology*, 37: 235–243.

Alvarez, A., Saez, J.M., Davila Costa, J.S., Colin, V.L., Fuentes, M.S., Cuozzo, S.A. et al. (2017). Actinobacteria: Current research and perspectives for bioremediation of pesticides and heavy metals. *Chemosphere*, 166: 41–62.

Asolkar, R.N., Kirkland, T.N., Jensen, P.R. and Fenical, W. (2010). Arenimycin, an antibiotic effective against rifampin- and methicillin-resistant *Staphylococcus aureus* from the marine actinomycete *Salinispora arenicola*. *Journal of Antibiotics*, 63: 37–39.

Barabote, R.D., Parales, J.V., Guo, Y.Y., Labavitch, J.M., Parales, R.E. and Berry, A.M. (2010). Xyn10A, a thermostable endoxylanase from *Acidothermus cellulolyticus* 11B. *Applied and Environmental Microbiology*, 76: 7363–7366.

Barabote, R.D., Xie, G., Leu, D.H., Normand, P., Necsulea, A., Daubin, V. et al. (2009). Complete genome of the cellulolytic thermophile *Acidothermus cellulolyticus* 11B provides insights into its ecophysiological and evolutionary adaptations. *Genome Research*, 19: 1033–1043.

Barka, E.A., Vatsa, P., Sanchez, L., Gaveau-Vaillant, N., Jacquard, C., Klenk, H.P., Clément, C., Ouhdouch, Y. and van Wezel, G.P. (2016). Taxonomy, physiology, and natural products of Actinobacteria. *Microbiology and Molecular Biology Reviews*, 80(1): 1–43.

Benammar, L., Bektaş, K.I., Menasria, T., Beldüz, A.O., Güler, H.I., Bedaida, I.K., Gonzalez, J.M. and Ayachi, A. (2020). Diversity and enzymatic potential of thermophilic bacteria associated with terrestrial hot springs in Algeria. *Brazilian Journal of Microbiology*, 51(4): 1987–2007.

Borgmeyer, J.R. and Crawford, D.L. (1985). Production and characterization of polymeric lignin degradation intermediates from two different *Streptomyces* spp. *Applied and Environmental Microbiology*, 49: 273–278.

Busarakam, K., Brown, R., Bull, A.T., Tan, G.Y., Zucchi, T.D., da Silva, L.J. et al. (2016). Classification of thermophilic actinobacteria isolated from arid desert soils, including the description of *Amycolatopsis deserti* sp. nov. *Antonie Van Leeuwenhoek*, 109: 319–334.

Carreto, L., Moore, E., Nobre, M.F., Wait, R., Riley, P.W., Sharp, R.J. et al. (1996). *Rubrobacter xylanophilus* sp. nov., a new thermophilic species isolated from a thermally polluted effluent. *International Journal of Systematic and Evolutionary Microbiology*, 46: 460–465.

Chaudhuri, B., Chowdhury, T. and Chattopadhyay, B. (2017). Comparative analysis of microbial diversity in two hot springs of Bakreshwar, West Bengal, India. *Genomics Data*, 12: 122–129.

Chen, C.Y., Huang, Y.C., Wei, C.M., Meng, M., Liu, W.H. and Yang, C.H. (2013). Properties of the newly isolated extracellular thermo-alkali-stable laccase from thermophilic actinomycetes, *Thermobifida fusca* and its application in dye intermediates oxidation. *AMB Express*, 3: 1–9.

Chen, M.Y., Wu, S.H., Lin, G.H., Lu, C.P., Lin, Y.T., Chang, W.C. et al. (2004). *Rubrobacter taiwanensis* sp. nov., a novel thermophilic, radiation-resistant species isolated from hot springs. *International Journal of Systematic and Evolutionary Microbiology*, 54: 1849–1855.

Chien, A., Edgar, D.B. and Trela, J.M. (1976). Deoxyribonucleic acid polymerase from the extreme thermophile *Thermus aquaticus*. *Journal of Bacteriology*, 127: 1550–1557.

Chitte, R.R. and Dey, S. (2002). Production of a fibrinolytic enzyme by thermophilic *Streptomyces* species. *World Journal of Microbiology and Biotechnology*, 18: 289–294.

Demir, T., Gübe, Ö., Yücel, M. and Hames-Kocabas, E.E. (2013). Increased alkalotolerant and thermostable ribonuclease (RNase) production from alkaliphilic *Streptomyces* sp. M49-1 by optimizing the growth conditions using response surface methodology. *World Journal of Microbiology and Biotechnology*, 29: 1625–1633.

Djokic, T., Van Kranendonk, M.J., Campbell, K.A., Walter, M.R. and Ward, C.R. (2017). Earliest signs of life on land preserved in ca. 3.5 Ga hot spring deposits. *Nature Communications*, 8: 15263.

Dong, Z.Y., Narsing Rao, M.P., Wang, H.F., Fang, B.Z., Liu, Y.H., Li, L. et al. (2019). Transcriptomic analysis of two endophytes involved in enhancing salt stress ability of *Arabidopsis thaliana*. *Science of the Total Environment*, 10: 107–117.

Duan, Y.Y., Ming, H., Dong, L., Yin, Y.R., Zhang, Y., Zhou, E.M. et al. 2014. *Streptomyces calidiresistens* sp. nov., isolated from a hot spring sediment. *Antonie Van Leeuwenhoek*, 106: 189–196.

Edwards, C. (1993). Isolation properties and potential applications of thermophilic actinomycetes. *Applied Biochemistry and Biotechnology*, 42(2): 161–179.

Fink, J.N., Resnick, A.J. and Salvaggio, J. (1971). Presence of thermophilic acintomycetes in residential heating systems. *Applied Microbiology*, 22: 730–731.

Gadkari, D., Schricker, K., Acker, G., Kroppenstedt, R.M. and Meyer, O. 1990. *Streptomyces thermoautotrophicus* sp. nov., a thermophilic CO- and H$_2$-oxidizing obligate chemolithoautotroph. *Applied and Environmental Microbiology*, 56: 3727–3734.

Gohel, S.D. and Singh, S.P. (2012). Purification strategies, characteristics, and thermodynamic analysis of a highly thermostable alkaline protease from a salt-tolerant alkaliphilic actinomycete, *Nocardiopsis alba* OK-5. *Journal of Chromatography B Analytical Technologies in the Biomedical and Life Sciences*, 889-890: 61–68.

Goodfellow, M. (2012). Phylum XXVI. Actinobacteria phyl. nov. pp. 33–34. *In*: Goodfellow, M., Kämpfer, P., Busse, H.J., Trujillo, M.E., Suzuki, K., Ludwig, W. and Whitman, W.B. (eds.). *Bergey's Manual of Systematic Bacteriology*, 2nd edn. Vol. 5, Springer, New York.

Habbeche, A., Saoudi, B., Jaouadi, B., Haberra, S., Kerouaz, B., Boudelaa, M. et al. (2014). Purification and biochemical characterization of a detergent-stable keratinase from a newly thermophilic actinomycete *Actinomadura keratinilytica* strain Cpt29 isolated from poultry compost. *Journal of BioScience and Biotechnology*, 117: 413–421.

Habib, N., Khan, I.U., Xiao, M., Li, S., Saqib, M., Xian, W.D. et al. (2020). *Marmoricola caldifontis* sp. nov., a novel actinobacterium isolated from a hot spring. *International Journal of Systematic and Evolutionary Microbiology*, 70: 2053–2058.

Hayakawa, M. and Nonomura, H. (1987). Humic acid-vitamin agar, a new medium for the selective isolation of soil actinomycetes. *Journal of Fermentation Technology*, 65: 501–509.

Hua, Z.S., Qu, Y.N., Zhu, Q., Zhou, E.M., Qi, Y.L., Yin, Y.R. et al. 2018. Genomic inference of the metabolism and evolution of the archaeal phylum Aigarchaeota. *Nature Communications*, 19; 9(1): 2832.

Huang, H., Yao, Y., He, Z., Yang, T., Ma, J., Tian, X. et al. (2011). Antimalarial β-carboline and indolactam alkaloids from *Marinactinospora thermotolerans*, a deep-sea isolate. *Journal of Natural Products*, 74: 2122–2127.

Hussein, E.I., Jacob, J.H., Shakhatreh, M.A.K., Abd Al-Razaq, M.A., Juhmani, A.F. and Cornelison, C.T. (2017). Exploring the microbial diversity in Jordanian hot springs by comparative metagenomic analysis. *MicrobiologyOpen*, 6(6): e00521. https://onlinelibrary.wiley.com/journal/20458827.

Itoh, T., Yamanoi, K., Kudo, T., Ohkuma, M. and Takashina, T. (2011). *Aciditerrimonas ferrireducens* gen. nov., sp. nov., an iron-reducing thermoacidophilic actinobacterium isolated from a solfataric field. *International Journal of Systematic and Evolutionary Microbiology*, 61: 1281–1285.

Jiang, H., Dong, C.Z., Huang, Q., Wang, G., Fang, B., Zhang, C. et al. (2012). Actinobacterial diversity in microbial mats of five hot springs in Central and Central-Eastern Tibet, China. *Geomicrobiology J.*, 29: 520–527.

Jiao, J.Y., Liu, L., Zhou, E.M., Wei, D.Q., Ming, H., Xian, W.D. et al. (2015). *Actinomadura amylolytica* sp. nov. and *Actinomadura cellulosilytica* sp. nov., isolated from geothermally heated soil. *Antonie Van Leeuwenhoek*, 108: 75–83.

Johnson, D.B., Bacelar-Nicolau, P., Okibe, N., Thomas, A. and Hallberg, K.B. (2009). *Ferrimicrobium acidiphilum* gen. nov., and *Ferrithrix thermotolerans* gen. nov., sp. nov.: Heterotrophic, iron-oxidizing, extremely acidophilic actinobacteria. *International Journal of Systematic and Evolutionary Microbiology*, 59: 1082–1089.

Kim, S.B., Falconer, C., Williams, E. and Goodfellow, M. (1998). *Streptomyces thermocarboxydovorans* sp. nov. and *Streptomyces thermocarboxydus* sp. nov., two moderately thermophilic carboxydotrophic species from soil. *International Journal of Systematic and Evolutionary Microbiology*, 48: 59–68.

Kim, B., Al-Tai, A.M., Kim, S.B., Somasundaram, P. and Goodfellow, M. (2000). *Streptomyces thermocoprophilus* sp. nov., a cellulase-free endo-xylanase-producing streptomycete. *International Journal of Systematic and Evolutionary Microbiology*, 50(2): 505–509.

Kim, B., Sahin, N., Minnikin, D.E., Zakrzewska-Czerwinska, J., Mordarski, M. and Goodfellow, M. (1999). Classification of thermophilic streptomycetes, including the description of *Streptomyces thermoalcalitolerans* sp. nov. *International Journal of Systematic and Evolutionary Microbiology*, 49: 7–17.

Kittisrisopit, S., Pittayakhajonwut, P., Tadtong, S. and Thawai, C. (2018). *Microbispora soli* sp. nov., isolated from soil of a hot spring. *International Journal of Systematic and Evolutionary Microbiology*, 68: 3863–3868.

Kumar, M., Yadav, A.N., Tiwari, R., Prasanna, R. and Saxena, A.K. (2013). Deciphering the diversity of culturable thermotolerant bacteria from Manikaran hot springs. *Annals of Microbiology*, 64: 741–751.

Kumar, M., Yadav, A.N., Tiwari, R., Prasanna, R. and Saxena, A.K. (2014). Evaluating the diversity of culturable thermotolerant bacteria from four hot springs of India. *Journal of Biodiversity, Bioprospecting and Development*, 1: 1–9.

Kurapova, I., Zenova, G.M., Sudnitsyn, I.I., Kizilova, A.K., Manucharova, N.A., Norovsuren, Z.H. et al. (2012). Thermotolerant and thermophilic actinomycetes from soils of Mongolia Desert Steppe Zone. *Microbiology*, 81: 98–108.

Lagier, J.C., Armougom, F., Million, M., Hugon, P., Pagnier, I., Robert, C. et al. (2012). Microbial culturomics: Paradigm shift in the human gut microbiome study. *Clinical Microbiology and Infection*, 18: 1185–1193.

Li, F., Liu, S., Lu, Q., Zheng, H., Osterman, I.A., Lukyanov, D.A. et al. (2019). Studies on antibacterial activity and diversity of cultivable actinobacteria isolated from mangrove soil in Futian and Maoweihai of China. *Evidence-Based Complementary and Alternative Medicine*, 9: 3476567.

Li, L. and Ma, Z. (2019). Global microbiome diversity scaling in hot springs with DAR (Diversity-Area Relationship) profiles. *Frontiers in Microbiology*, 22; 10: 118.

Liu, L., Salam, N., Jiao, J.Y., Jiang, H.C., Zhou, E.M., Yin. Y.R. et al. (2016). Diversity of culturable thermophilic actinobacteria in hot springs in Tengchong, China and studies of their biosynthetic gene profiles. *Microbial Ecology*, 72: 150–162.

Lobastova, T.G., Fokina, V.V., Bragin, E.Y., Shtratnikova, V.Y., Starodumova, I.P., Tarlachkov, S.V. et al. (2020). Draft genome sequence of the moderately thermophilic actinobacterial steroid-transforming *Saccharopolyspora hirsuta* subsp. *hirsuta* Strain VKM Ac-666[T]. *Microbiology Resource Announcements*, 9(1): e01327–19.

López-López, O., Cerdán, M.E. and González-Siso, M.I. (2013). Hot spring metagenomics. *Life (Basel)*, 3: 308–320.

Maiti, P.K. and Mandal, S. (2021). *Streptomyces cupreus* sp. nov., an antimicrobial producing actinobacterium isolated from Himalayan soil. *Archives of Microbiology*, Doi: 10.1007/s00203-020-02160-y.

Mangrola, A., Dudhagara, P., Koringa, P., Joshi, C.G., Parmar, M. and Patel, R. (2015). Deciphering the microbiota of Tuwa hot spring, India using shotgun metagenomic sequencing approach. *Genomics Data*, 4: 153–155.

Manikprabhu, D. and Li, W.J. (2015a). Antibiotics from discovery to journey. pp. 1–14. *In*: Dhanasekaran, D., Thajuddin, N. and Panneerselvam, A. (eds.). *Antimicrobials, Synthetic and Natural Compounds.* CRC Press. Boca Raton.

Manikprabhu, D. and Li, W.J. (2015b). Antimicrobial agents from actinomycetes: Chemistry and applications. pp. 99–115. *In*: Dhanasekaran, D., Thajuddin, N. and Panneerselvam, A. (eds.). *Antimicrobials, Synthetic and Natural Compounds.* CRC Press. Boca Raton.

Mehetre, G.T., Vinodh, J., Burkul, B.B., Desai, D., Santhakumari, B., Dharne, M.S. and Dastager, S.G. (2019). Bioactivities and molecular networking-based elucidation of metabolites of potent actinobacterial strains isolated from the Unkeshwar geothermal springs in India. *RSC Advances,* 9: 9850–9859.

Miller, S.R., Strong, A.L., Jones, K.L. and Ungerer, M.C. (2009). Bar-coded pyrosequencing reveals shared bacterial community properties along the temperature gradients of two alkaline hot springs in Yellowstone National Park. *Applied and Environmental Microbiology,* 75: 4565–4572.

Mohagheghi, A., Grohmann, K., Himmel, M., Leighton, L. and Updegraff, D.M. (1986). Isolation and characterization of *Acidothermus cellulolyticus* gen nov., sp. nov., a new genus of thermophilic, acidophilic, cellulolytic bacteria. *International Journal of Systematic and Evolutionary Microbiology,* 36: 435–443.

Najar, I.N., Sherpa, M.T., Das, S. and Thakur, N. (2020). Bacterial diversity and functional metagenomics expounding the diversity of xenobiotics, stress, defense and CRISPR gene ontology providing eco-efficiency to Himalayan hot springs. *Functional & Integrative Genomics,* 20: 479–496.

Ningsih, F., Yokota, A., Sakai, Y., Nanatani, K., Yabe, S., Oetari, A. et al. (2019). *Gandjariella thermophila* gen. nov., sp. nov., a new member of the family *Pseudonocardiaceae,* isolated from forest soil in a geothermal area. *International Journal of Systematic and Evolutionary Microbiology,* 69: 3080–3086.

Nishio, Y., Nakamura, Y., Kawarabayasi, Y., Usuda, Y., Kimura, E., Sugimoto, S. et al. (2003). Comparative complete genome sequence analysis of the amino acid replacements responsible for the thermostability of *Corynebacterium efficiens. Genome Research,* 13: 1572–1579.

Norman, J.S., King, G.M. and Friesen, M.L. (2017). *Rubrobacter spartanus* sp. nov., a moderately thermophilic oligotrophic bacterium isolated from volcanic soil. *International Journal of Systematic and Evolutionary Microbiology,* 67: 3597–3602.

Nouioui, I., Carro, L., García-López, M., Meier-Kolthoff, J.P., Woyke, T., Kyrpides, N.C., Pukall, R., Klenk, H.P., Goodfellow, M. and Göker, M. (2018). Genome-based taxonomic classification of the phylum *Actinobacteria. Frontiers in Microbiology,* Article 2007.

Oren, A. and Garrity, G.M. (2021). Valid publication of the names of forty-two phyla of prokaryotes. *International Journal of Systematic and Evolutionary Microbiology,* 71(10): 005056.

Palaniyandi, S.A., Yang, S.H., Zhang, L. and Suh, J.W. (2013). Effects of actinobacteria on plant disease suppression and growth promotion. *Applied Microbiology and Biotechnology,* 97: 9621–9636.

Prieto-Barajas, C.M., Alfaro-Cuevas, R., Valencia-Cantero, E. and Santoyo, G. (2017). Effect of seasonality and physicochemical parameters on bacterial communities in two hot spring microbial mats from Araró, Mexico. *Revista Mexicana de Biodiversidad,* 88: 616–624.

Purushe, S., Prakash, D., Nawani, N.N., Dhakephalkar, P. and Kapadnis, B. (2012). Biocatalytic potential of an alkalophilic and thermophilic dextranase as a remedial measure for dextran removal during sugar manufacture. *Bioresource Technology,* 115: 2–7.

Sabat, A.J., van Zanten, E., Akkerboom, V., Wisselink, G., van Slochteren, K., de Boer, R.F. et al. (2017). Targeted next-generation sequencing of the 16S-23S rRNA region for culture-independent bacterial identification-increased discrimination of closely related species. *Scientific Reports,* 7: 3434.

Salam, N., Jiao, J.Y., Zhang, X.T. and Li, W.J. (2020). Update on the classification of higher ranks in the phylum *Actinobacteria. International Journal of Systematic and Evolutionary Microbiology,* 70: 1331–1355.

Schuler, C.G., Havig, J.R. and Hamilton, T.L. (2017). Hot spring microbial community composition, morphology, and carbon fixation: Implications for interpreting the ancient rock record. *Frontiers in Earth Science,* 5: 97.

Shirling, E.B. and Gottlieb, D. (1966). Methods for characterization of *Streptomyces* species. *International Journal of Systematic and Evolutionary Microbiology,* 16: 313–340.

Shivlata, L. and Satyanarayana, T. (2015). Thermophilic and alkaliphilic actinobacteria: Biology and potential applications. *Frontiers in Microbiology*, 6: 1014.

Song, J., Weon, H.Y., Yoon, S.H., Park, D.S., Go, S.J. and Suh, J.W. (2001). Phylogenetic diversity of thermophilic actinomycetes and *Thermoactinomyces* spp. isolated from mushroom composts in Korea based on 16S rRNA gene sequence analysis. *FEMS Microbiology Letters*, 202: 97–102.

Song, Z., Zhi, X., Li, W., Jiang, H., Zhang, C. and Dong, H. (2009). Actinobacterial diversity in hot springs in Tengchong (China), Kamchatka (Russia), and Nevada (USA). *Geomicrobiology Journal*, 26: 256–263.

Stackebrandt, E., Rainey, F.A. and Ward-Rainey, N.L. (1997). Proposal for a new hierarchic classification system, *Actinobacteria* classis nov. *International Journal of Systematic and Evolutionary Microbiology*, 47: 479–491.

Stetter, K.O. (1994). The lesson of *Archaebacteria*. pp. 143–160. *In*: Bengtson, S. (ed.). Early Life on Earth. Columbia, New York.

Suyal, D.C., Joshi, D., Debbarma, P., Soni, R., Das, B. and Goel, R. (2019). Soil metagenomics: Unculturable microbial diversity and its function. In *Mycorrhizosphere and Pedogenesis* (pp. 355–362). Springer, Singapore.

Tan, G.Y.A., Ward, A.C. and Goodfellow, M. (2006). Exploration of *Amycolatopsis* diversity in soil using genus-specific primers and novel selective media. *Systematic and Applied Microbiology*, 29: 557–569.

Techapun, C., Charoenrat, T., Poosaran, N., Watanabe, M. and Sasaki, K. (2002). Thermostable and alkaline-tolerant cellulase-free xylanase produced by thermotolerant *Streptomyces* sp. Ab106. *Journal of BioScience and Bioengineering*, 93: 431–433.

Thawai, C., Thamsathit, W. and Kudo, T. (2013). *Planosporangium thailandense* sp. nov., isolated from soil from a Thai hot spring. *International Journal of Systematic and Evolutionary Microbiology*, 63: 1051–1055.

Tripathy, S., Padhi, S.K., Mohanty, S., Samanta, M. and Maiti, N.K. (2016). Analysis of the metatranscriptome of microbial communities of an alkaline hot sulfur spring revealed different gene encoding pathway enzymes associated with energy metabolism. *Extremophiles*, 20: 525–536.

Valverde, A., Tuffin, M. and Cowan, D.A. (2012). Biogeography of bacterial communities in hot springs: A focus on the actinobacteria. *Extremophiles*, 16: 669–679.

Waithaka, P.N., Mwaura, F.B., Wagacha, J.M. and Gathuru, E.M. (2017). Isolation of actinomycetes from geothermal vents of menengai crater in Kenya. *International Journal of Molecular Biology*, 2: 132–139.

Walker, J.J., Spear, J.R. and Pace, N.R. (2005). Geobiology of a microbial endolithic community in the Yellowstone geothermal environment. *Nature*, 434: 1011–1014.

Walter, M.R. (1996). Ancient hydrothermal ecosystems on Earth: A new palaeobiological frontier. *In Evolution of Hydrothermal Ecosystems on Earth* (and Mars?), Ciba Foundation Symposium 202. pp. 112–130. John Wiley & Sons, New York.

Woese, C.R. (1987). Bacterial evolution. *Microbiology Reviews*, 51: 221–271.

Wu, H., Liu, B. and Pan, S. (2015). *Thermoactinomyces guangxiensis* sp. nov., a thermophilic actinomycete isolated from mushroom compost. *International Journal of Systematic and Evolutionary Microbiology*, 65: 2859–2864.

Wu, H., Liu, B., Ou, X., Pan, S., Shao, Y. and Huang, F. (2018a). *Streptomyces thermoalkaliphilus* sp. nov., an alkaline cellulase producing thermophilic actinomycete isolated from tropical rainforest soil. *Antonie Van Leeuwenhoek*, 111: 413–422.

Wu, H., Liu, B., Shao, Y., Ou, X. and Huang, F. (2018b). *Thermostaphylospora grisealba* gen. nov., sp. nov., isolated from mushroom compost and transfer of *Thermomonospora chromogena* Zhang et al. 1998 to *Thermostaphylospora chromogena* comb. nov. *International Journal of Systematic and Evolutionary Microbiology*, 68: 602–608.

Wu, H., Wei, J. and Liu, B. (2019). *Thermomonospora catenispora* sp. nov., isolated from mushroom compost. *International Journal of Systematic and Evolutionary Microbiology*, 69: 2465–2470.

Yabe, S., Aiba, Y., Sakai, Y., Hazaka, M. and Yokota, A. (2011). *Thermasporomyces composti* gen. nov., sp. nov., a thermophilic actinomycete isolated from compost. *International Journal of Systematic and Evolutionary Microbiology*, 61: 86–90.

Yan, X., Yan, H., Liu, Z., Liu, X., Mo, H. and Zhang, L. (2011). *Nocardiopsis yanglingensis* sp. nov., a thermophilic strain isolated from a compost of button mushrooms. *Antonie Van Leeuwenhoek*, 100: 415–419.

You, Z.Q., Li, J., Qin, S., Tian, X.P., Wang, F.Z. and Zhang, S. (2013). *Georgenia sediminis* sp. nov., a moderately thermophilic actinobacterium isolated from sediment. *International Journal of Systematic and Evolutionary Microbiology*, 63: 4243–4247.

Zhang, L., Xi, L., Ruan, J. and Huang, Y. (2017). *Kocuria oceani* sp. nov., isolated from a deep-sea hydrothermal plume. *International Journal of Systematic and Evolutionary Microbiology*, 67: 164–169.

Zhou, E.M., Tang, S.K, Sjøholm, C., Song, Z.Q., Yu, T.T., Yang, L.L. et al. (2012a). *Thermoactinospora rubra* gen. nov., sp. nov., a thermophilic actinomycete isolated from Tengchong, Yunnan province, south-west China. *Antonie Van Leeuwenhoek*, 102: 177–185.

Zhou, E.M., Yang, L.L., Song, Z.Q., Yu, T.T., Nie, G.X., Ming, H. et al. (2012b). *Thermocatellispora tengchongensis* gen. nov., sp. nov., a new member of the family *Streptosporangiaceae*. *International Journal of Systematic and Evolutionary Microbiology*, 62: 2417–2423.

Zucchi, T.D., Tan, G.Y.A., Bonda, A.N.V., Frank, S., Kshetrimayum, J.D. and Goodfellow, M. (2012a). *Amycolatopsis granulosa* sp. nov., *Amycolatopsis ruanii* sp. nov. and *Amycolatopsis thermalba* sp. nov., thermophilic actinomycetes isolated from arid soils. *International Journal of Systematic and Evolutionary Microbiology*, 62: 1245–1251.

Zucchi, T.D., Tan, G.Y.A. and Goodfellow, M. (2012b). *Amycolatopsis thermophila* sp. nov. and *Amycolatopsis viridis* sp. nov., thermophilic actinomycetes isolated from arid soil. *International Journal of Systematic and Evolutionary Microbiology*, 62: 168–172.

# Chapter 5

# Entering Poorly Charted Waters
## The Biology of the Filamentous Acid-loving Actinomycetes and Acidimicrobia

*Patrycja Golińska,*[1,*] *Michael Goodfellow*[2] and *Vartul Sangal*[3]

## 1. Introduction

The phylum *Actinomycetota* (Oren and Garrity 2021), formerly *Actinobacteria sensu* Goodfellow (2012a), is one of the largest in the domain *Bacteria* as inferred from 16S rRNA gene sequencing studies (Ludwig et al. 2012a). Members of the phylum display an astonishing morphological, metabolic, physiological and phylogenomic diversity (Goodfellow et al. 2012, Barka et al. 2015, Nouioui et al. 2018). Actinomycetes, that is, members of the phylum *Actinomycetota*, are widely distributed in natural habitats where they are involved in the turnover of organic matter and in the transformation of recalcitrant molecules (Peczyńska-Czoch and Mordarski 1988, Buresova et al. 2019) but are best known as a unique source of specialized (secondary) metabolites of agricultural, industrial and medical significance, notably antibiotics (Takahashi and Nakashima 2018, Neuman and Cragg 2020).

The name *Actinomycetota* was validly published following the decision of the International Committee on Systematics of Prokaryotes to include the rank of phylum in the International Code of Nomenclature of Prokaryotes (Parker et al. 2019). This, in turn, was based on proposals by Whitman et al. (2018) and Oren et al. (2021). The phylum encompasses actinomycetes belonging to the classes *Acidimicrobiia* Norris 2012, "*Actinobacteria*" Stackebrandt et al. 1997 (a validly published, but illegitimate

---

[1] Department of Microbiology, Faculty of Biological and Veterinary Sciences, Nicolaus Copernicus University, Lwowska 1, 87 100 Toruń, Poland.
[2] School of Natural and Environmental Sciences, Ridley Building 2, Newcastle University, Newcastle upon Tyne, NE1 7RU, United Kingdom; michael.goodfellow@newcastle.ac.uk
[3] Faculty of Health and Life Sciences, Northumbria University, Newcastle upon Tyne, United Kingdom; vartul.sangal@northumbria.ac.uk
* Corresponding author: golinska@umk.pl

name that needs to be rephrased), *Coriobacteriia* König 2013 emend Gupta et al. (2013), *Nitriliruptoria* Ludwig et al. 2012b, *Rubrobacteria* Suzuki et al. 2012 emend Foesel et al. (2016) and *Thermoleophilia* Suzuki and Whitman 2013 emend Foesel et al. (2016).

Filamentous actinomycetes remain a rich source of new antibiotics of clinical significance (Goodfellow 2012b, Genilloud 2014, 2017) and hence have a role in combating emerging infectious diseases and in controlling multi-drug resistant microbial pathogens which are a major threat to global health (Tacconelli et al. 2018, Theuretzbacher et al. 2019). Nearly 70% of known antibiotics are produced by filamentous actinomycetes notably by members of the genus *Streptomyces* (Hopwood 2007, Barka et al. 2015). Streptomycetes are especially gifted *sensu* Baltz (2017, 2019) as they have large genomes (> 8 Mbp) rich in natural product-biosynthetic gene clusters (NP-BGCs) which encode for novel and uncharacterized antibiotics (Gomez-Escribano et al. 2015, Castro et al. 2018). Other filamentous actinomycetes with large genomes include rare, that is, historically poorly studied taxa, such as the genera *Amycolatopsis* (Sangal et al. 2018), *Frankia* (Nouioui et al. 2019), *Micromonospora* (Carro et al. 2018) and *Salinispora* (Jensen 2016). The discovery that the genomes of filamentous actinomycetes contain clade-, species- and strain-specific NP-bioclusters (Adamek et al. 2018, Carro et al. 2018, Martinet et al. 2020, Nicault et al. 2020) underlines the importance of reliable classification of actinomycetes in the search for novel compounds, especially with respect to taxonomic approaches to drug discovery (Goodfellow and Fiedler 2010, Goodfellow et al. 2018). Standard and innovative procedures are available for the selective isolation of taxon-specific actinomycetes from natural habitats (Goodfellow 2010, Goodfellow et al. 2018).

The search for new natural products that can be developed as resources for healthcare has shifted towards the isolation of filamentous actinomycetes from extreme ecosystems on the premise that extreme abiotic conditions favour the selection of novel strains with unexplored chemical diversity leading to the discovery of new bioactive compounds (Bull 2011, Kurtböke 2016, Bull and Goodfellow 2019). In practice, novel filamentous actinomycetes, especially streptomycetes, are a prolific source of new antibiotics, as exemplified by strains isolated from desert soils (Rateb et al. 2018, Djinni et al. 2019) and deep-sea sediments (Nweze et al. 2020, Subramani and Sipkema 2019). The extension of such studies to unexplored extreme habitats should yield additional drug leads as geographical location and biome type influence the distribution of streptomycete phenotypes associated with antibiotic production (Schlatter and Kinkel 2014, Andam et al. 2016). Geographical location and biome type are also critical features in the distribution of NP-BGCs in environmental samples collected on a global basis (Charlop-Powers et al. 2015, Hernandez et al. 2020).

Most actinomycetes, including streptomycetes, behave as neutrophiles in culture, with a growth range from about pH 5.0 to 9.0 and an optimum around pH 7.0 (Goodfellow and Williams 1983, Williams et al. 1984). Neutrophilic streptomycetes are a unique source of clinically significant antibiotics (Hopwood 2007, de Lima Procópio et al. 2012) and remain a target in the search for new bioactive compounds. In contrast, acidophilic (pH range 3.5–6.5, optimal growth around pH 4.5) and acidotolerant (pH range 4.5–7.5, optimal growth pH 5.0–5.5) actinomycetes with streptomycete-like properties (Williams and Mayfield 1971, Williams et al. 1971, Xu et al. 2006) have rarely featured in bioprospecting campaigns. This is surprising as Jensen noted in 1928 that novel isolates, designated "*Actinomyces* (*Streptomyces*) *acidophilus*", grew from pH 2.6 to 5.5 with little or no evidence of growth at pH 6.5. Such organisms are now known to be numerous and widespread in acidic habitats, notably in litter and mineral horizons of coniferous forests (Corke and Chase 1964, Williams et al. 1971, Khan and Williams 1975, Hagedorn 1976, Goodfellow and Dawson 1978, Goodfellow and Simpson 1987, Seong et al. 1993, Cho et al. 2006). They are readily isolated from such sites using acidified isolation media supplemented with antifungal antibiotics (Williams et al. 1971, Goodfellow and Dawson 1978). Acid-loving actinomycetes are of interest as they play a role in the turnover of organic matter in low pH habitats (Williams et al. 1971, 1984), are a source of acid stable enzymes (Williams and Flowers 1978, Williams and Robinson 1981), release compounds that promote plant growth (Poomthongdee et al. 2015) and are potential biocontrol agents (Guo et al. 2015, Niyasom et al. 2015, Lyu et al. 2017).

Acidophilic and neutrophilic streptomycetes display chemotaxonomic and morphological properties associated with the genus *Streptomyces* (Lonsdale 1985, Kämpfer 2012a) though isolates assigned to these taxa have different nutritional and physiological profiles (Khan and Williams 1975, Flowers and Williams 1977). Acidophilic actinomycetes with streptomycete-like properties have been assigned to several numerically defined clusters equated with taxospecies (Lonsdale 1985, Goodfellow and Simpson 1987, Seong 1992, Seong et al. 1993); a probabilistic frequency matrix was designed for identification of acidophilic actinomycetes by Seong et al. (1995). Apart from the proposal for "*S. acidophilus*" (Jensen 1928), no attempt has been made to assign acidophilic streptomycetes to validly published *Streptomyces* species though isolates from acidic rhizosphere soils were shown to be related to the type-strain of *Streptomyces misionensis* (Poomthongdee et al. 2015). In contrast, several *Streptomyces* species have been proposed to accommodate acidotolerant isolates, as exemplified by *S. guanduensis*, *S. paucisporeus*, *S. rubidus*, *S. yanglinensis* and *S. yeochonensis* (Kim et al. 2004, Xu et al. 2006).

The potential importance of filamentous, acidophilic and acidotolerant actinomycetes in agricultural and medical practices make it timely to explore their diversity and ecology in acidic environments and their potential as a source of new specialised (secondary) metabolites. Genome-based classifications of actinomycetes are bringing greater precision to such studies (Nouioui et al. 2018, Sangal et al. 2018), as are improved metrics for the detection of generic, species and subspecific boundaries (Nouioui et al. 2018, Sant'Anna et al. 2019, Thompson et al. 2020). Similarly, the distribution of stress-related genes in whole-genomes of actinomycetes isolated from extreme biomes provide an insight into how they adapt to harsh abiotic conditions that prevail in such environments (Busarakam et al. 2014, Abdel-Mageed et al. 2020). Whole-genome sequences of novel, extremely gifted actinomycetes from extreme habitats can also be scanned to detect the presence of uncharacterized, biosynthetic gene clusters using computer-based systems, such as the Antibiotic Resistant Target Seeker (ARTS) platform (Alanjary et al. 2017, Mungan et al. 2020).

## 2. Classification

Acid-loving actinomycetes which form substrate and aerial mycelia are common in acidic biomes, as shown by culture-dependent (Williams et al. 1971, Cho et al. 2006) and culture-independent (Janssen 2006, Dedysh et al. 2006) studies. Culturable strains include extreme acidophiles belonging to the class *Acidimicrobiia* (Norris 2012), these organisms grow optimally below pH 3.0 (Johnson and Quatrini 2016). Extremely acidophilic isolates from low pH sites are currently classified in the monospecific genera *Acidiferrimicrobium*, *Acidimicrobium*, *Aciditerrimonas*, *Acidothermus*, *Ferrimicrobium* and *Ferrithrix* (see Table 1). These taxa together with *"Acidithrix ferrooxidans"* (Jones and Johnson 2015) and the candidatus genus *"Acidithiomicrobium"* (Davis-Belmar and Norris 2009) belong to the family *Acidimicrobiaceae* (Stackebrandt et al. 1997) emend Zhi et al. (2009) which is classified in the order *Acidimicrobiales* (Stackebrandt et al. 1997) emend Zhi et al. (2009) of the class *Acidimicrobiia* (Norris 2012). This taxon also encompasses the order *Iamiales* (Salam et al. 2020) which is composed of the families *Iamiaceae* (Kurahashi et al. 2009) and *Ilumatobacteraceae* (Asem et al. 2018); these taxa include the genera *Aquihabitans* (Jin et al. 2013) and *Iamia* (Kurahashi et al. 2009), and *Desertimonas* (Asem et al. 2018) and *Ilumatobacter* (Matsumoto et al. 2009), respectively. None of the organisms classified in these taxa are acidophilic as they grow optimally at pH 7.0 but not below pH 6.0 (Norris 2012).

Extremely acidophilic actinomycetes have developed diverse lifestyles which allow them to thrive in highly acidic natural habitats such as hydrothermal sites and man-made environments, notably acid mine drainage and biomining operations (Méndez-García et al. 2015). Other environmental factors that shape extreme acidophilic communities include dissolved oxygen, high concentrations of dissolved metals, total organic carbon, and temperature. Extreme acidophiles can be distinguished from their mycelial counterparts as they form rod-shaped cells though some produce filaments of varying lengths (Clark and Norris 1996, Johnson et al. 2009). The two groups are found in well separated lineages in 16S rRNA gene trees with extreme acidophiles forming a clade in the most deeply rooted part of the tree (Ludwig et al. 2012b, Salam et al. 2020).

The focus of this article is on acidophilic and acidotolerant actinomycetes which form extensively branched substrate mycelia that bear aerial hyphae that differentiate into chains of spores. Detailed accounts of the ecology, metabolism, and physiology of the extreme acidophiles and their potential importance in biotechnology can be found elsewhere (Quatrini and Johnson 2016, Johnson and Quatrini 2020).

Following the pioneering work of Williams and his colleagues (Williams and Mayfield, 1971, Williams et al. 1971, 1972, Williams and Flowers 1978, Williams and Robinson 1981), acidophilic mycelial actinomycetes were isolated from acidic environmental samples and assigned to novel species classified in the genera *Actinocrinis* (Kim et al. 2017), *Actinospica* (Cavaletti et al. 2006), *Catenulispora* (Busti et al. 2006) and *Streptacidiphilus* (Kim et al. 2003), as shown in Table 1. *Actinocrinis* and *Actinospica* form the family *Actinospicaceae* (Cavaletti et al. 2006) and *Catenulispora*, the family *Catenulisporaceae* (Busti et al. 2006); the order *Catenulisporales* (Donadio et al. 2012) encompasses both families. The family *Streptomycetaceae* (Waksman and Henrici 1943), as conceived by Nouioui et al. (2018), includes *Streptomyces* (Waksman and Henrici 1943) emend Wellington et al. (1992), the type-genus, *Kitasatospora* (Ōmura et al. 1982) emend Nouioui et al. (2018) and *Streptacidiphilus* (Kim et al. 2003).

Kitasatospora* and S*treptacidiphilus* have had short, but eventful taxonomic histories. The former was seen to be a synonym of *Streptomyces* by Wellington et al. (1992), its generic status re-established by Zhang et al. (1997), underpinned by partial RNA polymerase β-subunit gene sequence data (Kim et al. 2004) and then questioned when *Kitasatospora* species fell within the evolutionary radiation of *Streptomyces* (Kämpfer 2012a, Labeda et al. 2012). In the present edition of Bergey's Manual of Systematic Bacteriology (Kämpfer 2012b), *Kitasatospora* and *Streptacidiphilus* are cited as genera *incertae sedis*. Currently, *Kitasatospora* is considered to be a sister taxon to *Streptomyces* based on conserved developmental gene sequences (Girard et al. 2013, 2014) and on concatenated sequences of conserved housekeeping genes (Labeda et al. 2017). Similarly, in comparative analyses of whole-genome sequences, *Kitasatospora* and S*treptacidiphilus* formed taxa equivalent in rank to *Streptomyces* (Nouioui et al. 2018). These studies transferred several *Streptomyces* species to the genus *Kitasatospora*, as exemplified by *Streptomyces aureofaciens* (Duggar 1948) and *Streptomyces misakiensis* (Nakamura 1961) and proposed the monospecific genera *Embleya* and *Yinghuangia* for species previously classified as *Streptomyces scabisporus* (Ping et al. 2004) and *Streptomyces aomiensis* (Nagai et al. 2011), respectively. Three of the 15 validly published *Streptacidiphilus* species, namely *S. albus*, *S. carbonis* and *S. neutrinimicus*, are composed of strains representing clusters defined in the comprehensive numerical phenetic study undertaken by Lonsdale (1985). The genus *Acidothermus*, which contains acidophilic, thermophilic and cellulolytic actinomycetes, unlike the genera assigned to the family *Acidimicrobiaceae*, is composed of strains which form slender rods and long slender filaments (Mohagheghi et al. 1986). This taxon is the sole member of the family *Acidothermaceae* (Rainey et al. 1997) (in Stackebrandt et al. 1997) emend Zhi et al. (2009) of the order *Acidothermales* (Sen et al. 2014).

Most aciditolerant actinomycetes from acidic soils have been assigned to the genera *Micromonospora* and *Streptomyces* (Zakalyukina et al. 2002) whereas aciditolerant streptomycetes have been classified into five validly published species (see Table 1). The type-strain of *Streptomyces yeochonensis* (Kim et al. 2004), the earliest described acidotolerant *Streptomyces* species, is the centrotype of numerically defined cluster 13 delineated by Lonsdale (1985). The type-strains of all of these *Streptomyces* species, apart from *Streptomyces guanduensis*, were found to form well supported lineages within the evolutionary ambit of the genus *Streptomyces* (Labeda et al. 2012, 2017). Similarly, *Streptomyces cocklensis* (Kim et al. 2012), which grows from pH 4.0 to 10.0, is clearly related to strains classified in the *S. yeochoensis* lineage (Labeda et al. 2017, Nouioui et al. 2018). Other acidophilic/ acidotolerant filamentous actinomycetes have been classified as novel species within the genera *Amycolatopsis* and *Nocardia*, as shown in Table 1.

**Table 1.** Type-strains of acidophilic and acidotolerant actinomycetes; growth conditions and the isolation source.

| Actinobacteria | Growth conditions | | Location of isolation | References |
|---|---|---|---|---|
| | pH range** | Temperature range (°C)** | | |
| **Amycelial strains:** | | | | |
| *Acidiferrimicrobium australe* DSM 106828[T] | 1.7–4.5 (3.0) | 20–39 (30) | An acidic, metal-contaminated stream draining from a former coal mine, near Curanilahue, Chile | González et al. (2020) |
| *Acidimicrobium ferrooxidans* DSM 10331[T] | ND (2.0) | ND (45–50) | Pyrite enrichment culture from an Icelandic geothermal site | Clark and Norris (1996) |
| *Aciditerrimonas ferrireducens* DSM 45281[T] | 2.0–4.5 (3.0) | 35–58 (50) | Solfataric soil, Ohwaku-dani, Hakone, Japan | Itoh et al. (2011) |
| *Acidothermus cellulolyticus* ATCC 43068[T] | 3.5–7.0 (5.0) | 37–65 (55) | Acidic hot springs at Yellowstone National Park, USA | Mohagheghi et al. (1986) |
| *Ferrimicrobium acidiphilum* DSM 19497[T] | Minimum 1.4 (2.0) | 20–37 (35) | Acid, mine water, abandoned Cae Coch sulfur mine, North Wales, UK | Johnson et al. (2009) |
| *Ferrithrix thermotolerans* DSM 19514[T] | Minimum 1.6 (1.8) | Maximum 50 | Mineral sample, Beryl Spring/Gibbon river area, Yellowstone National Park, Wyoming, USA | Johnson et al. (2009) |
| **Mycelial strains:** | | | | |
| *Actinocrinis puniceicyclus* DSM 45618[T] | 3.5–6.5 (5.5) | 13–30 (25) | Soil, Ochre Beds bog, Kootenay National Park, Canada | Kim et al. (2017) |
| *Actinospica robiniae* DSM 44927[T] | 4.8–6.2 (5.5) | 17–33 (22–28) | Temperate forest soil, Gerenzano, Italy | Cavaletti et al. (2006) |
| *Actinospica acidiphila* DSM 44926[T] | 4.2–6.0 (5.0) | 17–33 (28) | Temperate forest soil, Gerenzano, Italy | Cavaletti et al. (2006) |
| *Actinospica durhamensis* DSM 46820[T] | 4.0–6.0 (5.5) | 15–33 (28) | C-horizon soil under Sitka spruce (*Picea sitchensis*), Hamsterley Forest, County Durham, UK | Golinska et al. (2015) |
| *Amycolatopsis acidicola* NBRC 113896[T] | 4.0–7.0 (5.5–6.0) | 14–41 (28–30) | Kantulee peat swamp forest soil, Surat Thani Province, Thailand | Teo et al. (2020) |
| *Amycolatopsis acidiphila* JCM 30562[T] | 4.5–7.5 (5.5–6.0) | 15–40 (30) | Soil from coal mine, Nalaikh Province, Mongolia | Oyuntsetseg et al. (2017) |
| *Catenulispora acidiphila* DSM 44928[T] | 4.3–6.8 (6.0) | 11–37 (22–28) | Temperate forest soil, Gerenzano, Italy | Busti et al. (2006) |

*Table 1 contd. ...*

*...Table 1 contd.*

| Actinobacteria | Growth conditions | | Location of isolation | References |
|---|---|---|---|---|
| | pH range** | Temperature range (°C)** | | |
| *Catemulispora fulva* NBRC 110074[T] | 4.0–7.0 (7.0) | 15–42 (28) | Acid pine soil, Chungnam, Republic of Korea | Lee and Whang (2016) |
| *Catemulispora graminis* 45863[T] | 4.0–7.0 (5.5) | 15–37 (28) | Rhizosphere of *Phyllostachys nigro* var. *henonis*, Damyang, Jeonnam, Republic of Korea | Lee et al. (2012) |
| *Catemulispora pinisilvae* DSM 111109[T] | 5–7.5 (5.5) | 10–33 (28) | H-horizon, pine forest, near Toruń, Poland | Świecimska et al. (2021b) |
| *Catemulispora pinistramenti* DSM 111110[T] | 5–7.5 (5.5) | 15–33 (28) | L-horizon, pine forest, near Toruń, Poland | Świecimska et al. (2021a) |
| *Catemulispora rubra* DSM 44948[T] | 4.0–6.5 (5.0) | ND (20–40) | Forest soil, Oodate, Akita Prefecture, Japan | Tamura et al. (2007) |
| *Catemulispora subtropica* DSM 45249[T] | 5.0–8.0 (6.0–7.0) | 10–37 (25–30) | Paddy rice soil, Iriomote Island, Japan | Tamura et al. (2008) |
| *Catemulispora yoronensis* DSM 45250[T] | 5.0–7.0 (6.0–7.0) | 10–37 (25–30) | Forest soil from Yoro Valley, Chiba, Japan | Tamura et al. (2008) |
| *Kitasatospora acidiphila* JCM 32302[T] | 4.0–9.0 (5.0) | 15–37 (30) | Acid pine forest soil, Daejeon City, Republic of Korea | Kim et al. (2020) |
| *Kitasatospora arboriphila* DSM 44785[T] | 5.0–8.0 (5.0–7.0) | 15–40 (28–32) | Rhizosphere of *Maytenus aquifolia*, Ribeirao Preto, Brazil | Groth et al. (2004) |
| *Kitasatospora viridis* DSM 44826[T] | 4.0–7.0 (5.5) | 10–37 | Rhizosphere of *Camellia oleifera*, Jiangxi Province, China | Liu et al. (2005) |
| *Nocardia aciditolerans* DSM 45801[T] | 4.5–7.0 (5.5) | 10–30 (28) | C-horizon soil under Sitka spruce (*Picea sitchensis*), Hamsterley Forest, County Durham, UK | Golinska et al. (2013c) |
| *Nocardia jiangxiensis* DSM 17684[T] | 3.5–9.5 (5.5) | 17–37 | Rhizosphere of goose-grass (*Eleusine indica*) near copper mine, Wushan, Jiangxi Province, China | Cui et al. (2005) |
| *Nocardia miyunensis* DSM 17685[T] | 4.5–9.5 (5.5) | 17–37 | Acidic pine forest soil, Miyun County, Beijing, China | Cui et al. (2005) |
| *Nocardia acidivorans* DSM 45049[T] | ND | ND (25–30) | Soil from Stromboli Island, Italy | Kämpfer et al. (2007) |
| *Streptacidiphilus albus* DSM 41753[T] | 3.5–6.0 (5.0) | 10–25 | F-horizon soil under Sitka spruce (*Picea sitchensis*), Hamsterley Forest, County Durham, UK | Kim et al. (2003) |

| Species | pH range (optimum) | Temperature | Origin | Reference |
|---|---|---|---|---|
| *Streptacidiphilus anmyonensis* JCM 16223[T] | 3.0–8.0 (4.5–5.0) | 28–35 | Acidic pine soil, Anmyeon, Tae-An, Chungnam, Republic of Korea | Cho et al. (2008) |
| *Streptacidiphilus bronchialis* DSM 106435[T] | 5.0–7.0 | 20–40 | Bronchial lavage, 80-years old patient, Tennessee, USA | Nouioui et al. (2019) |
| *Streptacidiphilus carbonis* DSM 41754[T] | 3.5–6.0 (5.0) | 15–30 | Acidic coal mine waste, East Cramlington, Nortumber-land, UK | Kim et al. (2003) |
| *Streptacidiphilus durhamensis* DSM 45796[T] | 4.5–6.0 (5.5) | 10–30 (26) | F-horizon soil under Sitka spruce (*Picea sitchensis*), Hamsterley Forest, County Durham, UK | Golinska et al. (2013a) |
| *Streptacidiphilus griseoplanus* DSM 40009[T] | ND (5.0) | ND (28) | Grassland soil, Iowa, USA | Backus et al. (1957); Nouioui et al. (2019) |
| *Streptacidiphilus hamsterleyensis* DSM 45900[T] | 4.5–6.0 (5.5) | 10–30 (25) | H-horizon under Sitka spruce, Hamsterley Forest, County Durham, UK | Golinska et al. (2013b) |
| *Streptacidiphilus jiangxiensis* DSM 45096[T] | 3.5–6.5 (5.5) | 15–35 | Rhizosphere of wild tea plants, Jiangxi Agricultural University Campus, Jiangxi Province, China | Huang et al. (2004) |
| *Streptacidiphilus melanogenes* JCM 16224[T] | 3.0–8.0 (4.5–5.0) | 28–35 | Acidic pine soil, Sambong, Tae-An, Chungnam, Republic of Korea | Cho et al. (2008) |
| *Streptacidiphilus monticola* DSM 105744[T] | 5.0–9.0 (6.0) | 15–40 (28) | Acidic soil, Xianglu Mountain, Binxian, Heilongjiang Province, China | Song et al. (2018) |
| *Streptacidiphilus neutrinimicus* DSM 41755[T] | 3.5–5.5 (5.0) | 10–25 | H-horizon soil under Sitka spruce (*Picea sitchensis*), Hamsterley Forest, County Durham, UK | Kim et al. (2003) |
| *Streptacidiphilus oryzae* DSM 45098[T] | 3.0–6.5 (5.0) | 28–37 | Rice field soil, Nontaburi Province, Thailand | Wang et al. (2006) |
| *Streptacidiphilus pinicola* JCM 32300[T] | 5.0–8.0 (6.0) | 15–33 (30) | Soil from pine grove, Daejeon, Republic of Korea | Roh et al. (2018) |
| *Streptacidiphilus rugosus* JCM 16225[T] | 3.0–8.0 (4.5–5.0) | 28–35 | Acidic pine soil, Anmyeon, Tae-An, Chungnam, Republic of Korea | Cho et al. (2008) |
| *Streptacidiphilus torunensis* DSM 102291[T] | 4.0–6.5 (5.5) | 10–30 (27) | F-horizon soil under pine (*Pinus sylvestris*), near Torun, Poland | Golinska et al. 2016 |

*Table 1 contd. ...*

*...Table 1 contd.*

| Actinobacteria | Growth conditions | | Location of isolation | References |
|---|---|---|---|---|
| | pH range** | Temperature range (°C)** | | |
| *Streptomyces guanduensis* DSM 41944[T] | 4.5–7.0 (5.5) | 20–37 (28) | Acidic pine forest soil, Dashao, Guandu, Yunnan Province, China | Xu et al. (2006) |
| *Streptomyces paucisporeus* DSM 41946[T] | 4.5–7.5 (5.5) | 20–37 (28) | Acidic pine forest soil Dashao, Guandu, Yunnan Province, China | Xu et al. (2006) |
| *Streptomyces rubidus* DSM 41947[T] | 4.5–7.0 (5.5) | 20–37 (28) | Acidic pine forest soil, Yanglin, Yunnan Province, China | Xu et al. (2006) |
| *Streptomyces yanglinensis* DSM 41945[T] | 4.5–7.0 (5.5) | 20–37 (28) | Acidic pine forest soil, Yanglin, Yunnan Province, China | Xu et al. (2006) |
| *Streptomyces yeochonensis* DSM 41868[T] | 4.3–7.3 (5.5) | 20–37 | Acidic pine forest soil, Yeochon, Republic of Korea | Kim et al. (2004) |

**Footnote:***Strains included in the phylogenomic analyses, **Optimal pH and temperatures are shown in brackets, when available, ND; not determined.

# 3. Ecology

Actinomycetes mainly occur as saprophytes in diverse natural habitats, including soil. They are key elements of microbial communities but tend to be outpaced by fast-growing microorganisms when serial dilutions of environmental samples are plated onto standard nutrient media (Williams et al. 1984). This problem has been met by the addition of antibacterial and antifungal antibiotics to isolation media which favour the growth of filamentous actinomycetes, such as humic acid-vitamins (Hayakawa and Nonomura 1987) and starch-casein-nitrate (Küster and Williams 1964) agar. Cycloheximide and nystatin are commonly used to inhibit fungal growth (Williams and Davies 1965) and nalidixic acid to restrict that of fast-growing Gram-negative bacteria (Hayakawa et al. 1996). Similarly, chemical and physical pretreatment procedures are used to promote the isolation of actinomycetes and suppress that of unwanted microorganisms in environmental samples or in suspensions prepared from them prior to inoculation of selective media.

The effectiveness of pretreatment regimes reflects the biological properties of targeted taxa, as exemplified by those that produce spores. Spores of actinomycetes, for instance, are more resistant to dryness and heat than fungal spores and bacterial cells, hence drying and/or heating environmental samples promote the isolation of spore-forming strains on isolation plates (Williams et al. 1972). It is not possible to recommend a universal procedure for the selective isolation of the different kinds of actinomycetes found in environmental samples given their different nutritional and physiological requirements. Consequently, many approaches have been recommended for the isolation of actinomycetes, as considered in review articles (Goodfellow 2010, Jiang et al. 2016, Kurtböke 2017a). Most selective isolation methods involve the extraction of propagules from selected environmental samples, pretreatment(s) of samples, choice of selective media, selection of incubation conditions and choice of representative colonies from isolation plates for taxonomic studies.

The choice of selective isolation procedures is influenced by abiotic conditions at sampling sites, such as aeration, altitude, desiccation, pH, temperature, organic matter content, UV-radiation and extremes of salinity (Bull et al. 2018, Gómez-Silva 2018). Critically, physicochemical interactions between the propagules of actinomycetes and particulate matter determine the composition of inocula prepared from environmental samples. Traditional procedures used to address this problem, such as shaking samples in weak buffers, are not as effective as the dispersion and differential centrifugation procedure introduced by Hopkins et al. (1991). This multistage procedure, which combines several physico-chemical treatments, increases the yield and diversity of actinomycetes isolated from environmental samples (Goodfellow and Fiedler 2010).

pH is a major factor governing the distribution of filamentous actinomycetes in acidic soils (Goodfellow and Williams 1983, Kurtböke 2017b). Williams and his colleagues demonstrated that filamentous actinomycetes, almost exclusively streptomycetes, which grew between pH 3.5 and 6.5 could be isolated from acidic podsol and mine waste soils on acidified starch-casein and soil extract agar plates supplemented with antifungal antibiotics following incubation at 28°C for

14 to 21 days (Williams et al. 1971, Khan and Williams 1975). The presence of low numbers of neutrophilic streptomycetes in these habitats can be attributed to the localized occurrence of less acidic microsites (Williams and Mayfield 1971), coupled with the resistance of their spores to acidity (Flowers and Williams 1977). It has also been shown that the amendment of poorly buffered acidic soil with chitin of dead fungal mycelium results in a succession of acidophilic to neutrophilic streptomycetes that parallels ammonification and a consequent rise in pH (Williams and Robinson 1981).

Following the pioneering studies outlined above, acidophilic actinomycetes were isolated from acidic environments using a range of selective isolation procedures and assigned to *Actinocrinis, Actinospica, Catenulispora* and *Streptacidiphilus* species. Similarly, acidophilic and acidotolerant strains have been classified as new *Amycolatopsis, Kitasatospora, Nocardia* and *Streptomyces* species (see Table 1). Acidified starch-casein-nitrate broth (pH 4.5–5.0) set with either agar or gellan gum and supplemented with antifungal antibiotics has been used widely to isolate acidophilic filamentous actinomycetes from acidic habitats, notably from environmental samples taken from litter and mineral layers of coniferous forests (Kim et al. 2003, Cho et al. 2006, Golińska et al. 2013a–c, 2015, 2016, Świecimska et al. 2021a–b). Other acidified isolation formulations used for this purpose include GTV (Busti et al. 2006, Cavaletti et al. 2006), humic acid (Groth et al. 2004), inorganic salts-starch (Lee et al. 2012), peptone-yeast extract (Goodfellow and Dawson 1978) and yeast extract-malt extract (Cavaletti et al. 2006, Roh et al. 2018) media with agar or gellan gum used as gelling agents. In general, acidified isolation media are inoculated using aliquots of serial dilutions prepared by shaking environmental samples in ¼ strength Ringer's solution, as shown in Table 2. Soil suspensions prepared using the DDC procedure (Hopkins et al. 1991) have been used to good effect (Huang et al. 2004, Cui et al. 2005, Xu et al. 2006, Wang et al. 2006), notably by Wang et al. (2003). It is good practice to spread serial dilutions over dried isolation plates, as exemplified by Golińska et al. (2013c), as this reduces the spread of motile bacteria (Vickers and Williams 1987).

High viable counts of acid-loving filamentous actinomycetes have been recorded from acidic habitats, notably from litter and mineral horizons under spruce (Goodfellow and Dawson 1978), as exemplified in Table 3. Similar numbers of acidophilic filamentous isolates have been recorded from corresponding horizons under pine (*Pinus thunbergii*), from acidic coal waste sites (Seong 1992, Cho et al. 2006), and from the rhizospheres of rice plants and rubber trees (Poomthongdee et al. 2015). Goodfellow and Dawson (1978) noted that the number of acidophilic filamentous strains in litter and mineral horizons under pine (*Picea sitchensis*) varied throughout the year but were unable to correlate these fluctuations with changes in moisture content, pH, or temperature.

Extreme acidophilic acidophiles are found in low pH habitats such as geothermal sites and areas impacted by mining activities where they form complex associations with other extreme thermophiles belonging to other phyla, such as *Aquficae, Firmacutes, Nitrosporae, Proteobacteria* and *Verrucomicrobia* (Holanda et al. 2016, Rawat and Joshi 2019). In general, acidimicrobia isolated from such habitats are few

**Table 2.** Procedures used to selectively isolate acidophilic and acidotolerant actinomycetes.

| Taxa | Preparation of environmental suspensions | Pretreatment regimes | Selective media and incubation regimes | References |
|---|---|---|---|---|
| **Amycelial strains:** | | | | |
| *Af. australe* | Acid mine waters serially diluted in autotrophic basal salts media | None | Ferrous iron/tetrathionate/tryptic soy broth overlay plates | González et al. (2020) |
| *Ac. ferrooxidans* | Enriched sample from Icelandic site | Pyrite enrichment culture | Mineral salts medium incubated at 48°C | Clark and Norris (1996) |
| *At. ferrireducens* | Enriched solfataric soil sample | Yeast extract enrichment culture | *Sulfolobus* medium (pH 2.5) incubated at 55°C | Itoh et al. (2011) |
| *Ath. cellulolyticus* | Enriched sample from Norris Geyser basin | Yeast extract, D-cellobiose and cellulosic substrate enrichment culture | Low-phosphate basal salts medium incubated around 50°C for 1–2 weeks | Mohagheghi et al. (1986) |
| *F. acidiphilum* | Serial dilution of enrichment acid mine water | Pyrite enrichment culture | Ferrous iron-containing overlay media (pH 2.5) incubated at 30°C | Johnson et al. (2009) |
| *Fr. thermotolerans* | Serial dilution of enrichment sample | Iron-yeast extract enrichment culture | Ferrous iron-containing overlay media (pH 2.5) incubated at 45°C | Johnson et al. (2009) |
| **Mycelial strains:** | | | | |
| *A. puniceicyclus* | Suspensions of paddy field and forest soils | None | Minimal salts medium solidified with gellan gum (pH 3.9–4.0) incubated at room temperature for 24 days | Kim et al. (2017) |
| *Act. acidiphila,* *Act. robiniae,* *C. acidiphila* | Suspension of soil in citric acid/Na$_2$HPO$_4$ buffer (pH 7) | Soil dried under vacuum at 30°C for 7 days prior to preparation of soil suspension | GTV (pH 5) supplemented with CMM vitamin solution and cycloheximide (50 µg ml$^{-1}$) incubated at 28°C for 8 weeks | Busti et al. (2006), Cavaletti et al. (2006) |

*Table 2 contd. ...*

*...Table 2 contd.*

| Taxa | Preparation of environmental suspensions | Pretreatment regimes | Selective media and incubation regimes | References |
|---|---|---|---|---|
| *Act. durhamensis, C. pinisilvae, C. pinistramenti N. aciditolerans,* Str. albus, Str. carbonis, Str. neutrinimicus, Str. durhamensis, Str. hamsterleyensis, Str. torunensis* | Suspensions of litter and soil samples in ¼ Ringer's solution shaken on a reciprocal or rotary shaker for 20–30 minutes | None | Starch casein medium with agar or gellan gum as gelling agent (pH 4.5) supplemented with cycloheximide and nystatin (each of 50 µg ml⁻¹) incubated at 25–28°C for 14–28 days | Kim et al. (2003), Golinska et al. (2013a–c, 2015, 2016), Świecimska et al. (2021a-b) |
| *Am. acidicola* | Suspension of soil in sterilized soil extract solution | Suspension heated at 55°C for 10 minutes | Half-strength modified glycerol asparagine agar (pH 5.0) supplemented with nalidixic acid, nystatin and ketaconozol (25, 50 and 100 µg ml⁻¹), respectively, incubated at 28°C for 2 weeks | Teo et al. (2020) |
| *Am. acidiphila, Str. ammyonensis, Str. rugosus, Str. melanogenes* | Suspension of soil sample in ¼ Ringer's solution shaken on a reciprocal shaker for 20 minutes | Heat soil suspension at 50°C for 20 minutes | Starch casein agar (pH 4–5) supplemented with cycloheximide and nystatin (each of 50 µg ml⁻¹) incubated at 30°C for at least 7 days | Seong (1992), Cho et al. (2006), Oyuntsetseg et al. (2017) |
| *C. fulva* | Suspension of soil sample | None | 100-fold diluted nutrient medium incubated at 30°C for 20 days | Lee and Whang (2016) |
| *C. graminis* | Suspension of rhizosphere soil. | None | Inorganic salts–starch agar (pH 5.5) incubated at 28°C for 7–10 days | Lee et al. (2012) |
| *C. rubra* | Suspension of forest soil | None | Glucose-yeast extract agar (pH 4.0) | Tamura et al. (2007) |
| *C. subtropica, C. yoronensis* | Suspension of paddy field and forest soil | Air dried soil treated with yeast extract and sodium dodecyl sulphate | Humic acid vitamin agar supplemented with cycloheximide (50 µg ml⁻¹) and nalidixic acid (20 µg ml⁻¹) | Tamura et al. (2008) |

| | | | |
|---|---|---|---|
| *K. acidiphila* | Suspension of pine forest soil in ¼ Ringer's solution | None | Starch casein agar (pH 5) incubated at 30°C for 7 days | Kim et al. (2020) |
| *K. arboriphila* | Suspension of soil in aqueous sodium pyrophosphate, mixed on a rotary shaker for 30 min, supernatant decanted after soil particles settled. Second treatment of soil fraction with solvent and ultra-sonication for 5 minutes at 50 W. Combine supernatants prior to preparation serial dilutions | Air dried soil heated at 80°C for an hour prior to preparation of soil suspensions | Casein mineral, humic acid, PY-BHI agar and yeast extract-starch agar incubated at 28°C for 4 weeks | Groth et al. (2004) |
| *N. acidivorans* | Heated soil sample treated with 10 ml 0.1% (v/v) Tween 80 solution supplemented with ampicillin; serial dilutions plated onto selective medium | Soil sample heated for 10 seconds at 100°C | Mannitol-rifampicin agar supplemented with nystatin incubated at 27°C for 6 weeks | Kämpfer et al. (2007) |
| *K. viridis, N. jiangxiensis, N. miyunensis, S. guanduensis, S. paucisporeus, S. rubidus, S. yanglinensis* | Preparation of suspensions of pine forest soil using a modified DDC procedure* | None | Mineral salts agar (pH 4.5) supplemented with cycloheximide and nystatin (each of 50 μg ml$^{-1}$) incubated at 28 °C for 3 weeks | Cui et al. (2005), Liu et al. (2005), Xu et al. (2006) |
| *Str. bronchialis* | Bronchial wash from male patient | None | Trypticase soy agar supplemented with 5% sheep blood incubated at 35°C for 3–5 days | Nouioui et al. (2019) |
| *Str. jiangxiensis* | Suspensions of rhizosphere soils using a modified DDC procedure* | None | Mineral salts agar (pH 4.5) supplemented with cycloheximide and nystatin (each of 50 μg ml$^{-1}$) incubated 28°C for 3 weeks | Huang et al. (2004) |

*Table 2 contd. ...*

*...Table 2 contd.*

| Taxa | Preparation of environmental suspensions | Pretreatment regimes | Selective media and incubation regimes | References |
|---|---|---|---|---|
| *Str. monticola* | Suspension of soil sample in distilled water shaken on rotary shaker for 30 minutes at 28°C | Air dried soil | Cellulose–proline agar (pH 7.2) supplemented with cycloheximide and nalidixic acid and (each 50 µg ml⁻¹) incubated at 28°C for 15 days | Song et al. (2018) |
| *Str. oryzae* | Soil suspensions in ¼ Ringer's solution prepared using a modification of the DDC procedure** | Supernatant fractions heated at 55°C for 20 minutes | Starch casein agar (pH 4.5) supplemented with cycloheximide and nystatin (each of 50 µg ml⁻¹) incubated at 28°C for 14 days | Wang et al. (2006) |
| *Str. pinicola* | Suspension of pine grove soil | None | Yeast extract – malt extract agar (pH 5.0) supplemented with cycloheximide and nystatin (each at 50 µg ml⁻¹) incubated at 30°C | Roh et al. (2018) |

**Footnote:** Key to genera: A, *Actinocrinis*; Ac, *Acidimicrobium*; Act, *Actinospica*; Af, *Acidiferromicrobium*; Am, *Amycolatopsis*; At, *Aciditerrimonas*; Ath, *Acidothermus*; C, *Catenulispora*; F, *Ferromicrobium*; Fr, *Ferrithrix*; K, *Kitasatospora*; N, *Nocardia*; S, *Streptomyces* and Str, *Streptacidiphilus*.

DDC procedure (Hopkins et al. 1991), as modified by *Wang et al. (2003) and **Sembiring et al. (2000).

**Abbreviation:** DDC, dispersion and differential centrifugation procedure

**Table 3.** Numbers (cfu/g. dry weight of soil) of acidophilic/acidotolerant filamentous isolates from spruce (*Picea sitchensis*) litter and mineral samples from Hamsterley forest, County Durham, UK.

| Horizon | Sample 1 | | Sample 2 | |
|---|---|---|---|---|
| | Mean pH | Number x 10⁴ | Mean pH | Number x 10⁴ |
| L | $4.18 \pm 0.01$* | $31.0 \pm 7.5$* | $3.8 \pm 0.1$* | $86.8 \pm 51.3$* |
| F | $3.65 \pm 0.18$ | $43.2 \pm 13.6$ | $3.2 \pm 0.2$ | $43.2 \pm 24.9$ |
| H | $3.55 \pm 0.10$ | $3.7 \pm 0.5$ | $3.2 \pm 0.1$ | $207.0 \pm 32.6$ |
| A₁ | $3.33 \pm 0.12$ | $0.3 \pm 0.4$ | $3.1 \pm 0.1$ | $51.0 \pm 26.7$ |
| A₂ | $3.57 \pm 0.12$ | $2.5 \pm 1.1$ | $3.3 \pm 0.1$ | $26.8 \pm 4.3$ |

**Footnote:** Counts were done on acidified starch-casein agar (pH 4.5) supplemented with cycloheximide and nystatin (each at 50 µg ml⁻¹) and after incubation at 28°C for 21 days.

*Standard deviation. The samples were collected at a two-week interval.

in number but taxonomically diverse. They are generally isolated from enrichment cultures of acid mine waste waters, as illustrated in Table 2.

## 4. Phylogeny

The recovery of representatives of the extreme acidophilic taxa as a distinct lineage at the foot of the 16S rRNA actinomycete gene tree (Fig. 1) accords with results from previous studies (Ludwig et al. 2012, Salam et al. 2020). Indeed, these latter workers found that taxa forming the class *Acidimicrobiia* (Norris 2012) are more closely related to strains classified in the classes *Nitriliruptoria* (Ludwig et al. 2012b), *Rubrobacteria* (Suzuki 2012) emend Foesel et al. (2016), and *Thermoleophilia* (Suzuki and Whitman 2012) emend Foesel et al. (2016) than to those assigned to the recently proposed class *Actinomycetia* (Salam et al. 2020). These recently evolved actinomycetes include acidophilic strains belonging to the orders *Catenulisporales* and *Streptomycetales*.

With one exception, the phylogenetic tree shows that representatives of the acidophilic filamentous taxa form well supported lineages corresponding to the genera *Actinocrinis*, *Actinospica*, *Catenulispora* and *Streptacidiphilus*. The exception, *Actinospica acidiphila* DSM 44926ᵀ, formed a well-supported branch with the type-strains of *Streptomyces leeuwenhoekii*, *Streptomyces rocheii*, *Streptomyces somaliensis* and *Streptomyces violaceus* in a deeply rooted part of the *Streptomyces* clade. Similarly, with two exceptions, the *Kitasatospora*, *Streptacidiphilus* and *Streptomyces* strains were assigned to three clades which can be equated with the family *Streptomycetaceae*, a result in line with an earlier study (Takahashi 2017). One of the exceptions, *Kitasatospora albolonga* DSM 40570ᵀ forms a well-supported lineage with the type-strains of *Streptomyces anulatus* and *Streptomyces griseus* sharing a 16S rRNA gene similarity of 99.1% with each of them. This organism was initially classified as *Streptomyces abolongus* (Tsukiura et al. 1964) then reclassified in the genus *Kitasatospora* as *Kitasatospora albolonga* based on a MLSA sequence analysis of concatenated housekeeping genes (Labeda

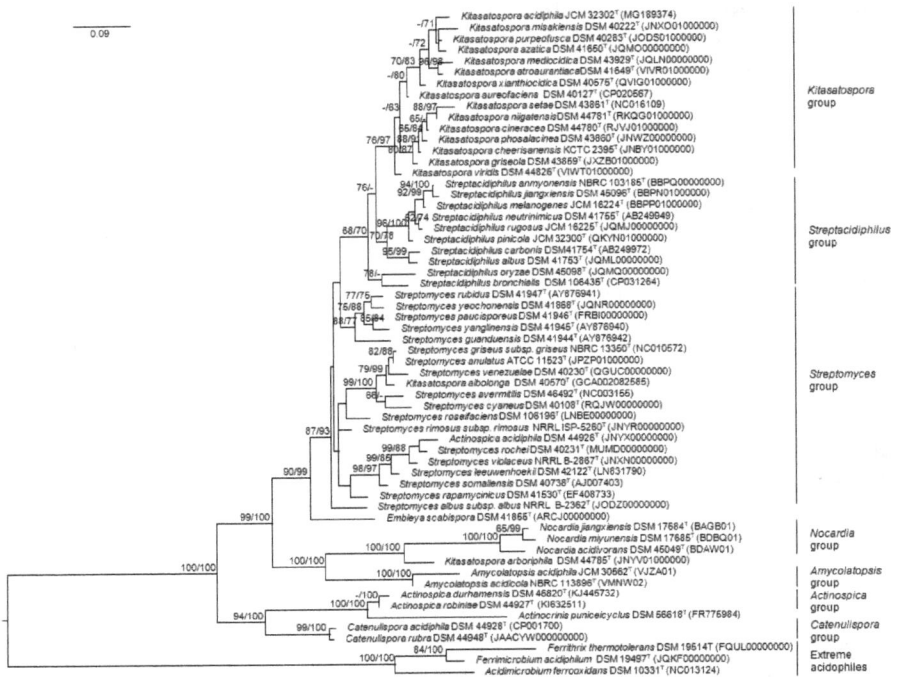

**Figure 1.** Maximum-likelihood and maximum-parsimony trees based on nearly complete 16S rRNA gene sequences (1285 to 1526 nucleotides) showing relationships between acidophilic/acidotolerant and neutrotolerant filamentous actinomycetes and between them and reference *Kitasatospora* and *Streptomyces* strains as well as their relationships with extreme acidophilic strains.

**Footnote:** The numbers at the branches are bootstrap support values when more than 60% for ML (left) and MP (right). Bar 0.09 substitutions per nucleotide position.

et al. 2017). The other exception, *Kitasatospora arboriphila* DSM 44785[T], forms a distinct branch in the phylogenetic tree which lies between lineages composed of the acidophilic *Amycolatopsis* and *Nocardia* strains (Fig. 1). Groth and her colleagues (2004) found that this strain could be distinguished from other *Kitasatospora* species though it was recovered at the edge of their *Kitasatospora* 16S rRNA gene tree, as was the case in a recent study (Takahashi 2017).

The acidotolerant *Streptomyces* species shown in Table 1 formed a well-supported lineage within the evolutionary radiation occupied by the genus *Streptomyces* (Fig. 1), a result in line with those from earlier studies (Xu et al. 2006, Labeda et al. 2012, 2017). Similarly, the acidophilic *Amycolatopsis* strains, *A. acidicola* NBRC 113896[T] and *A. acidiphila* JCM 30562[T], form a distinct branch in the 16S rRNA gene tree, as do *Nocardia acidivorans* DSM 45049[T], *Nocardia jiangxiensis* DSM 17684[T] and *Nocardia miyunensis* DSM 17685[T]; all of the *Nocardia* strains grow optimally at pH 5.5, but the latter two can be considered to be neutrotolerant as they grow at pH 9.5. Another acidophile, *Kitasatospora viridis* DSM 44826[T], was recovered at the periphery of the *Kitasatospora* clade, as is the case with

*Kitasatospora acidiphila* which was proposed by Kim et al. (2020) for a strain which grows optimally at pH 5.0.

In general, very good correlation has been found between 16S rRNA lineages and the distribution of chemotaxonomic and morphological properties of acidophilic and acidotolerant taxa (Goodfellow and O'Donnell 1989, Chun et al. 1996, Ludwig et al. 2012a), as was the case in this study. It can, for instance, be seen from Table 4 that the *Catenulispora* strains can be distinguished from all of the other acidophilic taxa as they form extensively branched, non-fragmenting substrate mycelia, aerial hyphae that differentiate into rod-shaped spores with smooth surfaces, have $LL$-A$_2$pm as the wall peptidoglycan, MK-9 (H$_4$, H$_6$, H$_8$) as the predominant isoprenologues and arabinose as a common sugar. Similarly, *Actinocrinis puniceicyclus* DSM 45618$^T$, another member of the order *Catenulisporales*, exhibits several discriminating features, it forms filaments that differentiate into chains of spores, has MK-11 as the predominant respiratory quinone and phosphohexose as the major polar lipid. The chemotaxonomic and morphological properties of the *Actinospica* species, which also belong to the *Catenulisporales*, were very similar even though *A. acidiphila* clustered with the *Streptomyces* strains. The *Actinospica* strains also show features that distinguished them from other acidophilic taxa, notably their ability to produce non-fragmenting substrate mycelia and, when formed, aerial hyphae that differentiate into long chains of cylindrical, rugose ornamented spores, as well as having 3-hydroxydiaminopimelic acid as the diamino acid of the wall peptidogycan. *Streptacidiphilus*, the largest taxon encompassing acidophilic actinomycetes, also has a distinctive profile; it forms extensively branched substrate mycelia, aerial hyphae which mature into long straight to flexuous chains of spores with smooth or warty surfaces, have $LL$-A$_2$pm as the major diamino acid of the wall peptidoglycan, MK-9 (H$_4$, H$_6$, H$_8$) as predominant isoprenologues and a varied sugar pattern, as shown in Table 4.

The type-strains of the acidotolerant *Streptomyces* (Table 1) have chemotaxonomic and morphological features typical of streptomycetes (Kämpfer 2012a, Kusuma et al. 2020). They formed an extensively branched substrate mycelium, aerial hyphae that carried smooth surfaced spores in flexuous spore chains, produced $LL$-A$_2$pm as the major diamino acid in the wall peptidoglycan, MK-9 (H$_6$, H$_8$) as predominant isoprenologues and complex mixtures of straight chain and *iso*- and *anteiso*-branched fatty acids. This profile is very similar to that shown by the *Streptomyces* strains (Table 4). However, the two groups can be distinguished based on sugar and polar lipid composition though these data are not available for *S. yeochonensis* DSM 41868$^T$. Unlike the streptacidiphili, the streptomycetes lacked diagnostic sugars and phosphatidylinositol and produced phosphatidylmethylethanolamine. The chemotaxonomic and morphological properties of the acidophilic/acidotolerant *Kitasatospora* strains are similar to those shown by the *Streptacidiphilus* and *Streptomyces* strains though only the former show mixtures of $LL$- and *meso*-A$_2$pm in whole cell hydrolysates, as the substrate mycelia of kitasatosporae contain *meso*-A$_2$pm and aerial and substrate spores $LL$-A$_2$pm (Kämpfer 2012b). Further, the kitasatosporae, unlike the streptacidophili,

**Table 4.** Key chemotaxonomic and morphological properties of acidophilic and acidotolerant actinomycetes

| Actinobacteria | Wall diamino acid (peptidoglycan type) | Menaquinone profile (MK) | Predominant fatty acids | Cell sugars | Polar lipids | DNA G+C content (mol%) | Morphology | References |
|---|---|---|---|---|---|---|---|---|
| **Amycelial strains:** | | | | | | | | |
| *Acidiferrimicrobium australe* DSM 106828[T] | *meso*-A$_2$pm | MK-9 (H$_8$); minor amounts MK-9 (H$_2$, H$_4$, H$_6$) | i-C16:0, ai-C17:0 | ND | DPG, PI, AGL, GL, PGL | 74.1 | Motile, rod-shaped cells. | González et al. (2020) |
| *Acidimicrobium ferrooxidans* DSM 10331[T] | ND | ND | i-C16:0 | ND | ND | ND | Small rod-shaped cells which may be in filaments of varying lengths. | Clark and Norris (1996), González et al. (2020) |
| *Aciditerrimonas ferrireducens* DSM 45281[T] | *meso*-A$_2$pm | MK-9 (H4, H8) | i-C16:0, ai-C17:0, i-C18:0 | ND | N-methyl PE, PGL | 74.1 | Motile, rod-shaped cells. | Itoh et al. (2011) |
| *Acidothermus cellulolyticus* ATCC 43068[T] | A$_2$pm glucosamine, muramic acid, serine, alanine | ND | ND | ND | ND | 60.7 | Slender rods and long, slender filaments. | Mohagheghi et al. (1986) |
| *Ferrimicrobium acidiphilum* DSM 19497[T] | *meso*-A$_2$pm (A1γ) | MK-8 (H$_{10}$); minor amounts MK-8 and MK-8 (H$_{10}$) with one or two methyl groups | i-C14:0, i-C15:0, ai-C15:0, i-C16:0, C17:1ω6c | ND | ND | 55 | Gram-negative, motile rods. Spores are not formed. | Johnson et al. (2009) |
| *Ferrithrix thermotolerans* DSM 19514[T] | *meso*-A$_2$pm (A1γ) | ND | i-C16:0 | ND | ND | 50.2 | Filaments forming visible flocs, single, rod-shaped cells occasionally observed. | Johnson et al. (2009) |

**Mycelial strains:**

| | | | | | | G+C | | |
|---|---|---|---|---|---|---|---|---|
| *Actinocrinis puniceicyclus* DSM 45618$^T$ | *meso*-A$_2$pm (A1γ) | MK-11 | i-C15:0, ai-C15:0, i-C16:0 | ND | Phex | 70.2 | Long filaments which differentiate into chains of spores. | Kim et al. (2017) |
| *Actinospica acidiphila* DSM 44926$^T$ | 3-OH A$_2$pm; *meso*-A$_2$pm (trace) | MK-9 (H$_4$, H$_6$, H$_8$) | i-C15:0, ai-C15:0, i-C16:0, | Ara, Man, Rha, Xyl | DPG, PE, PI, PME | 69.2 | Non-fragmenting vegetative mycelium. Aerial hyphae differentiate into long chains of cylindrical, rugose ornamented spores. | Cavaletti et al. (2006) |
| *Actinospica durhamensis* DSM 46820$^T$ | 2,6-diamino-3-hydroxy and *meso*-A$_2$pm | MK-9 (H$_2$, H$_4$, H$_6$, H$_8$) | i-C15:0, ai-C15:0, i-C16:0, ai-C17:0 | Ara, Gal, Man, Rha (trace) | DPG, PE, PG, PI | 68.0 | Extensively branched substrate mycelium, aerial hyphae not formed. | Golinska et al. (2015) |
| *Actinospica robiniae* DSM 44927$^T$ | 3-OH A$_2$pm; *meso*-A$_2$pm (trace) | MK-9 (H$_4$, H$_6$, H$_8$) | i-C15:0, ai-C15:0, i-C16:0, | Gal, Man, Rha | DPG, PE, PI, PME, | 70.8 | Non-fragmenting vegetative mycelium. Aerial hyphae divide into chains of squat to slightly rugose cylindrical spores. | Cavaletti et al. (2006) |
| *Amycolatopsis acidicola* NBRC 113896$^T$ | *meso*-A$_2$pm | MK-9 (H$_2$, H$_4$, H$_6$) | C16:0, i-C16:0, C17:0 cyclo, | Ara, Gal, Rib | DPG, PE, Hydroxy-PE, PG, PI, 2 PL | 69.7 | Extensively branched mycelium that fragments into rod-shaped elements. | Teo et al. (2020) |
| *Amycolatopsis acidiphila* JCM 30562$^T$ | *meso*-A$_2$pm | MK-9 (H$_4$) | i-C15:0, ai-C15:0, i-C16:0, C17:1 cyclo | Ara, Gal, Glu, Rib | PE, PG, PI, PIDM, PME | 72.0 | Extensively branched aerial and substrate hyphae. | Oyuntsetseg et al. (2017) |

*Table 4 contd.…*

*...Table 4 contd.*

| Actinobacteria | Wall diamino acid (peptidoglycan type) | Menaquinone profile (MK) | Predominant fatty acids | Cell sugars | Polar lipids | DNA G+C content (mol%) | Morphology | References |
|---|---|---|---|---|---|---|---|---|
| *Catenulispora acidiphila* DSM 44928[T] | LL-A$_2$pm (A3γ) | MK-9 (H$_4$, H$_6$, H$_8$) | i-C16:0, ai-C17:0 | Ara, Glu, Rha, Rib, Xyl | DPG, PG, PI, PIMs, 2 PL | 71.9 | Branched, non-fragmenting, vegetative mycelium, straight to flexuous aerial hyphae differentiate into chains of cylindrical spores. | Busti et al. (2006) |
| *Catenulispora fulva* NBRC 110074[T] | LL-A$_2$pm | MK-9 (H$_4$, H$_6$, H$_8$) | ai-C15:0, i-C16:0, i-C17:0, ai-C17:0 | Ara, Gal, Glu, Man, Rha, Rib | DPG, PC, PI, PIM, 2AL, GL, L, PL | 71.0 | Extensively branched substrate mycelium forms flexuous or straight chains of rod-shaped, smooth surfaced spores. Aerial mycelium not formed. | Lee and Whang (2016) |
| *Catenulispora graminis* 45863[T] | LL-A$_2$pm | MK-9 (H$_4$, H$_6$, H$_8$) | i-C16:0 | Ara, Xyl | PG, PIM, PS, 3PL | 72.8 | Extensively branched substrate mycelium forms flexuous or straight spore chain of rod-shaped spores. Aerial mycelium not formed. | Lee et al. (2012) |
| *Catenulispora pinisilvae* DSM 111109[T] | LL-A$_2$pm | MK-9 (H$_4$, H$_6$, H$_8$) | i-C16:0, ai-C17:0 | Ara, Man, Rha, Xyl, Rib (trace) | DPG, PG, PI, PIM, 3L | 69.9 | Extensively branched substrate mycelium. Aerial mycelium not formed. | Świecimska et al. (2021b) |
| *Catenulispora pinistramenti* DSM 111110[T] | LL-A$_2$pm | MK-9 (H$_6$, H$_8$) | i-C16:0, ai-C17:0 | Ara, Man, Rha, Xyl, Rib | DPG, PG, PI, PIM, 4L | 70.0 | Extensively branched substrate mycelium. Aerial mycelium not formed. | Świecimska et al. (2021a) |

| | | | | | | | | |
|---|---|---|---|---|---|---|---|---|
| *Catemulispora rubra* DSM 44948[T] | LL-A$_2$pm | MK-9 (H$_4$, H$_6$, H$_8$) | i-C16:0, ai-C17:0 | Ara, Glu, Man, Rib | PG, PI | 69.1 | Extensively branched, non-fragmenting substrate mycelium. Aerial hyphae differentiate into rod-shaped, smooth surfaced spores borne in hook-like or flexuous chains. | Tamura et al. (2007) |
| *Catemulispora subtropica* DSM 45249[T] | LL-A$_2$pm (A1γ) | MK-9(H$_8$); minor amounts MK-9 (H$_4$, H$_6$, H$_{10}$) | i-C16:0, ai-C17:0 | Ara, Gal, Man | DPG | 70–71 | Extensively branched, non-fragmenting substrate mycelium bearing sparse aerial hyphae. | Tamura et al. (2008) |
| *Catemulispora yoronensis* DSM 45250[T] | LL-A$_2$pm (A1γ) | MK-9(H$_8$); minor amounts MK-9 (H$_4$, H$_6$, H$_{10}$) | i-C16:0, ai-C17:0 | Ara, Gal, Man | DPG | 69 | Extensively branched, non-fragmenting substrate mycelium bearing sparse aerial hyphae. | Tamura et al. (2008) |
| *Kitasatospora acidiphila* JCM 32302[T] | *meso*-/LL-A$_2$pm | MK-9 (H$_6$,H$_8$) | i-C15:0, ai-C15:0, ai-C15:1 | Gal, Glu, Man | DPG, PE, PG, PI | ND | Extensively branched substrate mycelium. Forms long straight chains of cylindrical and smooth-surfaced spores. | Kim et al. (2020) |
| *Kitasatospora arboriphila* DSM 44785[T] | *meso*-/LL-A$_2$pm | MK-9 (H$_6$,H$_8$) | i-C15:0, ai-C15:0, i-C16:0 | Gal, Glu, Man, Rib | DPG, PE, PG, PI, PIMs | 64.5 | Extensively branched substrate mycelium. Forms long straight to spiral chains together with hooks and loops. Spores are cylindrical and smooth-surfaced. | Groth et al. (2004), Nouioui et al. 2018 |

*Table 4 contd. ...*

*...Table 4 contd.*

| Actinobacteria | Wall diamino acid (peptidoglycan type) | Menaquinone profile (MK) | Predominant fatty acids | Cell sugars | Polar lipids | DNA G+C content (mol%) | Morphology | References |
|---|---|---|---|---|---|---|---|---|
| *Kitasatospora viridis* DSM 44826[T] | *meso-*/LL-A$_2$pm | MK-9 (H$_6$ H$_8$) | i-C15:0, ai-C15:0, C16:0, i-C16:0, ai-C17:0 | Gal, Glu | DPG, PE, PI, PIMs | ND | Extensively branched substrate mycelium. Aerial hyphae differentiate into long, spiral chains of smooth-surfaced, cylindrical spores. | Liu et al. (2005) |
| *Nocardia aciditolerans* DSM 45801[T] | *meso-*A$_2$pm | MK-8 (H$_6$), MK8 (H$_4$ cyclo) | C16:0, C18:1 ω9c, 10-methyl C18:0 | Ara, Gal, Rib | DPG, PE, PG, PI, PIMs | 71.3 | Extensively branched, substrate mycelium fragment into irregular rod- and coccoid-like elements. Aerial hyphae formed. | Golinska et al. (2013c) |
| *Nocardia acidivorans* DSM 45049[T] | ND | MK-8 (H$_4$, ω-cycl), MK8 (H$_4$) trace | C15:0, C16:0, C17:1 ω8c, 10-methyl C17:0 | ND | DPG, PE, PI, PIMs | | Extensively branched, substrate mycelium. Aerial hyphae formed. | Kämpfer et al. (2007) |
| *Nocardia jiangxiensis* DSM 17684[T] | *meso-*A$_2$pm | MK-8 (H$_6$), MK8 (H$_4$ cyclo) | C16:0, C16:1 ω7t, C18:0, C18:1 ω9c, 10-methyl C18:0 | Ara Gal | DPG, PE, PI, PIMs | ND | Extensively branched, substrate mycelium fragments into rod-shaped elements. Aerial hyphae formed. | Cui et al. (2005) |
| *Nocardia miyunensis* DSM 17685[T] | *meso-*A$_2$pm | MK-8 (H$_6$), MK8 (H$_4$ cyclo) | C16:0, C16:1 ω7c, C18:0, C18:1 ω9c, 10-methyl C18:0 | Ara Gal | DPG, PE, PI, PIMs | ND | Extensively branched, substrate mycelium fragments into rod-shaped elements. Aerial hyphae formed. | Cui et al. (2005) |

| Species | Diaminopimelic acid | Menaquinones | Fatty acids | Sugars | Polar lipids | G+C (%) | Morphology | Reference |
|---|---|---|---|---|---|---|---|---|
| *Streptacidiphilus albus* DSM 41753[T] | LL-A$_2$pm | MK-9 (H$_6$, H$_8$) | i-C15:0, ai-C15:0, C16:0, i-C16:0, ai-C17:0 | Gal, Rha | DPG, PE, PI, PIMs | 70-72 | Extensively branched, non-fragmenting substrate mycelium; aerial mycelium differentiates into long flexuous chains of smooth surfaced spores. | Kim et al. (2003) |
| *Streptacidiphilus anmyonensis* JCM 16223[T] | LL-A$_2$pm; meso-A$_2$pm (trace) | MK-9 (H$_6$, H$_8$) | i-C15:0, ai-C15:0, C16:0, i-C16:0, i-C17:0, ai-C17:0 | Gal, Rha | DPG, PE, PI, PIMs | ND | Extensively branched, non-fragmenting substrate mycelium; aerial mycelium differentiates into long flexuous chains of smooth surfaced spores. | Cho et al. (2008) |
| *Streptacidiphilus bronchialis* DSM 106435[T] | LL-A$_2$pm | MK-9 (H$_8$) | ai-C15:0, i-C16:0 | Glu, Man, Rib | DPG, PE, PI, GPL, AGL, L | 72.6 | Extensively branched, non-fragmenting substrate mycelium; aerial mycelium differentiates into flexuous chains of spores with warty surfaces. | Nouioui et al. (2019) |
| *Streptacidiphilus carbonis* DSM 41754[T] | LL-A$_2$pm | MK-9 (H$_6$, H$_8$) | i-C15:0, ai-C15:0, C16:0, i-C16:0, ai-C17:0 | Gal, Rha | DPG, PE, PI, PIMs | 70-72 | Extensively branched, non-fragmenting substrate mycelium; aerial mycelium differentiates into long flexuous chains of spores with smooth surfaces. | Kim et al. (2003) |
| *Streptacidiphilus durhamensis* DSM 45796[T] | LL-A$_2$pm | MK-9 (H$_6$, H$_8$) | i-C15:0, ai-C15:0, i-C16:0 | Gal, Rha | DPG, PE, PI, PIMs | 71.0 | Extensively branched, non-fragmenting substrate mycelium; aerial mycelium differentiates into long, straight to flexuous chains of spores with smooth surfaces. | Golinska et al. (2013a) |

*Table 4 contd. ...*

...*Table 4 contd.*

| Actinobacteria | Wall diamino acid (peptidoglycan type) | Menaquinone profile (MK) | Predominant fatty acids | Cell sugars | Polar lipids | DNA G+C content (mol%) | Morphology | References |
|---|---|---|---|---|---|---|---|---|
| *Streptacidiphilus griseoplanus* DSM 40009[T] | LL-A$_2$pm | MK-9 (H$_6$) | ai-C15:0, C16:0 | Glu, Man (trace), Rib | DPG, PE, PI, GPL, AGL, AL, GL, L, 3PL | 72.5 | Extensively branched, non-fragmenting substrate mycelium; aerial mycelium differentiates into spiral chains of spores with warty surfaces. | Backus et al. (1957), Nouioui et al. (2019) |
| *Streptacidiphilus hamsterleyensis* DSM 45900[T] | LL-A$_2$pm | MK-9 (H$_6$, H$_8$) | i-C15:0, ai-C15:0, C16:0, i-C16:0 | Gal, Rha | DPG, PE, PI, PIMs | 71.0 | Extensively branched, non-fragmenting substrate mycelium; aerial mycelium differentiates into long, straight to flexuous spore chains with smooth surfaces. | Golinska et al. (2013b) |
| *Streptacidiphilus jiangxiensis* DSM 45096[T] | LL-A$_2$pm | MK-9 (H$_6$, H$_8$) | i-C15:0, ai-C15:0, C16:0, i-C16:0 | Gal, Rha | DPG, PE, PI, PIMs | 70.8–71.7 | Extensively branched, non-fragmenting substrate mycelium; aerial mycelium differentiates into long flexuous chains of spores with smooth surfaces. | Huang et al. (2004) |
| *Streptacidiphilus melanogenes* JCM 16224[T] | LL-A$_2$pm | MK-9 (H$_6$, H$_8$) | i-C15:0, ai-C15:0, C16:0, i-C16:0, i-C17:0, ai-C17:0 | Gal, Rha | DPG, PE, PI, PIMs | ND | Extensively branched, non-fragmenting substrate mycelium; aerial mycelium differentiates into long flexuous chains of spores with smooth surfaces. | Cho et al. (2008) |

| Species | Diaminopimelic acid | Menaquinone | Fatty acids | Sugars | Phospholipids | G+C (%) | Morphology | Reference |
|---|---|---|---|---|---|---|---|---|
| *Streptacidiphilus monticola* DSM 105744$^T$ | LL-A$_2$pm | MK-9 (H$_4$, H$_6$, H$_8$) | C14:0, C15:0, C16:0, ai-C17:0 | Gal, Glu, Rha, Rib | DPG, PE, PI, PIMs, 2PL | 71.0 | Extensively branched, non-fragmenting substrate mycelium; aerial mycelium differentiates into flexuous to straight chains of cylindrical spores with smooth surfaces. | Song et al. (2018) |
| *Streptacidiphilus neutrinimicus* DSM 41755$^T$ | LL-A$_2$pm | MK-9 (H$_6$, H$_8$) | i-C15:0, ai-C15:0, C16:0, i-C16:0, ai-C17:0 | Gal, Rha | DPG, PE, PI, PIMs | 70–72 | Extensively branched, non-fragmenting substrate mycelium; aerial mycelium differentiates into long chains of spores with smooth surfaces. | Kim et al. (2003) |
| *Streptacidiphilus oryzae* DSM 45098$^T$ | LL-A$_2$pm; meso-A$_2$pm (trace) | MK-9 (H$_6$, H$_8$) | i-C15:0, ai-C15:0, C16:0, i-C16:0 | Gal, Glu, Man, Rib | DPG, PE, PG, PI, PIMs | ND | Extensively branched, non-fragmenting substrate mycelium;aerial mycelium differentiates into long flexuous chains of spores with smooth surfaces. | Wang et al. (2006) |
| *Streptacidiphilus pinicola* JCM 32300$^T$ | LL-A$_2$pm | MK-9(H$_4$, H$_6$, H$_8$) | i-C15:0, ai-C15:0, C16:0, i-C16:0 | Gal | DPG, PE, PG, PI, PIMs | 68.6 | Extensively branched, non-fragmenting substrate mycelium; aerial mycelium differentiates into straight chains of spores with smooth surfaces. | Roh et al. (2018) |
| *Streptacidiphilus rugosus* JCM 16225$^T$ | LL-A$_2$pm; meso-A$_2$pm (trace) | MK-9 (H$_6$, H$_8$) | i-C15:0, ai-C15:0, C16:0, i-C16:0, i-C17:0, ai-C17:0 | Gal, Rha | DPG, PE, PI, PIMs | ND | Extensively branched, non-fragmenting substrate mycelium; aerial mycelium differentiates into long flexuous chains of smooth surfaced spores. | Cho et al. (2008) |

*Table 4 contd. ...*

...*Table 4 contd.*

| Actinobacteria | Wall diamino acid (peptidoglycan type) | Menaquinone profile (MK) | Predominant fatty acids | Cell sugars | Polar lipids | DNA G+C content (mol%) | Morphology | References |
|---|---|---|---|---|---|---|---|---|
| *Streptacidiphilus torunensis* DSM 10229$^T$ | LL-A$_2$pm | MK-9 (H$_4$, H$_6$, H$_8$) | ai-C15:0, C16:0, i-C16:0, ai-C17:0 | Gal, Rha | DPG, PE, PG, PI, PIMs | 72.3 | Extensively branched, non-fragmenting substrate mycelium; aerial mycelium differentiates into long straight to flexuous chains of spores with smooth surfaces. | Golinska et al. 2016 |
| *Streptomyces guanduensis* DSM 41944$^T$ | LL-A$_2$pm; *meso*-A$_2$pm (trace) | MK-9 (H$_6$, H$_8$) | C16:0, i-C14:0, i-C16:0, i-C18:0, ai-C15:0, ai-C17:0 | None | DPG, PE, PIMs, PME | 72.7 | Extensively branched, non-fragmenting substrate mycelium; aerial mycelium differentiate into flexuous chains of spores with smooth surfaces. | Xu et al. (2006) |
| *Streptomyces paucisporeus* DSM 41946$^T$ | LL-A$_2$pm; *meso*-A$_2$pm (trace) | MK-9 (H$_6$, H$_8$) | C16:0, i-C14:0, i-C16:0, i-C18:0, ai-C15:0, ai-C17:0 | None | DPG, PE, PIMs, PME | 74.8 | Extensively branched, non-fragmenting substrate mycelium; aerial mycelium differentiates into flexuous chains of spores with smooth surfaces. | Xu et al. (2006) |
| *Streptomyces rubidus* DSM 41947$^T$ | LL-A$_2$pm; *meso*-A$_2$pm (trace) | MK-9 (H$_6$, H$_8$) | C16:0, i-C14:0, i-C16:0, i-C18:0, ai-C15:0, ai-C17:0 | None | DPG, PE, PIMs, PME | 70.6 | Extensively branched, non-fragmenting substrate mycelium; aerial mycelium differentiates into flexuous chains of spores with smooth surfaces. | Xu et al. (2006) |

| | | | | | | | | |
|---|---|---|---|---|---|---|---|---|
| *Streptomyces yanglinensis* DSM 41945[T] | LL-A$_2$pm; *meso*-A$_2$pm (trace) | MK-9 (H$_6$,H$_8$) | C16:0, i-C14:0, i-C16:0, i-C18:0, ai-C15:0, ai-C17:0 | None | DPG, PE, PIMs, PME | 74.8 | Extensively branched, non-fragmenting substrate mycelium; aerial mycelium differentiates into flexuous chains of spores with smooth surfaces. | Xu et al. (2006) |
| *Streptomyces yeochonensis* DSM 41868[T] | LL-A$_2$pm | MK-9 (H$_6$,H$_8$) | C16:0, i-C14:0, i-C16:0, i-C18:0, ai-C15:0, ai-C17:0 | None | ND | ND | Extensively branched, non-fragmenting substrate mycelium; aerial mycelium differentiates into straight to flexuous chains of spores with smooth surfaces. | Kim et al. (2004) |

**Footnote:** *Only the *Nocardia* strains contained mycolic acids.

ND; not determined, A$_2$pm; diaminopimelic acid, i; iso, ai; anteiso, Ara; arabinose, Gal; galactose, Man; mannose, Rha; rhamnose, Rib; ribose, Xyl; xylose.

DPG; diphosphatidylglycerol, PC; phosphatidylcholine, PE; phosphatidylethanolamine, PGL; phosphoglycolipid, Phex; phosphohexose, PI; phosphatidylinositol, PIMs phosphatidylinositol mannosides; PME; phosphatidylmethylethanolamine, PS; phosphatidylserine, AGL; aminoglycolipid, AL; aminolipid, GL; glycolipid, GPL; glycophospholipid, L; unknown lipid, PGL; phosphoglycolipid, PL; unknown phospholipid.

form spiral chains of spores and have a different sugar composition (Table 4). In contrast, both the acidophilic and netrotolerant *Nocardia* strains can be distinguished from all of the other taxa as they form an extensively branched substrate mycelium which fragments into rod- and coccoid-like elements; contain mycolic acids, produce cyclized menaquinones with eight isoprene units and have fatty acid profiles which include tuberculostearic acid (10-methyl C18:0); all of these properties are typical of *Nocardia* strains (Goodfellow and Maldonado 2012).

## 5. Phylogenomics

The genome sequences of 63 actinomycetes representing 12 genera (Tables 1 and 5) were obtained from GenBank and annotated using Prokka (Seemann 2014). Protein sequences were used to generate a phylogenomic tree using PhyloPhlAn (Asnicar et al. 2020) and the tree was rerooted on the longest branch. The relationships between the strains in the phylogenomic tree (Fig. 2) is consistent with those found

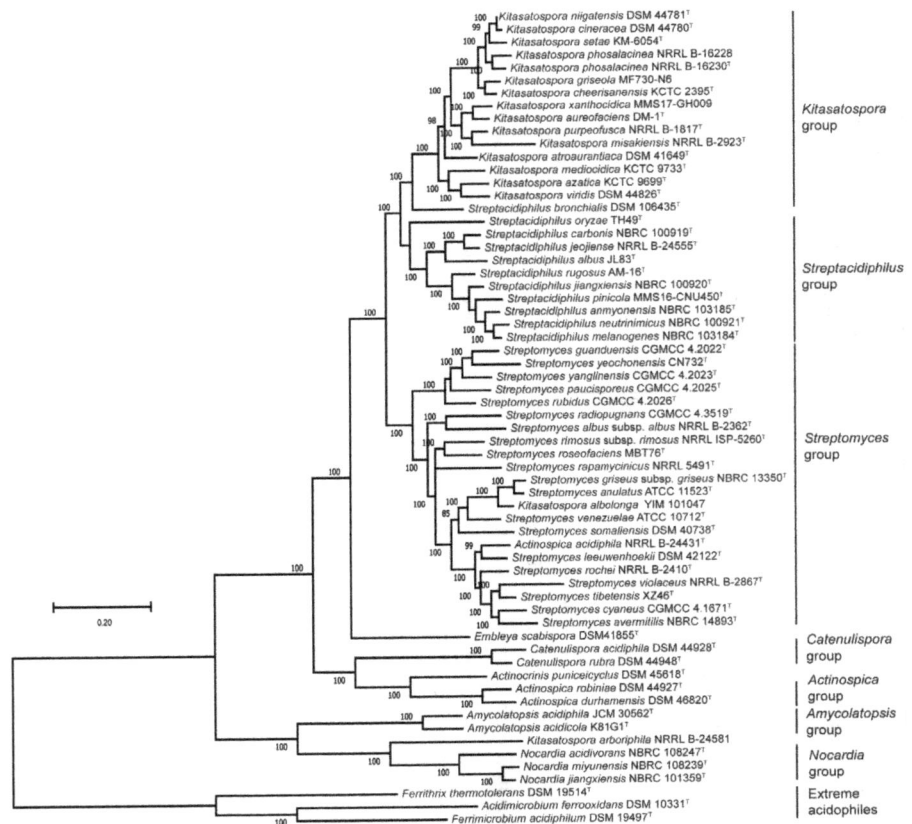

**Figure 2.** A phylogenomic tree constructed by PhyloPhlAn 3.0 showing relationships between acidophilic/acidotolerant and neutrotolerant filamentous actinomycetes and between them and extreme acidophilic strains.

in the 16S rRNA gene tree (Fig. 1) though lineages in the former are supported by much higher bootstrap values, a result which reflects the fact that phylogenomic trees are based on many more nucleotides than single gene trees. It is for this reason that classifications based on whole-genome sequences and associated software tools have led to marked improvements in the systematics of actinomycetes (Nouioui et al. 2018, Sangal et al. 2018, 2019).

The phylogenetic tree (Fig. 2) shows that the extreme acidophilic strains form a well separated taxon though *F. thermotolerans* DSM 19514[T] is sharply separated from the other two strains. It is also clear from the tree that with few exceptions the *Kitasatospora, Streptacidiphilus* and *Streptomyces* strains can be assigned to three well supported lineages. The exceptions include *S. bronchialis* DSM 100435[T] and *S. oryzae* TH49[T], the former is closely related to the *Kitasatospora* strains, and the latter forms a branch just outside the *Streptacidophilus* lineage, as found in a previous study (Nouioui et al. 2018). Another singleton, *K. arboriphila* NRRL B-24581[T], clusters next to the *Nocardia* strains, as in the 16S rRNA gene tree. Again, as before, the *A. acidophila* and *K. albolonga* strains were recovered in the *Streptomyces* lineage though the former was most closely related to *Streptomyces leeuwenhoekii* (Busarakam et al. 2014). The type-strain of *E. scabispora* lay at the periphery of the *Kitasatospora/Streptacidiphilus/Streptomyces* clades, as found by Nouioui et al. (2018). It is also interesting that the strict acidophiles representing the *Actinocrinis, Actinospica* and *Catenulispora* lineages are more closely related to the strains belonging to the family *Streptomycetaceae* than they are to the *Amycolatopsis* and *Nocardia* strains.

Relatively little attention has been paid to designing metrics for the recognition of generic boundaries (Sant'Anna et al. 2019). In the present study, pairwise percentages of conserved proteins (POCP) were calculated between the strains using the scripts data_file_4.sh (Moose 2017) and runPOCP.sh (Pantiukh and Grouzdev 2017) which are based on the work of Qin et al. (2014) who compared the proteomes of strains in bidirectional BLAST searches. POCP similarities of 50% or above have been used to establish whether members of closely related taxa merit generic status (Qin et al. 2014, Sangal et al. 2018). It is evident from Fig. 3 that on this basis the generic status of the *Kitasatospora, Streptacidiphilus* and *Streptomyces* can be questioned. In contrast, the generic status of the *Actinocrinis, Actinospica* (minus the *A. acidophila* strain), *Catenulispora* and *Embleya* strains are underpinned by POCP similarities, as are those of the extreme acidophiles assigned to the genera *Acidimicrobium, Ferrimicrobium* and *Ferrothrix*. POCP similarities also show that *K. arboriphila* (Groth et al. 2004) and *K. misakiensis* (Nakamura 1961, Labeda et al. 2017) may merit generic status.

Pairwise nucleotide index (ANI) values between the strains were determined using the FastANI program (Jain et al. 2018) to determine whether any of the strains shared ANI similarities above the 95–96 % threshold used to assign closely related strains to the same species (Chun et al. 2018). Figure 4 shows that *Kitasatospora cineracea* DSM 44780[T] and *Kitasatospora niigatensis* DSM 44781[T] share an ANI value well above this threshold. Tajima et al. (2001), who proposed these species,

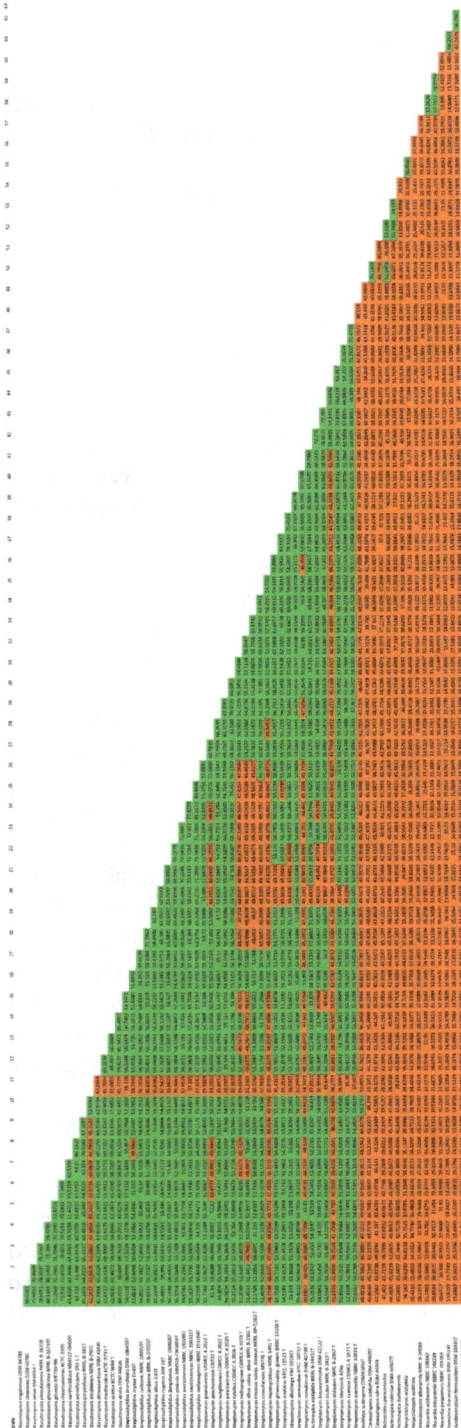

**Figure 3.** Pairwise POCP values between the genomes in the dataset. POCP values ≥ 50 are highlighted in green and those below 50% in orange.

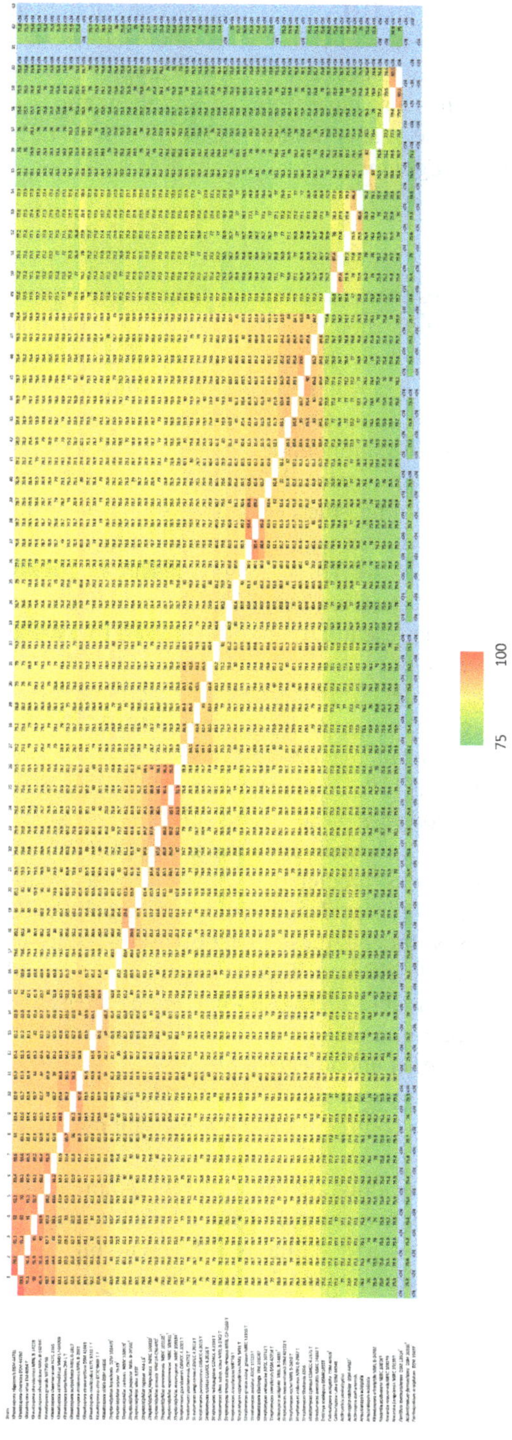

**Figure 4.** A heatmap from pairwise ANI values among the genomes. FastANI program does not return any values for those pairs where ANI is below 74%. Such pairs are highlighted in blue.

distinguished between the type-strains based on DNA homology and physiological data. Nevertheless, these strains have been repeatedly shown to be closely related based on 16S rRNA gene sequence data (Groth et al. 2003, 2004, Kämpfer 2012b, Labeda et al. 2012), as found in this study. It is also clear from the phylogenomic tree that they form a well-supported lineage. These strains should be classified as a single species; *K. cineracea* has priority over *K. niigatensis*.

## 6. Biotechnological Potential

The search for novel specialised metabolites from acid-loving, filamentous spore-forming actinomycetes is in its infancy even though early studies showed that *Streptomyces*-like strains from acidic soils were a potential source of antifungal antibiotics and acid-stable enzymes (Williams and Khan 1974, Williams and Flowers 1978, Williams and Robinson 1981). However, it has been shown that *C. acidophila* DSM 44928[T] produces catenulipeptin, a class III lantipeptide which partially restores aerial hyphal growth when applied to surfactin treated *Streptomyces coelicolor* (Wang and van der Donk 2012). Genome mining studies on this strain led to the discovery of a novel actinocoumarin gene cluster which encoded for two novel coumarins, calcibiocin A and B (Zettler et al. 2014). These compounds are relatively small molecules compared with other aminocoumarins and unlike them do not show antibiotic activity suggesting that they may represent intermediates in a pathway that forms a more complex biologically active agent.

The present study provides further evidence that acidophilic actinomycetes are gifted in the sense that they have large genomes rich in NP-BGCs, as exemplified by *Actinocrinis* (6.77 Mbp), *Actinospica* (9.63–9.92 Mbp), *Catenulispora* (10.47–12.8 Mbp) and *Streptacidiphilus* (7.91–9.72 Mbp) strains, as is the case with the moderately acidotolerant *Streptomyces* type-strains (7.82–9.89 Mbp). Biosystematic analyses of the genomes of representatives of the acidophilic genera revealed the presence of many NP-bioclusters predicted to encode for a broad-range of specialised metabolites, notably lanthipeptides, non-ribosomal polyketide synthetases and type I polyketide synthases. Indeed, these metabolic profiles are as diverse as those of the neutrophilic strains classified as *E. scabispora*, *K. setae* and *S. albus* subsp. *albus*. These results show that acid-loving sporoactinomycetes have a much greater potential to synthesise specialised metabolites, including antibiotics, than previously realised. Consequently, novel members of these taxa and their acidotolerant relatives should feature in bioprospecting campaigns designed to detect new classes of bioactive compounds needed to control multi-drug resistant microbial pathogens.

Neutrophilic streptomycetes and their acidophilic/acidotolerant counterparts have long been seen as prospective agents for the control of fungal root pathogens despite the lack of direct evidence (Goodfellow and Williams 1983, Williams et al. 1984, Goodfellow and Simpson 1987). Recently, acidophilic actinomycetes were reported to be more effective in inhibiting fungal growth than neutrophilic strains under acidic conditions (Zakalyukina and Zenova 2007, Guo et al. 2015, Tamreihao

**Table 5.** Genomic and physiological features of the representative acidophilic, acidotolerant and neutrophilic strains following the order of the clades in the genomic tree.

| Genus, species and strain ID | Accession | Size (Mbp) | GC (mol%) | #Contigs | Optimal pH | Temperature range |
|---|---|---|---|---|---|---|
| *Kitasatospora atroaurantiaca* DSM 41649[T] | VIVR01000000 | 7.87 | 71.2 | 1 | 5.0–8.0* | 10–37 |
| *Kitasatospora aureofaciens* DM-1[T] | NZ_CP020567 | 6.82 | 72.6 | 1 | 7.0–8.0 | 10–45 |
| *Kitasatospora azatica* KCTC 9699[T] | NZ_JQMO00000000 | 8.27 | 71.6 | 3 | 5.0–8.0* | 11–34 |
| *Kitasatospora cheerisanensis* KCTC 2395[T] | JNBY01000000 | 8.17 | 73.5 | 178 | 7.0–8.0 | 20–37 |
| *Kitasatospora cineracea* DSM 44780[T] | RJVJ01000000 | 8.76 | 73.9 | 5 | 7.0 | 15–37 |
| *Kitasatospora griseola* MF730-N6 | JXZB01000000 | 7.97 | 72.7 | 8 | 5.5–9.0* | 15–37 |
| *Kitasatospora mediocidica* KCTC 9733[T] | NZ_JQLN00000000 | 8.68 | 71.9 | 7 | 5.0–7.0* | 15–30 |
| *Kitasatospora misakiensis* NRRL B-2923[T] | JNXO01000000 | 8.16 | 73 | 6906 | 7.0 | ND |
| *Kitasatospora niigatensis* DSM 44781[T] | RKQG01000000 | 8.58 | 74 | 7 | 7.0 | 15–41 |
| *Kitasatospora phosalacinea* NRRL B-16228 | GCA_000717185 | 8.62 | 74.1 | 340 | ND | ND |
| *Kitasatospora phosalicinea* NRRL B-16230[T] | NZ_JNWZ00000000 | 7.62 | 74.1 | 227 | 5.5–9.0* | 15–42 |
| *Kitasatospora purpeofusca* NRRL B-1817[T] | JODS01000000 | 9.34 | 73.6 | 138 | 7.0 | 10–37 |
| *Kitasatospora setae* KM-6054[T] | NC_016109 | 8.78 | 74.2 | 1 | ND | 15–37 |
| *Kitasatospora viridis* DSM 44826[T] | VIWT01000000 | 10.09 | 73.1 | 9 | 5.5 | 10–37 |
| *Kitasatospora xanthocidicus* MMS17-GH009 | QVIG01000000 | 9.69 | 73.5 | 4 | ND | ND |
| *Streptacidiphilus bronchialis* DSM 106435[T] | CP031264 | 7.09 | 72.7 | 1 | 5.0–7.0* | 20–40 |
| *Streptacidiphilus albus* JL83[T] | JQML00000000 | 9.92 | 71.8 | 1 | 5.0 | 10–25 |
| *Streptacidiphilus anmyonensis* NBRC 103185[T] | BBPQ00000000 | 9.39 | 72.2 | 384 | 4.5–5.0 | 28–35 |
| *Streptacidiphilus carbonis* NBRC 100919[T] | BBPM01000000 | 8.46 | 71.5 | 316 | 5.0 | 15–30 |
| *Streptacidiphilus jeojiense* NRRL B-24555[T] | JOEH01000000 | 9.16 | 71.5 | 144 | ND | ND |

*Table 5 contd. ...*

...*Table 5 contd.*

| Genus, species and strain ID | Accession | Size (Mbp) | GC (mol%) | #Contigs | Optimal pH | Temperature range |
|---|---|---|---|---|---|---|
| *Streptacidiphilus jiangxiensis* NBRC 100920<sup>T</sup> | BBPN01000000 | 9.53 | 72.2 | 225 | 5.5 | 15–30 |
| *Streptacidiphilus melanogenes* NBRC 103184<sup>T</sup> | BBPP01000000 | 8.77 | 72 | 282 | 4.5–5.0 | 28–35 |
| *Streptacidiphilus neutrinimicus* NBRC 100921<sup>T</sup> | BBPO01000000 | 8.42 | 72.2 | 476 | 5.0 | 10–25 |
| *Streptacidiphilus oryzae* TH49<sup>T</sup> | JQMQ00000000 | 7.81 | 73.4 | 6 | 5.0 | 28–37 |
| *Streptacidiphilus pinicola* MMS16-CNU450<sup>T</sup> | QKYN01000000 | 8.43 | 71.8 | 342 | 6.0 | 15–33 |
| *Streptacidiphilus rugosus* AM-16T | JQMJ00000000 | 9.00 | 71.8 | 4 | 4.5–5.0 | 28–35 |
| *Actinospica acidiphila* NRRL B-24431<sup>T</sup> | JNYX00000000 | 7.31 | 72.6 | 98 | 5.0 | 17–33 |
| *Kitasatospora albolonga* YIM 101047<sup>T</sup> | GCA_002082585 | 8.03 | 72 | 1 | 6.5–8.0* | ND |
| *Streptomyces albus* NRRL B-2362<sup>T</sup> | NZ_JODZ00000000 | 7.85 | 72.7 | 98 | 6.5–8.0* | ND |
| *Streptomyces amulatus* ATCC 11523<sup>T</sup> | NZ_JPZP01000000 | 8.85 | 71.7 | 8 | 6.5–8.0* | ND |
| *Streptomyces avermitilis* NBRC 14893<sup>T</sup> | NC_003155 | 9.03 | 70.7 | 1 | 6.5–8.0* | ND |
| *Streptomyces cyaneus* CGMCC 4.1671<sup>T</sup> | NZ_RQJW00000000 | 9.83 | 70.4 | 36 | 6.5–8.0* | ND |
| *Streptomyces griseus* NBRC 13350<sup>T</sup> | NC_010572 | 8.55 | 72.2 | 1 | 6.5–8.0* | ND |
| *Streptomyces guanduensis* CGMCC 4.2022<sup>T</sup> | NZ_FNIE00000000 | 8.22 | 73.1 | 35 | 5.5 | 20–37 |
| *Streptomyces leeuwenhoekii* DSM 42122<sup>T</sup> | NZ_LN831790 | 7.9 | 72.8 | 1 | 7.0 | 4–50 |
| *Streptomyces paucisporeus* CGMCC 4.2025<sup>T</sup> | NZ_FRBI00000000 | 8.16 | 72.3 | 91 | 5.5 | 20–37 |
| *Streptomyces radiopugnans* CGMCC 4.3519<sup>T</sup> | NZ_FOET00000000 | 6.07 | 73.4 | 43 | 6.5–8.0* | 20–50 |
| *Streptomyces rapamycinicus* NRRL 5491<sup>T</sup> | CP006567 | 12.7 | 71.1 | 394 | 6.5–8.0* | 10–37 |
| *Streptomyces rimosus* NRRL ISP-5260<sup>T</sup> | NZ_JNYR00000000 | 9.59 | 71.9 | 129 | 6.5–8.0* | ND |
| *Streptomyces rochei* NRRL B-2410<sup>T</sup> | NZ_MUMD00000000 | 7.68 | 72.5 | 917 | 6.5–8.0* | ND |
| *Streptomyces roseofaciens* MB176<sup>T</sup> | NZ_LNBE00000000 | 8.64 | 72.1 | 18 | 7.0 | 20–50 |

| | | | | | | |
|---|---|---|---|---|---|---|
| *Streptomyces rubidus* CGMCC 4.2026[T] | NZ_FODD00000000 | 9.01 | 72.9 | 115 | 5.5 | 20–37 |
| *Streptomyces somaliensis* DSM 40738[T] | NZ_AJJM00000000 | 5.18 | 74.1 | 243 | 6.5–8.0* | ND |
| *Streptomyces tibetensis* XZ46[T] | SZVR00000000 | 8.99 | 71.2 | 79 | 4.0–9.0* | 4–40 |
| *Streptomyces venezuelae* ATCC 10712[T] | NZ_QGUC00000000 | 8.39 | 72.5 | 38 | 7.0 | ND |
| *Streptomyces violaceus* NRRL B-2867[T] | JNXN00000000 | 9.16 | 70.6 | 5150 | 6.5–8.0* | ND |
| *Streptomyces yanglinensis* CGMCC 4.2023[T] | NZ_FNVU00000000 | 9.59 | 72.6 | 48 | 5.5 | 20–37 |
| *Streptomyces yeochonensis* CN732[T] | NZ_JQNR00000000 | 7.82 | 73.6 | 6 | 5.5 | 20–37 |
| *Embleya scabispora* DSM 41855[T] | ARCJ00000000 | 11.39 | 70.9 | 199 | 7.0 | 18–36 |
| *Catenulispora acidiphila* DSM 44928[T] | CP001700 | 10.47 | 69.8 | 1 | 6.0 | 11–37 |
| *Catenulispora rubra* DSM 44948[T] | JAACYW00000000 | 12.8 | 70 | 888 | 5.0 | ND |
| *Actinocrinis puniceicyclus* DSM 45618[T] | – | 6.77 | 70.3 | 434 | 5.5 | 13–30 |
| *Actinospica durhamensis* DSM 46820[T] | – | 9.63 | 71.1 | 953 | 5.5 | 15–33 |
| *Actinospica robiniae* DSM 44927[T] | K1632511 | 9.92 | 70.3 | 3 | 5.5 | 17–33 |
| *Amycolatopsis acidicola* K81G1[T] | VMNW02 | 9.47 | 69.7 | 195 | 5.5–6.0 | 14–41 |
| *Amycolatopsis acidiphila* JCM 30562[T] | VIZA01 | 8.08 | 70.5 | 212 | 5.5–6.0 | 15–40 |
| *Kitasatospora arboriphila* NRRL B-24581[T] | JNYV01000000 | 5.75 | 64.5 | 45 | 5.0–7.0 | 15–40 |
| *Nocardia acidivorans* NBRC 108247[T] | BDAW01 | 7.57 | 66.9 | 161 | ND | ND |
| *Nocardia jiangxiensis* NBRC 101359[T] | BAGB01 | 10.45 | 66.8 | 174 | 5.5 | 17–37 |
| *Nocardia miyunensis* NBRC 108239[T] | BDBQ01 | 10.52 | 67 | 208 | 5.5 | 17–37 |
| *Ferrithrix thermotolerans* DSM 19514[T] | FQUL00000000 | 2.49 | 51.1 | 101 | 1.8 | 50 |
| *Acidimicrobium ferrooxidans* DSM 10331[T] | NC_013124 | 2.16 | 68.3 | 1 | 2.0 | 45–50 |
| *Ferrimicrobium acidiphilum* DSM 19497[T] | JQKF00000000 | 2.93 | 55.3 | 115 | 2.0 | 37 |

*In the absence of an optimal pH value the pH range is given. ND, not determined; -, not yet available.

et al. 2018a), as exemplified by the antagonistic activity of strains against the rice pathogens, *Fusarium moniliforme*, *Helminthosporium oryzae* and *Rhizoctonia solani* (Poomthongdee et al. 2015). Similarly, an acidotolerant *Streptomyces* strain was found to inhibit the growth of rice fungal pathogens such as *Curvularia oryzae*, *Fusarium oxysporum* and *Pyricularia oryzae* (Tamreihao et al. 2018b). Two antibiotics, reveromycins A and B, produced by an acidophilic *Streptomyces* strain effectively suppressed the growth of several fungal root pathogens, including *Botrytis cinerea*, *Rhizopus stolonifer* and *Sclerotinia sclerotiorum* under acidic conditions (Lyu et al. 2017). The antifungal activity of these compounds was higher at pH 4.5 than 7.0 suggesting that the application of the production strain to acidic soils may achieve high antifungal efficacy in the suppression of pathogenic fungi. In this context an extract of reveromycin repressed strawberry fruit rot caused by several fungi, including *B. cinerea* and *M. hiemalis* (Lyu et al. 2017).

A time-honoured biocontrol strategy involves amending soil with organic substrates on the premise that this promotes the growth of saprophytic antagonists, especially streptomycetes, at the expense of fungal pathogens (Williams et al. 1984, Goodfellow and Simpson 1987). From the onset, chitin amendments were seen to be especially effective (Buxton et al. 1965, Vruggink 1976). Acidophilic/acidotolerant actinomycetes are considered to play a significant role in the turnover of chitin in acidic soil and litter (Williams and Robinson 1981, Guo et al. 2015). Streptomycetes, including an acidotolerant strain, have been shown to produce cell-wall degrading chitinases that inhibit the growth of fungal pathogens (Srividya et al. 2012, Tamreihao et al. 2018b). Application of chitinase-producing acidophilic/acidotolerant actinomycetes with the capacity to synthesise bioactive compounds may prove to be an effective way of controlling fungal phytopathogens in acidic soils thereby providing an attractive alternative to the use of synthetic fungicides.

Acid-loving and neutrophilic filamentous actinomycetes have potential as biofertilizers, as they can promote plant growth directly and indirectly (Poomthongdee et al. 2015, Tamreihao et al. 2018a, Chaiharn et al. 2020). Direct contributions to plant welfare can be achieved through the production of phytohormones, such as cytokinins and indole-3-acetic acid, phosphate solubilization and the synthesis of iron-chelating siderophores, as explained by Nouioui et al. (2019). Such studies on acidophilic/acidotolerant actinomycetes are at a pioneering stage though these organisms, notably streptomycetes, can synthesise siderophores and solubilize phosphate (Poomthongdee et al. 2015, Tamreihao et al. 2018a). It has been shown that rice plants benefited following inoculation of rice seeds with an acidotolerant *Streptomyces* strain under greenhouse conditions (Tamreihao et al. 2016, 2018b). The genome of the type-strain of *Streptacidiphilis oryzae* harbours genes associated with siderophore production (Kim et al. 2015). Indirect ways of promoting plant growth, such as the production of antibiotics and cell-wall degrading extracellular enzymes, was touched upon earlier.

It can be concluded that acidophilic/acidotolerant actinomycetes have potential as a source of new specialised metabolites, notably novel antibiotics, and as inoculants designed to promote plant growth and a greening of agricultural practices.

In contrast, the chemolithotrophic lifestyles of their extreme acidophilic counterparts are well suited for metal recovery and have potential in mining for copper and gold (Harrison 2016).

## 7. Biosynthetic Potential of Representative Strains

Seven representative genome sequences were analysed using version 2 of the Antibiotic Resistant Target Seeker (ARTS) platform (Mungan et al. 2020) which detects bioclusters predicted to encode for uncharacterized BGCs based on the presence of resistant target genes, as described by Alanjary et al. (2017). The number of BGCs in the genomes of the strains ranged from 25 in *C. acidophila* DSM 44928[T] to 50 in *E. scabispora* DSM 41855[T] (Table 6). The major BGC types included bacteriocin, butyrolactones, lanthipeptides, non-ribosomal peptide synthetases (NRPS), type 1, 2 and 3 polyketide synthases (T1, T2 and T3 PKS), terpenes and hybrids of different cluster types. The genomes of most of the strains contained 3–5 bioclusters predicted to encode for known products with high identities (> 75%). *Actinocrinis puniceicyclus* DSM 45618[T] was considered to encode for the NRPS antibiotics, icosalides A and B (100%); and *Actinospica robiniae* DSM 44927[T] for the terpenes, geosmin (100%) and 2-methylisoborneol (75%), and a T2 PKS/NRPS spore pigment (75%). Similarly, the genome of *C. acidophila* DSM 44928[T] contained genes expressing for the aminocoumarin, cacibiocin B (92%), the class III lanthipeptide, catenulipeptin (100%), the siderophore, desferrioxamin B (100%), geosmin (100%), and a NPAA-RiPP-like terpene, 2 methylisoborneol (100%) whereas the genome of *Streptacidiphilus albus* DSM 41753[T] harboured genes encoding for the T3 PKS alkylresorcinol (100%), the NRPS-T1 PKS, antimycin and geosmin. In turn, the genomes of *K. setae* KM-6054[T] and *E. scabispora* DSM 41855[T] were equipped with bioclusters predicted to encode for alkylresorcinol (100%), the T1 PKS, bafilomycin B1 (100%), geosmin (100%) and a T2 PKS spore pigment (75%), and alkylresorcinol (100%), the T1 PKS-terpene, hitachimycin (77%), the T3 PKS, naringenin (100%), the T1 PKS, spectinabilin/orinocin/ SNF4435C/SNF4435D (90%) and a T2 PKS spore pigment (83%), respectively. Finally, the genome of *S. albus* subsp. *albus* NRL B-2362[T] harboured genes predicted to encode for desferrioxamine E (100%), ectoin (100%) and geosmin (100%). In contrast, 20–45 bioclusters detected in the strains showed low similarities to known BGCs and can be considered to encode for novel biomolecules.

Few of the detected bioclusters were conserved amongst all of the strains. The only exceptions were a siderophore cluster found in the genomes of *K. setae* KM-6054[T] and *S. albus* subsp. *albus* NRRL B-2362[T], and a terpene biocluster identified in the type-strains of *Actinocrinis puniceicyclus* and *Actinospica robiniae*; three genes were common to each of these bioclusters. Multiple butyrolactone and terpene bioclusters were predicted in the genome of *K. setae* KM-6054[T] (Table 6). Two butyrolactone bioclusters and two terpene BGCs showed significant similarities to each other based on BIG-SCAPE (Biosynthetic Genes Similarity Clustering and

**Table 6.** Biosynthetic gene clusters detected in the genomes of four representative acidophilic and three neutrophilic actinomycetes.

| BGC type | Acidophiles | | | | Neutrophiles | | |
|---|---|---|---|---|---|---|---|
| | ap1 | ar32 | ca35 | sal52 | es3 | ks5 | sal10 |
| Aminocoumarin | | | 1 | | | | |
| Bacteriocin | 1 | 1 | | | 2 | 3 | 1 |
| Bacteriocin.butyrolactone.PKS-like | | | | 1 | | | |
| Bacteriocin.NRPS | 1 | | | | | | |
| Bacteriocin.terpene.NRPS-like | | | 1 | | | | |
| Betalactone.NRPS | | 1 | | | 1 | | |
| Betalactone.terpene | | | | | 1 | | |
| Butyrolactone | | 1 | | 2 | 1 | 6 | |
| Butyrolactone.NRPS | | | | | | 1 | |
| Butyrolactone.PKS-like | 1 | | | | | | |
| CDPS | | | | 1 | 1 | | 1 |
| CDPS.terpene | | | | | | | 1 |
| Ectoine | | | | | | | 1 |
| Hgle-KS | | 1 | | | 1 | | |
| Indole | | | | | | 1 | |
| Indole.NRPS | | | | | 1 | | |
| Ladderane | 1 | | | | | 1 | |
| Lanthipeptide | 1 | 6 | 5 | 6 | 4 | 5 | 2 |
| Lanthipeptide.betalactone.T3PKS.PKS-like.NRPS | | | | 1 | | | |
| Lanthipeptide.lipolanthine | | | 1 | | | | |
| Lanthipeptide.NRPS | | | | | 1 | | |
| Lanthipeptide.NRPS-like | | | | 1 | | | |
| Lanthipeptide.T2PKS.T1PKS | | | | | | | 1 |
| LAP | 1 | | | | | 1 | |
| Lassopeptide | | | | | | 2 | |
| NRPS | 10 | 5 | 3 | 4 | 10 | 2 | 3 |
| NRPS-like | | 2 | | 3 | 1 | 1 | |
| NRPS.ladderane | | | | | | 1 | |
| Nucleoside | | | | | | | 1 |
| Oligosaccharide.T1PKS | | | | | | | 1 |
| Other | | 1 | | 1 | 1 | 1 | |
| Phosphonate.NRPS | | | | | | 1 | |
| Phosphonate.NRPS-like | 1 | | | | | | |
| PKS-like.NRPS | | | | 1 | | | |
| Siderophore | 1 | 1 | 2 | 2 | 3 | 2 | 2 |

*Table 6 contd. ...*

*...Table 6 contd.*

| BGC type | Acidophiles | | | | Neutrophiles | | |
|---|---|---|---|---|---|---|---|
| | ap1 | ar32 | ca35 | sal52 | es3 | ks5 | sal10 |
| T1PKS | 7 | 2 | 1 | | 8 | 1 | 7 |
| T1PKS.arylpolyene.NRPS | | | | | | 1 | |
| T1PKS.butyrolactone.PKS-like | | | | 1 | | | |
| T1PKS.hglE-KS | | | | | 1 | | |
| T1PKS.hglE-KS.NRPS | | | | 1 | | | |
| T1PKS.LAP.PKS-like | | | | | | 1 | |
| T1PKS.NRPS | 1 | 1 | 1 | 1 | | | 1 |
| T1PKS.NRPS-like | | | | | | | 1 |
| T1PKS.terpene | | | | | 1 | | |
| T2PKS | | | | 3 | 1 | 1 | |
| T2PKS.NRPS | | 1 | | | | | |
| T2PKS.oligosaccharide.PKS-like | | | | 1 | | | |
| T3PKS | | 1 | 1 | | 3 | 1 | |
| T3PKS.NRPS | | | | 1 | | | |
| T3PKS.terpene | | | | 1 | | | |
| Terpene | 4 | 5 | 2 | 3 | 4 | 6 | 5 |
| Terpene.butyrolactone | | | | 1 | | | |
| Terpene.NRPS | | | | | 1 | | |
| Terpene.NRPS-like | | | | | 1 | | |
| Terpene.T1PKS | | | | 1 | | | |
| Terpene.T1PKS.NRPS | | | | | 1 | | 1 |
| Thiopeptide.LAP | | | 2 | 2 | 1 | | |
| Thiopeptide.tfua-related | | 1 | | | | | |
| Transat-PKS.T1PKS.NRPS | | | | | | 1 | |
| Transat-PKS.T1PKS.PKS-like.NRPS | | | | 1 | | | |

**Note:** ap1 – *Actinocrinis puniceicyclus* DSM 45618[T]; ar32 – *Actinospica robiniae* DSM 44927[T]; ca35 – *Catenulispora acidiphila* DSM 44928[T]; es3 – *Embleya scabispora* DSM41855[T]; ks5 - *Kitasatospora setae* KM-6054[T]; sal10 - *Streptomyces albus* subsp. *albus* NRRL B-2362[T]; sal52 - *Streptacidiphilus albus* JL83[T].

Prospecting Engine) analyses. A large number of bioclusters, most notably NRPS and PKS type clusters, were detected among these strains, they showed very limited similarities to characterised BGCs that produce known antimicrobials. These results taken together show that strains representing each of the lineages in the phylogenomic tree produce different specialised metabolites. Additional studies are needed to explore the full biosynthetic potential of these strains.

## 8. Adaptation to Acidic Environments

Ninety-four genes comprising 44 sigma factors, 7 molecular chaperones, 10 heat shock proteins together with 16 genes involved in osmotic stress and 17 in oxidative stress responses were overexpressed in *Streptomyces coelicolor* when grown under acidic conditions (Kim et al. 2008). These genes were sought in the genomes of the 63 representative strains using protein BLAST (Camacho et al. 2009). The searches resulted in 24,055 hits with an e-value < 1e-05. Target hits that were smaller in length (< 70% of the query sequences; 1,450 hits), had alignment lengths < 50% of these sequences (1,482 hits) and sequence identities < 30% (2,269 hits) were deleted from the dataset, as were 12,124 duplicate hits to the same target; only single hits with the best bit-scores were kept. The remaining 6,731 hits were binned into five categories, namely heat shock, molecular chaperones, osmotic stress, oxidative stress, and sigma factors.

The BLAST searches revealed 51–76 copies of sigma factors among strains within the *Kitasatospora* lineage (35 in *K. misakiensis* NRRL B-2923[T]), 56–84 amongst those comprising the *Streptacidiphilus* clade, and 44-100 copies among the *Streptomyces* strains (Table 7). The *Catenulispora* strains have by far the highest number of copies of sigma factors, namely 115–123. In contrast, only 8–12 sigma factors were identified in the extreme acidophiles, *Acidimicrobium ferrooxidans* DSM 10331[T], *Ferrimicrobium acidiphilum* DSM 19497[T] and *Ferrithrix thermotolerans* DSM 19514[T], indicating that in these cases they may not be directly involved in adaptation to acidic environments.

Five to 13 chaperones, 7–21 heat shock response proteins and 7–21 proteins involved in responses to oxidative stress were detected in the genomes of the strains, the smallest number, 5–8, were found in the genomes of the three extreme acidophiles mentioned above. Interestingly, most of the *Streptomyces* strains have 7–32 genes involved in responses to osmotic stress; these genes were relatively rare amongst the other strains. Most of these genes are annotated to putative gas vesicle synthesis proteins. Ectoine and trehalose biosynthetic pathways are known to help acidophilic bacteria survive osmotic stress (Rivera-Araya et al. 2019). The ectoine BGC was only identified in the genome of *S. albus* subsp. *albus* NRRL B-2362[T]. In contrast, multiple copies of genes involved in trehalose biosynthesis (8–23 genes) were present in the genomes of most of the strains though only five such genes were detected in the *A. ferrooxidans* DSM 10331[T] and *F. thermotolerans* DSM 19514[T] genomes (data not shown).

The ability to metabolise chitin is associated with adaptation of acidophilic/acidotolerant actinomycetes to acidic habitats (Williams and Robinson 1981, Nawani et al. 2002, Guo et al. 2015). The genomes of the tested strains contained between 2–8 copies of genes predicted to express for chitinases (exochitinases, chitinases or putative bifunctional chitinases/lysozymes), except for *K. arboriphila* NRRL B-24581[T], *K. aureofaciens* DM-1[T], *N. acidivorans* NBRC 108247[T], *N. jiangxiensis* NBRC 101359[T], *N. miyunensis* NBRC 108239[T], *Streptacidiphilus oryzae* TH49[T], *Streptomyces somaliensis* DSM 40738[T], and the extreme acidophiles *Acidimicrobium ferrooxidans* DSM 10331[T], *Ferrithrix thermotolerans* DSM 19514[T] and *Ferrimicrobium acidiphilum* DSM 19497[T] (data not shown).

**Table 7.** Genes detected in the genomes of representative actinomycetes potentially involved in adaptation to acidic conditions.

| Organisms | Sigma factors | Chaperones | Heat shock proteins | Oxidative stress | Osmotic stress |
|---|---|---|---|---|---|
| **Kitasatospora lineage** | | | | | |
| K. misakiensis NRRL B-2923[T] | 35 | 11 | 7 | 12 | |
| K. aureofaciens DM-1[T] | 51 | 10 | 13 | 13 | |
| K. atroaurantiaca DSM 41649[T] | 53 | 8 | 12 | 13 | |
| K. azatica KCTC 9699[T] | 53 | 7 | 11 | 12 | |
| K. viridis DSM 44826[T] | 55 | 8 | 12 | 14 | |
| K. xanthocidica MMS17-GH009[T] | 58 | 9 | 14 | 19 | |
| K. griseola MF730-N6[T] | 62 | 9 | 12 | 15 | |
| K. mediocidica KCTC 9733[T] | 65 | 7 | 11 | 15 | |
| K. cheerisanensis KCTC 2395[T] | 67 | 9 | 12 | 14 | |
| K. phosalicinea NRRL B-16230[T] | 68 | 9 | 13 | 17 | 8 |
| K. phosalacinea NRRL B-16228[T] | 73 | 9 | 13 | 20 | 8 |
| K. cineracea DSM 44780[T] | 74 | 9 | 14 | 16 | 8 |
| K. setae KM-6054[T] | 74 | 10 | 13 | 16 | |
| K. niigatensis DSM 44781[T] | 76 | 9 | 13 | 20 | 8 |
| K. purpeofusca NRRL B-1817[T] | 76 | 12 | 21 | 18 | |
| **Streptacidiphilus lineage** | | | | | |
| S. bronchialis DSM 106435[T] | 41 | 10 | 16 | 11 | 8 |
| S. oryzae TH49[T] | 56 | 9 | 16 | 15 | 15 |
| S. carbonis NBRC 100919[T] | 67 | 7 | 18 | 15 | |
| S. albus JL83[T] | 68 | 7 | 15 | 14 | |

*Table 7 contd. ...*

...Table 7 contd.

| Organisms | Sigma factors | Chaperones | Heat shock proteins | Oxidative stress | Osmotic stress |
|---|---|---|---|---|---|
| S. ammyonensis NBRC 103185^T | 68 | 9 | 12 | 14 | 8 |
| S. neutrinimicus NBRC 100921^T | 68 | 9 | 21 | 18 | 8 |
| S. pinicola MMS16-CNU450^T | 69 | 9 | 13 | 12 | |
| S. jeojiense NRRL B-24555^T | 73 | 7 | 13 | 16 | |
| S. rugosus AM-16^T | 76 | 9 | 13 | 14 | |
| S. melanogenes NBRC 103184^T | 78 | 9 | 14 | 16 | 8 |
| S. jiangxiensis NBRC 100920^T | 84 | 9 | 14 | 14 | 8 |
| **Streptomyces lineage** | | | | | |
| S. somaliensis DSM 40738^T | 44 | 7 | 10 | 13 | 8 |
| S. yeochonensis CN732^T | 47 | 8 | 10 | 10 | |
| S. roseofaciens MBT76^T | 49 | 8 | 10 | 19 | 8 |
| S. radiopugnans CGMCC 4.3519^T | 50 | 9 | 11 | 11 | 7 |
| S. albus subsp. albus NRRL B-2362^T | 54 | 11 | 8 | 13 | 8 |
| S. guanduensis CGMCC 4.2022^T | 55 | 8 | 15 | 16 | 16 |
| K. albolonga  YIM 101047^T | 56 | 11 | 11 | 14 | |
| S.  violaceus NRRL B-2867^T | 59 | 7 | 16 | 13 | 13 |
| S. amulatus ATCC 11523^T | 62 | 13 | 12 | 18 | 8 |
| S. venezuelae ATCC 10712^T | 63 | 9 | 9 | 19 | 16 |
| S. rapamycinicus NRRL 5491^T | 67 | 10 | 14 | 17 | 32 |
| S. griseus subsp. griseus NBRC 13350^T | 69 | 12 | 12 | 15 | |
| S. leeuwenhoekii DSM 42122^T | 75 | 9 | 12 | 15 | 8 |

| | | | | | |
|---|---|---|---|---|---|
| S. rimosus subsp. rimosus NRRL ISP-5260[T] | 76 | 9 | 12 | 14 | 8 |
| S. avermitilis NBRC 14893[T] | 78 | 8 | 17 | 18 | 24 |
| A. acidiphila NRRL B-2443[T] | 79 | 10 | 12 | 18 | 9 |
| S. rochei NRRL B-2410[T] | 79 | 10 | 11 | 20 | 17 |
| S. yanglinensis CGMCC 4.2023[T] | 84 | 9 | 19 | 15 | 8 |
| S. cyaneus CGMCC 4.1671[T] | 85 | 9 | 14 | 15 | 8 |
| S. paucisporeus CGMCC 4.2025[T] | 95 | 9 | 20 | 16 | 9 |
| S. tibetensis XZ46[T] | 95 | 9 | 15 | 16 | 9 |
| S. rubidus CGMCC 4.2026[T] | 100 | 8 | 15 | 21 | 8 |
| Embleya scabispora DSM 41855[T] | 86 | 9 | 13 | 14 | |
| **Catenulispora lineage** | | | | | |
| C. rubra DSM 44948[T] | 115 | 8 | 13 | 19 | |
| C. acidiphila DSM 44928[T] | 123 | 8 | 18 | 18 | |
| Actinocrinis puniceicyclus DSM 45618[T] | 41 | 6 | 12 | 11 | 8 |
| **Actinospica lineage** | | | | | |
| A. durhamensis DSM 46820[T] | 70 | 7 | 14 | 18 | |
| A. robiniae DSM 44927[T] | 90 | 7 | 17 | 17 | 11 |
| **Amycolatopsis lineage** | | | | | |
| A. acidicola NBRC 113896[T] | 54 | 11 | 11 | 21 | 7 |
| A. acidiphila JCM 30562[T] | 68 | 8 | 13 | 17 | |
| Kitasatospora arboriphila NRRL B-24581[T] | 24 | 6 | 9 | 13 | |

*Table 7 contd. ...*

*...Table 7 contd.*

| Organisms | Sigma factors | Chaperones | Heat shock proteins | Oxidative stress | Osmotic stress |
|---|---|---|---|---|---|
| **Nocardia lineage** | | | | | |
| *N. acidivorans* NBRC 108247[T] | 43 | 9 | 14 | 16 | |
| *N. miyunensis* NBRC 108239[T] | 44 | 7 | 15 | 20 | 8 |
| *N. jiangxiensis* NBRC 101359[T] | 49 | 8 | 17 | 24 | |
| **Extreme acidophilic lineage** | | | | | |
| *Ferrithrix thermotolerans* DSM 19514[T] | 12 | 5 | 8 | 7 | |
| *Acidimicrobium ferrooxidans* DSM 10331[T] | 9 | 5 | 8 | 7 | |
| *Ferrimicrobium acidiphilum* DSM 19497[T] | 8 | 5 | 8 | 7 | |

These results taken together indicate that majority of the acidophilic and acidotolerant strains are well equipped to thrive in acidic habitats.

## 9. Conclusions and Future Perspectives

New approaches to systematics of actinomycetes show that acidophilic filamentous, spore-forming members belong to evolutionary lineages which correspond to the genera *Actinocrinis*, *Actinospica*, *Catenulispora* and *Streptacidiphilus*. Similarly, acidophilic and acidotolerant isolates have been assigned to species in several genera, notably *Amycolatopsis*, *Kitasatospora*, *Nocardia* and *Streptomyces*; species assigned to these taxa tend to accommodate neutrophilic strains. In general, confidence can be placed in the taxonomic integrity of species encompassing acidophilic/acidotolerant strains especially when underpinned by data drawn from whole-genome sequences, notably thorough the use of ANI and digital DNA:DNA hybridization (dDDH) values (Chun et al. 2018). However, it is not unusual for these metrics to give conflicting results, as exemplified by studies on closely related *Micromonospora* and *Rhodococcus* species (Riesco et al. 2018, Thompson et al. 2020). In such instances, biological characteristics, such as ecological and phenotypic properties, should be used to refine conclusions drawn from ANI and dDDH values (Li et al. 2015, Riesco et al. 2018, Thompson et al. 2020). Metrics recently introduced for the recognition of generic boundaries also need to be interpreted with care, especially between closely related lineages. In the present study, lineages corresponding to the genera *Kitasatospora*, *Streptacidiphilus* and *Streptomyces* generally shared POCP similarities above 50%, a threshold that has been considered somewhat arbitrary (Barco et al. 2020). Interestingly, raising the POCP threshold a little allows these taxa to be distinguished, a result that correlates with the discontinuous distribution of chemotaxonomic and physiological markers between them (Kim et al. 2003).

This study provides further evidence that acid-loving filamentous sporoactinomycetes are gifted as they have large genomes (6.77–12.80 Mbp) which contain many NP-BGCs predicted to express for a diverse range of specialised metabolites, most of them novel and uncharacterised. It is particularly encouraging that the genomes of the representative acidophilic strains contained a similar range of bioclusters as their neutrophilic counterparts, especially given their potential as a source of non-ribosomal peptide synthetases and type I polyketide synthases which are critical driving forces for the generation of chemical diversity and structural complexity of natural products. Consequently, strains classified in the genera *Actinocrinis*, *Actinospica*, *Catenulispora* and *Streptacidiphilus* should be included in the search for lead structures needed for the development of antibiotic, anticancer and immunosuppresive compounds. Current strategies designed for this purpose are based on the application of technical advances in microbiology, molecular biology, bioinformatics and natural product chemistry. It is understandable that natural product pipelines tend to be focused on outcomes, notably the discovery for new and effective antibiotics. However, in the wider context the isolation, dereplication and screening of novel filamentous actinomycetes from poorly studied habitats are key elements in culture-dependent bioprospecting strategies (Goodfellow and Fiedler 2010, Goodfellow et al. 2018).

This survey of properties of acidophilic and acidotolerant filamentous actinomycetes provides evidence that the extension of the taxonomic approach to drug discovery to poorly studied acidic habitats will lead to the discovery of many drug leads, as found in drug discovery campaigns involving the isolation of actinomycetes from unexplored extreme biomes (Bull and Goodfellow 2019, Sayed et al. 2020). The continued application of this approach to coniferous forest soils should be especially effective as communities of actinomycetes in litter and mineral horizons are likely to show marked differences given different combinations of ecological drivers that operate within them. However, it is imperative to combine traditional cultivation approaches with culture-independent 'omics' technologies to realize the full extent of the diversity of acid-loving filamentous actinomycetes in acidic habitats. This dual approach to revealing the extent of actinomycete diversity is a fitting objective at a time when the capacity to acquire metagenomic data outstrips the ability to culture representative actinomycetes from extreme habitats (Idris et al. 2017, Bull and Goodfellow 2019). Further, the cultivation of representative elements of actinomycete dark matter form acidic habitats is essential for screening programmes and for investigating their ecological and physiological properties. Eco-physiological studies are an essential precursor for the objective selection of biocontrol agents against phytopathogenic fungi and for the development of inoculants to promote plant growth in acidic soils. These objectives represent the tip of the iceberg as more extensive work is needed to determine the genetic mechanisms that underlie the behaviour of acid-loving filamentous sporoactinomycetes in acidic habitats, the role that they play in natural recycling and how they interact with other elements of the indigenous microbiota. In short, progress has been made but much remains to be done.

Finally, it is our earnest hope that this article will encourage others to study the biology of a sadly neglected, but fascinating group of spore-forming actinomycetes and thereby help to map what remain poorly charted waters.

# References

Abdel-Mageed, W.M., Lehri, B., Jarmusch, S.A., Miranda, K., Al-Wahaibi, L.H., Stewart, H.A. et al. (2020). Whole genome sequencing of four bacterial strains from South Shetland Trench revealing biosynthetic and environmental adaptation gene clusters. *Marine Genomics*, 54: 100782.

Adamek, M., Alanjary, M., Sales-Ortells, H., Goodfellow, M., Bull, A.T., Winkler, A. et al. (2018). Comparative genomics reveals phylogenetic distribution patterns of secondary metabolites in *Amycolatopsis* species. *BMC Genomics*, 19: 426.

Alanjary, M., Kronmiller, B., Adamek, M., Blin, K., Weber, T., Huson, D. et al. (2017). The Antibiotic Resistant Target Seeker (ARTS), an exploration engine for antibiotic cluster prioritization and novel drug target discovery. *Nucleic Acids Research*, 45: W42–W48.

Andam, C.P., Doroghazi, J.R., Campbell, A.N., Kelly, P.J., Choudoir, M.J. and Buckley, D.H. (2016). A latitudinal diversity gradient in terrestrial bacteria of the genus *Streptomyces*. *mBio*, 7: e02200–15.

Asem, M.D., Shi, L., Jiao, J.Y., Wang, D., Han, M.X., Dong, L. et al. (2018). *Desertimonas flava* gen. nov., sp. nov. isolated from a desert soil, and proposal of *Ilumatobacteraceae* fam. nov. *International Journal of Systematic and Evolutionary Microbiology*, 68: 3593–3599.

Asnicar, F., Thomas, A.M., Beghini, F., Mengoni, C., Manara, S., Manghi, P. et al. (2020). Precise phylogenetic analysis of microbial isolates and genomes from metagenomes using PhyloPhlAn 3.0. *Nature Communications*, 11: 2500.

Backus, E.J., Tresner, H.D. and Campbell, T.H. (1957). The nucleocidin and alazopetin producing organisms: two new species of *Streptomyces*. *Antibiotics and Chemotherapy (Northfield III)*, 7: 532–541.

Baltz, R.H. (2017). Gifted microbes for genome mining and natural product discovery. *Journal of Industrial Microbiology and Biotechnology*, 44: 573–588.

Baltz, R.H. (2019). Natural product drug discovery in the genomic era: Realities, conjectures, misconceptions, and opportunities. *Journal of Industrial Microbiology and Biotechnology*, 46: 281–299.

Barco, R.A., Garrity, G.M., Scott, J.J., Amend, J.P., Nealson, K.H. and Emerson, D. (2020). A genus definition for *Bacteria* and *Archaea* based on a standard genome relatedness index. *mBio*, 11: e02475–19.

Barka, E.A., Vatsa, P., Sanchez, L., Gaveau-Vaillant, N., Jacquard, C., Meier-Kolthoff, J.P. et al. (2015). Taxonomy, physiology, and natural products of *Actinobacteria*. *Microbiology and Molecular Biology Reviews*, 80: 1–43.

Bull, A.T. (2011). Actinobacteria of the extremobiosphere. pp. 1204–1240. *In*: Horikoshi, K. (ed.). *Extremophiles Handbook*. Springer Verlag: Tokyo.

Bull, A.T. and Goodfellow, M. (2019). Dark, rare and inspirational microbial matter in the extremobiosphere: 16000 m of bioprospecting campaigns. *Microbiology (Reading)*, 165: 1252–1264.

Bull, A.T., Idris, H., Sanderson, R., Asenjo, J., Andrews, B. and Goodfellow, M. (2018). High altitude, hyper-arid soils of the Central-Andes harbor mega-diverse communities of actinobacteria. *Extremophiles*, 22: 47–57.

Buresova, A., Kopecky, J., Hrdinkova, V., Kamenik, Z., Omelka, M. and Sagova-Mareckova, M. (2019). Succession of microbial decomposers is determined by litter type, but site conditions drive decomposition rates. *Applied and Environmental Microbiology*, 85: e01760–19.

Busarakam, K., Bull, A.T., Girard, G., Labeda, D., van Wezel, G.P. and Goodfellow, M. (2014). *Streptomyces leeuwenhoekii* sp. nov., the producer of chaxalactins and chaxamycins, form a distinct branch in *Streptomyces* gene trees. *Antonie van Leeuwenhoek*, 105: 849–861.

Busti, E., Cavaletti, L., Monciardini, P., Schumann, P., Rohde, M., Sosio, M. et al. (2006). *Catenulispora acidiphila* gen. nov., sp. nov., a novel, mycelium-forming actinomycete, and proposal of *Catenulisporaceae* fam. nov. *International Journal of Systematic and Evolutionary Microbiology*, 56: 1741–1746.

Buxton, E.E., Khalifa, O. and Ward, V. (1965). Effect of soil amendment with chitin on pea wilt caused by *Fusarium oxysporum* f. *pisi*. *Annals of Applied Biology*, 55: 83–88.

Camacho, C., Coulouris, G., Avagyan, V., Ma, N., Papadopoulos, J., Bealer, K. et al. (2009). BLAST+: architecture and applications. *BMC Bioinformatics*, 10: 421.

Carro, L., Nouioui, I., Sangal, V., Meier-Kolthoff, J.P., Trujillio, M.E., Montero-Calasanz, M.C. et al. (2018). Genome-based classification of micromonosporae with a focus on their biotechnological and ecological potential. *Scientific Reports*, 8: 525.

Castro, J.F., Razmilic, V., Gomez-Escribano, J.P., Andrews, B., Asenjo, J.A. and Bibb, M.J. (2018). The 'gifted' actinomycete *Streptomyces leeuwenhoekii*. *Antonie van Leeuwenhoek*, 111: 1433–1448.

Cavaletti, L., Monciardini, P., Schumann, P., Rohde, M., Bamonte, R. Busti, E. et al. (2006). *Actinospica robiniae* gen. nov., sp. nov. and *Actinospica acidiphila* sp. nov.: proposal for *Actinospicaceae* fam. nov. and *Catenulisporinae* subord. nov. in the order *Actinomycetales*. *International Journal of Systematic and Evolutionary Microbiology*, 56: 1747–1753.

Chaiharn, M., Theantana, T. and Pathom-aree, W. (2020). Evaluation of biocontrol activities of *Streptomyces* spp. against rice blast disease fungi. *Pathogens (Basel, Switzerland)*, 9: 126.

Charlop-Powers, Z., Owen, J.G., Reddy, B.V., Ternei, M.A., Guimarães, D.O., de Frias, U.A. et al. (2015). Global biogeographic sampling of bacterial secondary metabolism. *eLife*, 4: e05048.

Cho, S.H., Han, J.H., Seong, C.N. and Kim, S.B. (2006). Phylogenetic diversity of the acidophilic sporoactinobacteria isolated from various soils. *Journal of Microbiology*, 44: 600–606.

Cho, S.H., Han, J.H., Ko, H.Y. and Kim, S.B. (2008). *Streptacidiphilus anmyonensis* sp. nov., *Streptacidiphilus rugosus* sp. nov. and *Streptacidiphilus melanogenes* sp. nov., acidophilic actinobacteria isolated from *Pinus* soils. *International Journal of Systematic and Evolutionary Microbiology*, 58: 1566–1570.

Chun, J., Kang, S.O., Hah, Y.C. and Goodfellow, M. (1996). Phylogeny of mycolic acid-containing actinomycetes. *Journal of Industrial Microbiology and Biotechnology*, 17: 205–213.

Chun, J., Oren, A., Ventosa, A., Christensen, H., Arahal, D.R., da Costa, M.S. et al. (2018). Proposed minimal standards for the use of genome data for the taxonomy of prokaryotes. *International Journal of Systematic and Evolutionary Microbiology*, 68: 461–466.

Clark, D.A. and Norris, P.R. (1996). *Acidimicrobium ferrooxidans* gen. nov., sp. nov.: mixed-culture ferrous iron oxidation with *Sulfobacillus* species. *Microbiology*, 141: 785–790.

Corke, C.T. and Chase, F.E. (1964). Comparative studies of actionomycete populations in acid podzolic and neutral mull forest soils. *Soil Science Society of America, Proceedings*, 28: 68–70.

Cui, Q., Wang, L., Huang, Y., Liu, Z. and Goodfellow, M. (2005). *Nocardia jiangxiensis* sp. nov. and *Nocardia miyunensis* sp. nov., isolated from acidic soils. *International Journal of Systematic and Evolutionary Microbiology*, 55: 1921–1925.

Davis-Belmar, C.S. and Norris, P.R. (2009). Ferrous iron and pyrite oxidation by "*Acidithiomicrobium*" species. *Advanced Materials Research*, 71-73: 271–274.

De Lima Procópio, R.E., Silva, I.R., Martins, M.K., Azevedo, J.L. and Araújo, J.M. (2012). Antibiotics produced by *Streptomyces*. *Brazilian Journal of Infectious Diseases*, 16: 466–741.

Dedysh, S.N., Pankratov, T.A., Belova, S.E., Kulichevskaya, I.S. and Liesack, W. (2006). Phylogenetic analysis and *in situ* identification of bacteria community composition in an acidic sphagnum peat bog. *Applied and Environmental Microbiology*, 72: 2110–2117.

Djinni, I., Defant, A., Kecha, M. and Mancini, I. (2019). Actinobacteria derived from Algerian ecosystems as a prominent source of antimicrobial molecules. *Antibiotics*, 8: 172.

Donadio, S., Cavaletti, L. and Monciardini, P. (2012). Order IV. *Catenulisporales* ord. nov. p. 225. *In*: Goodfellow, M., Kämpfer, P., Busse, H.J., Trujillo, M.E., Suzuki, K.I., Ludwig, W. et al. (eds.). *Bergey's Manual of Systematic Bacteriology*, Second Edition, Volume 5, Part A. Springer, New York.

Duggar, B. (1948). Aureomycin: A product of the continuing search for new antibiotics. *Annals of the New York Academy of Science*, 51: 177–181.

Flowers, S.T.H. and Williams, S.T. (1977). The influence of pH on the growth rate and viability of neutrophilic and acidophilic streptomycetes. *Microbios*, 18: 223–228.

Foesel, B.U., Geppert, A., Rohde, M. and Overmann, J. (2016). *Parviterribacter kavangonensis* gen. nov., sp. nov. and *Parviterribacter multiflagellatus* sp. nov., novel members of *Parviterribacteraceae* fam. nov. within the order *Solirubrobacterales*, and emended descriptions of the classes *Thermoleophilia* and *Rubrobacteria* and their orders and families. *International Journal of Systematic and Evolutionary Microbiology*, 66: 652–665.

Genilloud, O. (2014). The re-emerging role of microbial natural products in antibiotic discovery. *Antonie van Leeuwenhoek*, 106: 173–188.

Genilloud, O. (2017). Actinomycetes: Still a source of novel antibiotics. *Nat. Prod. Rep.*, 34: 1203–1232.

Girard, G., Traag, B.A., Sangal, V., Mascini, N., Hoskisson, P.A., Goodfellow, M. et al. (2013). A novel taxonomic marker that discriminates between morphologically complex actinomycetes. *Open Biology*, 3: 130073.

Girard, G., Willemse, J., Zhu, H., Claessen, D., Bukarasam, K., Goodfellow, M. et al. (2014). Analysis of novel kitasatosporae reveals significant evolutionary changes in conserved developmental genes between *Kitasatospora* and *Streptomyces*. *Antonie van Leeuwenhoek*, 106: 365–380.

Golińska, P., Ahmed, L., Wang, D. and Goodfellow, M. (2013a). *Streptacidiphilus durhamensis* sp. nov., isolated from a spruce forest soil. *Antonie van Leeuwenhoek*, 104: 199–206.

Golińska, P., Dahm, H. and Goodfellow, M. (2016). *Streptacidiphilus toruniensis* sp. nov., isolated from a pine forest soil. *Antonie van Leeuwenhoek*, 109: 1583–1591.

Golińska, P., Kim, B.Y., Dahm, H. and Goodfellow, M. 2013b. *Streptacidiphilus hamsterleyensis* sp. nov., isolated from a spruce forest soil. *Antonie van Leeuwenhoek*, 104: 965–972.

Golińska, P., Wang, D. and Goodfellow, M. (2013c). *Nocardia aciditolerans* sp. nov., isolated from a spruce forest soil. *Antonie van Leeuwenhoek*, 103: 1079–88.

Golińska, P., Zucchi, T.D., Silva, L., Dahm, H. and Goodfellow, M. (2015). *Actinospica durhamensis* sp. nov., isolated from a spruce forest soil. *Antonie van Leeuwenhoek*, 108: 435–442.

Gomez-Escribano, J.P., Castro, J.F., Razmilic, V., Chandra, G., Andrews, B., Asenjo, J.A. et al. (2015). The *Streptomyces leeuwenhoekii* genome: *de novo* sequencing and assembly in single contigs of the chromosome, circular plasmid pSLE1 and linear plasmid pSLE2. *BMC Genomics*, 16: 485.

Gómez-Silva, B. (2018). Lithobiontic life: "Atacama rocks are well and alive". *Antonie van Leeuwenhoek*, 111: 1333–1343.

González, D., Huber, K.J., Tindall, B., Hedrich, S., Rojas-Villalobos, C., Quatrini, R. et al. (2020). *Acidiferrimicrobium australe* gen. nov., sp. nov., an acidophilic and obligately heterotrophic, member of the *Actinobacteria* that catalyses dissimilatory oxido-reduction of iron isolated from metal-rich acidic water in Chile. *International Journal of Systematic and Evolutionary Microbiology*, 70: 3348–3354.

Goodfellow, M. (2010). Selective isolation of actinobacteria. pp. 13–27. *In*: Baltz, R.H., Davies, J.E., Demain, A.L., Bull, A.T., Junker, B., Katz, L. et al. (eds.). *Manual of Industrial Microbiology and Biotechnology*, American Society for Microbiology Press, Washington, USA.

Goodfellow, M. (2012a). Phylum XXVI. *Actinobacteria* phyl. nov., pp. 33–34. *In*: Goodfellow, M., Kämpfer, P., Busse, H.-J., Trujillo, M.E., Suzuki, K., Ludwig, W. et al. (eds.). *Bergey's Manual of Systematic Bacteriology*, Second Edition, Volume 5, The *Actinobacteria*, Part A. Springer, New York.

Goodfellow, M. (2012b). Class I. *Actinobacteria* Stackebrandt, Rainey and Ward-Rainey 1997, 483. pp. 34–35. *In*: Goodfellow, M., Kämpfer, P., Busse, H.-J., Trujillo, M.E., Suzuki, K., Ludwig, W. et al. (eds.). *Bergey's Manual of Systematic Bacteriology*, Second Edition, Volume 5, Part A. Springer, New York.

Goodfellow, M. and Dawson, D. (1978). Qualitative and quantitative studies of bacteria colonizing *Picea sitchensis* litter. *Soil Biology and Biochemistry*, 10: 303–307.

Goodfellow, M. and Fiedler, H.P. (2010). A guide to successful bioprospecting: Informed by actinobacterial systematics. *Antonie van Leeuwenhoek*, 98: 119–142.

Goodfellow, M. and Maldonado, L.A. (2012). Genus I. *Nocardia* Trevisan 1889. pp. 376–419. *In*: Goodfellow, M., Kämpfer, P., Busse, H.-J., Trujillo, M.E., Suzuki, K., Ludwig, W. et al. (eds.). *Bergey's Manual of Systematic Bacteriology*. Second Edition, Volume 5. The *Actinobacteria*, Part B. Springer, New York.

Goodfellow, M. and O'Donnell, A.G. (1989). Search and discovery of industrially significant actinomycetes. pp. 343–83. *In*: Baumberg, S., Hunter, I.S. and Rhodes, P.M. (eds.). *Microbial Products: New Approaches*. Cambridge University Press, Cambridge, UK. 1989.

Goodfellow, M. and Simpson, K.E. (1987). Ecology of streptomycetes. *Frontier in Applied Microbiology*, 2: 97–125.

Goodfellow, M. and Williams, S.T. (1983). Ecology of actinomycetes. *Annual Review of Microbiology*, 37: 189–216.

Goodfellow, M., Kämpfer, P., Busse, H.J., Trujillo, M.E., Suzuki, K.I., Ludwig, W. et al. (2012). Bergey's *Manual of Systematic Bacteriology, Second Edition*, Volume 5, The *Actinobacteria*, Parts A and B. Springer, New York.

Goodfellow, M., Nouioui, I., Sanderson, R., Xie, F. and Bull, A.T. (2018). Rare taxa and dark microbial matter: Novel bioactive actinobacteria abound in Atacama Desert soils. *Antonie van Leeuwenhoek*, 111: 1315–1332.

Groth, I., Schütze, B., Boettcher, T., Pullen, C.B., Rodriguez, C., Leistner, E. et al. (2003). *Kitasatospora putterlickiae* sp. nov., isolated from rhizosphere soil, transfer of *Streptomyces kifunensis* to the genus *Kitasatospora* as *Kitasatospora kifunensis* comb. nov., and emended description of *Streptomyces aureofaciens* Duggar 1948. *International Journal of Systematic and Evolutionary Microbiology*, 53: 2033–2040.

Groth, I., Rodriguez, C., Schutze, B., Schmitz, P., Leistner, E. and Goodfellow, M. (2004). Five novel *Kitasatospora* species from soil: *Kitasatospora arboriphila* sp. nov., *K. gansuensis* sp. nov., *K. nipponensis* sp. nov., *K. paranensis* sp. nov. and *K. terrestris* sp. nov. *International Journal of Systematic and Evolutionary Microbiology*, 2121–2129.

Guo, X., Liu, N., Li, X., Ding, Y., Shang, F., Gao, Y. et al. (2015). Red soils harbor diverse culturable actinomycetes that are promising sources of novel secondary metabolites. *Applied and Environmental Microbiology*, 81: 3086–3103.

Gupta, R.S., Chen, W.J., Adeolu, M. and Chai, Y. (2013). Molecular signatures for the class *Coriobacteriia* and its different clades; proposal for division of the class *Coriobacteriia* into the emended order *Coriobacteriales*, containing the emended family *Coriobacteriaceae* and *Atopobiaceae* fam. nov., and *Eggerthellales* ord. nov., containing the family *Eggerthellaceae* fam. nov. *International Journal of Systematic and Evolutionary Microbiology*, 63: 3379–3397.

Hagedorn, C. (1976). Influences of soil acidity on *Streptomyces* populations inhabiting forest soils. *Applied and Environmental Microbiology*, 32: 368–75.

Harrison, S.T.L. (2016). Biotechnologies that utilise acidophiles. pp. 265–284. *In*: Quatrini, R. and Johnson, D.B. (eds.). *Acidophiles: Life in Extremely Acidic Environments*. Caister Academic Press, Norfolk, UK.

Hayakawa, M. and Nonomura, H. (1987). Humic acid–vitamin agar, a new medium for the selective isolation of soil actinomycetes. *Journal of Fermentation Technology*, 65: 501–509.

Hayakawa, M., Momose, Y., Yamazaki, T. and Nonomura, H. (1996). A method for the selective isolation of *Microtetraspora glauca* and related four-spored actinomycetes from soil. *Journal of Applied Bacteriology*, 80: 375–386.

Hernandez, A., Nguyen, L.T., Dhakal, R. and Murphy, B.T. (2020). The need to innovate sample collection and library generation in microbial drug discovery: A focus on academia. *Natural Products Reports*. Doi: 10.1039/d0np00029a.

Holanda, R., Hedrich, S., Ñancucheo, I., Oliveira, G., Grail, B.M. and Johnson, D.B. (2016). Isolation and characterisation of mineral-oxidising "*Acidibacillus*" spp. from mine sites and geothermal environments in different global locations. *Research in Microbiology*, 167: 613–23.

Hopkins, D.W., Macnaughton, S.J. and O'Donnell, A.G. (1991). A dispersal and differential centrifugation technique for representatively sampling microorganisms from soil. *Soil Biology and Biochemistry*, 23: 217–225.

Hopwood, D.A. (2007). *Streptomyces in Nature and Medicine*. Oxford University Press.

Huang, Y., Cui, Q., Wang, L., Rodriguez, C., Quintana, E., Goodfellow, M. et al. (2004). *Streptacidiphilus jiangxiensis* sp. nov., a novel actinomycete isolated from acidic rhizosphere soil in China. *Antonie van Leeuwenhoek*, 86: 159–165.

Idris, H., Goodfellow, M., Sanderson, R., Asenjo, J.A. and Bull, A.T. (2017). Actinobacterial rare biospheres and dark matter revealed in habitats of the Chilean Atacama Desert. *Scientific Reports*, 7: 8373.

Itoh, T., Yamanoi, K., Kudo, T., Ohkuma, M. and Takashina, T. (2011). *Aciditerrimonas ferrireducens* gen. nov., sp. nov., an iron-reducing thermoacidophilic actinobacterium isolated from a solfataric field. *International Journal of Systematic and Evolutionary Microbiology*, 61: 1281–1285.

Jain, C., Rodriguez, R.L., Phillippy, A.M., Konstantinidis, K.T. and Aluru, S. (2018). High throughput ANI analysis of 90 K prokaryotic genomes reveals clear species boundaries. *Nature Communications*, 9: 5114.

Janssen, P.H. (2006). Identifying the dominant soil bacterial taxa in libraries of 16S rRNA and 16S rRNA Genes. *Applied and Environmental Microbiology*, 72: 1719–1728.

Jensen, H.L. (1928). *Actinomyces acidophilus* n. sp.-a group of acidophilus actinomycetes isolated from the soil. *Soil Science*, 25: 225–236.

Jensen, P.R. (2016). Natural products and the gene cluster revolution. *Trends Microbiology*, 24: 968–977.

Jiang, Y., Li, Q., Chen, X. and Jiang, C. (2016). Isolation and cultivation methods of actinobacteria.pp. 39–57. *In*: Dhanasekaran, D. and Jiang, Y. (eds.). *Actinobacteria - Basics and Biotechnological Applications*. Intech Open, Ltd, London, UK.

Jin, L., Huy, H., Kim, K.K., Lee, H.G., Kim, H.S., Ahn, C.Y. et al. (2013). *Aquihabitans daechungensis* gen. nov., sp. nov., an actinobacterium isolated from reservoir water. *International Journal of Systematic and Evolutionary Microbiology*, 63: 2970–2974.

Johnson, D.B. and Quatrini, R. (2016). Acidophile microbiology in space and time. pp. 3–16. *In*: Quatrini, R. and Johnson, B. (eds.). *Acidophiles: Life in Extremely Acidic Environments*. Caister Academic Press, Norfolk.

Johnson, D.B. and Quatrini, R. (2020). Acidophile microbiology in space and time. *Current Issues in Molecular Biology*, 39: 63–76.

Johnson, D.B., Bacelar-Nicolau, P., Okibe, N., Thomas, A. and Hallberg, K.B. (2009). *Ferrimicrobium acidiphilum* gen. nov., sp. nov. and *Ferrithrix thermotolerans* gen. nov., sp. nov.: heterotrophic, iron-oxidizing, extremely acidophilic actinobacteria. *International Journal of Systematic and Evolutionary Microbiology*, 59: 1082–1089.

Jones, R.M. and Johnson, D.B. (2015). *Acidithrix ferrooxidans* gen. nov., sp. nov.; a filamentous and obligately heterotrophic, acidophilic member of the Actinobacteria that catalyzes the dissimilatory oxido-reduction of iron. *Research in Microbiology*, 166: 111–120.

Kämpfer, P. (2012a). Genus *Streptomyces*. pp. 1455–1767. *In*: Goodfellow, M., Kämpfer, P., Busse, H.J., Trujillo, M.E., Suzuki, K.I., Ludwig, W. et al. (eds.). *Bergey's Manual of Systematic Bacteriology*, Second Edition, Volume 5, The *Actinobacteria*, Part B. Springer, New York.

Kämpfer, P. (2012b). Family I. *Streptomycetaceae* Waksman and Herrici 1943, 339[AL] emend. Rainey, Ward-Rainey and Stackebrandt 1997, 486 emend. Kim, Lonsdale, Seong and Goodfellow 2003b. 113 emend. Zhi, Li and Stackebrandt 2009, 600, pp. 1446–1454. *In*: Goodfellow, M., Kämpfer, P., Busse, H.J., Trujillo, M.E., Suzuki, K.I., Ludwig, W. et al. (eds.). *Bergey's Manual of Systematic Bacteriology*, Second Edition, Volume 5, The *Actinobacteria*, Part B, Springer, NewYork.

Kämpfer, P., Huber, B., Buczolits, S., Thummes, K., Grün-Wollny, I. and Busse, H.J. (2007). *Nocardia acidivorans* sp. nov., isolated from soil of the island of Stromboli. *International Journal of Systematic and Evolutionary Microbiology*, 57: 1183–1187.

Khan, M.R. and Williams, S.T. (1975). Studies on the ecology of actinomycetes in soil. VIII: Distribution and characteristics of acidophilic actinomycetes. *Soil Biology and Biochemistry*, 7: 345–348.

Kim, B.J., Kim, C.J., Chun, J., Koh, Y.H., Lee, S.H., Hyun, J.W. et al. (2004). Phylogenetic analysis of the genera *Streptomyces* and *Kitasatospora* based on partial RNA polymerase beta-subunit gene (rpoB) sequences. *International Journal of Systematic and Evolutionary Microbiology*, 54: 593–598.

Kim, B.Y., Zucchi, T.D., Fiedler, H.P. and Goodfellow, M. (2012). *Streptomyces cocklensis* sp. nov., a dioxamycin-producing actinomycete. *International Journal of Systematic and Evolutionary Microbiology*, 62: 279–283.

Kim, J.J., Marjerrison, C.E., Cornish Shartau, S.L., Brady, A.L., Sharp, C.E., Rijpstra, W.I.C. et al. (2017). *Actinocrinis puniceicyclus* gen. nov., sp. nov., an actinobacterium isolated from an acidic spring. *International Journal of Systematic and Evolutionary Microbiology*, 67: 602–609.

Kim, M.J., Roh, S.G., Kim, M.K., Park, C., Kim, S. and Kim, S.B. (2020). *Kitasatospora acidiphila* sp. nov., isolated from pine grove soil, exhibiting antimicrobial potential. *International Journal of Systematic and Evolutionary Microbiology*, 70: 5567–5575.

Kim, S.B., Lonsdale, J., Seong, C.N. and Goodfellow, M. (2003). *Streptacidiphilus* gen. nov., acidophilic actinomycetes with wall chemotype I and emendation of the family *Streptomycetaceae* (Waksman and Henrici (1943)[AL]) emend. Rainey et al. 1997. *Antonie Van Leeuwenhoek*, 83: 107–116.

Kim, S.B., Seong, C.N., Jeon, S.J., Bae, K.S. and Goodfellow, M. (2004). Taxonomic study of neutrotolerant acidophilic actinomycetes isolated from soil and description of *Streptomyces yeochonensis* sp. nov. *International Journal of Systematic and Evolutionary Microbiology*, 54: 211–214.

Kim, Y.J., Moon, M.H., Song, J.Y., Smith, C.P., Hong, S.K. and Chang, Y.K. (2008). Acidic pH shock induces the expression of a wide range of stress-response genes. *BMC Genomics*, 9: 604.

Kim, Y.R., Park, S., Kim, T.S., Kim, M.K., Han, J.H., Joung, Y. et al. (2015). Draft genome sequence of *Streptacidiphilus oryzae* TH49[T], an acidophilic actinobacterium isolated from soil. *Genome Announcements*, 3: e00703-15.

Kurahashi, M., Fukunaga, Y., Sakiyama, Y., Harayama, S. and Yokota, A. (2009). *Iamia majanohamensis* gen. nov., sp. nov., an actinobacterium isolated from sea cucumber *Holothuria edulis*, and proposal of *Iamiaceae* fam. nov. *IInternational Journal of Systematic and Evolutionary Microbiology*, 59: 869–873.

Kurtböke, D.İ. (2016). Actinomycetes in biodiscovery: Genomic advances and new horizons. Chapter 35, pp. 567–590. *In*: Gupta, V.K., Sharma, G.D., Tuohy, M.G. and Gaur, R. (eds.). *The Handbook of Microbial Resources*. CAB International Publications, Oxfordshire, UK.

Kurtböke, D.İ. (2017a). Ecology and habitat distribution of actinobacteria. pp. 123–149. *In*: Wink, J., Mohammadipanah, F. and Hamedi, J. (eds.). *Ecology and Habitat Distribution of Actinobacteria*. Biology and Biotechnology of Actinobacteria. Springer, Cham.

Kurtböke, D.İ. (2017b). Bioactive actinomycetes: Reaching rarity through sound understanding of selective culture and molecular diversity. *In*: Kurtböke, D.I. (ed.). *Microbial Resources: From Functional Existence in Nature to Applications*. Elsevier, Academic Press, ISBN: 9780128047651.

Kusuma, A.B., Nouioui, I., Klenk, H.-P. and Goodfellow, M. (2020). *Streptomyces harenosi* sp. nov., a home for a gifted strain isolated from Indonesian sand dune soil. *International Journal of Systematic and Evolutionary Microbiology*, 70: 4874–4882.

Küster, E. and Williams, S.T. (1964). Selection of media for isolation of streptomycetes. *Nature*, 202: 928–929.

Labeda, D.P., Goodfellow, M., Brown, R., Ward, A.C., Lanoot, B., Vancanneyt, M. et al. (2012). Phylogenetic study of the species within the family *Streptomycetaceae*. *Antonie van Leeuwenhoek*, 101: 73–104.

Labeda, D.P., Dunlap, C.A., Rong, X., Huang, Y., Doroghazi, J.R., Ju, K.S. et al. (2017). Phylogenetic relationships in the family *Streptomycetaceae* using multi-locus sequence analysis. *Antonie van Leeuwenhoek*, 110: 563–583.

Lee, H.J. and Whang, K.S. (2016). *Catenulispora fulva* sp. nov., isolated from forest soil. *International Journal of Systematic and Evolutionary Microbiology*, 66: 271–275.

Lee, H.J., Han, S.I. and Whang, K.S. (2012). *Catenulispora graminis* sp. nov., a rhizobacterium from bamboo (*Phyllostachys nigro* var. *henonis*) rhizosphere soil. *International Journal of Systematic and Evolutionary Microbiology*, 62: 2589–2592.

Li, X., Huang, Y. and Whitman, W.B. (2015). The relationship of the whole genome sequence identity to DNA hybridization varies between genera of prokaryotes. *Antonie van Leeuwenhoek*, 107: 241–249.

Liu, Z., Rodriguez, C., Wang, L., Cui, Q., Huang, Y., Quintana, E.T. et al. (2005). *Kitasatospora viridis* sp. nov., a novel actinomycete from soil. *International Journal of Systematic and Evolutionary Microbiology*, 55: 707–711.

Lonsdale, J.T. (1985). *Aspects of the Biology of Acidophilic Actinomycetes*. Ph.D. thesis, University of Newcastle, Newcastle upon Tyne, UK.

Ludwig, W., Euzéby, J., Schumann, P., Busse, H.-J., Trujillo, M.E., Kämpfer, P. et al. (2012a). Road map of the phylum *Actinobacteria*. pp. 1–28. *In*: Goodfellow, M., Kämpfer, P., Busse, H.J., Trujillo, M.E., Suzuki, K.I., Ludwig, W. et al. (eds.). *Bergey's Manual of Systematic Bacteriology*, Second Edition, Volume 5, Part A, Springer, New York.

Ludwig, W., Euzéby, J. and Whitman, W.B. (2012b). Class IV. Nitriliruptoria class nov., pp. 2000–2001. *In*: Goodfellow, M., Kämpfer, P., Busse, H.J., Trujillo, M.E., Suzuki, K.I., Ludwig, W. et al. (eds.). *Bergey's Manual of Systematic Bacteriology*, Second Edition, Volume 5, The Actinobacteria, Part B. Springer, New York.

Lyu, A., Liu, H., Che, H.J., Yang, L., Zhang, J., Wu, M.D. et al. (2017). Reveromycins A and B from *Streptomyces* sp. 3-10: Antifungal activity against plant pathogenic fungi *in vitro* and in a strawberry food model system. *Frontiers in Microbiology*, 8: 550.

Martinet, L., Naômé, A., Baiwir, D., De Pauw, E., Mazzucchelli, G. and Rigali, S. (2020). On the risks of phylogeny-based strain prioritization for drug discovery: *Streptomyces lunaelactis* as a case study. *Biomolecules*, 10: 1027.

Matsumoto, A., Kasai, H., Matsuo, Y., Omura, S., Shizuri, Y. and Takahashi, Y. (2009). *Ilumatobacter fluminis* gen. nov., sp. nov., a novel actinobacterium isolated from the sediment of an estuary. *Journal of General and Applied Microbiology*, 55: 201–205.

Méndez-García, C., Peláez, A.I., Mesa, V., Sánchez, J., Golyshina, O.V. and Ferrer, M. (2015). Microbial diversity and metabolic networks in acid mine drainage habitats. *Frontiers in Microbiology*, 6: 475.

Mohagheghi, A., Grohmann, K., Himmel, M., Leighton, L. and Updegraff, D.M. (1986). Isolation and characterization of *Acidothermus cellulolyticus* gen. nov., sp. nov., a new genus of thermophilic, acidophilic, cellulolytic bacteria. *International Journal of Systematic Bacteriology*, 36: 435–443.

Moose, A. (2017). POCP calculation for two genomes. (FigShare).

Mungan, M.D., Alanjary, M., Blin, K., Weber, T., Medema, M.H. and Ziemert, N. (2020). ARTS 2.0: Feature updates and expansion of the Antibiotic Resistant Target Seeker for comparative genome mining. *Nucleic Acids Research*, 48: W546–W552.

Nagai, A., Khan, S.T., Tamura, T., Takagi, M. and Shinya, T. (2011). *Streptomyces aomiensis* sp. nov., isolated from a soil sample using the membrane-filter method. *International Journal of Systematic and Evolutionary Microbiology*, 61: 947–950.

Nakamura, G. (1961). Studies on antibiotic actinomycetes. I. On *Streptomyces* producing a new antibiotic tubermycin. *Journal of Antibiotics (Tokyo) Series A*, 14: 86–89.

Nawani, N.N., Kapadnis, B.P., Das, A.D., Rao, A.S. and Mahajan, S.K. (2002). Purification and characterization of a thermophilic and acidophilic chitinase from Microbispora sp. V2. *Journal of Applied Microbiology*, 93: 965–975.

Neuman, D.J. and Cragg, G.M. (2020). Natural products as sources of new drugs over the nearly four decades from 01/1981 to 09/2019. *Journal of Natural Products*, 83: 770–803.

Nicault, M., Tidjani, A.-R., Gauthier, A., Dumarcay, S., Gelhaye, E., Bontemps, C. et al. (2020). Mining the biosynthetic potential for specialized metabolism of a *Streptomyces* soil community. *Antibiotics*, 9: 271.

Niyasom, C., Boonmak, S. and Meesri, N. (2015). Antimicrobial activity of acidophilic actinomycetes isolated from acidic soil. *KMITL Science and Technology*, 15: 62–69.

Norris, P.R. (2012). Class II. *Acidimicrobiia* class. nov. p. 1968. *In*: Goodfellow, M., Kämpfer, P., Busse, H.J., Trujillo, M.E., Suzuki, K.I., Ludwig, W. et al. (eds.). *Bergey's Manual of Systematic Bacteriology*, Second Edition, Volume 5, The *Actinobacteria*, Part B. Springer, New York.

Nouioui, I., Carro, L., Garcia-Lopez, M., Meier-Kolthoff, J.P., Woyke, T., Kyrpides, N.C. et al. (2018). Genome-based taxonomic classification of the phylum *Actinobacteria*. *Frontiers in Microbiology*, 9: 2007.

Nouioui, I., Cortés-albayay, C., Carro, L., Castro, J.F., Gtari, M., Ghodhbane-Gtari, F. et al. (2019). Genomic insights into plant-growth-promoting potentialities of the genus *Frankia*. *Frontiers in Microbiology*, 10: 1457.

Nweze, J.A., Mbaoji, F.N., Huang, G., Li, Y., Yang, L., Zhang, Y. et al. (2020). Antibiotics development and the potentials of marine-derived compounds to stem the tide of multidrug-resistant pathogenic bacteria, fungi, and protozoa. *Marine Drugs*, 18: 145.

Ōmura, S., Takahashi, Y., Iwai, Y. and Tanaka, H. (1982). *Kitasatosporia*, a new genus of the order *Actinomycetales*. *Journal of Antibiotics (Tokyo)*, 35: 1013–1019.

Oren, A. and Garrity, G.M. (2021). Valid publication of the names of forty-two phyla of prokaryotes. *International Journal of Systematic and Evolutionary Microbiology*, 71: 005056.

Oren, A., Arahal, D.R., Rosselló-Móra, R., Sutcliffe, I.C. and Moore, E.R.B. (2021). Emendation of rules 5b, 8, 15 and 22 of the International Code of Nomenclature of Prokaryotes to include the rank of phylum. *International Journal of Systematic and Evolutionary Microbiology*, 71: 004851.

Oyuntsetseg, B., Cho, S.H., Jeon, S.J., Lee, H.B., Shin, K.S., Kim, I.S. et al. (2017). *Amycolatopsis acidiphila* sp. nov., a moderately acidophilic species isolated from coal mine soil. *International Journal of Systematic and Evolutionary Microbiology*, 67: 3387–3392.

Pantiukh, K. and Grouzdev, D. (2017). POCP-matrix calculation for a number of genomes. (FigShare).

Parker, C.T., Tindall, B.J. and Garrity, G.M. (2019). International Code of Nomenclature of Prokaryotes. *International Journal of Systematic and Evolutionary Microbiology*, 69: S1–S111.

Peczyńska-Czoch, W. and Mordarski, M. (1988). Actinomycete enzymes. pp. 220-283. *In*: Goodfellow, M., Williams, S.T. and Mordarski, M. (eds.). *Actinomycetes in Biotechnology*, Pergamon Press, Oxford, UK.

Ping, X., Takahashi, Y., Seino, A., Iwai, Y. and Omura, S. (2004). *Streptomyces scabrisporus* sp. nov. *International Journal of Systematic and Evolutionary Microbiology*, 54: 577–581.

Poomthongdee, N., Duangmal, K. and Pathom-aree, W. (2015). Acidophilic actinomycetes from rhizosphere soil: Diversity and properties beneficial to plants. *Journal of Antibiotics*, 68: 106–114.

Qin, Q.L., Xie, B.B., Zhang, X.Y., Chen, X.L., Zhou, B.C., Zhou, J. et al. (2014). A proposed genus boundary for the prokaryotes based on genomic insights. *Journal of Bacteriology*, 196: 2210–2215.

Quatrini, R. and Johnson, D.B. (2016). *Acidophiles: Life in Extremely Acidic Environments*. Caister Academic Press, UK.

Rateb, M.E., Ebel, R. and Jaspars, M. (2018). Natural product diversity of actinobacteria in the Atacama Desert. *Antonie van Leeuwenhoek*, 111: 1467–1477.

Rawat, N. and Joshi, G.K. (2019). Bacterial community structure analysis of a hot spring soil by next generation sequencing of ribosomal RNA. *Genomics*, 111: 1053–1058.

Riesco, R., Carro, L., Román-Ponce, B., Prieto, C., Blom, J., Klenk, H.P. et al. (2018). Defining the species *Micromonospora saelicesensis* and *Micromonospora noduli* under the framework of genomics. *Frontiers in Microbiology*, 9: 1360.

Rivera-Araya, J., Pollender, A., Huynh, D., Schlomann, M., Chavez, R. and Levican, G. (2019). Osmotic imbalance, cytoplasm acidification and oxidative stress induction support the high toxicity of chloride in acidophilic bacteria. *Frontiers in Microbiology*, 10: 2455.

Roh, S.G., Kim, M.K., Park, S., Yun, B.R., Park, J., Kim, M.J. et al. (2018). *Streptacidiphilus pinicola* sp. nov., isolated from pine grove soil. *International Journal of Systematic and Evolutionary Microbiology*, 68: 3149–3155.

Sant'Anna, F.H., Bach, E., Porto, R.Z., Guella, F., Sant'Anna, H.E. and Passaglia, L.M.P. (2019). Genomic metrics made easy: What to do and where to go in the new era of bacterial taxonomy. *Critical Reviews in Microbiology*, 45: 182–200.

Salam, N., Jiao, J.Y., Zhang, X.T. and Li, W.J. (2020). Update on the classification of higher ranks in the phylum *Actinobacteria*. *International Journal of Systematic and Evolutionary Microbiology*, 70: 1331–1355.

Sangal, V., Goodfellow, M., Blom, J., Tan, G.Y.A., Klenk, H.-P. and Sutcliffe, I.C. (2018). Revisiting the taxonomic status of the biomedically and industrially important genus *Amycolatopsis*, using a phylogenomic approach. *Frontiers in Microbiology*, 9: 2281.

Sangal, V., Goodfellow, M., Jones, A.L., Sevoir, R.J. and Sutcliffe, I.C. (2019). Refined systematics of the genus *Rhodococcus* based on whole genome analyses. pp. 22–31. *In*: Alvarez, H. (ed.). *Biology of Rhodococcus*. Second edition, Springer, Cham.

Sayed, A.M., Hassan, M.H.A., Alhadrami, H.A., Hassan, H.M., Goodfellow, M. and Rateb, M.E. (2020). Extreme environments: Microbiology leading to specialized metabolites. *Journal of Applied Microbiology*, 128: 630–657.

Schlatter, D.C. and Kinkel, L.L. (2014). Global biogeography of *Streptomyces* antibiotic inhibition, resistance, and resource use. *FEMS Microbiology and Ecology*, 88: 386–397.

Seemann, T. (2014). Prokka: Rapid prokaryotic genome annotation. *Bioinformatics*, 30: 2068–2069.

Sembiring, L., Ward, A.C. and Goodfellow, M. (2000). Selective isolation and characterisation of members of the *Streptomyces violaceusniger* clade associated with the roots of *Paraserianthes falcataria*. *Antonie Van Leeuwenhoek*, 78: 353–366.

Sen, A., Daubin, V., Abrouk, D., Gifford, I., Berry, A.M. and Normand, P. (2014). Phylogeny of the class *Actinobacteria* revisited in the light of complete genomes. The orders '*Frankiales*' and *Micrococcales* should be split into coherent entities: proposal of *Frankiales* ord. nov., *Geodermatophilales* ord. nov., *Acidothermales* ord. nov. and *Nakamurellales* ord. nov. *International Journal of Systematic and Evolutionary Microbiology*, 64: 3821–3832.

Seong, C.N. (1992). *Numerical Taxonomy of Acidophilic and Neutrotolerant Actinomycetes Isolated from Acid Soil in Korea*. PhD thesis. Seoul National University, Seoul, Republic of Korea.

Seong, C.N., Goodfellow, M., Ward, A.C. and Hah, Y.C. (1993). Numerical classification of acidophilic actinomycetes isolated from acid soil in Korea. *Korean Journal of Microbiology*, 31: 355–363.

Seong, C.N., Park, S.K., Goodfellow, M., Kim, S.B. and Hah, Y.C. (1995). Construction of probability identification matrix and selective medium for acidophilic actinomycetes using numerical classification data. *Journal of Microbiology*, 33: 95–102.

Song, W., Duan, L., Jin, L., Zhao, J., Jiang, S., Sun, T. et al. (2018). *Streptacidiphilus monticola* sp. nov., a novel actinomycete isolated from soil. *International Journal of Systematic and Evolutionary Microbiology*, 68: 1757–1761.

Srividya, S., Thapa, A., Bhat, D.V., Golmei, K. and Dey, N. (2012). *Streptomyces* sp. 9p as effective biocontrol against chilli soilborne fungal phytopathogens. *European Journal of Experimental Biology*, 2: 163–173.

Stackebrandt, E., Rainey, F.A. and Ward-Rainey, N.L. (1997). Proposal for a new hierarchic classification system, *Actinobacteria* classis nov. *International Journal of Systematic Bacteriology*, 47: 479–491.

Subramani, R. and Sipkema, D. (2019). Marine rare actinomycetes: A promising source of structurally diverse and unique novel natural products. *Marine Drugs*, 26: 17–249.

Suzuki, K.-I. (2012). Class V. *Rubrobacteria* class. nov. pp. 2004–2005. *In*: Goodfellow, M., Kämpfer, P., Busse, H.-J., Trujillo, M.E., Suzuki, K.-I., Ludwig, W. et al. (eds.). *Bergey's Manual of Systematic Bacteriology*, Second Edition, Volume 5, The *Actinobacteria*, Part B, Springer, New York.

Suzuki, K.-I. and Whitman, W.B. (2012). Class VI. *Thermoleophilia* class. nov. p. 2010. *In*: Goodfellow, M., Kämpfer, P., Busse, H.-J., Trujillo, M.E., Suzuki, K.-I., Ludwig, W. et al. (eds.). *Bergey's Manual* of *Systematic Bacteriology*, Second Edition, Volume 5, The *Actinobacteria*, Part B, Springer, New York.

Świecimska, M., Golińska, P., Wypij, M. and Goodfellow, M. (2021a). *Catenulispora pinistramenti* sp. nov., novel actinobacteria isolated from pine forest soil in Poland. *International Journal of Systematic and Evolutionary Microbiology*, 71: 5063.

Świecimska, M., Golińska, P., Wypij, M. and Goodfellow, M. (2021b). Genomic-based classification of *Catenulispora pinisilvae* sp. nov., novel actinobacteria isolated from a pine forest soil in Poland and emended description of *Catenulispora rubra*. *Systematic and Applied Microbiology*, 44: 126164.

Tacconelli, E., Carrara, E., Savoldi, A., Harbarth, S., Mendelson, M. Monnet, D.L. et al. (2018). WHO Pathogens Priority List Working Group. Discovery, research, and development of new antibiotics: The WHO priority list of antibiotic-resistant bacteria and tuberculosis. *Lancet Infectious Diseases*, 18: 318–327.

Tajima, K., Takahashi, Y., Seino, A., Iwai, Y. and Ōmura, S. (2001). Description of two novel species of the genus *Kitasatospora* Ōmura et al. 1982, *Kitasatospora cineracea* sp. nov. and *Kitasatospora niigatensis* sp. nov. *International Journal of Systematic and Evolutionary Microbiology*, 51: 1765–1771.

Takahashi, Y. (2017). Continuing fascination of exploration in natural substances from microorganisms. *Bioscience, Biotechnology and Biochemistry*, 81: 6–12.

Takahashi, Y. and Nakashima, T. (2018). Actinomycetes, an inexhaustible source of naturally occurring antibiotics. *Antibiotics (Basel)*, 7: 45.

Tamreihao, K., Ningthoujam, D.S., Nimaichand, S., Singh, E.S., Reena, P., Singh, S.H. et al. (2016). Biocontrol and plant growth promoting activities of a *Streptomyces corchorusii* strain UCR3-16 and preparation of powder formulation for application as biofertilizer agents for rice plant. *Microbiology Research*, 192: 260–270.

Tamreihao, K., Nimaichand, S. and Ningthoujam, D.S. (2018a). Use of acidophilic or acidotolerant actinobacteria for sustainable agricultural production in acidic soils. pp. 453–464. *In*: Egamberdieva, D., Birkeland, N.K., Panosyan, H. and Li, W.J. (eds.). *Extremophiles in Eurasian Ecosystems: Ecology, Diversity, and Applications. Microorganisms for Sustainability*. Volume 8. Springer, Singapore.

Tamreihao, K., Nimaichand, S., Chanu, S.B., Devi, K.A., Lynda, R., Jeeniita, N. et al. (2018b). Acidotolerant *Streptomyces* sp. MBRL 10 from limestone quarry site showing antagonism against fungal pathogens and growth promotion in rice plants. *Journal of King Saud University – Science*, 30: 143–152.

Tamura, T., Ishida, Y., Sakane, T. and Suzuki, K. (2007). *Catenulispora rubra* sp. nov., an acidophilic actinomycete isolated from forest soil. *International Journal of Systematic and Evolutionary Microbiology*, 57: 2272–2274.

Tamura, T., Ishida, Y., Otoguro, M. and Suzuki, K. (2008). Catenulispora subtropica sp. nov. and Catenulispora yoronensis sp. nov. *International Journal of Systematic and Evolutionary Microbiology*, 58: 1552–1555.

Teo, W.F.A., Srisuk, N. and Duangmal, K. (2020). *Amycolatopsis acidicola* sp. nov., isolated from peat swamp forest soil. *International Journal of Systematic and Evolutionary Microbiology*, 70: 1547–1554.

Theuretzbacher, U., Gottwalt, S., Beyer, P., Butler, M., Czaplewski, L., Lienhardt, C. et al. (2019). Analysis of the clinical antibacterial and antituberculosis pipeline. *Lancet Infectious Diseases*, 19: e40–e50.

Thompson, D., Cognat, V., Goodfellow, M., Koechler, S., Heintz, D., Carapito, C. et al. (2020). Phylogenomic classification and biosynthetic potential of the fossil fuel-biodesulfurizing *Rhodococcus* strain IGTS8. *Frontiers in Microbiology*, 11: 1417.

Tsukiura, H., Okanishi, M., Koshiyama, H., Ohmori, T., Miyaki, T. and Kawaguchi, H. (1964). Proceomycin, a new antibiotic. *Journal of Antibiotics (Tokyo) Series A*, 17: 223–229.

Vickers, J.C. and Williams, S.T. (1987). An assessment of plate inoculation procedures for the enumeration and isolation of soil streptomycetes. *Microbios Letters*, 35: 113–117.

Vruggink, H. (1976). Influence of agricultural crops on the actinomycetes flora in soil. *Plant and Soil*, 44: 639–654.

Waksman, S.A. and Henrici, A.T. (1943). The nomenclature and classification of the actinomycetes. *Journal of Bacteriology*, 46: 337–341.

Wang, H. and van der Donk, W.A. (2012). Biosynthesis of the class III lantipeptide catenulipeptin. *ACS Chemical Biology*, 7: 1529–1535.

Wang, L., Huang, Y., Cui, Q., Xie, Q., Zhang, Y. and Liu, Z. (2003). Isolation of acidophilic and acidoduric streptomycetes using a dispersion and differential centrifugation approach. *Microbiology (English traanslation of Microbiologia)*, 30: 104–106.

Wang, L., Huang, Y., Liu, Z., Goodfellow, M. and Rodriguez, C. (2006). *Streptacidiphilus oryzae* sp. nov., an actinomycete isolated from rice-field soil in Thailand. *International Journal of Systematic and Evolutionary Microbiology*, 56: 1257–1261.

Wellington, E.M.H., Stackebrandt, E., Sanders, D., Wolstrup, J. and Jorgensen, N.O.G. (1992). Taxonomic status of *Kitasatosporia*, and proposed unification with *Streptomyces* on the basis of phenotypic and 16S rRNA analysis and emendation of *Streptomyces* Waksman and Henrici 1943, 339AL. *International Journal of Systematic Bacteriology*, 42: 156–160.

Whitman, W.B., Oren, A., Chuvochina, M., da Costa, M.S., Garrity, G.M., Rainey, F.A. et al. (2018). Proposal of the suffix -ota to denote phyla. Addendum to Proposal to include the rank of phylum in the International Code of Nomenclature of Prokaryotes. *International Journal of Systematic and Evolutionary Microbiology*, 68: 967–969.

Williams, S.T. and Davies, F.L. (1965). Use of antibiotics for selective isolation and enumeration of actinomycetes in soil. *Journal of General Microbiology*, 38: 251–261.

Williams, S.T. and Flowers, T.H. (1978). The influence of pH on starch hydrolysis by neutrophilic and acidophilic actinomycetes. *Microbios*, 20: 99–106.

Williams, S.T. and Khan, M.R. (1974). Antibiotics—A soil microbiologist's view-point. *Postepy Higieny i Medycyny Doswiadczalnej*, 28: 395–408.

Williams, S.T. and Mayfield, C.I. (1971). Studies on the ecology of actinomycetes in soil. III. The behaviour of neutrophilic streptomycetes in soil. *Soil Biology and Biochemistry*, 3: 197–208.

Williams, S.T. and Robinson, C.S. (1981). The role of streptomycetes in decomposition of chitin in acidic soils. *Journal of General Microbiology*, 127: 55–63.

Williams, S.T., Davies, F.L., Mayfield, C.I. and Khan, M.R. (1971). Studies on the ecology of actinomycetes in soil. II. The pH requirements of streptomycetes in acid soils. *Soil Biology and Biochemistry*, 3: 187–195.

Williams, S.T., Lanning, S. and Wellington, E.M.H. (1984). Ecology of actinomycetes. pp. 481–528. *In*: Goodfellow, M., Mordarski, M. and Williams, S.T. (eds.). *The Biology of the Actinomycetes*. Academic Press, London.

Williams, S.T., Shameemullah, M., Watson, E.T. and Mayfield, C.I. (1972). Studies on the ecology of actinomycetes in soil—VI. The influence of moisture tension on growth and survival. *Soil Biology and Biochemistry*, 4: 215–225.

Xu, C., Wang, L., Cui, Q., Huang, Y., Liu, Z., Zheng, G. et al. (2006). Neutrotolerant acidophilic *Streptomyces* species isolated from acidic soils in China: *Streptomyces guanduensis* sp. nov., *Streptomyces paucisporeus* sp. nov., *Streptomyces rubidus* sp. nov. and *Streptomyces yanglinensis* sp. nov. *International Journal of Systematic and Evolutionary Microbiology*, 56: 1109–1115.

Zakalyukina, Y.V. and Zenova, G.M. (2007). Antagonistic activity of soil acidophilic actinomycetes. *Biology Bulletin of Russian Academy of Science*, 34: 329–332.

Zakalyukina, Y.V., Zenova, G.M. and Zvyagintsev, D.G. (2002). Acidophilic soil actinomycetes. *Microbiology*, 71: 342–345.

Zettler, J., Xia, H., Burkard, N., Kulik, A., Grond, S., Heide, L. et al. (2014). New aminocoumarins from the rare actinomycete *Catenulispora acidiphila* DSM 44928: Identification, structure elucidation, and heterologous production. *Chembiochem*, 15: 612–621.

Zhang, Z., Wang, Y. and Ruan, J. (1997). A proposal to revive the genus *Kitasatospora* (Omura, Takahashi, Iwai, and Tanaka 1982). *International Journal of Systematic Bacteriology*, 47: 1048–1054.

Zhi, X.Y., Li, W.J. and Stackebrandt, E. (2009). An update of the structure and 16S rRNA gene sequence-based definition of higher ranks of the class *Actinobacteria*, with the proposal of two new suborders and four new families and emended descriptions of the existing higher taxa. *International Journal of Systematic and Evolutionary Microbiology*, 59: 589–608.

# Chapter 6

# A Retrospect on Actinomycete Diversity, Novelty and Secondary Metabolites Isolated from Deserts of China

*Cheng-hang Sun,\* Shao-wei Liu, Fei-na Li, Zhong-ke Jiang, Ting Wang* and *Qin-pei Lu*

## 1. Introduction

Deserts are characterized by various harsh environmental conditions, "mainly soil deficiency in water and nutrients, high salinity and pH, low precipitation, high temperatures and UV irradiation", as stated by Soussi et al. (Soussi et al. 2016). Since the beginning of the 20th century, very little research on the examination of microbial characteristics of deserts has been carried out, because deserts as one of the extreme environments are considered to be abiotic and sterile.

In 1912, Lipman (Lipman 1912, Whitford 2002) explored the microbiological characteristics of soil samples collected from deserts in California and first demonstrated that desert soils were inhabited by ammonifying organisms, nitrifying and nitrogen-fixing organisms (*Azotobacter* organisms). Actinomycetes are keystone species in terrestrial ecosystems and are acknowledged as a source of bioactive compounds (Idris et al. 2017). To the best of our knowledge, the first report on bioprospecting for actinomycetes and their bioactive compounds in the desert appears to be in the Whaley's doctoral dissertation (Whaley 1964) followed by a subsequent paper by Whaley and Boyl (1967) on the isolation and identification of antagonistic actinomycetes from desert plant rhizosphere in Arizona, the United States to discover antibiotics against soil-borne plant pathogens during the Golden Age of Antibiotics Discovery (Walsh and Wencewicz 2013).

Department of Microbial Chemistry and Beijing Key Laboratory of Antimicrobial Agents, Institute of Medicinal Biotechnology, Chinese Academy of Medical Sciences & Peking Union Medical College, Beijing 100050, China.
* Corresponding author: sunchenghang@imb.pumc.edu.cn

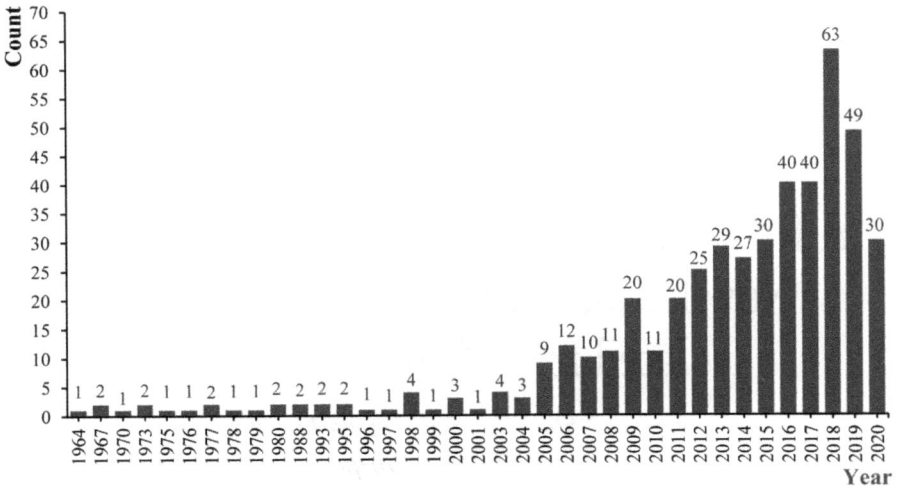

**Figure 1.** Number of papers retrieved in PubMed by searching actinomycete and related words and Desert.

By searching PubMed (https://pubmed.ncbi.nlm.nih.gov/) using the keywords *Actinomycetia* (or *actinomycetes* or *actinobacteria*) and desert (or deserts), the number of papers retrieved (Fig. 1) indicated that until the end of the 1990s, only sporadic investigations on examination of actinomycetes in different deserts were reported by researchers in countries with vast areas of deserts, such as Egypt (Taha et al. 1973, Zayed et al. 1973), Kuwait (Diab and Al Zaidan 1976), the former Union of Soviet Socialist Republics (Zenova et al. 1996), Algeria (Sabaou et al. 1992), the United States (Garrity et al. 1996), and Australia (Holmes et al. 2000), etc. Although the search results cannot represent the whole picture of research data on desert derived actinomycetes in detail, to some extent, they can still reflect the trends and characteristics in different periods. Desert-derived actinomycetes has received much more attention in the past two decades though.

In the 21st century, to cope with two worldwide crises, desertification (Le Houérou 1996) and antimicrobial resistance, resulting in the loss of massive croplands (Emadodin et al. 2019) and generating 'superbugs' (Infectious Diseases Society of America 2010, Bassetti et al. 2013), microorganism surveys in deserts have been boosted. On the one hand, to clarify microorganisms and their function in the form of biological soil crusts, which plays a vitally important role in maintaining soil stability and halting soil erosion, research on biodiversity of microorganisms including actinomycetes from the biotope of deserts has increased quickly. In parallel, to discover new therapeutic drugs, especially new antibiotics against drug-resistant pathogens, research on bioprospecting for untapped actinomycete strains from extreme environments has been steadily gathering appreciation. In exploring actinomycetes and its secondary metabolites as potential drugs from deserts, especially the Atacama Desert (Goodfellow et al. 2018, Rateb et al. 2018) and Sahara Desert (Djinni et al. 2019), much progress has been achieved in the past two decades, which has encouraged researchers worldwide to endeavor more efforts to explore actinomycetes and their bioactive compounds.

In terms of desert area, China is the third largest country, and deserts in China occur over a wide range of elevations and tectonic settings, spanning many ecosystems and climatic zones, and therefore leading to rich actinomycete diversity, novelty and abundant chemical diversity. This is the first review to fully summarize the diversity and novelty of actinomycetes as well as nearly all known and new secondary metabolites produced by desert-derived ones from Chinese deserts.

## 2. Overview of Major Sandy Deserts in China

It is well known that deserts are the outcome of arid environments. "Approximately one-third of the earth's land surface is desert, arid land with meager rainfall that supports only sparse vegetation" as stated by A.S. Walker (Walker 1992). After Australia and Saudi Arabia, China is the third largest country in terms of desert area. Updated in 2015, China had a desert area of 1.08 million km² (Chang et al. 2020). These deserts extend for 4,500 km from east to the west and 600 km from north to south between 75°–125° east longitude and 35°–50° north latitude (Yang et al. 2010) as shown in Fig. 2. They stretch as an arc across Northwest, North, and Northeast China including nine provinces and regions, including Xinjiang Uygur Autonomous Region, Qinghai Province, Gansu Province, Ningxia Hui Autonomous Region, Inner Mongolia Autonomous Region, Shaanxi Province, Liaoning Province, Jilin Province, and Heilongjiang Province.

In Chinese terminology, 'desert' has both broad and narrow meanings. In a broad sense, deserts ('Huang-Mo' in Chinese) mean various kinds of territories including gravel, rock, mud, salty and sandy deserts etc. The word 'gobi' ('Ge-Bi' in Chinese) "describes specifically deserts and semideserts paved with gravel or rock debris" as stated by Chao Sung-chiao (Chao 1984). In a narrow sense, deserts only refer to two landscapes: sandy desert and sandy land. Geomorphologically,

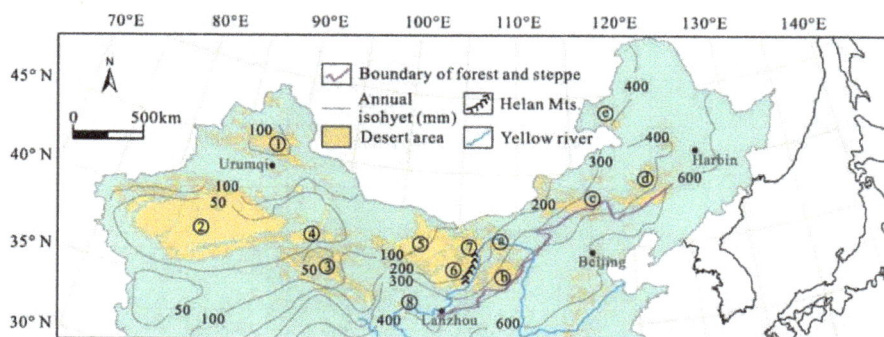

**Figure 2.** Distribution of major sandy deserts and sandy lands in northern China (Zhang 2007, Li et al. 2014).

**Footnote:** 1: Gurbantunggut, 2: Taklamakan, 3: Qaidam (or Chaidamu), 4: Kumtag, 5: Badain Jaran, 6: Tengger, 7: Wulanbuhe (or Ulan Buh), 8: Gonghe, a: Kubuqi (or Hobq), b: Mu Us, c: Hunshandake (or Onqin Daga), d: Horqin, and e: Hulunbeier (or Hulun Buir).

the sandy desert ('Sha-Mo' in Chinese) means areas mainly with mobile dunes, whereas the sandy land ('Sha-Di' in Chinese) means fields that are predominantly stable dunes (Chao 1984, Yang et al. 2004, Wang 2011). As shown in Fig. 2, the Helan Mountains, as a divided line at approximately 106° east longitude, delineate Chinese deserts as west sandy desert and east sandy land. From the Helan Mountains westward, the annual precipitation is below 200 mm, and eight major sandy deserts, Gurbantunggut, Taklamakan, Qaidam, Kumtag, Badain Jaran, Tengger, Wulanbuhe, and Kubuqi Desert, are concentrated in these arid or hyperarid areas. While the annual precipitation amounts east of the Helan Mountains are approximately 200–400 mm, four major sandy lands, Mu Us, Hunshandake, Horqin, and Hulunbeier Sandy Land are scattered in semiarid regions (Chao 1984, Li et al. 2014, Salam et al. 2018). Table 1 summarizes the eight major sandy deserts and four sandy lands in China.

"Unlike tropical and sub-tropical deserts, the deserts in China are cold deserts in the temperate zone" as stated by Wang Tao (Wang 2011). Most Chinese deserts are far from the ocean and separated by mountains such as the Himalayas, Tianshan, Altay, Altun, Qilian, Kunlun and Karakoram Mountains, etc. From the Turpan

**Table 1.** Information on the major eight sandy deserts and four sandy lands in China.

| Sandy deserts or sandy lands | Coordinates and location[1,2] | Altitude (m)[1] | Area (10⁴ km²)[2] |
|---|---|---|---|
| Taklamakan desert | 36°15′–42°03′ N, 76°14′–90°04′E Tarim basin in Xinjiang | 800–1400 | 34.69 |
| Gurbantunggut desert | 44°08′–48°25′ N, 82°38′–91°41′E Junggar Basin of Xinjiang | 300–600 | 4.99 |
| Badain Jaran desert | 39°20′–42°15′ N, 99°23′–104°27′E The west of Alxa (or Alashan) Plateau | 1300–1800 | 4.91 |
| Tengger desert | 37°26′–40°02′ N, 102°25′–105°43′E The southeast of Alxa Plateau | 1400–1600 | 3.91 |
| Kumtag desert | 39°08′–40°40′ N, 90°31′–94°53′E The north of Altun Mountains | 1000–1200 | 2.08 |
| Qaidam desert | 35°50′–38°52′ N, 90°10′–98°34′E Qaidam basin of Qinghai | 2600–3400 | 1.35 |
| Kubuqi desert | 39°34′–40°48′ N, 107°03′–111°23′E The north of Erdos Plateau | 1000–1200 | 1.30 |
| Wulanbuhe desert | 39°07′–40°54′ N, 105°33′–107°01′E The southeast of Alxa Plateau | 1000 | 0.98 |
| Horqin sandy land | 42°33′–45°44′ N, 117°48′–124°29′E The downstream of Xiliao River | 100–300 | 3.50 |
| Mu Us sandy land | 37°25′–39°43′ N, 107°07′–110°35′E The middle south of Erdos Plateau | 1300–1600 | 3.80 |
| Hunshandake sandy land | 42°52′–44°11′ N, 111°42′–117°46′E The east of Inner Mongolia | 1000–1400 | 3.33 |
| Hulunbeier sandy land | 47°22′–49°34′ N, 117°06′–120°38′E The northeast of Inner Mongolia | 600 | 0.78 |

**Note:** 1) Quoting from (Chen 2001, Wang 2011), 2) Quoting from (Yang 2018).

Depression 155 m below sea level to the intramontane basins in the western Tibetan Plateau at altitudes of more than 5,000 m above sea level (Laity 2008), deserts in China occur over a wide range of elevations and tectonic settings (Yang et al. 2010), spanning many ecosystems and climatic zones. For example, both the Taklamakan and Gurbantunggut deserts lie in the inland Tarim and Junggar basins, respectively. Meanwhile, the Badain Jaran, Tengger, and Wulanbuhe Deserts are located on the Alxa Plateau and both the Kubuqi Desert and Mu Us Sandy Land are on the Erdos Plateau. Notably, the Qaidam Desert lies in the Qaidam Basin of the Qinghai-Tibet Plateau, as listed in Table 1. The remarkable differences in regional physical geographical factors lead to Chinese deserts with distinctive characteristics.

In terms of desert area, the Xinjiang Uygur Autonomous Region is the largest in China, and the Taklamakan Desert, as the largest sandy desert in China and the second-largest shifting desert in the world, alone reaches 346,904.97 km$^2$ (Yang 2018). Separated by the Tianshan Mountains, Xingjiang is divided into south, north, and east regions. The average annual precipitation in the Gurbantunggut Desert in northern Xinjiang ranges between 70 and 150 mm, making it much more humid than Taklamakan southern Xinjiang. Melted snow in spring in the Gurbantunggut Desert helps plants such as Haloxylon, Ephedra, Artemisia and Carex grow and makes a higher percentage of vegetation in deserts of the north than that of the south (Yang et al. 2004, Wang 2011). In addition, deserts are also scattered in the Turpan and Hami (or Kumul) Depression in eastern Xinjiang. The extremely hot and dry climate in Turpan is so unique that it is known as "the fire continent" in China (the maximum air temperature once reached 48.9°C, and the sand surface temperature unexpectedly reached 82.3°C) (Wang 2011).

Aside from the hyperarid deserts in Tarim and Turpan, the desert in the Qaidam Basin is also unique. With an average elevation of ~ 2,800 m, the Qaidam Desert is one of the highest deserts in the world. The annual precipitation is less than 20–50 mm in the northwestern Qaidam Basin, while annual evaporations are as high as ~ 2,000–3,600 mm, enabling the Qaidam Desert to be one of the driest regions on Earth (Wang et al. 2018a).

## 2.1 Diversity of Actinomycetes in Deserts of China

The study of microbiology in Chinese deserts can be traced back to the 1960s. Hsien-Wu Chang, a pioneer of agricultural microbiology in China and his colleagues finished a preliminary study on the microbiological properties of sand dunes in the Tengger Desert, Ningxia Hui Autonomous Region in 1962 (Zhang and Xu 1962). Since then, various microbiological studies in Chinese deserts, such as the composition of microflora in sand dunes, identification of *Streptomyces* in extreme environments and the effect of the ecological environment including drought, high temperature, alkaline and poor organic matter on the distribution of microorganisms (Chen et al. 1983, Zhou and Wang 1983, Pan et al. 1990, Pan et al. 2010), etc., have been found in academic journals and have gradually increased since 1990 (Pan et al. 2010).

Bioprospecting for actinomycetes from Chinese deserts as unexplored extreme environments has been focused since 21st century, not only by local universities and research institutes such as Xinjiang University; Xinjiang Normal University; Xinjiang Agricultural University; Tarim University; Kashi University; Shihezi University; Xinjiang Academy of Agricultural Sciences; Xinjiang Institute of Ecology and Geography, Chinese Academy of Sciences; Inner Mongolia University; Lanzhou University; North Minzu University, Northwest A&F University; Northwest Institute of Eco-Environment and Resource, Chinese Academy of Sciences; etc., but also by other research institutions far from desert regions in China such as Yunnan University; Wuhan University; Sichuan Agricultural University; Chinese Academy of Forestry; Institute of Microbiology, Chinese Academy of Sciences; Institute of Medicinal Biotechnology, Chinese Academy of Medical Sciences, etc.

Reports on the diversity of actinomycetes in Chinese deserts from the 1960s to September 2020, were scattered in different papers and dissertations. The literatures (Luo et al. 2009, Tohty et al. 2009, Gu et al. 2011, Hao et al. 2011, Hu et al. 2012, Zhang et al. 2012, Tuo et al. 2012a, Guan et al. 2013a, Hu et al. 2013, Xu 2013, Zhang 2013a, Ding et al. 2013, Guan et al. 2014, Li et al. 2015a, Lv 2015a, Qi et al. 2015, Sun et al. 2015a, Tang 2016, Li 2016a, Zhang et al. 2016a, Li et al. 2016b, Li and Zhang 2017, Sun et al. 2018, Li et al. 2018a, Liu et al. 2019a, Xing et al. 2019, Jiang 2020, Li et al. 2020a, Li et al. 2020b) was carefully selected, analyzed and summarized to present the whole pictures of actinomycetes diversity in Chinese deserts. Actinomycete diversity in various samples, such as soils, plants and biological soil crusts, analyzed by culture-dependent and culture-independent methods, was covered by these papers. Airborne microbial flora in sandstorm original source areas in Chinese deserts are also covered. Research samples were mainly collected from the Taklamakan, Lop Nor, Gurbantunggut, Kumtag, Tengger, Badain Jaran, Qaidam, Hunshandake Deserts, and the Color Dsert in Dengpa, Tibet.

As exemplified by the Actinomycete diversity of the Taklamakan Desert, researchers at Xinjiang University have explored the diversity of the airborne microbial flora in the Taklamakan Desert (Duan 2011, Bai 2013, Wang et al. 2018b). Researchers at the Institute of Medicinal Biotechnology, Chinese Academy of Medical Sciences, explored the diversity of soil and psammophytes collected from both the southern edge of the Taklamakan Desert and the hinterland along the Alar-Hetian Desert Highway of the Taklamakan Desert (Tuo 2012b, Dong et al. 2013, Liu 2015a, Liu et al. 2021, Wang et al. 2021). The principles for paper citation or selection are largely dependent on the fact that as many genera are reported in a single paper as possible, or reported in English due to shortened reference citations in the review and convenience for international readers. In summary, 138 genera affiliated with 49 families in 24 orders of five classes in the phylum *Actinomycetota* have already been identified in Chinese deserts from the 1960s to September 2020 (Table 2).

## 2.2 Novelty of Actinomycetes in Deserts of China

As early as the late 1990s, a novel species, *Saccharomonospora xinjiangensis* XJ-54[T] (Jin et al. 1998), was claimed to be isolated from Xinjiang. However, no ultimate evidence in the published literature has indicated that the strain was recovered from deserts. To the best of our knowledge, *Jiangella gansuensis* is the first novel taxon in

**Table 2.** Actinomycetel diversity in Chinese deserts from the 1960s to September 2020.

| Class (5) | Order (24) | Family (49) | Genus (138) |
|---|---|---|---|
| *Actinomycetia* | *Actinomycetales* | *Actinomycetaceae* | *Actinomyces* |
| | *Actinopolysporales* | *Actinopolysporaceae* | *Actinopolyspora* |
| | *Cryptosporangiales* | *Cryptosporangiaceae* | *Cryptosporangium* |
| | *Frankiales* | *Frankiaceae* | *Frankia* |
| | *Geodermatophilales* | *Geodermatophilaceae* | *Blastococcus* |
| | | | *Geodermatophilus* |
| | | | *Modestobacter* |
| | *Glycomycetales* | *Glycomycetaceae* | *Glycomyces* |
| | *Jatrophihabitantales* | *Jatrophihabitantaceae* | *Jatrophihabitans* |
| | *Jiangellales* | *Jiangellaceae* | *Jiangella* |
| | | | *Phytoactinopolyspora* |
| | *Kineosporiales* | *Kineosporiaceae* | *Kineococcus* |
| | | | *Kineosporia* |
| | *Micrococcales* | *Bogoriellaceae* | *Georgenia* |
| | | *Brevibacteriaceae* | *Brevibacterium* |
| | | *Cellulomonadaceae* | *Actinotalea* |
| | | | *Cellulomonas* |
| | | | *Pseudactinotalea* |
| | | *Dermabacteraceae* | *Brachybacterium* |
| | | *Dermatophilaceae* | *Arsenicicoccus* |
| | | *Intrasporangiaceae* | *Aquipuribacter* |
| | | | *Janibacter* |
| | | | *Ornithinicoccus* |
| | | | *Phycicoccus* |
| | | *Jonesiaceae* | *Jonesia* |
| | | *Microbacteriaceae* | *Agreia* |
| | | | *Agrococcus* |
| | | | *Agromyces* |
| | | | *Clavibacter* |
| | | | *Cryobacterium* |
| | | | *Curtobacterium* |
| | | | *Diaminobutyricimonas* |
| | | | *Frigoribacterium* |
| | | | *Labedella* |
| | | | *Leifsonia* |
| | | | *Leucobacter* |

*Table 2 contd. ...*

*...Table 2 contd.*

| Class (5) | Order (24) | Family (49) | Genus (138) |
|---|---|---|---|
| | | | *Galbitalea* |
| | | | *Microbacterium* |
| | | | *Microcella* |
| | | | *Mycetocola* |
| | | | *Naasia* |
| | | | *Okibacterium* |
| | | | *Planctomonas* |
| | | | *Plantibacter* |
| | | | *Pseudoclavibacter* |
| | | | *Pseudolysinimonas* |
| | | | *Rathayibacter* |
| | | | *Salinibacterium* |
| | | *Micrococcaceae* | *Arthrobacter* |
| | | | *Citricoccus* |
| | | | *Glutamicibacter* |
| | | | *Kocuria* |
| | | | *Micrococcus* |
| | | | *Neomicrococcus* |
| | | | *Nesterenkonia* |
| | | | *Paenarthrobacter* |
| | | | *Pseudarthrobacter* |
| | | | *Rothia* |
| | | | *Zhihengliuella* |
| | | *Ornithinimicrobiaceae* | *Ornithinimicrobium* |
| | | *Promicromonosporaceae* | *Cellulosimicrobium* |
| | | | *Isoptericola* |
| | | | *Krasilnikoviella* |
| | | | *Myceligenerans* |
| | | | *Oerskovia* |
| | | | *Promicromonospora* |
| | | *Ruaniaceae* | *Haloactinobacterium* |
| | *Micromonosporales* | *Micromonosporaceae* | *Actinoplanes* |
| | | | *Allorhizocola* |
| | | | *Asanoa* |
| | | | *Couchioplanes* |
| | | | *Micromonospora* |
| | | | *Phytohabitans* |

*Table 2 contd. ...*

*...Table 2 contd.*

| Class (5) | Order (24) | Family (49) | Genus (138) |
|---|---|---|---|
| | | | *Plantactinospora* |
| | | | *Polymorphospora* |
| | | | *Spirilliplanes* |
| | | | *Virgisporangium* |
| | | | *'Wangella'* |
| | *Mycobacteriales* | *Corynebacteriaceae* | *Corynebacterium* |
| | | *Dietziaceae* | *Dietzia* |
| | | *Gordoniaceae* | *Gordonia* |
| | | | *Williamsia* |
| | | *Mycobacteriaceae* | *Mycobacterium* |
| | | | *Mycolicibacterium* |
| | | *Nocardiaceae* | *Nocardia* |
| | | | *Rhodococcus* |
| | | *Tsukamurellaceae* | *Tsukamurella* |
| | *Nakamurellales* | *Nakamurellaceae* | *Nakamurella* |
| | *Propionibacteriales* | *Actinopolymorphaceae* | *Actinopolymorpha* |
| | | *Kribbellaceae* | *Kribbella* |
| | | *Nocardioidaceae* | *Aeromicrobium* |
| | | | *Marmoricola* |
| | | | *Nocardioides* |
| | | *Propionibacteriaceae* | *Auraticoccus* |
| | | | *Desertihabitans* |
| | | | *Friedmanniella* |
| | | | *Granulicoccus* |
| | | | *Mariniluteicoccus* |
| | | | *Microlunatus* |
| | | | *Propionibacterium* |
| | *Pseudonocardiales* | *Pseudonocardiaceae* | *Actinoalloteichus* |
| | | | *Actinokineospora* |
| | | | *Actinomycetospora* |
| | | | *Actinophytocola* |
| | | | *Actinosynnema* |
| | | | *Amycolatopsis* |
| | | | *Crossiella* |
| | | | *Kibdelosporangium* |
| | | | *Lechevalieria* |
| | | | *Lentzea* |

*Table 2 contd. ...*

*...Table 2 contd.*

| Class (5) | Order (24) | Family (49) | Genus (138) |
|---|---|---|---|
| | | | *Prauserella* |
| | | | *Pseudonocardia* |
| | | | *Saccharomonospora* |
| | | | *Saccharopolyspora* |
| | | | *Saccharothrix* |
| | *Streptomycetales* | *Streptomycetaceae* | *Streptomyces* |
| | *Streptosporangiales* | *Nocardiopsaceae* | *Nocardiopsis* |
| | | | *Streptomonospora* |
| | | *Streptosporangiaceae* | *Nonomuraea* |
| | | | *Streptosporangium* |
| | | *Thermomonosporaceae* | *Actinocorallia* |
| | | | *Actinomadura* |
| | | | *Spirillospora* |
| | | | *Thermomonospora* |
| Acidimicrobiia | *Acidimicrobiales* | *Acidimicrobiaceae* | *Acidimicrobium* |
| Coriobacteriia | *Coriobacteriales* | *Atopobiaceae* | *Atopobium* |
| | | *Coriobacteriaceae* | *Collinsella* |
| | *Eggerthellales* | *Eggerthellaceae* | *Cryptobacterium* |
| | | | *Denitrobacterium* |
| | | | *Eggerthella* |
| | | | *Slackia* |
| Rubrobacteria | *Gaiellales* | *Gaiellaceae* | *Gaiella* |
| | *Rubrobacterales* | *Rubrobacteraceae* | *Rubrobacter* |
| Thermoleophilia | *Solirubrobacterales* | *Conexibacteraceae* | *Conexibacter* |
| | | *Parviterribacteraceae* | *Parviterribacter* |
| | | *Patulibacteraceae* | *Patulibacter* |
| | | *Solirubrobacteraceae* | *Solirubrobacter* |
| | *Thermoleophilales* | *Thermoleophilaceae* | *Thermoleophilum* |

**Note:** strains marked with single quotes mean the nomenclatural status are invalid and unpublished in List of Prokaryotic names with Standing in Nomenclature (LPSN, https://www.bacterio.net/).

the phylum *Actinomycetota* clearly described as isolated from the Chinese desert. The type-strain *J. gansuensis* YIM 002$^T$ was isolated from a desert soil sample collected in Sunan County, located in Hexi Corridor, Gansu Province (Song et al. 2005). It was classified as a novel species in a new genus, which was originally affiliated with the family *Nocardioidaceae* within the suborder *Propionibacterineae*. Currently, genus *Jiangella* is placed in the family *Jiangellaceae* of the order *Jiangellales* within the recently proposed class *Actinomycetia* (Tang et al. 2011). Until September 2020, 73 new actinomycete species including two new orders, two new families and 12 new genera, were discovered from samples collected in the Chinese desert (Table 3).

**Table 3.** Novel Actinomycete taxa discovered in Chinese deserts from 2005 to September 2020.

| No. | Name (Level of new taxa) | Family | Desert or location | Sample | References |
|-----|--------------------------|--------|--------------------|--------|-----------|
| 1. | *Jiangella gansuensis* (S, G) | *Jiangellaceae* | Gansu | Soil | (Song et al. 2005) |
| 2. | *Nocardiopsis quinghaiensis* (S) | *Nocardiopsaceae* | Qaidam Basin | Soil | (Chen et al. 2008) |
| 3. | *Nocardiopsis ganjiahuensis* (S) | *Nocardiopsaceae* | Xinjiang | Soil | (Zhang et al. 2008) |
| 4. | *Haloglycomyces albus* (S, G) | *Glycomycetaceae* | Lop Nur | Soil | (Guan et al. 2009) |
| 5. | *Zhihengliuella alba* (S) | *Micrococcaceae* | Xinjiang | Soil | (Tang et al. 2009a) |
| 6. | *Kocuria halotolerans* (S) *Rothia halotolerans* (HS) | *Micrococcaceae* | Xinjiang | Soil | (Tang et al. 2009b) (Nouioui et al. 2018) |
| 7. | *Kineococcus xinjiangensis* (S) | *Kineosporiaceae* | Xinjiang | Soil | (Liu et al. 2009) |
| 8. | *Actinopolyspora xinjiangensis* (S) | *Actinopolysporaceae* | Lop Nur | Soil | (Guan et al. 2010a) |
| 9. | *Nocardiopsis terrae* (S) | *Nocardiopsaceae* | Qaidam Basin | Soil | (Chen et al. 2010) |
| 10. | *Brevibacterium salitolerans* (S) | *Brevibacteriaceae* | Lop Nur | Soil | (Guan et al. 2010b) |
| 11. | *Microbacterium radiodurans* (S) | *Microbacteriaceae* | Xinjiang | Soil | (Zhang et al. 2010) |
| 12. | *Saccharopolyspora lacisalsi* (S) | *Pseudonocardiaceae* | Lop Nur | Soil | (Guan et al. 2011a) |
| 13. | *Glycomyces halotolerans* (S) | *Glycomycetaceae* | Lop Nur | Soil | (Guan et al. 2011b) |
| 14. | *Yuhushiella deserti* (S, G) *Amycolatopsis arida* (HS) | *Pseudonocardiaceae* | Xinjiang | Soil | (Mao et al. 2011) (Nouioui et al. 2018) |
| 15. | *Mycetocola manganoxydans* (S) | *Microbacteriaceae* | Taklamakan | Soil | (Luo et al. 2012) |
| 16. | *Corynebacterium deserti* (S) | *Corynebacteriaceae* | Xinjiang | Soil | (Zhou et al. 2012) |
| 17. | *Amycolatopsis salitolerans* (S) *Haloechinothrix salitolerans* (HS) | *Pseudonocardiaceae* | Lop Nur | Soil | (Guan et al. 2012) (Nouioui et al. 2018) |
| 18. | *Streptomyces fukangensis* (S) | *Streptomycetaceae* | Gurbantunggut | Soil | (Zhang et al. 2013b) |
| 19. | *'Myceligenerans salitolerans'* (S) | *Promicromonosporaceae* | Lop Nur | Soil | (Guan et al. 2013b) |

*Table 3 contd. ...*

*...Table 3 contd.*

| No. | Name (Level of new taxa) | Family | Desert or location | Sample | References |
|---|---|---|---|---|---|
| 20. | *Actinopolyspora lacussalsi* (S) | *Actinopolysporaceae* | Lop Nur | Soil | (Guan et al. 2013c) |
| 21. | *Actinophytocola gilvus* (S) | *Pseudonocardiaceae* | Tengger | Soil | (Sun et al. 2014) |
| 22. | *Prauserella shujinwangii* (S) | *Pseudonocardiaceae* | Taklamakan | Soil | (Liu et al. 2014) |
| 23. | *Streptomyces lopnurensis* (S) | *Streptomycetaceae* | Lop Nur | Soil | (Zheng et al. 2014) |
| 24. | *Nesterenkonia rhizosphaerae* (S) | *Micrococcaceae* | Gurbantunggut | Rhizosphere soil | (Wang et al. 2014) |
| 25. | *Saccharopolyspora halotolerans* (S) | *Pseudonocardiaceae* | Lop Nur | Soil | (Lv et al. 2014) |
| 26. | *Glycomyces albus* (S) | *Glycomycetaceae* | Lop Nur | Soil | (Han et al. 2014a) |
| 27. | *Glycomyces fuscus* (S) | *Glycomycetaceae* | Lop Nur | Soil | (Han et al. 2014a) |
| 28. | *Haloactinopolyspora alkaliphila* (S) | *Jiangellaceae* | Gurbantunggut | Soil | (Zhang et al. 2014) |
| 29. | *Prauserella endophytica*(S) | *Pseudonocardiaceae* | Taklamakan | Halophyte (endophyte) | (Liu et al. 2015b) |
| 30. | *Saccharothrix lopnurensis* (S) | *Pseudonocardiaceae* | Lop Nur | Soil | (Li et al. 2015b) |
| 31. | *Actinotalea suaedae* (S) *Pseudactinotalea suaedae* (HS) | *Cellulomonadaceae* | Gurbantunggut | Halophyte (endophyte) | (Zhao et al. 2015) (Cho et al. 2017) |
| 32. | *Allosalinactinospora lopnorensis* (S, G) | *Nocardiopsaceae* | Lop Nur | Rhizosphere soil | (Guo et al. 2015) |
| 33. | *Arthrobacter liuii* (S) | *Micrococcaceae* | Taklamakan | Soil | (Yu et al. 2015) |
| 34. | *Glycomyces tarimensis* (S) | *Glycomycetaceae* | Taklamakan | Soil | (Lv et al. 2015b) |
| 35. | *Nocardioides deserti* (S) | *Nocardioidaceae* | Taklamakan | Rhizosphere soil | (Tuo et al. 2015) |
| 36. | *Nesterenkonia populi* (S) | *Micrococcaceae* | Taklamakan | Halophyte (endophyte) | (Liu et al. 2015c) |
| 37. | *Tenggerimyces mesophilus* (S, G) | *Actinopolymorphaceae* | Tengger | Soil | (Sun et al. 2015b) |
| 38. | *Paraglycomyces xinjiangensis* (S, G) *Salininema proteolyticum* (HR) | *Glycomycetaceae* | Lop Nur | Soil | (Luo et al. 2015) (Nikou et al. 2015) (Li et al. 2016c) |
| 39. | '*Aeromicrobium halotolerans*' (S) | *Nocardioidaceae* | Xinjiang Turpan | Soil | (Yan et al. 2016) |

| 40. | *Nocardiopsis ansamitocini* (S) | *Nocardiopsaceae* | Gurbantunggut | Soil | (Zhang et al. 2016b) |
|---|---|---|---|---|---|
| 41. | *Egibacter rhizosphaerae* (S, G, F, O) | *Egibacteraceae* | Gurbantunggut | Rhizosphere soil | (Zhang et al. 2016c) |
| 42. | *Egicoccus halophilus* (S, G, F, O) | *Egicoccaceae* | Xinjiang | Soil | (Zhang et al. 2016d) |
| 43. | *Ornithinicoccus halotolerans* (S) | *Intrasporangiaceae* | Gurbantunggut | Soil | (Zhang et al. 2016e) |
| 44. | *Arthrobacter deserti* (S) | *Micrococcaceae* | Xinjiang Turpan | Soil | (Hu et al. 2016) |
| 45. | *Streptomyces indoligenes* (S) | *Streptomycetaceae* | Taklamakan | Rhizosphere soil | (Luo et al. 2016) |
| 46. | *Phytoactinopolyspora alkaliphila* | *Jiangellaceae* | Gurbantunggut | Soil | (Zhang et al. 2016f) |
| 47. | *Lipingzhangella halophila* (S, G) | *Nocardiopsaceae* | Gurbantunggut | Soil | (Zhang et al. 2016g) |
| 48. | *Nocardiopsis rhizosphaerae* (S) | *Nocardiopsaceae* | Gurbantunggut | Rhizosphere soil | (Zhang et al. 2016h) |
| 49. | *'Streptomyces luozhongensis'* (S) | *Streptomycetaceae* | Lop Nur | Soil | (Zhang et al. 2017) |
| 50. | *Streptomyces luteus* (S) | *Streptomycetaceae* | Taklamakan | Soil | (Luo et al. 2017) |
| 51. | *Kribbella deserti* (S) | *Kribbellaceae* | Mu Us | Rhizosphere soil | (Sun et al. 2017) |
| 52. | *Actinomadura deserti* (S) | *Thermomonosporaceae* | Taklamakan | Soil | (Cao et al. 2018) |
| 53. | *Streptomyces dengpaensis* (S) | *Streptomycetaceae* | colour desert, Tibet | Soil | (Li et al. 2018c) |
| 54. | *Desertimonas flava* (S, G) | *Ilumatobacteraceae* | Gurbantunggut | Soil | (Asem et al. 2018) |
| 55. | *Microbacterium halophytorum* (S) | *Microbacteriaceae* | Gurbantunggut | Halophyte (endophyte) | (Li et al. 2018b) |
| 56. | *Actinoplanes deserti* (S) | *Micromonosporaceae* | Xinjiang Turpan | Soil | (Habib et al. 2018) |
| 57. | *'Streptomyces qaidamensis'* (S) | *Streptomycetaceae* | Qaidam Basin | Soil | (Zhang et al. 2018) |
| 58. | *Streptomyces desertarenae* (S) | *Streptomycetaceae* | Gurbantunggut | Soil | (Li et al. 2019c) |
| 59. | *'Lentzea isolaginshaensis'* (S) | *Pseudonocardiaceae* | Isolaginsha, Ningxia | Soil | (Wang et al. 2019) |
| 60. | *Labedella populi* (S) | *Microbacteriaceae* | Taklamakan | Halophyte (endophyte) | (Li et al. 2019a) |
| 61. | *Labedella phragmitis* (S) | *Microbacteriaceae* | Taklamakan | Halophyte (endophyte) | (Li et al. 2019a) |
| 62. | *Blastococcus deserti* (S) | *Geodermatophilaceae* | Gurbantunggut | Soil | (Yang et al. 2019a) |

*Table 3 contd. ...*

*...Table 3 contd.*

| No. | Name (Level of new taxa) | Family | Desert or location | Sample | References |
|---|---|---|---|---|---|
| 63. | 'Aeromicrobium endophyticum' (S) | *Nocardioidaceae* | Taklamakan | Halophyte (endophyte) | (Li et al. 2019c) |
| 64. | *Nakamurella deserti* (S) | *Nakamurellaceae* | Taklamakan | Soil | (Liu et al. 2019a) |
| 65. | *Nocardia mangyaensis* (S) | *Nocardiaceae* | Qaidam Basin | Soil | (Yang et al. 2019b) |
| 66. | *Microbacterium suaedae* (S) | *Microbacteriaceae* | Gurbantunggut | Halophyte (endophyte) | (Zhu et al. 2019) |
| 67. | *Planctomonas deserti* (S, G) | *Microbacteriaceae* | Taklamakan | Rhizosphere soil | (Liu et al. 2019c) |
| 68. | *Desertihabitans aurantiacus* (S, G) | *Propionibacteriaceae* | Badain Jaran | Soil | (Sun et al. 2019) |
| 69. | *Nocardioides vastitatis* (S) | *Nocardioidaceae* | Taklamakan | Soil | (Liu et al. 2020a) |
| 70. | *Cellulomonas telluris* (S) | *Cellulomonadaceae* | Badain Jaran | Soil | (Shi et al. 2020) |
| 71. | *Desertihabitans brevis* (S) | *Propionibacteriaceae* | Taklamakan | Soil | (Liu et al. 2020b) |
| 72. | *Saccharothrix deserti* (S) | *Pseudonocardiaceae* | Taklamakan | Soil | (Liu et al. 2020c) |
| 73. | 'Streptomyces taklimakanensis' (S) | *Streptomycetaceae* | Taklamakan | Soil | (Yuan et al. 2020) |

**Note:** strains marked with single quotes mean the nomenclatural status are invalid and unpublished in List of Prokaryotic names with Standing in Nomenclature (LPSN, https://www.bacterio.net/). S, new species; G, new genus; F, new family; O, new order; HS, heterotypic synonym.

Among the 73 new taxa, in addition to *J. gansuensis* YIM 002[T], the taxonomic status of the other five strains was reclassified as follows: (1) The *Yuhushiella deserti* RA45[T] Strain, a novel species of a new genus in the family *Pseudonocardiaceae* isolated from desert soil in Xinjiang (Mao et al. 2011) was later transferred into the genus *Amycolatopsis* and identified as *Amycolatopsis arida* comb. nov. by comparison of chemotaxonomic and whole-genome phylogenies data (Nouioui et al. 2018). (2) The strain *Amycolatopsis salitolerans* TRM F103[T], a new species of the genus *Amycolatopsis* (Guan et al. 2012), was transferred into the genus *Haloechinothrix* and revised to *Haloechinothrix salitolerans* comb. nov. (Nouioui et al. 2018). (3) Strain *Kocuria halotolerans* as a new species of the genus *Kocuria* (Tang et al. 2009b), was transferred into the genus *Rothia* and reclassified as *Rothiahalotolerans* comb. nov (Nouioui et al. 2018). (4) The novel genus *Paraglycomyces* and its type-strain *Paraglycomyces xinjiangen*sis TRM 49201[T] (Luo et al. 2015) were reclassified as a later heterotypic synonym of *Salininema proteolyticum* (Nikou et al. 2015) and the genus *Salininema* and species *Salininema proteolyticum* were further emended (Li et al. 2016c). (5) The strain *Actinotalea suaedae* (Zhao et al. 2015) as a new species isolated from halophyte plants in Xinjiang, was reclassified as *Pseudactinotalea suaedae* comb. nov. (Cho et al. 2017).

According to the current taxonomic classification, the 73 new species are affiliated with 46 genera in 26 families. Nine of them were discovered in the genus *Streptomyces* and account for 12.3% of the total novel species. The remaining 64 new species, accounted for 87.7% of all 73 species, are affiliated with rare genera. Five, four, and three species were assigned to the genera *Nocardiopsis*, *Glycomyces*, and *Microbacterium*, respectively, accounting for 7.8%, 6.3% and 4.7% of the 64 species in rare genera. At the family level, most of the novel species are affiliated with *Pseudonocardiaceae* (ten species), followed by *Streptomycetaceae* (nine species) and *Microbacteriaceae* (seven species).

Most new species were isolated from soil or rhizosphere soil in different deserts, but strains *Prauserella endophytica* SP28S-3[T] (Liu et al. 2015b), *Pseudactinotalea suaedae* EGI 60002[T] (Zhao et al. 2015, Cho et al. 2017), *Nesterenkonia populi* GP10-3[T] (Liu et al. 2015c), *Microbacterium halophytorum* YJYP 303[T] (Li et al. 2018b), *Labedella phragmitis* 11W25H-1[T] (Li et al. 2019a), *Labedella populi* 8H24J-4-2[T] (Li et al. 2019a), *Aeromicrobium endophyticum* 9W16Y-2[T] (Li et al. 2019b), and *Microbacterium suaedae* YZYP 306[T] (Zhu et al. 2019) were endophytic actinomycetes obtained from halophyte plants. Interestingly, a piece of bark with 'Tears of *Populus euphratica*', from which *Nesterenkonia populi* GP10-3[T] was isolated, was a very special niche during our bioprospecting for actinomycetes to discover new antibiotics from the Taklamakan Desert in 2011. The bark was close to a wound, from which briny water leaked out and when it evaporated out, the bark was covered by a thin layer of white salt as shown in Fig. 3. This instance demonstrated that the special niches in desert ecosystems are noteworthy for careful investigation.

Sixty-one out of 73 new species were discovered from deserts in Xinjiang. Among them, 19, 16, and 15 species were isolated from samples collected in the Taklamakan, Lop Nur and Gurbantunggut Deserts, respectively, and the other 11 species were

**Figure 3.** Strain *Nesterenkonia populi* GP10-3ᵀ isolated from a piece of bark covered by a thin layer of white salt coming from briny water leaked by a wound (A) in *Populuse uphratica* (B) grown in the Taklamakan Desert.

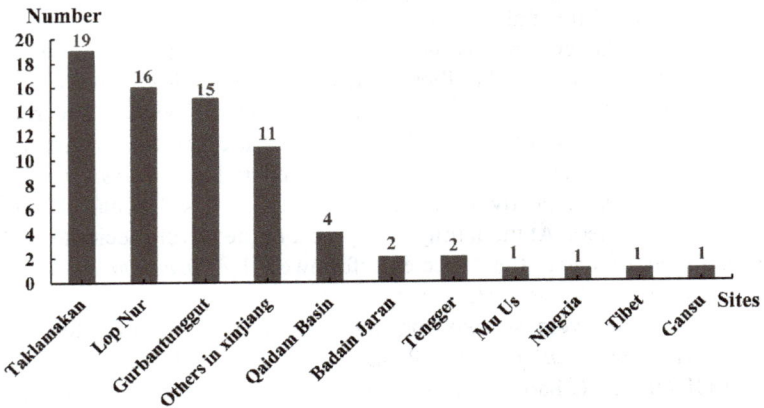

**Figure 4.** Distribution of novel Actinomycete species in Chinese deserts from 2005 to September 2020.

isolated from different deserts scattered in different sites of Xinjiang. The remaining 12 new species were isolated either from Qaidam, Badain Jaran, Tengger deserts, Mu Us sandy land or different deserts in the Ningxia Hui Autonomous Region, Tibet Autonomous Region, and Gansu Province, as shown in Fig. 4.

The overall tendency for the discovery of new actinomycete taxa from Chinese deserts is presented in Fig. 5. New species of actinomycetes were found and published sporadically over nearly a decade from 2005 to 2013, but a sharp increase occurred from 2014 to 2016, and a total of 28 new species were reported within three years. Compared with large desert areas in China, the number of new species discovered is very low, even though a fluctuation can be observed in the past few years. it is expected that the discovery of new actinomycete taxa from Chinese deserts will continue to increase as researches expanded from Taklamakan and Lop Nur to more deserts such as the Gurbantunggut and Tengger Deserts to explore new species.

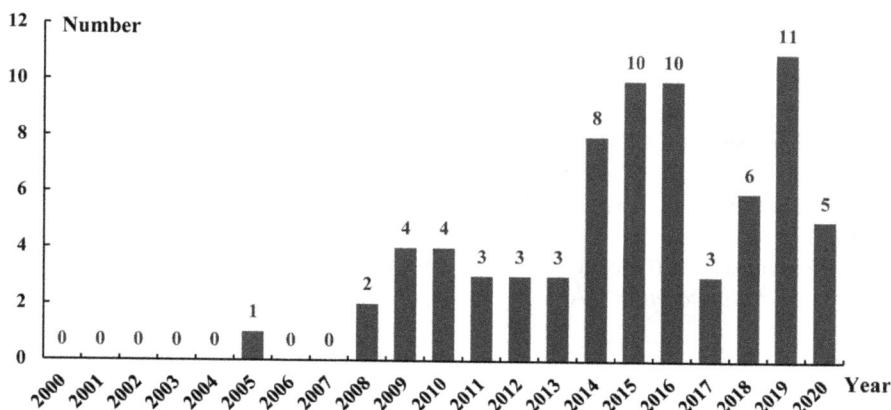

**Figure 5.** New actinomycete taxa from Chinese deserts published from 2000 to September 2020.

## 2.3 Compounds from Desert Derived Actinomycetes in China

Sixty-two compounds including 21 new, 21 known, and 20 predicted by mass spectra, as shown in Fig. 6 to Fig. 10, were reported from 15 actinomycete strains isolated from Chinese deserts, including five novel species. These compounds are composed of angucyclines, $\alpha$-pyrones, indoles, macrolides, naphthoquinones, nucleotides, polypeptides, tetracyclines, etc. They were divided into two sections according to the novelty of producing strains. Compounds in each section were summarized in order of the years, in which the producing strains were isolated. It must be mentioned that all chemical structures predicted by bioinformatics analysis such as antiSMASH based on the genomic sequences were excluded. Even 20 compounds, predicted by UPLC-HRMS or HPLC-MS from the culture broths of *A. lopnorensis* CA15-2$^T$, *Streptomyce* sp. Sd-31, and *Saccharothrix* sp. 16Sb2-4, did not concretely obtain and structurally identified by NMR, etc. but their chemical structures of compounds were included. Above all, a new thiopeptide antibiotic identified in our group from the culture broth of an endophytic *Micromonospra* sp. strain isolated from psammophytes in the Taklamakan Desert was included, but its chemical structure could not be shown in this review, due to patent application.

## 2.4 Compounds from Novel Actinomycete Species

*J. gansuensis* YIM 002$^T$ was published in 2005 (Song et al. 2005) and is the type-strain of the type species of the genus *Jiangella* in the family *Jiangellaceae* of the order *Jiangellales*. Eight compounds (Fig. 6) were isolated from its culture broth including five new pyrrol-2-aldehyde derivatives, jiangrines A-E (**1–5**), a new indolizine derivative, jiangrine F (**6**), and a new glycolipid, jiangolide (**7**), along with the known pyrrolezanthine (**8**) (Han et al. 2014b). Compounds **3** and **4** were obtained as an inseparable mixture in a nearly 1:1 ratio. All the compounds displayed no cytotoxicity at 100 μM by the MTT method. Meanwhile, compounds **1**, **2**, **3/4**, **5**, **7**, and **8** demonstrated anti-inflammatory activities by inhibiting NO production in LPS-treated RAW 264.7 macrophage cells, with $IC_{50}$ values of 97.8, 60.7, 30.1, 54.9, 61.4, and 58.8 μM, respectively. In addition, antiSMASH analysis showed that strain YIM 002 has the potential to produce pristinamycin and other new antibiotic compounds (Jiao et al. 2017).

    *A. lopnorensis* CA15-2[T] published in 2015, is the type-strain of the type species of the genus *Allosalinactinospora* in the family *Nocardiopsaceae* (Guo et al. 2015). The strain was isolated from soil samples collected in the Lop Nor region, Xinjiang, and identified as a novel species and type-strain of the new genus *Allosalinactinospora* (Guo et al. 2015). Bioinformatics analysis showed that the genome of *A. lopnorensis* CA15-2[T] has 17 putative biosynthetic gene clusters for

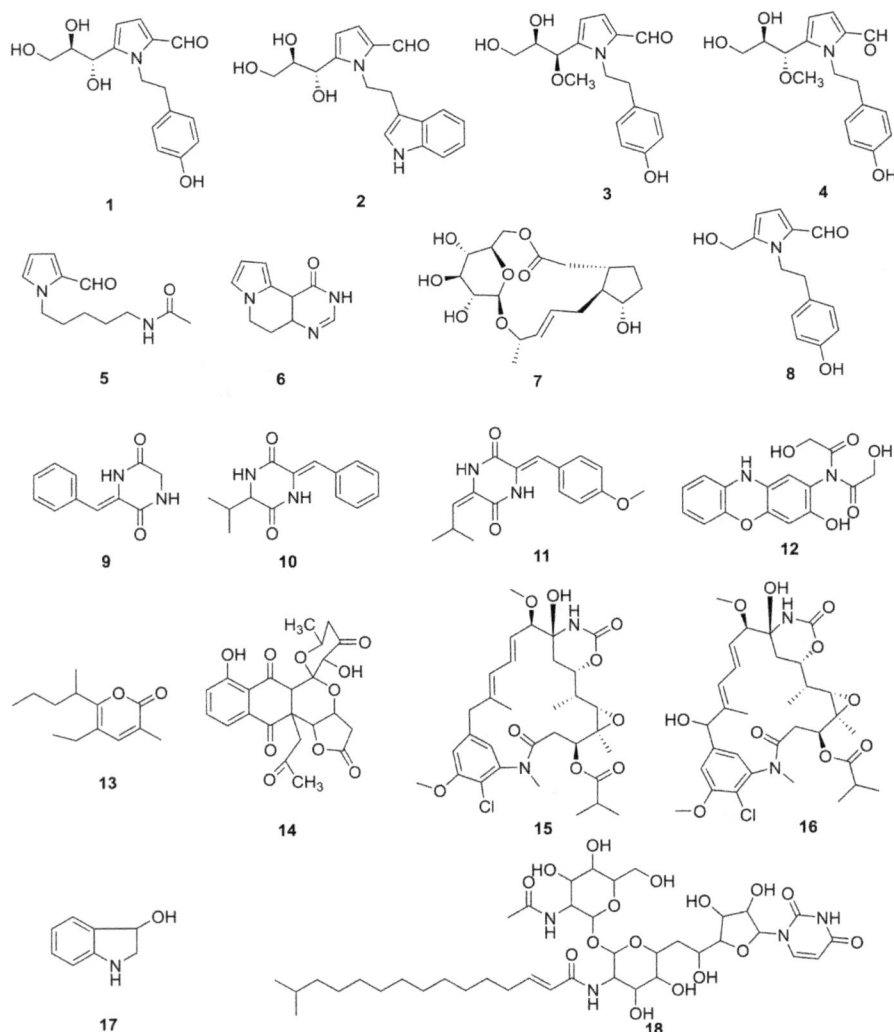

**Figure 6.** Chemical structures of compound 1–18.

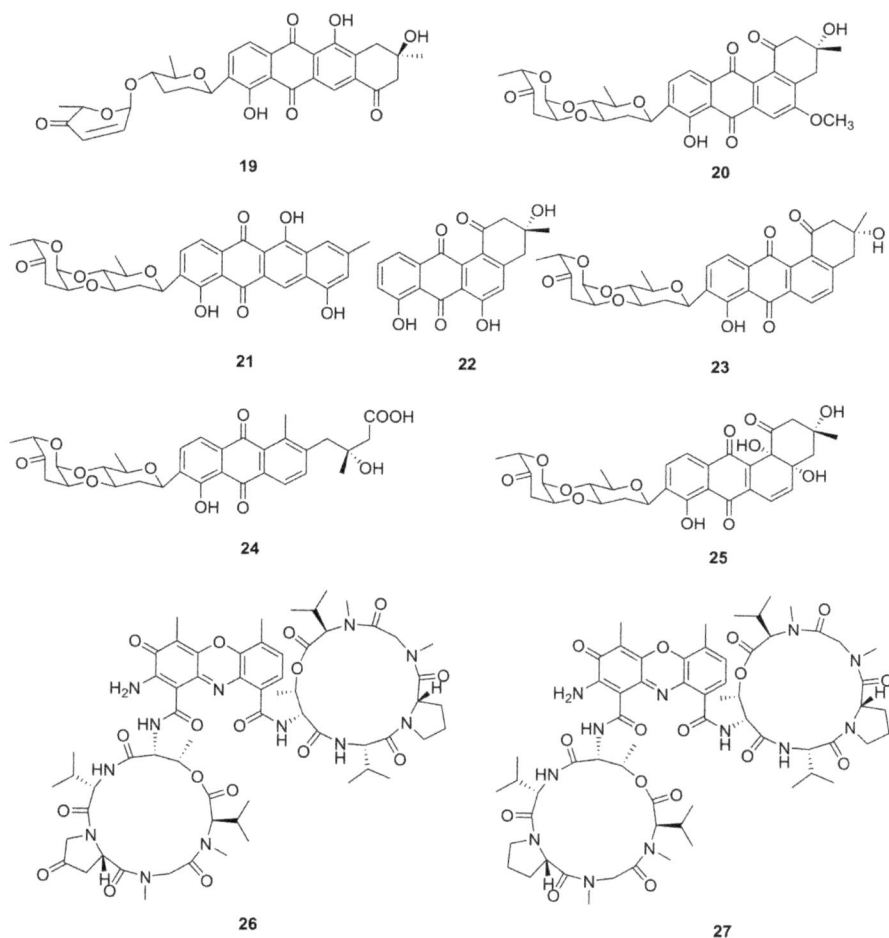

**Figure 7.** Chemical structures of compound 19–27.

known or potential novel secondary metabolites. Prediction by UPLC-MS against the database indicated the strain might produce six known compounds (Fig. 6), including three diketopiperazine group compounds: (S, Z)-3-benzylidene-piperazine-2, 5-dione (**9**), (S, Z)-3-benzylidene-6-isopropylpiperazine-2,5-dione (**10**), (3S, 6Z)-3-(4-methoxybenzylidene)-6-(2-methylpropylidene)-piperazine-2,5-dione (**11**), and one phenoxazine derivative: 2-hydroxy-N-(3-hydroxy-10H-phenoxazin-2-yl)-N-(2-hydroxyacetyl) acetamide (**12**), one alpha-pyrone compound, nocardiopyrones A (**13**) and one pyranonaphthoquinone antibiotic, Griseusins G (**14**) (Huang et al. 2016).

*Nocardiopsis* sp. EGI 80425[T] was isolated in 2015 from an arid-saline soil sample in a desert in Xinjiang (Liu 2015d) and identified as a novel species, *Nocardiopsis ansamitocini* (Zhang et al. 2016b). From the culture broth of the strain, two known ansamitocins (Fig. 6), ansamitocin P-3 (**15**) and 15-hydroxyansamitocin P-3 (**16**) were found in 2015.

*Streptomyces luozhongensis* TRM 49605$^T$ published in 2017 is the type-strain of a novel species of the genus *Streptomyces* (Zhang et al. 2017). It was isolated in a soil sample collected from Lop Nur, Luozhong, Xinjiang. From its culture broth, two known compounds (Fig. 6), 3-hydroxy-tetrahydro-indole (**17**) and tunicamycin VII (**18**) were found. Compound **18** exhibited antifungal activities and antibacterial activities (Zhang 2016i).

*Streptomyces dengpaensis* XZHG99$^T$ published in 2018 is the type-strain of a novel species of the genus *Streptomyces* (Li et al. 2018c). It was isolated in a desert soil sample collected from the Color Desert, Dengpa District, Tibet Autonomous Region. Nine compounds (Fig. 7), including three new angucycline group antibiotics, grincamycins L–N (**19–21**) and four known angucycline group antibiotics, rabelomycin (**22**), moromycin B (**23**), fridamycin D (**24**), saquayamycin B1 (**25**), along with the known actinomycin X$_2$ (**26**), and actinomycin D (**27**) were found (Bao et al. 2018). All compounds displayed significant cytotoxicity against a panel of human A549, H157, MCF7, MDA-MB-231, and HepG2 cancer cell lines by using the SRB method, with IC$_{50}$ values ranging from 0.09 nM to 17.30 µM. Compounds **22**, **26**, and **27** showed decent antibacterial activities against *Mycobacterium smegmatis* ATCC 607 and *Staphylococcus aureus* ATCC 25923, by the twofold serial dilution method in 96-well microplates, with IC$_{50}$ values from 0.12 to 23.1 µM.

## 2.5  Compounds from Known Actinomycete Species

*Streptomyces* sp. Sd-31 was isolated from a soil sample collected in 2010 from the Qinghai-Lake Desert Island at an elevation of 3,304 m in the northeastern Qinghai-Tibet Plateau. The culture broth of strain Sd-31 showed very strong inhibitory activities against *S. aureus* and *Bacillus subtilis* by the paper-disc diffusion method. Two known naphthoquinone antibiotics (Fig. 8), granatomycin A (**28**) and C (**29**), were predicted by HPLC-MS, indicating that *Streptomyces* sp. Sd-31 might be able to produce these compounds (Ding et al. 2013).

*Nocardiopsis* sp. 38-7L-1 and a *Micromonospra* strain were isolated in research on bioprospecting for antibiotics from psammophytes collected in the Taklamakan Desert in 2011 (Liu 2015a). A known macrolide antibiotic (Fig. 8), rosamicin (**30**), was found in the culture broth of *Nocardiopsis* sp. 38-7L-1 (Chen et al. 2015). Meanwhile, a new thiopeptide antibiotic with very strong antibacterial activities, *in vitro*, against gram-positive bacteria was isolated and identified from the culture broth of an endophytic *Micromonospra* strain, the structure did not show due to patent application.

*Streptomyces* sp. CA14-2 was isolated from a soil sample of *Tamarix chinensis Lour* rhizosphere from Lop Nor in 2012, and a known nucleotide antibiotic (Fig. 8), toyocamycin (**31**) with strong antifungal activities was obtained from the culture broth of strain CA14-2 (Tuo et al. 2012a, Tuo 2012b).

*Streptomyces thermolilacinus* SPC6 was formerly identified as *Streptomyces violaceusniger* SPC6 and reported in 2013 (Chen et al. 2013). The strain was isolated from the Linze Desert, part of the Badain Jaran Desert, located in Gansu Province. The known cyclic lipopeptide antibiotic marihysin A (**32**), as shown in Fig. 8, with good antifungal activity was found in the culture broth of the strain. Bioinformatics

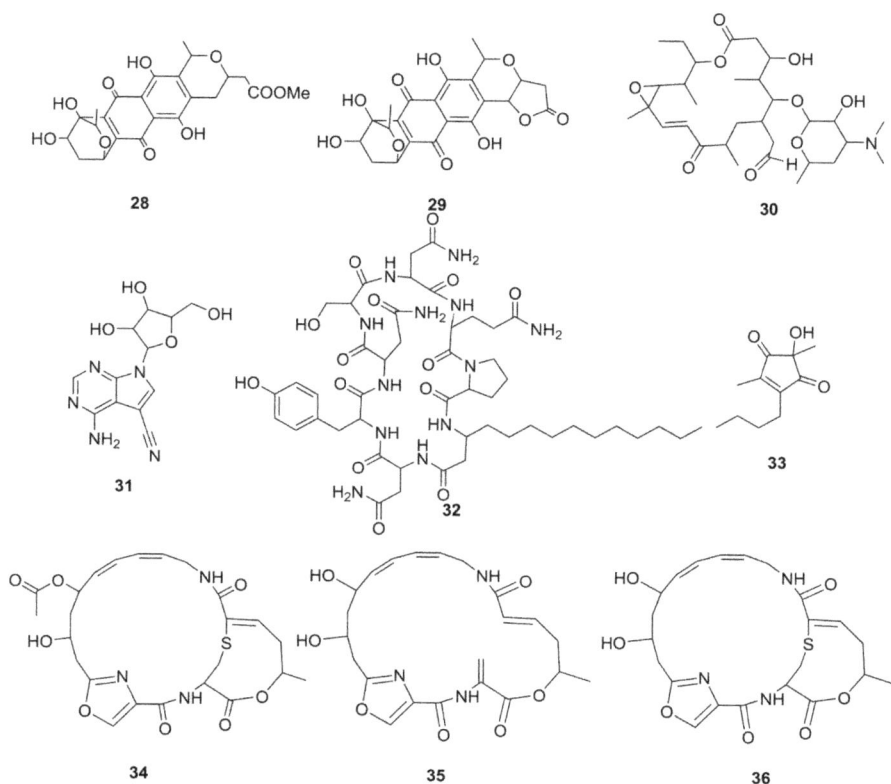

**Figure 8.** Chemical structures of compound 28–36.

analysis showed that the genome of *Streptomyces thermolilacinus* SPC6 has 52 putative biosynthetic gene clusters for a wealth of secondary metabolites (Xu 2018).

*Streptomyces* sp. LQ13 was isolated from Xinjiang arid-saline desert environments in 2015. A new polyketide (LQ13I) (**33**), as shown in Fig. 8, was obtained from the culture broth of strain LQ13 (Liu 2015d).

*Streptomyces* sp. 8P21H-1 was isolated in 2016 from psammophytes collected in the Taklamakan Desert (Li 2020c). A novel streptogramin-type antibiotic, acetyl-griseoviridin (**34**), along with two known compounds, desulphurizing griseoviridin (**35**) and griseoviridin (**36**), as shown in Fig. 8, were found from the culture broth of the strain. Both compounds **35** and **36** displayed antibacterial activity by inhibiting protein translation (Wang et al. 2021).

Two *Nocardiopsis* strains HDN154-146 and HDN154-168 were isolated in 2018 from soil samples collected in theTaklamakan Desert. Six compounds (Fig. 9) including five new α-pyrone derivatives, nocahypyrones A-E (**37–41**), and one known analog germicidin G (**42**) were isolated from the culture broth of strain HDN154-146, and three new α-pyrone derivatives nocahypyrones F-H (**43–45**), were isolated from the culture broth of strain HDN154-168 (Zhao et al.

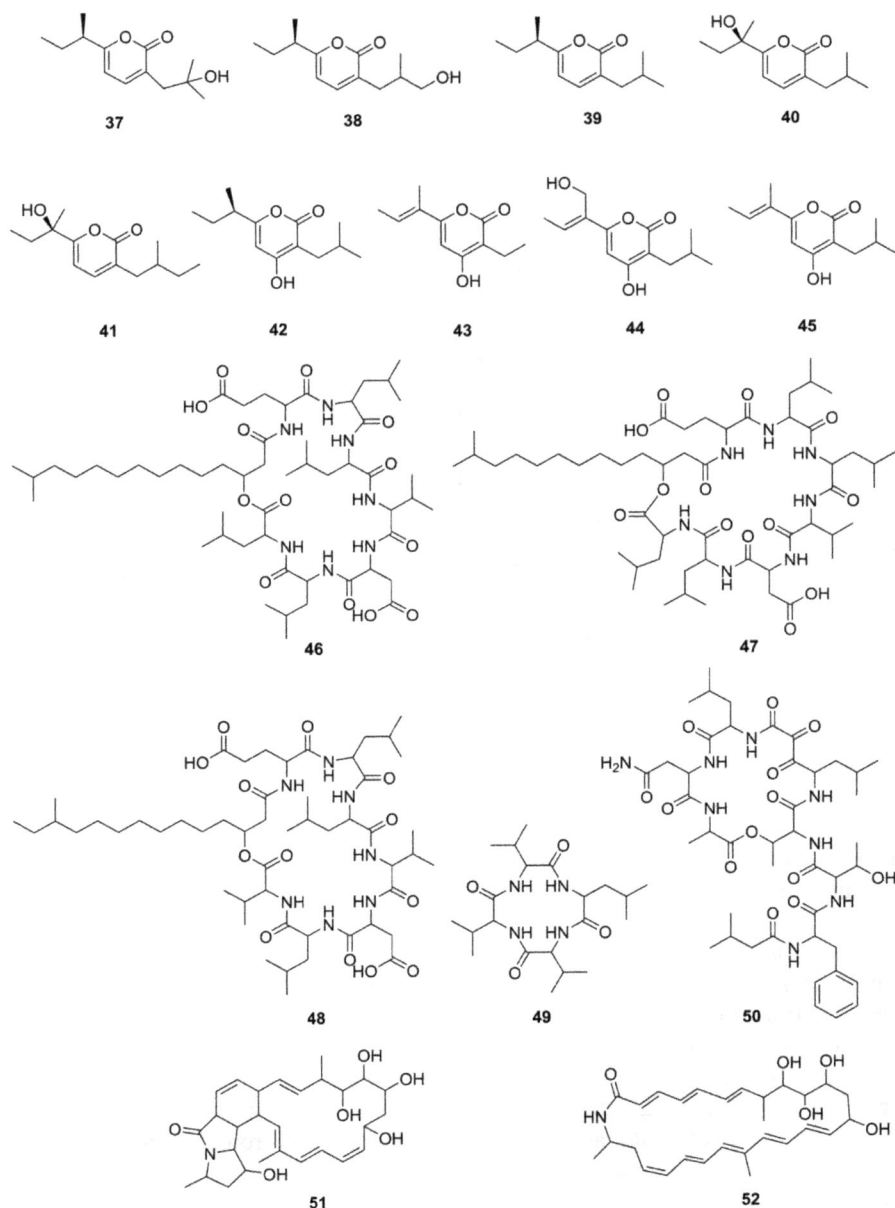

**Figure 9.** Chemical structures of compound 37–52.

2019). **All** compounds were tested for their cytoprotective activities, compounds **41** and **44** showed activities to induce the expression of phase II detoxifying enzymes in HaCaT cells such as superoxidedismutase 2 (SOD2) and heme oxygenase 1 (HO-1) to protect mammalian cells from over inflammatory reactions and oxidative damage.

*Saccharothrix* sp. 16Sb2-4 was isolated from a soil sample collected in Taklamakan Desert (Liu et al. 2021). Through UPLC-QToF-MS/MS based molecular networking analysis and identification against the various databases, a total of 16 compounds (Figs. 9–10) in four families were predicted. These compounds are surfactin C (**46**), surfactin C14 (**47**), [Val7]-surfactin C15 (**48**), xenotetrapeptide (**49**), YM-47142 (**50**), dracolactam A (**51**), micromonolactam (**52**), aldgamycin G (**53**), aldgamycin H (**54**), aldgamycin K (**55**), swalpamycin B (**56**), aldgamycin M (**57**), mycinamicin II

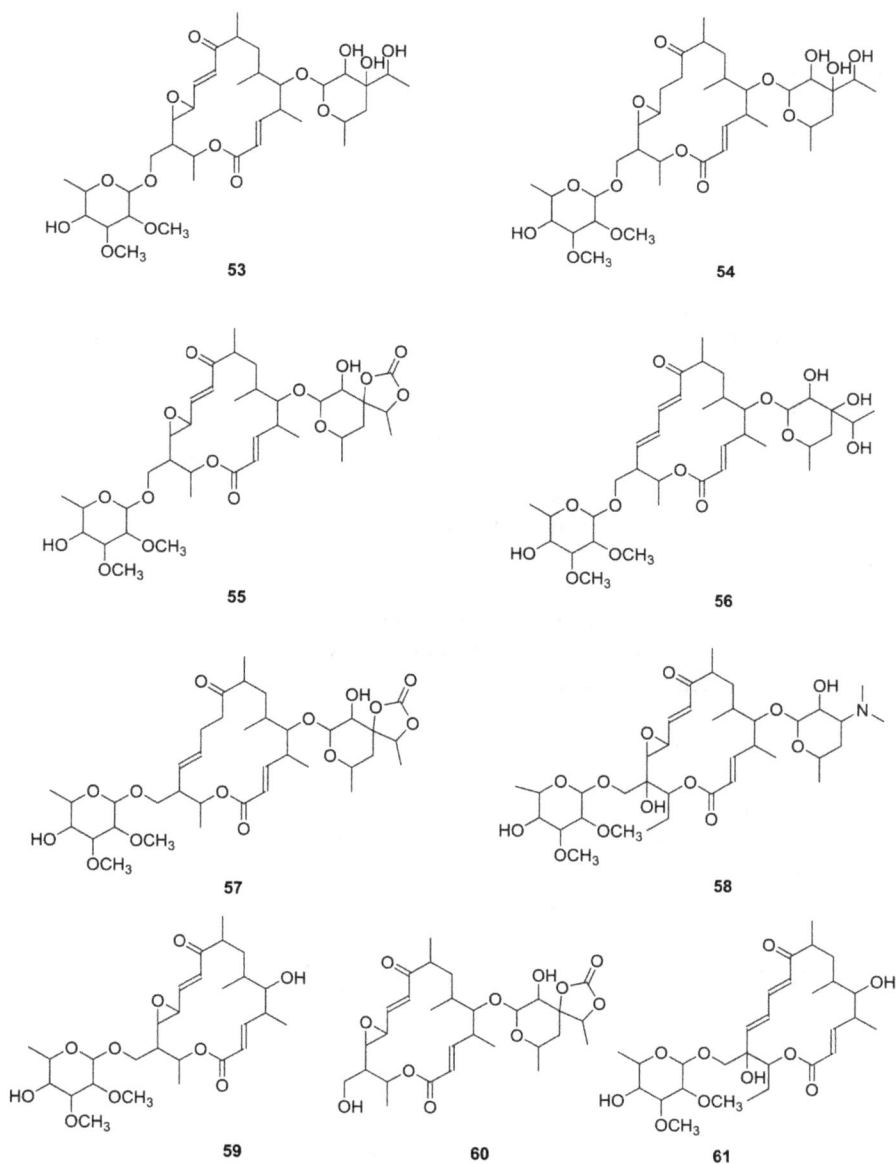

**Figure 10.** Chemical structures of compound 53–61.

(**58**), 5-*O*-dealdgarosyl-aldgamycin G (**59**), 14-*O*-demycinosyl-aldgamycin G (**60**), and *O*-dedesosaminyl-mycinamicin V (**61**). Four known 16-membered macrolides, aldgamycin G (**53**), H (**54**), K (**55**), swalpamycin B (**56**) were finally purified and identified by NMR from the culture broth of the strain. Compounds **53–56** all displayed antibacterial activity by inhibiting protein translation (Liu et al. 2021).

## 3. Discussion and Conclusion

Due to their extreme environmental conditions, deserts have not received enough attention from microbiologists as a resource of novel species of microorganisms and secondary metabolites despite accounting for approximately one-third of the Earth's land surface and being distributed on most continents in the world. Even two decades ago, research on bioprospecting for bioactive secondary metabolites such as antibiotics from actinomycetes in deserts was almost blank. However, the situation has been gradually changed since the beginning of the 21st century. Currently, progress achieved by exploring cultivable actinomycete diversity and secondary metabolites in biomes of the Atacama Desert in northern Chile (Bull and Goodfellow 2019) and the Sahara Desert (Djinni et al. 2019) has encouraged researchers in multiple countries to invest more efforts and resources in similar studies. Studies in this field were carried out not only in China as presented in this review, but also in many other Asian countries such as India (Ibeyaima et al. 2016) and Pakistan (Fatima et al. 2019) in South Asia, Saudi Arabia (Nithya et al. 2015) in the Arabian Peninsula of Western Asia and Kazakhstan (Ziyat et al. 2019) in Central Asia and Mongolia (Tsetseg et al. 1999) in East Asia.

After analyzing almost all international and domestic papers including dissertations and proceeding papers on actinomycetes and their secondary metabolites in Chinese deserts from the 1960s until now, the diversity and new taxa of actinomycetes found in the Chinese desert were summarized. It needs to be mentioned that the diversity of strains consists of two parts: the known species including 138 genera in 49 families of 24 orders in five classes, and the novel species including 46 genera in 26 families of 15 orders in three classes. After a combination of two groups of data, a total of 149 genera affiliated with 52 families in 26 orders of six classes in the phylum *Actinomycetota* were found in the Chinese deserts. As reported in 2020, there are currently 425 genera with validly published names in the phylum *Actinomycetota*, which were affiliated with 79 families in 46 orders of six classes (Salam et al. 2020). By comparison of the above three groups of data (Fig. 11), it can be found that over one third of genera in the phylum *Actinomycetota* have already been found in Chinese deserts, indicating that Chinese deserts are abundant with actinomycetes. Most of 73 new species from Chinese deserts were identified in the past decade, of which 11 new species including two new genera were identified by a polyphasic approach (Guo et al. 2015, Liu 2015a, Liu et al. 2015b and 2015c, Tuo et al. 2012a, Tuo 2012b, 2015, Li et al. 2019a and 2019b, 2020c, Liu et al. 2019b and 2019c, Liu et al. 2020a and 2020b) during our bioprospecting for new antibiotics from actinomycetes isolated in soils and psammophytes of the Taklamakan Desert and the Lop Nur region. In addition, many potential new species

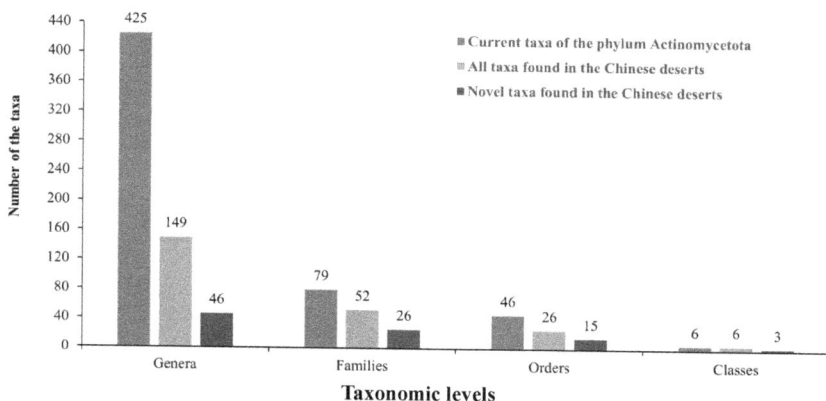

**Figure 11.** Biodiversity and novelty analysis of actinomycetes distributed in Chinese deserts.

from the Taklamakan desert are currently waiting to be identified in our group (Liu et al. 2021). Considering the unexplored area in vast Chinese deserts, the potential of discovering new actinomycete species with biodiversity and novelty has never been as promising as now.

Until now, only 15 actinomycete strains isolated from Chinese deserts have been studied for their capacity to produce bioactive metabolites, which accounted for only a very small proportion of strains discovered. Even with such a small fraction of strain samples, it has been demonstrated that the bioactive secondary metabolites produced are quite diverse (Figs. 6–10). These findings also supported that actinomycetes are still one of the most important sources of chemical diversity and a reservoir for mining metabolites with novel structures. Similar to other environments, deserts are open spaces, therefore, rediscovery of known actinomycete strains and known bioactive metabolites produced from these strains are inevitable. This implied the importance of earlier dereplication of strains and compounds. Extreme harsh ecosystems in most deserts provide unique opportunities for the discovery of new strains. For example, a new genus strain, *A. lopnorensis* CA15-2$^{T}$ was isolated from a hyperarid desert in the Lop Nur region. On the other hand, by applying genomic sequencing, analysis of gene clusters for potential novel metabolites, and analysis with UPLC-UV-HRESIMS/MS against the various databases can make earlier and more precise dereplication of known compounds possible. The combination of multiple strategies based on merging multiple disciplines such as bioinformatics, structural biology, synthetic biology, analytic chemistry, and microbiology. will drive forward the exploration of desert-derived unique actinomycete strains for more novel bioactive metabolites.

# Acknowledgments

The work was supported by grants from the CAMS Innovation Fund for Medical Sciences (CIFMS 2021-I2M-1-028, CAMS 2017-I2M-B&R-08 and 2017-I2M-1-012), and the National Natural Science Foundation of China (grant numbers 81373308 and 81621064). The authors are very grateful to Dr. Rong-feng Li from

Department of Chemistry at Johns Hopkins University, Baltimore, Maryland, US for his assistance with the writing of the review. The authors are also grateful to Ms. Qin Yang and Li-fang Liu for data proofreading.

# References

Asem, M.D., Shi, L., Jiao, J., Wang, D., Han, M., Dong, L. et al. (2018). *Desertimonas flava* gen. nov., sp. nov. isolated from a desert soil, and proposal of *Ilumatobacteraceae* fam. nov. *International Journal of Systematic and Evolutionary Microbiology*, 68: 3593–3599.

Bai, X. (2013). *Biodiversity of Airborne Microbial Flora during Dust Events in West Takalamakan Desert*. M.D., Xinjiang University. Urumqi, China.

Bao, J., He, F., Li, Y., Fang, L., Wang, K., Song, J. et al. (2018). Cytotoxic antibiotic angucyclines and actinomycins from the *Streptomyces* sp. XZHG99$^T$. *Journal of Antibiotics*, 71: 1018–1024.

Bassetti, M., Merelli, M., Temperoni, C. and Astilean, A. (2013). New antibiotics for bad bugs: Where are we? *Annals of Clinical Microbiology and Antimicrobials*, 12: 22.

Bull, A.T. and Goodfellow, M. (2019). Dark, rare and inspirational microbial matter in the extremobiosphere: 16000 m of bioprospecting campaigns. *Microbiology*, 165: 1252–1264.

Cao, C., Xu, T., Liu, J., Cai, X., Sun, Y., Qin, S. et al. (2018). *Actinomadura deserti* sp. nov., isolated from desert soil. *International Journal of Systematic and Evolutionary Microbiology*, 68: 2930–2935.

Chang, Q., Lu, H., Lv, N., Cui, M. and Li, H. (2020). Remote sensing monitoring and climate impact analysis of desert area change in China from 1992 to 2015. *Journal of Desert Research*, 40: 57–63.

Chao, S. (1984). The sandy deserts and the gobi of China. pp. 95–113. *In*: El-Baz, F. (eds.). *Deserts and Arid Lands*. Springer, Dordrecht.

Chen, C., Liu, J., Jiang, Z., Li, X., Liu, S., Tuo, L. et al. (2015). Endophytic actinomycetes from psammophytes of Taklamakan Desert against *Pseudomonas aeruginosa* and study on bioactive product of strain 38-7L-1. *Chinese Journal of Antibiotics*, 40: 81–87.

Chen, G. (2001). Origins of arguments on the area of desertified lands in china. *Deserts in China*, 21: 209–212.

Chen, X., Zhang, B., Zhang, W., Wu, X., Zhang, M., Chen, T. et al. (2013). Genome sequence of *streptomyces violaceusniger* strain SPC6, a halotolerant streptomycete that exhibits rapid growth and development. *Genome Announcements*, 1: e00494-13.

Chen, Y., Zhang, Y., Tang, S., Liu, Z., Xu, L., Zhang, L. et al. (2010). *Nocardiopsis terrae* sp. nov., a halophilic actinomycete isolated from saline soil. *Antonie van Leeuwenhoek*, 98: 31–38.

Chen, Y.G., Cui, X.L., Kroppenstedt, R.M., Stackebrandt, E., Wen, M.L., Xu, L.H. et al. (2008). *Nocardiopsis quinghaiensis* sp. nov., isolated from saline soil in China. *International Journal of Systematic and Evolutionary Microbiology*, 58: 699–705.

Chen, Z., Zhang, J. and Li, D. (1983). The Microbiological characteristics in various types of sand dunes in southeastern Tengri desert. *Journal of Desert Research*, 03: 24–30.

Cho, H., Hamada, M., Ahn, J., Weon, H., Joa, J. and Suzuki, K. et al. (2017). *Pseudactinotalea terrae* gen. nov., sp. nov., isolated from greenhouse soil, and reclassification of *Actinotalea suaedae* as *Pseudactinotalea suaedae* comb. nov. *International Journal of Systematic and Evolutionary Microbiology*, 67: 704–709.

Diab, A. and Al Zaidan, A. (1976). Actinomycetes in the desert of Kuwait. *Zentralblatt fur Bakteriologie, Mikrobiologie, und Hygiene*, 131: 545.

Ding, D., Chen, G., Wang, B., Wang, Q., Liu, D., Peng, M. et al. (2013). Culturable actinomycetes from desert ecosystem in northeast of Qinghai-Tibet Plateau. *Annals of Microbiology*, 63: 259–266.

Djinni, I., Defant, A., Kecha, M. and Mancini, I. (2019). Actinobacteria derived from Algerian ecosystems as a prominent source of antimicrobial molecules. *Antibiotics (Basel)*, 8: 172.

Dong, Y., Guo, L., Habden, X., Liu, J., Chen, C., Jiang, Z. et al. (2013). Study on diversity and bioactivity of actinomycetes isolated from the south edge of the Taklamakan Desert. *Chinese Journal of Antibiotics*, 38: 241–247.

Duan, W. (2011). *The Diversity of Air Microbial Communities from Sandstorm Source Areas of the Taklamakan Desert in Xinjiang*. M.D., College of Science and Technology of Xinjiang University. Urumqi, China.

Emadodin, I., Reinsch, T. and Taube, F. (2019). Drought and desertification in Iran. *Hydrology*, 6: 66.

Fatima, A., Aftab, U., Shaaban, K.A., Thorson, J.S. and Sajid, I. (2019). Spore forming Actinobacterial diversity of Cholistan Desert Pakistan: Polyphasic taxonomy, antimicrobial potential and chemical profiling. *BMC Microbiology*, 19: 49.

Garrity, G.M., Heimbuch, B.K. and Gagliardi, M. (1996). Isolation of zoosporogenous actinomycetes from desert soils. *Indian Journal of Microbiology and Biotechnology*, 17: 260–267.

Goodfellow, M., Nouioui, I., Sanderson, R., Xie, F. and Bull, A.T. (2018). Rare taxa and dark microbial matter: Novel bioactive actinobacteria abound in Atacama Desert soils. *Antonie van Leeuwenhoek*, 111: 1315–1332.

Gu, P., Hao, L., Xu, R., Ma, H. and Hu, M. (2011). Diversity analysis of endophytic actinomycetes isolated from *Sophora alopecuroides* L. of Lingwu Baijitan National Nature Reserve in Ningxia. *Journal of Ningxia University (Natural Science Edition)*, 32: 380–385.

Guan, T., Guan, T., Wu, N., Wu, N., Tang, S., Tang, S. et al. (2013b). *Myceligenerans salitolerans* sp. nov., a halotolerant actinomycete isolated from a salt lake in Xinjiang, China. *Extremophiles*, 17: 147–152.

Guan, T., Liu, Y., Zhao, K., Xia, Z., Zhang, X. and Zhang, L. (2010a). *Actinopolyspora xinjiangensis* sp. nov., a novel exteremely halophilic actinomycete isolated from a salt lake in Xinjiang, China. *Antonie van Leeuwenhoek*, 98: 447–453.

Guan, T., Teng, Y., Che, Z., Zhang, L., Zhang, X. Xing, Y. et al. (2013a). Comparison of isolation media for actinobacteria from Lop Nur salt lake. *Biotechnology*, 23: 56–60.

Guan, T., Wei, B., Zhang, Y., Xia, Z., Che, Z., Chen, X. et al. (2013c). *Actinopolyspora lacussalsi* sp. nov., an extremely halophilic actinomycete isolated from a salt lake. *International Journal of Systematic and Evolutionary Microbiology*, 63: 3009–3013.

Guan, T., Wu, N., Xia, Z., Ruan, J., Zhang, X., Huang, Y. et al. (2011a). *Saccharopolyspora lacisalsi* sp. nov., a novel halophilic actinomycete isolated from a salt lake in Xinjiang, China. *Extremophiles*, 15: 373–378.

Guan, T., Xia, Z., Tang, S., Wu, N., Chen, Z., Huang, Y. et al. (2012). *Amycolatopsis salitolerans* sp. nov., a filamentous actinomycete isolated from a hypersaline habitat. *International Journal of Systematic and Evolutionary Microbiology*, 62: 23–27.

Guan, T., Yun, T., Ren, D., Liu, S., Jiang, H., Zhao, H. et al. (2014). Diversity and *PKS* II genes on the cultured actinobacterial in Lop Nur. *Biotechnology*, 24: 67–71.

Guan, T., Zhao, K., Xiao, J., Liu, Y., Xia, Z., Zhang, X. et al. (2010b). *Brevibacterium salitolerans* sp. nov., an actinobacterium isolated from salt-lake sediment. *International Journal of Systematic and Evolutionary Microbiology*, 60: 2991.

Guan, T.W., Tang, S.K., Wu, J.Y., Zhi, X.Y., Xu, L.H., Zhang, L.L. et al. (2009). *Haloglycomyces albus* gen. nov., sp. nov., a halophilic, filamentous actinomycete of the family *Glycomycetaceae*. *International Journal of Systematic and Evolutionary Microbiology*, 59: 1297–1301.

Guan, T.W., Xia, Z.F., Xiao, J., Wu, N., Chen, Z.J., Zhang, L.L. et al. (2011b). *Glycomyces halotolerans* sp. nov., a novel actinomycete isolated from a hypersaline habitat in Xinjiang, China. *Antonie Van Leeuwenhoek*, 100: 137–143.

Guo, L., Tuo, L., Habden, X., Zhang, Y., Liu, J., Jiang, Z. et al. (2015). *Allosalinactinospora lopnorensis* gen. nov., sp. nov., a new member of the family *Nocardiopsaceae* isolated from soil. *International Journal of Systematic and Evolutionary Microbiology*, 65: 206–213.

Habib, N., Khan, I.U., Chu, X., Xiao, M., Li, S., Fang, B. et al. (2018). *Actinoplanes deserti* sp. nov., isolated from a desert soil sample. *Antonie van Leeuwenhoek*, 111: 2303–2310.

Han, L., Gao, C., Jiang, Y., Guan, P., Liu, J., Li, L. et al. (2014b). Jiangrines A–F and jiangolide from an actinobacterium, *Jiangella gansuensis*. *Journal of Natural Products*, 77: 2605–2610.

Han, X., Luo, X. and Zhang, L. (2014a). *Glycomyces fuscus* sp. nov. and *Glycomyces albus* sp. nov., actinomycetes isolated from a hypersaline habitat. *International Journal of Systematic and Evolutionary Microbiology*, 64: 2437–2441.

Hao, L., Hu, M., Ma, H., Xu, R. and Gu, P. (2011). Isolation and identification of endophytic actinomycetes from *Sophora alopecurodies* L. *Journal of Agricultural Sciences*, 32: 18–21.

Holmes, A.J., Bowyer, J., Holley, M.P., O'Donoghue, M., Montgomery, M. and Gillings, M.R. (2000). Diverse, yet-to-be-cultured members of the *Rubrobacter* subdivision of the actinobacteria are widespread in Australian arid soils. *FEMS Microbiology Ecology*, 33: 111–120.

Hu, M., Fan, Y. and Gu, P. (2012). Studies on the distribition of endophytic actinomyces and antimicrobal activity from *Sophora alopecuroides* of arid and desolate areas in Ningxia. *Journal of Agricultural Sciences*, 33: 30–34.

Hu, M., Zhou, X., Wang, L., Zhang, W., Jia, Q. and Gu, P. (2013). Producing enzyme activity and identification about endophytic actinomycetes isolated from *Sophora alopecuroides* L. *Northern Horticulture*, 98–102.

Hu, Q., Chu, X., Xiao, M., Li, C., Yan, Z., Hozzein, W.N. et al. (2016). *Arthrobacter deserti* sp. nov., isolated from a desert soil sample. *International Journal of Systematic and Evolutionary Microbiology*, 66: 2035–2040.

Huang, C., Leung, R.K., Guo, M., Tuo, L., Guo, L. and Yew, W.W. et al. (2016). Genome-guided investigation of antibiotic substances produced by *Allosalinactinospora lopnorensis* CA15-2$^T$ from Lop Nor region, China. *Scientific Reports*, 6: 20667.

Ibeyaima, A., Rana, J., Dwivedi, A., Gupta, S., Sharma, S., Saini, N. et al. (2016). Characterization of *Yuhushiella* sp. TD-032 from the Thar Desert and its antimicrobial activity. *Journal of Advanced Pharmaceutical Technology and Research*, 7: 32–36.

Infectious Diseases Society of America. (2010). The 10 × '20 initiative: pursuing a global commitment to develop 10 new antibacterial drugs by 2020. *Clinical Infectious Diseases*, 50: 1081–1083.

Idris, H., Goodfellow, M., Sanderson, R., Asenjo, J.A. and Bull, A.T. (2017). Actinobacterial rare biospheres and dark matter revealed in habitats of the chilean atacama desert. *Scientific Reports*, 7: 8311–8373.

Jiang, M. (2020). *Study on the Diversity and Biological Characteristics of Actinomycetes, Geodermatophilaceae in the Desert Environment*. M.D., Peking Union Medical College. Beijing, China.

Jiao, J., Carro, L., Liu, L., Gao, X., Zhang, X., Hozzein, W.N. et al. (2017). Complete genome sequence of *Jiangella gansuensis* strain YIM 002$^T$ (DSM 44835$^T$), the type species of the genus *Jiangella* and source of new antibiotic compounds. *Standards in Genomic Sciences*, 12: 21.

Jin, X., Xu, L.H., Mao, P.H., Hseu, T.H. and Jiang, C.L. (1998). Description of *Saccharomonospora xinjiangensis* sp. nov. based on chemical and molecular classification. *International Journal of Systematic and Evolutionary Microbiology*, 48: 1095–1099.

Laity, J. (2008). *Deserts and Desert Environments*. Wiley-Blackwell Press, London.

Le Houérou, H.N. (1996). Climate change, drought and desertification. *J. Arid Environ.*, 34: 133–185.

Li, F. (2020c). *Bioprospecting of Actinobacteria as Pharmaceutical Resource from Mangrove and Desert, and Identification of New Species from Special Habitats*. Ph.D., Peking Union Medical College. Beijing, China.

Li, F., Lu, Q., Liao, S., Jin, T., Li, W. and Sun, C. (2019a). *Labedella phragmitis* sp. nov. and *Labedella populi* sp. nov., two endophytic actinobacteria isolated from plants in the Taklamakan Desert and emended description of the genus *Labedella*. *Systematic and Applied Microbiology*, 42: 126004.

Li, F., Liao, S., Liu, S., Jin, T. and Sun, C. (2019b). *Aeromicrobium endophyticum* sp. nov., an endophytic actinobacterium isolated from reed (*Phragmites australis*). *Journal of Microbiology*, 57: 725–731.

Li, H. (2016a). *Diversity Analysis of Culturable Bacteria from Xerophyte* Haloxylon ammodendron *Rhizosphere*. M.D., Lanzhou University. Lanzhou, China.

Li, J., Jin, X., Zhang, X., Chen, L., Liu, J., Zhang, H. et al. (2020a). Comparative metagenomics of two distinct biological soil crusts in the Tengger Desert, China. *Soil Biology and Biochemistry*, 140: 107637.

Li, J. and Zhang, X. (2017). Microbial diversity analysis of different biological soil crusts in Tengger desert. *Ecological Science*, 36: 36–42.

Li, J., Zhang, X., Sun, M. and Zhang, Y. (2016b). Analysis of soil microbial diversity in shapotou area of Tengger Desert. *Journal of Ecology and Rural Environment*, 32: 780–787.

Li, J., Zhang, X., Chen, Y., Jin, X., Ma, Z., Ji, D. et al. (2020b). Potential functions of actinobacteria diversity in cyanobacteria and moss crusts in the southeastern Tengger Desert. *Acta Ecologica Sinica*, 40: 1–12.

Li, K., Bai, Z. and Zhang, H. (2015). Community succession of bacteria and eukaryotes in dune ecosystems of Gurbantünggüt Desert, Northwest China. *Extremophiles*, 19: 171–181.

Li, L., Yang, Z., Asem, M.D., Fang, B., Salam, N., Alkhalifah, D.H.M. et al. (2019c). *Streptomyces desertarenae* sp. nov., a novel actinobacterium isolated from a desert sample. *Antonie van Leeuwenhoek*, 112: 367–374.

Li, Q., Wu, H., Guo, Z., Yu, Y., Ge, J., Wu, J. et al. (2014). Distribution and vegetation reconstruction of the deserts of northern China during the mid-Holocene. *Geophysical Research Letters*, 41: 5184–5191.

Li, X., Liu, J., Wu, Y., Zhang, W., Li, J., Liu, S. et al. (2016c). Description of *Salilacibacter albus* gen. nov., sp. nov., isolated from a dried salt lake, and reclassification of *Paraglycomyces xinjiangensis* Luo et al. 2015 as a later heterotypic synonym of *Salininema proteolyticum* Nikou et al. 2015 with emended descriptions of the genus *Salininema* and *Salininema proteolyticum*. *International Journal of Systematic and Evolutionary Microbiology*, 66: 2558–2565.

Li, Y., Buheliqihan Baikeli, Bao, J., Gao, J., Wang, J. and Li, Y. (2018a). Bioprospecting and bioactivity screening of culturable actinobacteria from color desert in Dengpa, Tibet. *Microbiology (China)*, 45: 1651–1660.

Li, Y., Li, Y., Wang, L.W. and Bao, J. (2018c). *Streptomyces dengpaensis* sp. nov., an actinomycete isolated from desert soil. *International Journal of Systematic and Evolutionary Microbiology*, 68: 3322–3326.

Li, Y., Liu, L., Cheng, C., Shi, X., Lu, C., Dong, Z. et al. (2015b). *Saccharothrix lopnurensis* sp. nov., a filamentous actinomycete isolated from sediment of Lop Nur. *Antonie van Leeuwenhoek*, 108: 975–981.

Li, Y., Zhu, Z., Li, Y., Xiao, M., Han, M., Wadaan, M.A.M. et al. (2018b). *Microbacterium halophytorum* sp. nov., a novel endophytic actinobacterium isolated from halophytes. *International Journal of Systematic and Evolutionary Microbiology*, 68: 3928–3934.

Lipman, C.B. (1912). The distribution and activities of bacteria in soils of the arid region. *University of California Publications in Agricultural Sciences*, 1: 1–20.

Liu, J. (2015a). *Bioprospecting of Endophytic Actinobacteria as Pharmaceutical Resource from Plants Collected in Taklamakan Desert*. Ph.D., Peking Union Mediacal College. Beijing, China.

Liu, J., Habden, X., Guo, L., Tuo, L., Jiang, Z., Liu, S. et al. (2015b). *Prauserella endophytica* sp. nov., an endophytic actinobacterium isolated from *Tamarix taklamakanensis*. *Antonie van Leeuwenhoek*, 107: 1401–1409.

Liu, J., Sun, Y., Liu, J., Wu, Y., Cao, C., Li, R. et al. (2020d). *Saccharothrix deserti* sp. nov., an actinomycete isolated from desert soil. *International Journal of Systematic and Evolutionary Microbiology*, 70: 1882–1887.

Liu, J., Tuo, L., Habden, X., Guo, L., Jiang, Z., Liu, X. et al. (2015c). *Nesterenkonia populi* sp. nov., an actinobacterium isolated from *Populus euphratica*. *International Journal of Systematic and Evolutionary Microbiology*, 65: 1474–1479.

Liu, M., Peng, F., Wang, Y., Zhang, K., Chen, G. and Fang, C. (2009). *Kineococcus xinjiangensis* sp. nov., isolated from desert sand. *International Journal of Systematic and Evolutionary Microbiology*, 59: 1090.

Liu, M., Zhang, L., Ren, B., Yang, N., Yu, X., Wang, J. et al. (2014). *Prauserella shujinwangii* sp. nov., from a desert environment. *International Journal of Systematic and Evolutionary Microbiology*, 64: 3833–3837.

Liu, Q. (2015d). *Study on the Secondary Metabolites Produced by Three Actinobacterial Strains from Extreme Environments*. M.D., Yunnan University, China.

Liu, S., Li, F., Liu, H., Yu, L. and Sun, C. (2020c). *Desertihabitans brevis* sp. nov., an actinobacterium isolated from sand of the Taklamakan desert, and emended description of the genus *Desertihabitans*. *International Journal of Systematic and Evolutionary Microbiology*, 70: 1166–1171.

Liu, S., Li, F., Qi, X., Xie, Y. and Sun, C. (2019b). *Nakamurella deserti* sp. nov., isolated from rhizosphere soil of *Reaumuria* in the Taklamakan desert. *International Journal of Systematic and Evolutionary Microbiology*, 69: 214–219.

Liu, S., Li, F., Zheng, H., Qi, X., Huang, D., Xie, Y. et al. (2019c). *Planctomonas deserti* gen. nov., sp. nov., a new member of the family *Microbacteriaceae* isolated from soil of the Taklamakan desert. *IInternational Journal of Systematic and Evolutionary Microbiology*, 69: 616–624.

Liu, S., Wang, T., Lu, Q., Li, F., Wu, G., Jiang, Z. et al. (2021). Bioprospecting of soil-derived actinobacteria along the Alar-Hotan Desert Highway in the Taklamakan Desert. *Frontiers in Microbiology*, 12: 604999.

Liu, S., Xue, C., Li, F. and Sun, C. (2020b). *Nocardioides vastitatis* sp. nov., isolated from Taklamakan desert soil. *International Journal of Systematic and Evolutionary Microbiology*, 70: 77–82.

Liu, Y., Wang, Z., Zhao, L., Wang, X., Liu, L., Hui, R. et al. (2019a). Differences in bacterial community structure between three types of biological soil crusts and soil below crusts from the Gurbantunggut Desert, China. *European Journal of Soil Science*, 70: 630–643.

Luo, M., Han, J., Jiang, P. and Wu, H. (2009). Diversity of culturable halophilic bacteria isolated from Lop Nur region in Xinjiang. *Biodiversity Science*, 17: 288–295.

Luo, X., Han, X., Zhang, F., Wan, C. and Zhang, L. (2015). *Paraglycomyces xinjiangensis* gen. nov., sp. nov., a halophilic actinomycete. *International Journal of Systematic and Evolutionary Microbiology*, 65: 4263–4269.

Luo, X., Kai, L., Wang, Y., Wan, C. and Zhang, L. (2017). *Streptomyces luteus* sp. nov., an actinomycete isolated from soil. *IInternational Journal of Systematic and Evolutionary Microbiology*, 67: 543–547.

Luo, X., Sun, Y., Xie, S., Wan, C. and Zhang, L. (2016). *Streptomyces indoligenes* sp. nov., isolated from rhizosphere soil of *Populus euphratica*. *International Journal of Systematic and Evolutionary Microbiology*, 66: 2424–2428.

Luo, X., Wang, J., Zeng, X., Wang, Y., Zhou, L., Nie, Y. et al. (2012). *Mycetocola manganoxydans* sp. nov., an actinobacterium isolated from the Taklamakan desert. *International Journal of Systematic and Evolutionary Microbiology*, 62: 2967–2970.

Lv, J. (2015a). Analysis of microbial species diversity in Kumtag Desert and screening of antagonistic strains against *Colletotrichum gloeosporioides*. M.D., Hebei Agricultural University. Heibei, China.

Lv, L., Zhang, Y., Xia, Z., Zhang, J. and Zhang, L. (2014). *Saccharopolyspora halotolerans* sp. nov., a halophilic actinomycete isolated from a hypersaline lake. *International Journal of Systematic and Evolutionary Microbiology*, 64: 3532–3537.

Lv, L., Zhang, Y. and Zhang, L. (2015b). *Glycomyces tarimensis* sp. nov., an actinomycete isolated from a saline-alkali habitat. *International Journal of Systematic and Evolutionary Microbiology*, 65: 1587–1591.

Mao, J., Wang, J., Dai, H., Zhang, Z., Tang, Q., Ren, B. et al. (2011). *Yuhushiella deserti* gen. nov., sp. nov., a new member of the suborder *Pseudonocardineae*. *International Journal of Systematic and Evolutionary Microbiology*, 61: 621–630.

Nikou, M.M., Ramezani, M., Ali Amoozegar, M., Rasouli, M., Abolhassan Shahzadeh Fazeli, S., Schumann, P. et al. (2015). *Salininema proteolyticum* gen. nov., sp. nov., a halophilic rare actinomycete isolated from wetland soil, and emended description of the family *Glycomycetaceae*. *International Journal of Systematic and Evolutionary Microbiology*, 65: 3727–3733.

Nithya, K., Muthukumar, C., Duraipandiyan, V., Dhanasekaran, D. and Thajuddin, N. (2015). Diversity and antimicrobial potential of culturable actinobacteria from desert soils of saudi arabia. *Journal of Pharmaceutical Sciences and Research*, 7: 117–122.

Nouioui, I., Carro, L., García-López, M., Meier-Kolthoff, J.P., Woyke, T., Kyrpides, N.C. et al. (2018). Genome-based taxonomic classification of the phylum *Actinobacteria*. *Frontiers in Microbiology*, 9: 2007.

Pan, H., Cheng, Z., Zhang, Y., Mu, S. and Qi, X. (2010). Research progress and developing trends on microorganisms of Xinjiang specific environments. *Journal of Arid Land*, 2: 51–56.

Pan, H., Wang, X. and Wang, L. (1990). Effect of the ecological environments on the distribution of microorganism. *Arid Zone Research*, 07: 44–49.

Qi, H., Zhou, X., Hu, M., Gao, Y., Chen, Y., Zhang, Q. et al. (2015). Diversity and distribution of endophytic actinomycetes strains in *Sophora alopecuroides* L. from Baijitan Nature Reserve of Ningxia. *Microbiology (China)*, 42: 990–1000.

Rateb, M.E., Ebel, R. and Jaspars, M. (2018). Natural product diversity of actinobacteria in the Atacama Desert. *Antonie van Leeuwenhoek*, 111: 1467–1477.

Sabaou, N., Hacene, H., Bennadji, A. and Bennadji, H. (1992). Distribution quantitative et qualitative des actinomyctes dans les horizons de sol de surface et profonds d'une palmeraie algerienne. *Canadian Journal of Microbiology*, 38: 1066–1073.

Salam, N., Jiao, J., Zhang, X. and Li, W. (2020). Update on the classification of higher ranks in the phylum *Actinobacteria*. *International Journal of Systematic and Evolutionary Microbiology*, 70: 1331–1355.

Salam, N., Yang, Z., Asem, M.D., Hozzein, W.N. and Li, W. (2018). Microbial diversity in asian deserts: Distribution, biotechnological importance, and environmental impacts. pp. 365–387. *In*: Egamberdieva, D., Birkeland, N., Panosyan H., and Li, W. (eds.). *Extremophiles in Eurasian Ecosystems: Ecology, Diversity, and Applications*. Springer, Singapore.

Shi, Y., Sun, Y., Ruan, Z., Su, J., Yu, L. and Zhang, Y. (2020). *Cellulomonas telluris* sp. nov., an endoglucanase-producing actinobacterium isolated from Badain Jaran desert sand. *International Journal of Systematic and Evolutionary Microbiology*, 70: 631–635.

Song, L., Li, W., Wang, Q., Chen, G., Zhang, Y. and Xu, L. (2005). *Jiangella gansuensis* gen. nov., sp. nov., a novel actinomycete from a desert soil in north-west China. *International Journal of Systematic and Evolutionary Microbiology*, 55: 881–884.

Soussi, A., Ferjani, R., Marasco, R., Guesmi, A., Cherif, H., Rolli, E. et al. (2016). Plant-associated microbiomes in arid lands: Diversity, ecology and biotechnological potential. *Plant and Soil*, 405: 357–370.

Sun, H., Zhang, T., Yu, L., Lu, X., Mou, X. and Zhang, Y. (2014). *Actinophytocola gilvus* sp. nov., isolated from desert soil crusts, and emended description of the genus *Actinophytocola* Indananda et al. 2010. *International Journal of Systematic and Evolutionary Microbiology*, 64: 3120–3125.

Sun, H., Zhang, T., Yu, L., Sen, K. and Zhang, Y. (2015a). Ubiquity, diversity and physiological characteristics of *Geodermatophilaceae* in Shapotou National Desert Ecological Reserve. *Frontiers in Microbiology*, 6: 1059.

Sun, H., Zhang, T., Wei, Y., Liu, H., Yu, L. and Zhang, Y. (2015b). *Tenggerimyces mesophilus* gen. nov., sp. nov., a member of the family *Nocardioidaceae*. *International Journal of Systematic and Evolutionary Microbiology*, 65: 3359–3364.

Sun, J., Xu, L., Guo, Y., Li, W., Shao, Z., Yang, Y. et al. (2017). *Kribbella deserti* sp. nov., isolated from rhizosphere soil of *Ammopiptanthus mongolicus*. *International Journal of Systematic and Evolutionary Microbiology*, 67: 692–696.

Sun, Y., Shi, Y., Wang, H., Zhang, T., Yu, L., Sun, H. et al. (2018). Diversity of bacteria and the characteristics of actinobacteria community structure in badain jaran desert and tengger desert of china. *Frontiers in Microbiology*, 9: 1068.

Sun, Y., Wang, H., Zhang, T., Liu, W., Liu, H., Yu, L. et al. (2019). *Desertihabitans aurantiacus* gen. nov., sp. nov., a novel member of the family *Propionibacteriaceae*. *International Journal of Systematic and Evolutionary Microbiology*, 69: 2486–2491.

Taha, S.M., Zayed, M.N., Moawad, H. and Khalaf-Allah, M. (1973). Studies in actinomycetes in Egyptian soils II. The occurrence of antibiotic-producing actinomycetes. *Zentralblatt fur Bakteriologie, Mikrobiologie, und Hygiene*, 128: 110–115.

Tang, K. (2016). *Community Structure and Diversity of Bacteria in Biological Soil Crusts of Hunshandake Deserts*. M.D., Inner Mongolia Agricultural University. Inner Mongolia, China.

Tang, S., Wang, Y., Lou, K., Mao, P., Xu, L., Jiang, C. et al. (2009a). *Kocuria halotolerans* sp. nov., an actinobacterium isolated from a saline soil in China. *International Journal of Systematic and Evolutionary Microbiology*, 59: 1316.

Tang, S., Zhi, X., Wang, Y., Shi, R., Lou, K., Xu, L. et al. (2011). *Haloactinopolyspora alba* gen. nov., sp. nov., a halophilic filamentous actinomycete isolated from a salt lake, with proposal of *Jiangellaceae* fam. nov. and *Jiangellineae* subord. nov. *International Journal of Systematic and Evolutionary Microbiology*, 61: 194–200.

Tang, S.K., Wang, Y., Chen, Y., Lou, K., Cao, L.L., Xu, L.H. et al. (2009b). *Zhihengliuella alba* sp. nov., and emended description of the genus *Zhihengliuella*. *International Journal of Systematic and Evolutionary Microbiology*, 59: 2025–2032.

Tohty, D., Habdin, X., Chen, J., Zhu Hasanjan, Y., An, D. et al. (2009). Biological characteristics and active substances of alkalophilic actinomycetes from extreme environment of Lop Nur peripherals in Xinjiang, China. *The Chinese Journal of Applied and Environmental Biology*, 15: 835–839.

Tsetseg, B., Ganbaatar, T. and Puntsag, T. (1999). *Evaluation of Actinomycetes Isolated form the Gobi Desert Soils as Producers of Extracellular Enzymes*. Kunming, China.

Tuo, L. (2012b). Studies on actinomycetes from rhizospheric soil of psammophyte in Lop Nor region and the compound produced by strain CA14-2. M.D., China Pharmaceutical University. Nanjing, China.

Tuo, L., Dong, Y., Habden, X., Liu, J., Guo, L., Liu, X. et al. (2015). *Nocardioides deserti* sp. nov., an actinobacterium isolated from desert soil. *International Journal of Systematic and Evolutionary Microbiology*, 65: 1604–1610.

Tuo, L., Xugela, H., Guo, L., Zhang, Y., Tao, L., Wang, F. et al. (2012a). Studies on diversity and bioactivity of rhizosphere actinomycetes from psammophyte in Lop Nor region. *Chinese Journal of Antibiotics*, 37: 21–26.

Walker, A.S. (1992). *Deserts: Geology and Resources*. General Interest Publication, USA.

Walsh, C.T. and Wencewicz, T.A. (2013). Prospects for new antibiotics: A molecule-centered perspective. *Journal of Antibiotics*, 67: 7–22.

Wang, D., Lin, Q., Otkur, M., Yang, H. and Lou, K. (2018b). Diversity of airborne bacterial communities in sandstorm area of Taklimakan. *Acta Microbiologica Sinica*, 58: 1420–1430.

Wang, H., Zhang, Y., Chen, J., Hozzein, W.N., Li, L., Wadaan, M.A.M. et al. (2014). *Nesterenkonia rhizosphaerae* sp. nov., an alkaliphilic actinobacterium isolated from rhizosphere soil in a saline-alkaline desert. *International Journal of Systematic and Evolutionary Microbiology*, 64: 4021–4026.

Wang, J., Xiao, L., Reiss, D., Hiesinger, H., Huang, J., Xu, Y. et al. (2018a). Geological features and evolution of yardangs in the qaidam basin, tibetan plateau (NW china): A terrestrial analogue for mars. *The Journal of Geophysical Research: Planets*, 123: 2336–2364.

Wang, L., Li, Y. and Li, Y. (2019). *Lentzea isolaginshaensis* sp. nov., an actinomycete isolated from desert soil. *Antonie van Leeuwenhoek*, 112: 633–639.

Wang, T. (2011). *Deserts and Aeolian Desertification in China*. Science Press, China.

Wang, T., Li, F., Lu, Q., Wu, G., Jiang, Z., Liu, S. et al. (2021). Studies on diversity, novelty, antimicrobial activity, and new antibiotics of cultivable endophytic actinobacteria isolated from psammophytes collected in Taklamakan Desert. *Journal of Pharmaceutical Analysis*, 11: 241–250.

Whaley, J. and Boyle, A. (1967). Antibiotic production by *Streptomyces* species from the rhizosphere of desert plants. *Phytopathology*, 57: 347.

Whaley, J.W. (1964). *Physiological Studies of Antagonistic Actinomycetes from the Rhizosphere of Desert Plants*. Ph.D., The University of Arizona. Tucson, Arizona, USA.

Whitford, W.G. (2002). *Ecology of Desert Systems*. Academic Press, New York.

Xing, R., Gao, Q.B., Zhang, F.Q., Wang, J.L. and Chen, S.L. (2019). Large-scale distribution of bacterial communities in the Qaidam Basin of the Qinghai-Tibet Plateau. *Microbiology Open*, 00: e909.

Xu, L. (2013). *The Study of Soil Microbial Diversity in Kumtag Desert*. M.D., Chinese Academy of Agricultural Sciences, Beijing, China.

Xu, T.T. (2018). *Discovery of Cyclic Lipopeptide Marihysin A and Study on its Biosynthetic Gene Cluster in Streptomyces thermolilacinus SPC6*. M.D., Lanzhou University. Lanzhou, China.

Yan, Z., Lin, P., Chu, X., Kook, M., Li, C. and Yi, T. (2016). *Aeromicrobium halotolerans* sp. nov., isolated from desert soil sample. *Archives of Microbiology*, 198: 423–427.

Yang, R., Zhang, B., Sun, H., Zhang, G., Li, S., Liu, G. et al. (2019b). *Nocardia mangyaensis* sp. nov., a novel actinomycete isolated from crude-oil-contaminated soil. *International Journal of Systematic and Evolutionary Microbiology*, 69: 397–403.

Yang, W. (2018). *Atlas of Sandy Deserts in China*. Science Press, China.

Yang, X., Zhou, H., Li, S. and Zhang, K. (2010). Sandy deserts, gobi, sandlands and sandified land in dryland. pp. 177–216. *In*: Ci, L. and Yang, X. (eds.). *Desertification and its Control in China*. Springer, Berlin, Heidelberg.

Yang, X., Rost, K.T., Lehmkuhl, F., Zhenda, Z. and Dodson, J. (2004). The evolution of dry lands in northern China and in the Republic of Mongolia since the Last Glacial Maximum. *Quaternary International*, 118-119: 69–85.

Yang, Z., Asem, M.D., Li, X., Li, L., Salam, N., Alkhalifah, D.H.M. et al. (2019a). *Blastococcus deserti* sp. nov., isolated from a desert sample. *Archives of Microbiology*, 201: 193–198.

Yu, X., Zhang, L., Ren, B., Yang, N., Liu, M., Liu, X. et al. (2015). *Arthrobacter liuii* sp. nov., resuscitated from Xinjiang desert soil. *International Journal of Systematic and Evolutionary Microbiology*, 65: 896–901.

Yuan, L., Zhang, L., Luo, X., Xia, Z., Sun, B. and Zeng, H. (2020). *Streptomyces taklimakanensis* sp. nov., an actinomycete isolated from the Taklimakan desert. *Antonie van Leeuwenhoek*, 113: 1023–1031.

Zayed, M.N., Taha, S.M., Moawad, H. and Khalf-Allah, M. (1973). Studies in actinomycetes in Egyptian soils. I. The distribution of actinomycetes. *Zentralblatt fur Bakteriologie, Mikrobiologie, und Hygiene*, 128: 101–109.

Zenova, G.M., Chernov, I.Y., Gracheva, T.A. and Zvyagintsev, D.G. (1996). Structure of actinomycete complexes in deserts. *Mikrobiologiya*, 65: 704–709.

Zhang, B., Kong, W., Wu, N. and Zhang, Y. (2016a). Bacterial diversity and community along the succession of biological soil crusts in the Gurbantunggut Desert, Northern China. *J. Basic Microbiol.*, 56: 670–679.

Zhang, B., Tang, S., Chen, X., Zhang, G., Zhang, W., Chen, T. et al. (2018). *Streptomyces qaidamensis* sp. nov., isolated from sand in the Qaidam Basin, China. *The Journal of Antibiotics*, 71: 880–886.

Zhang, J. (2013a). *Diversity of Related Bacteria of Kumtag Desert Plants*. M.D., Chinese Academy of Agricultural Sciences. Beijing, China.

Zhang, R. (2016i). *Study on Polyphasic Taxonomy and its Secondary Metabolites of Two New Species of Streptomyces*. M.D., Tarim University. Tarim, China.

Zhang, R., Han, X., Xia, Z., Luo, X., Wan, C. and Zhang, L. (2017). *Streptomyces luozhongensis* sp. nov., a novel actinomycete with antifungal activity and antibacterial activity. *Antonie van Leeuwenhoek*, 110: 195–203.

Zhang, W., Zhang, G., Liu, G., Wang, L., Dong, X., Yue, J. et al. (2012). Characteristics of cultivable microbial community number and structure at the southeast edge of Tengger Desert. *Acta Ecologica Sinica*, 32: 567–577.

Zhang, W., Zhu, H.H., Yuan, M., Yao, Q., Tang, R., Lin, M. et al. (2010). *Microbacterium radiodurans* sp. nov., a UV radiation-resistant bacterium isolated from soil. *International Journal of Systematic and Evolutionary Microbiology*, 60: 2665–2670.

Zhang, X., Zhang, L.P., Yang, R., Shi, N., Lu, Z., Chen, W.X. et al. (2008). *Nocardiopsis ganjiahuensis* sp. nov., isolated from a soil from Ganjiahu, China. *International Journal of Systematic and Evolutionary Microbiology*, 58: 195–199.

Zhang, X.S. (2007). *Vegetation Map of China and its Geographic Pattern*. Beijing: Geological Publishing House: Beijing.

Zhang, Y., Wang, H., Yang, L., Guo, J., Xiao, M., Huang, M. et al. (2016e). *Ornithinicoccus halotolerans* sp. nov., and emended description of the genus *Ornithinicoccus*. *International Journal of Systematic and Evolutionary Microbiology*, 66: 1894–1899.

Zhang, Y., Wang, H., Yang, L., Zhou, X., Zhi, X., Duan, Y. et al. (2016c). *Egibacter rhizosphaerae* gen. nov., sp. nov., an obligately halophilic, facultatively alkaliphilic actinobacterium and proposal of *Egibacteraceae* fam. nov. and *Egibacterales* ord. nov. *International Journal of Systematic and Evolutionary Microbiology*, 66: 283–289.

Zhang, Y., Wang, H., Liu, Q., Hozzein, W.N., Wadaan, M.A.M., Cheng, J. et al. (2013b). *Streptomyces fukangensis* sp. nov., a novel alkaliphilic actinomycete isolated from a saline-alkaline soil. *Antonie van Leeuwenhoek*, 104: 1227–1233.

Zhang, Y., Chen, J., Wang, H., Xiao, M., Yang, L., Guo, J. et al. (2016d). *Egicoccus halophilus* gen. nov., sp. nov., a halophilic, alkalitolerant actinobacterium and proposal of *Egicoccaceae* fam. nov. and *Egicoccales* ord. nov. *International Journal of Systematic and Evolutionary Microbiology*, 66: 530–535.

Zhang, Y., Liu, Q., Wang, H., Zhang, D., Zhang, Y., Park, D. et al. (2014). *Haloactinopolyspora alkaliphila* sp. nov., and emended description of the genus *Haloactinopolyspora*. *International Journal of Systematic and Evolutionary Microbiology*, 64: 1945–1951.

Zhang, Y., Lu, X., Ding, Y., Wang, S., Zhou, X., Wang, H. et al. (2016g). *Lipingzhangella halophila* gen. nov., sp. nov., a new member of the family *Nocardiopsaceae*. *International Journal of Systematic and Evolutionary Microbiology*, 66: 4071–4076.

Zhang, Y., Lu, X., Ding, Y., Zhou, X., Wang, H., Guo, J. et al. (2016h). *Nocardiopsis rhizosphaerae* sp. nov., isolated from rhizosphere soil of *Halocnermum strobilaceum* (Pall.) Bieb. *International Journal of Systematic and Evolutionary Microbiology*, 66: 5129–5133.

Zhang, Y., Lu, X., Ding, Y., Zhou, X., Li, L., Guo, J. et al. (2016f). *Phytoactinopolyspora alkaliphila* sp. nov., an alkaliphilic actinomycete isolated from a saline-alkaline soil. *International Journal of Systematic and Evolutionary Microbiology*, 66: 2058–2063.

Zhang, Y., Zhang, Y., Park, D., Wang, H., Kim, C., Guo, J. et al. (2016b). *Nocardiopsis ansamitocini* sp. nov., a new producer of ansamitocin P-3 of the genus *Nocardiopsis*. *International Journal of Systematic and Evolutionary Microbiology*, 66: 230–235.

Zhang, Z. and Xu, G. (1962). Studies on the microbiological properties of the sand dunes in Tengkeli desert of Ningsia Hui Autonomous Region. *Acta Pedologica Sinica*, 227–234.

Zhao, S., Li, L., Li, S., Wang, H., Hozzein, W.N., Zhang, Y. et al. (2015). *Actinotalea suaedae* sp. nov., Isolated from the halophyte *Suaeda physophora* in Xinjiang, Northwest China. *Antonie van Leeuwenhoek*, 107: 1–7.

Zhao, T., Chang, Y., Zhu, T., Li, J., Gu, Q., Li, D. et al. (2019). α-pyrone derivatives with cyto-protective activity from two Takla Makan desert soil derived actinomycete *Nocardiopsis* strains recovered in seawater based medium. *Nat. Prod. Res.*, 33: 2498–2506.

Zheng, B., Han, X., Xia, Z., Wan, C. and Zhang, L. (2014). *Streptomyces lopnurensis* sp. nov., an actinomycete isolated from soil. *International Journal of Systematic and Evolutionary Microbiology*, 64: 4179–4183.

Zhou, P. and Wang, D. (1983). Identification of strain of *subhalophilic streptomyces*. *Acta Microbiologica Sinica*, 23: 216–219.

Zhou, Z., Yuan, M., Tang, R., Chen, M., Lin, M. and Zhang, W. (2012). *Corynebacterium deserti* sp. nov., isolated from desert sand. *International Journal of Systematic and Evolutionary Microbiology*, 62: 791–794.

Zhu, Z., Li, Y., Li, Y., Xiao, M., Han, M., Wadaan, M.A.M. et al. (2019). *Microbacterium suaedae* sp. nov., Isolated from *Suaeda aralocaspica*. *International Journal of Systematic and Evolutionary Microbiology*, 69: 411–416.

Ziyat, A., Goddfellow, M., Nurgozhina, A., Sergazy, S. and Nurgaziev, M. (2019). Novel actinobacterial diversity in Kazakhstan deserts soils as a source of new drug leads. *J. Microbiol. Biotech. Food Sci.*, 8: 1057.

# Chapter 7

# Multi-Metal Tolerant Actinomycetes from Tin Tailings of an Ex-Mining Area

*Getha-Krishnasamy** and *Hema-Thopla Govender*

## 1. Introduction

Heavy metals and metalloids constitute a group of inorganic chemical hazards which are generally present in trace amounts in the environment. While most metals such as iron (Fe), zinc (Zn), cobalt (Co), chromium (Cr) and copper (Cu), to name a few, are essential micronutrients that can be toxic at high concentrations, others like mercury (Hg), lead (Pb), cadmium (Cd), and the metalloid arsenic (As) are toxic even at low concentrations (El Baz et al. 2015). Due to their non-degradative nature, most metal(loid)s persist in soil for a long time and can enter life forms through air, water, and food chain, exerting long-term effects on the ecosystem. The toxicity and mobility of heavy metal(loid)s in soil are strongly dependent on their chemical form and specific binding properties (Fashola et al. 2016). Metal(loid) ions have a strong electrostatic attraction and high binding affinity with the same sites that essential metal ions bind to in various cellular structures. This causes destabilization of structures and biomolecules such as enzymes, DNA and RNA, thus inducing replication defects and consequent mutagenesis, hereditary genetic disorders and other adverse effects (Wuana and Okieimen 2011, Coelho et al. 2015).

Various natural processes such as mineral weathering, soil erosion and emission of toxic gases from volcanic activities, forest fires and particles released from vegetation, are responsible for the continuous release and distribution of heavy metal(loid) into the environment (Garrett 2000, Nagajyoti et al. 2010). Additionally, emissions from industrial and electronic wastes, wood preservatives, pesticide and fertilizer applications, effluent and municipal sewage sludge applications, and ore mining activities are some of the anthropogenic processes that cause heavy metal(loid) accumulation in soil (Kapahi and Sachdeva 2019).

Biomolecules Research Laboratory, Bioactivity Programme, Natural Products Division, Forest Research Institute Malaysia (FRIM), 52109 Kepong, Selangor, Malaysia.
* Corresponding author: getha@frim.gov.my

Industrial processes such as metalliferous mining and smelting activities are identified as one of the major anthropogenic sources of metal(loid)s in the environment (Limcharoensuk et al. 2015). During mining, mineral processing and metallurgical extraction, metal(loid)s are extracted from the ores. Heavier and larger particles known as mine tailings generated during these activities are discharged as wastes and contribute to soil contamination (Zibret et al. 2018, Wuana and Okieimen 2011). The release of metal waste residues and other pollutants from mining sites to soil drainage represent a potential source of contamination to surrounding areas especially to agricultural lands (Sheoran et al. 2010).

Heavy metal contamination is a serious problem which not only impacts the environment but also causes detrimental effects on all life forms, including microorganisms (Bhakta et al. 2014). Although bacterial communities have shown a reduced diversity in metal contaminated soils (de Quadros et al. 2016), many are observed to possess corresponding physiological functions related to heavy metal(loid) tolerance to cope with the pollution (Feng et al. 2018, Fernandes et al. 2018). Several tolerant species of bacteria have been reported, among others the Gram-negative *Cupriavidus metallidurans* isolated from metal processing sludge with multi-tolerance to Co, Cr, Hg, Ni, Cd, Cu, Pb and Zn to survive under diverse metal stress (Lal et al. 2013). Similarly, metal tolerance have been demonstrated in bacteria isolated from metal-contaminated environments belonging to the genera *Desulfovibrio, Tsukamurella, Pseudomonas, Brevundimonas, Bacillus, Corynebacterium, Arthrobacter, Ralstonia, Alcaligenes* and *Burkholderia* (Karnachuk et al. 2008, Limcharoensuk et al. 2015, Divakar et al. 2018, Soto et al. 2019). Investigations on heavy metal(loid) tolerance in bacteria, especially when associated with their ability to sequester or detoxify metal ions, have raised much interest because of the practical application in bioremediation (Amoroso et al. 1998, Coelho et al. 2015). Others have demonstrated that heavy metal resistance-encoding genetic systems in bacteria can provide critical tools to develop metal biosensors to detect bioavailable metal concentrations in the environment. One example is the multi-metal tolerant *Achromobacter* sp. A022 from lead-contaminated soil in Australia showed potential applications in copper biosensor development (Ng et al. 2012).

Phylum *Actinomycetota* (Nouioui et al. 2018) covers the former Order of *Actinomycetales*, currently defined in the recently proposed Class *Actinomycetia*, https://lpsn.dsmz.de/class/actinomycetia (Salam et al. 2020) and commonly known as "Actinomycetes". Actinomycetes are a group of metabolically versatile Gram-positive filamentous bacteria found in diverse habitats, including extreme environments (Nafis et al. 2019). Their mycelial form of growth and rapid colonization ability make them as potential agents for bioremediation of both organic and inorganic compounds (Alvarez et al. 2017, El Baz 2017). They are known as natural components of the microbiota in metal-rich acidic ecosystems and have a range of mechanisms that provide tolerance to metal toxicants (Wakelin et al. 2011). The effect of a wide range of heavy metal(loid) on different species of actinomycetes were reported by various researchers (Abbas and Edwards 1989, Amoroso et al.

1998, Albarracin et al. 2005, 2010, Polti et al. 2007, Guo et al. 2009, Schmidt et al. 2005, 2007, 2009, Lin et al. 2011, 2012, Remenar et al. 2014, El Baz et al. 2015). Table 1 briefly shows the types of metal resistance/tolerance reported over the last 15 years in actinomycetes isolated from contaminated environments.

**Table 1.** Heavy metal(loid) tolerance in different species of actinomycetes isolated from contaminated environments.

| Isolation source | Species | Metal (MTC/MIC)* [Mechanism of tolerance] | References |
|---|---|---|---|
| Ex-uranium mining site, Germany | *Streptomyces acidiscabies* E13 | $Ni^{2+}$ (5 mM) | Schmidt et al. 2005, Dimkpa et al. 2008 |
| Ex-uranium mining site, Germany | *S. tendae* F4 | $Cd^{2+}$ (1.5 mM) [biosorption 41.7 mg/g cell dry weight after 7 days] | Schmidt et al. 2005, Sineriz et al. 2009 |
| Riparian sediments contaminated with Ni and Uranium, Steed Pond USA | *S. aureofaciens* NR-3 *Kitasatospora cystarginea* NR-4 | $Ni^{2+}$ (42.1 mM, at pH 6) $Ni^{2+}$ (17.0 mM, at pH 6) | Van Nostrand et al. 2007 |
| Soil contaminated with Cr(III) and Cr(VI), Colombia | *Streptomyces* sp. CG252 | $Cr^{6+}$ (0.5 g/l); $Cu^{2+}$ (1 g/l) | Morales et al. 2007 |
| Copper processing plant wastewater, Tucuman, Argentina | *Streptomyces* sp. M3 | $Cr^{6+}$ (10 mM) [specific removal rate 75.5 mg/g cells] | Polti et al. 2007 |
| Uranium mining site in Eastern Thuringia, Germany | *S. acidiscabies* E13 | $Ni^{2+}$ (Biomineral production] | Haferburg et al. 2008 |
| Copper-contaminated sediments, Tucuman, Argentina | *Amycolatopsis tucumanensis* DSM 45259ᵀ | $Cu^{2+}$ (3 mM) [bioaccumulation 25 mg/g cells] | Albarracın et al. 2008, 2010 |
| Sugarcane plant wastewater (contaminated), Argentina | *Streptomyces* sp. MC1 | $Cr^{6+}$ [30% removal after 96 h of incubation at initial concentration 10 mg/l] | Polti et al. 2009, 2010 |
| Lead-contaminated soil, Gansu province, China | *S. plumbiresistens* CCNWHX 13-160ᵀ | $Pb^{2+}$ (4 mM) | Guo et al. 2009 |
| Acid mine drainage (ex-uranium mine), Germany | *S. mirabilis* P10A-3 & P16A-1 | $Ni^{2+}$ (> 100 mM); Zn (100 mM) | Schmidt et al. 2009 |
| Marine sediment, Marakkanam, India | *Streptomyces* spp. VITSVK5 | $Cr^{3+}$ (1 g/l); $Cr^{6+}$ (1 g/l) | Saurav and Kannabiran 2009 |
| Industrial metal mine | *Streptomyces* sp. K33 | $Cd^{2+}$ [biosorption 38.5 mg/g dry cells] [70% recovery of Cd(2+) at pH ≤ 3] | Yuan et al. 2009 |

*Table 1 contd. ...*

*...Table 1 contd.*

| Isolation source | Species | Metal (MTC/MIC)* [Mechanism of tolerance] | References |
|---|---|---|---|
| Soil from mining area, Urumqi, China | *Kocuria flava* CR1 | $Cu^{2+}$ [97% removal at initial concentration 1 g/l] | Achal et al. 2011 |
| Zinc & copper mine, Shaanxi province, China | *S. zinciresistens* CCNWNQ 0016$^T$ | $Zn^{2+}$ (35 mM) [bioaccumulation 26.8 mg/g cells]; $Cd^{2+}$ (22 mM) [bioaccumulation 85 mg/g cells] | Lin et al. 2012 |
| Extreme habitats and metal contaminated sites, India | *Streptomyces* sp. NK404 | $Cd^{2+}$ (5 mM) [59.49% removal by cells & 61.18% removal by spores] | Prithviraj et al. 2012 |
| Gold mine, Riyadh, Saudi Arabia | *S. diastaticus* and *S. albus* | $Al^{3+}$, $Ag^+$, $Co^{2+}$ (1 mM/25 ml agar medium) | Al-Kadeeb et al. 2012 |
| Heavy metal contaminated farmland, Slovakia | *Streptomyces* sp. | Ni (100 mM); Co (100 mM); Cd (100 mM) | Remenar et al. 2014 |
| Copper iron sulfide ore, Sossego mine, Brazil | *K. marina* SO9-6 | $Cu^{2+}$; $Hg^{2+}$; $As^{2+}$; $Zn^{2+}$ | Castro et al. 2015 |
| Ex-mining area Bir Nhass, Morocco | *Streptomyces* sp. BN3 | $Pb^{2+}$ (0.3 mg/ml) [bioaccumulation 600 mg/g cells] | El Baz et al. 2015 |
| Mining area, Iran | *Promicromonospora* sp. UTMC 2243 | $Cd^{2+}$ (9.2 mM) [96.5% removal after 10 days at initial concentration 5.5 mM] | Hamedi et al. 2015 |
| Zinc mine, Tak Province, Thailand | *Tsukamurella paurometabola* A155 | $Zn^{2+}$ (22 mM) [max. biosorption capacity at 1 6.75 mg/g] | Limcharoensuk et al. 2015 |
| Industrial effluents, Egypt | *Nocardiopsis* sp. MORSY1948 | $Ni^{2+}$, $Cr^{6+}$, $Zn^{2+}$, $Cu^{2+}$, $Co^{2+}$, $Mn^{2+}$ (> 1 g/l) [60.1% $Ni^{2+}$ removal at pH 6] | El-Gendy and El-Bondkly 2016 |
| | *Nocardia* sp. MORSY2014 | $Ni^{2+}$, $Cr^{6+}$, $Zn^{2+}$, $Fe^{3+}$, $Cd^{2+}$, $Ar^{2+}$ (> 1 g/l) [50.5% $Ni^{2+}$ removal at pH 6] | |
| Marine sediments, Chile | *Streptomyces* sp. H-KF8 | $Cu^{2+}$ (0.75 mM); $Co^{2+}$ (6 mM); $Hg^{2+}$ (60 µM); $Cr^{4+}$ (20 mM); $Ni^{2+}$ (15 mM) | Undabarrena et al. 2017 |
| Highly alkaline Al brown mud disposal site, Slovakia | *Streptomyces* sp. K11 | $Zn^{2+}$ (> 150 mM) [bioaccumulation > 4.4 mmol/g cells] | Sedlakova-Kadukova et al. 2019 |

*MTC (maximum tolerable concentration); MIC (minimum inhibitory concentration).

Building on the above presented information, this chapter presents findings from our study on metal(loid) tolerant actinomycetes from an abandoned tin mining site in Malaysia, with a special focus on multi-metal tolerance and the ability of a tolerant *Streptomyces* strain to bioaccumulate $Cd^{2+}$ from solution. Published information is also included to provide a broad view on past research carried out on heavy metal(loid) tolerant actinomycetes, their adaptive mechanisms, and the prospects of using them in industrial applications.

## *1.1 Adaptations to Tolerate Heavy Metal(loid) in Actinomycetes*

In response to high concentrations of heavy metal(loid) ions, many bacteria have evolved different mechanisms of resistance to adapt and survive in the presence of these toxicants in their surroundings (Nies 1999). The genetic characteristics of tolerant species are responsible for various mechanisms involved to control intracellular concentration to reduce the toxicity caused by metal(loid) ions (Nies and Silver 1995). These mechanisms include intracellular metal(loid) sequestration, extracellular metal(loid) sequestration, metal(loid) exclusion by permeability barriers, exclusion from cell by active efflux transporters, bioprecipitation, biotransformation and enzymatic detoxification to a less toxic form (Ramasamy et al. 2006). Figure 1 illustrates the main interactions of metal(loid) ions with bacterial cell.

Metal(loid) uptake by bacteria is categorized as either biosorption or bioaccumulation. Biosorption is a metabolism-independent process in which metal(loid) ions are passively adsorbed onto the cell wall by a number of ways which include adsorption reactions, surface complexation and ion exchange reactions. Various cell wall components such as polysaccharides, proteins, glucans, chitin, mannans and phospholipids contain metal-binding functional groups which are capable of binding to heavy metal(loid) (El Baz 2017, Igiri et al. 2018). Besides living cells, the process is also carried out by dead cell biomass through physicochemical interaction between the ions and cell surface functional groups (Fashola et al. 2016). Efficient metal(loid) biosorption capability was reported in *Streptomyces* sp. K33. The strain showed $Cd^{2+}$ uptake rate of 38.49 mg/g dry cells and used surface functional groups such as amino, carboxyl, hydroxyl and carbonyl groups for its high $Cd^{2+}$ adsorption capacity (Yuan et al. 2009). Similarly, high $Cd^{2+}$ biosorption was observed in *S. tendae* F4 which showed a maximum specific biosorption rate of

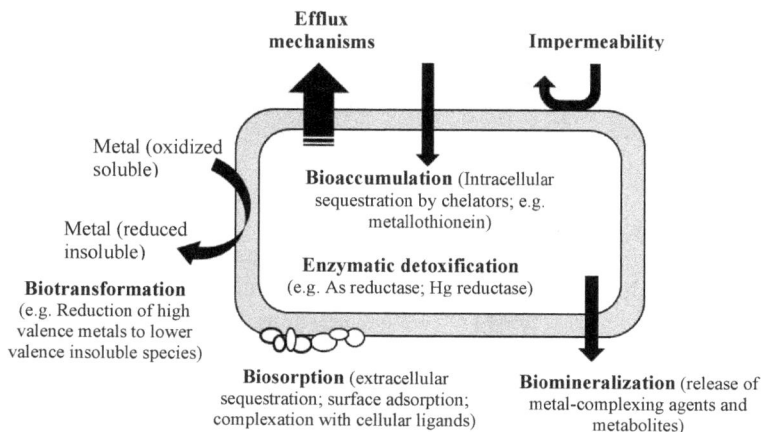

**Figure 1.** Different mechanisms of interaction of metal(loid) ions on surface or inside of bacterial cell (adapted from El Baz 2017 and Gadd 2010).

41.7 mg $Cd^{2+}$/g dry cells after 7 days of growth, with mostly adsorbed in cell wall (Sineriz et al. 2009). Rho and Kim (2002) reported of a high $Pb^{2+}$ adsorption capacity in *S. viridochromogenes* (164 mg/mg cells). However, they also observed that strains showing strongest $Pb^{2+}$ biosorption capacity need not be those with highest tolerance to the metal. Indicating that in some actinomycetes, metal biosorption in cell material, live or dead, may not necessarily be the only factor in metal tolerance.

Metal bioaccumulation is an active process involving energy from cell metabolism to transport metal(loid) ions into cytoplasm through cell membrane. The pollutant is later sequestered by complexation with specific oligopeptides (glutathione) or cysteine-rich protein (metallothionein) chelators and removed through the cell metabolic cycle (Limcharoensuk et al. 2015). The high intracellular uptake of $Cu^{2+}$ demonstrated in *Amycolatopsis tucumanensis* DSM 45259$^T$ (25 mg/g cells), was associated with the efflux system of copper P-Type ATPases (Albarracin et al. 2008, 2010). Sedlakova-Kadukova et al. (2019) observed that the high tolerance to $Zn^{2+}$ in *Streptomyces* sp. K1 was contributed to its ability to uptake metal by both biosorption and bioaccumulation processes. The maximum $Zn^{2+}$ bioaccumulation capacity of 4.4 mmol/g live cells shown by this strain was far higher than its biosorption capacity (0.75 mmol/g dead cells), indicating that high metal bioaccumulating ability is responsible for its extremely high $Zn^{2+}$ tolerance.

Biotransformation in actinomycetes is brought about by a number of enzymatic activities that resulted in formation of less toxic form of heavy metal(loid) by oxidation, reduction, methylation and demethylation (Igiri et al. 2018). Biomethylation involves the transfer of methyl group ($CH_3$) to transform the more toxic inorganic forms of Hg, As, Cd and Pb to methylated species (Fashola et al. 2016). Detoxification by metal(loid) reduction is carried out by some strains of actinomycetes. *Streptomyces griseus* NCIM 2020 (Poopal and Laxman 2009) and *Streptomyces* sp. MC1 (Polti et al. 2010) can effectively reduce the mutagenic chromate to Cr(III) using cell-associated NAD(P)H-dependent chromate reductases. Since the reduction product Cr(III) is quite insoluble, efflux systems which include ATPases and chemiosmotic ion/proton exchangers assist in exporting out the reduced products from cell (Nies 1999, Silver and Phung 2005).

Other adaptation strategies are also used by actinomycetes and these include production of organic compounds such as pigments, biosurfactants (Lakshmipathy et al. 2010) and low-molecular mass chelating secondary metabolites known as siderophores (Dimkpa et al. 2009) which can indirectly increase the mobility of heavy metal(loid) ions. Besides $Fe^{3+}$, siderophores also form complexes with $Cd^{2+}$, $Cr^{3+}$ and other metal(loid)s to transport them across cell membranes (Limcharoensuk et al. 2015). Biomineralization is carried out by some strains of actinomycetes where cells or associated extracellular polymers serve as nucleation surfaces and alter the chemical composition of fluids to raise the supersaturation of solution with a mineral phase. For example, the biomineral nickel struvite $Ni(NH_4)(PO_4).6H_2O$ is formed by viable cells of *S. acidiscabies* E13. The capacity to induce biomineralization was postulated as the resistance mechanism which provided adaptations in this strain to tolerate high Ni concentrations in its habitat, an ex-uranium mining site (Haferburg et al. 2008).

## 1.2 Metal Resistance-encoding Genes in Actinomycetes

Heavy metal(loid) resistance determinants in bacteria can be chromosomal or plasmid-mediated, and they involve many operons. The resistance systems differ from one metal(loid) to another and can be directed against a group of chemically related ions. For instance, arsenate [As(V)] resistance is conferred by the *ars* operon in bacteria and uptake of this toxic oxyanions of arsenic into cytoplasm is mediated by transporters such as phosphate transporters (Silver and Phung 2005). Intracellular As(V) is reduced to arsenite [Ar(III)] by ArsC enzyme encoded by the *arsC* gene, and Ar(III) is transported out either by ATPase or chemiosmotic transporter (Fashola et al. 2016). In resistance to mercury, two different *mer* operons are responsible, namely the *merRTPA* gene for inorganic $Hg^{2+}$ and the *merB* gene which protects cells against organomercurial compounds (Undabarrena et al. 2017). Some of the major metal(loid) resistance operons found in bacteria are shown in Table 2.

Although metal(loid) tolerance genes have been widely studied in Gram negative bacteria, a number of studies on tolerance genes in actinomycetes have been reported. Genes homologous to the mercuric reductase and organomercurial lyase of *mer* operon was reported in *Streptomyces* sp., strains CHR3 and CHR28 isolated from metal-contaminated marine sediments. Both strains showed high resistance to mercuric chloride and phenylmercuric acetate (Ravel et al. 1998). Nickel resistance genes were reported in two actinomycete strains isolated from $Ni^{2+}$ contaminated riparian sediments (Van Nostrand et al. 2007). *Streptomyces aureofaciens* NR-3 and *Kitasatospora cystarginea* NR-4 showed high resistance to $Ni^{2+}$ (42.6 mM at pH 6), $Co^{2+}$ and $Zn^{2+}$. Presence of *nreB* gene which encodes for an efflux transporter specific to $Ni^{2+}$ resistance was detected in both strains. The extremely high level of $Ni^{2+}$ resistance observed in them, however, suggested that other mechanisms of resistance in addition to *nreB* may be present. These include genes for $Ni^{2+}$ uptake transporter nikB, and genes for $Ni^{2+}$ containing metalloproteins. *Kitasatospora*

**Table 2.** Heavy metal(loid) resistance operons and mechanisms in bacteria (adapted from Silver and Phung 2005).

| Metal(loid) ions | Gene mnemonic | Protein function |
|---|---|---|
| $AsO_4^{3-}$ and $As(OH)_3$ | *ars* | Arsenate reductase and transport |
| $AsO_4^{3-}$ | *arr* | Respiratory arsenate reductase |
| $As(OH)_3$ | *aso* | Arsenite oxidase and transport |
| $Hg^{2+}$ and organomercurials | *mer* | Mercuric reductase and transport |
| $Cd^{2+}$ | *cad* | P-type efflux ATPase |
| $Ni^{2+}$ | *nre* | CBA efflux permease |
| $Co^{2+}$, $Ni^{2+}$ | *cnr* | CBA efflux permease |
| $Ni^{2+}$, $Co^{2+}$, $Cd^{2+}$ | *ncc* | CBA efflux permease |
| $Cd^{2+}$, $Zn^{2+}$, $Co^{2+}$ | *czc* | CBA efflux permease |
| $Pb^{2+}$ | *pbr* | Lead resistance and efflux |
| $Cu^{2+}$, $Cu^+$ | *cop* and *pco* | Copper resistance and transport |

*cystarginea* NR-4 also showed the presence of a Zn$^{2+}$ exporter (ZiaA) which can explain the higher level of Zn$^{2+}$ resistance observed compared to *S. aureofaciens* NR-3. In Gram-positive bacteria, accumulation of Cd$^{2+}$ leads to the expression of the *cadA* resistance gene located on plasmids which mediates resistance to Cd$^{2+}$ and Zn$^{2+}$ by active ion efflux systems (Nies and Silver 1995, Oger et al. 2003).

## 1.3 Metallomorphic Habitat in Abandoned Tin Mining Area in Malaysia

Tin mining dictates one of the oldest industries in Malaysia which started in the early 18th century. The state of Perak on west coast of Peninsular Malaysia was the leading venue for tin production. Most of the resources have been exhausted and these areas are left abandoned (Ahmad and Jones 2013). An ex-tin mining area located in the district of Bidor (4°06'N latitude, 101°16'E longitude) in Perak was selected as the metallomorphic site in our study (Fig. 2).

Bidor has a long history of being one of the major tin mining sites during the 1950s but operations ceased by late 1980s. Tin deposits are mainly discovered in alluvial mine pits in the form of cassiterite (SnO$_2$) or tin ore which is associated with other heavy metals, most commonly iron. During the processing of alluvial deposits, a large volume of water is used to suspend fine soil fractions to recover the tin ore. This generated large amount of waste slurry, namely tailings, generally characterized by low fertility and poor physical properties (Yong and Abdul-Talib 2017). Tin tailings in Bidor consisted of sand, slime, and sandy slime tailings. Sand tailings which comprise > 90% coarse sand, is poorer in fertility than slime which consists of > 90% of clay and silt (Ang et al. 2014). Besides the physical properties of low water retention capacity and soil water deficit for plant uptake in the sand tailings, harsh microclimate on the tailing surface with a maximum temperature of 40°C and lowest relative humidity at 32% contributed to the stress conditions in Bidor (Ho et al. 2008). In addition, by-products of long-term mining activities inherently contributed to the high contents of four potentially toxic elements (PTEs) in the Bidor tailings. High levels of Cd (0.02–0.36 and 0.03–0.58 mg/kg), Pb (3.91–21.82 and 13.97–18.27 mg/kg), Hg (0.03–0.07 and 0.08–0.64 mg/kg) and As (0.02–4.48 and 0.03–3.18 mg/kg) were reported in both sand and slime tailings, respectively (Ang et al. 2010). Recent reports showed that high levels of Ni (21.9 mg/kg) and Cu (79.1 mg/kg) were observed in another abandoned tin mine in Perak, even at a soil depth of 100 cm (Ahmad et al. 2018).

Efforts have been taken to rehabilitate and convert tin tailing areas in Bidor to agricultural land. Agricultural products from some of these sites, however, showed high contents of Cd, Pb, Hg and As (Ang and Ng 2000). The identification of PTEs in the tailings, metal uptake by plants and high concentration of each element in agricultural products indicated the impact of metal(loid) contamination resulting from mining activities in Bidor tailings. Environments enriched with high concentration of heavy metal(loid)s is known to act as metallomorphic habitats which harbor microorganisms with remarkable metal-resistance attributes (Pal et al. 2005). According to Wakelin et al. (2011), the metallomorphic environment in mine

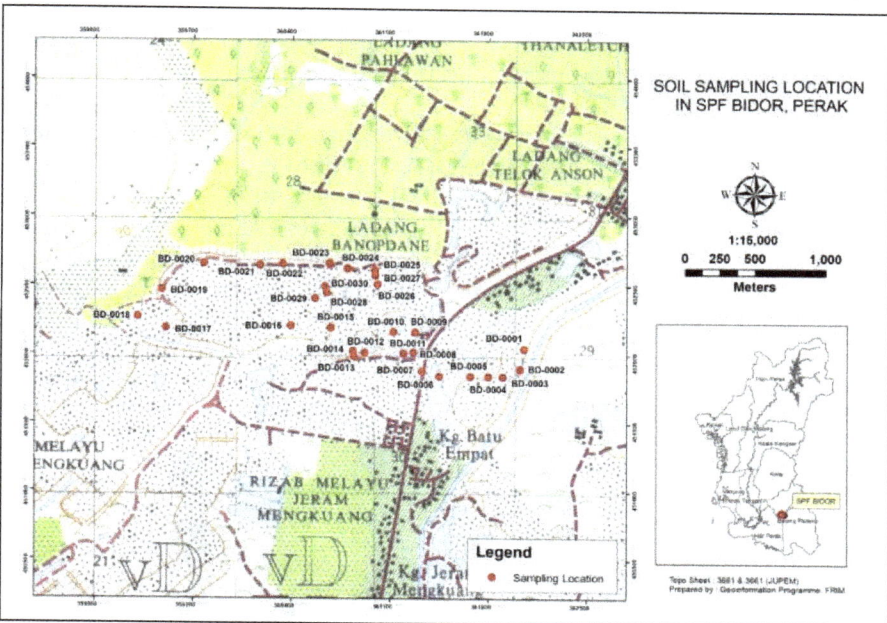

**Figure 2:** (A) Map of Peninsular Malaysia showing state of Perak; (B) Location of sampling points at ex-tin mining area in Bidor (tin tailing samples collected at a depth of 5–10 cm) (Courtesy of the FRIM Geoinformation Programme).

tailings provides an excellent experimental platform to explore geomicrobiological interactions because it produces considerable changes in microbial communities and their activities. Thus, Bidor tin tailings provided an ideal location to examine the adaptation of actinomycetes isolated from this anthropogenically-contaminated environment.

## 1.4 Multi-metal(loid) Tolerant Actinomycetes from Bidor Tailings

Sand and slime tailing samples for isolation of actinomycetes were pretreated using dry heat (60°C) and chemical (yeast extract-sodium dodecyl sulfate) methods (Jiang et al. 2016). Various selective media were used for the isolation using procedures as described earlier (Getha et al. 2014). The isolates were cryopreserved as glycerol suspensions (20%, w/v) and maintained at the Microbial Culture Collection of Forest Research Institute Malaysia (FRIM-MCC; WDCM No. 1129). We employed a two-stage screening technique to select isolates with multi-metal tolerance and high capability to tolerate the toxicants, as described by Hema et al. (2014). In the primary screening, a rapid assay was carried out to evaluate isolates for tolerance to six heavy metal(loid) salts at the following concentration: $Hg^{2+}$ (0.4 mM), $Cd^{2+}$ (0.6 mM), $Pb^{2+}$ (4 mM), $As^{3+}$ (25 mM), $Ni^{2+}$ (39 mM) and $Zn^{2+}$ (37 mM). The assay was based on direct agar diffusion method on Minimal Medium (MM) agar, where use of MM minimizes complexation of the added heavy metal(loid) with ingredients in the medium (Albarracin et al. 2005). In the secondary screening, selected tolerant isolates were grown for 10 days at 28°C in MM broth amended with serially diluted concentrations of the respective metal solutions to which tolerance was observed. The maximum tolerable concentration (MTC) of metal(loid)s that allowed growth was recorded (Hema et al. 2014).

A total of 238 isolates of actinomycetes were isolated from Bidor tailing samples. A remarkable proportion of these isolates (52.1%) were grouped as *Streptomyces*-like based on macromorphological and cultural characteristics (Hema et al. 2014). Results from primary screening indicated that the isolates exhibited a high incidence of tolerance to As, Zn, Cd and Pb (> 20%) at the test concentration. Nickel tolerance was observed in about 15% of isolates, while only less than 1% was tolerant to Hg (Fig. 3). Although high concentrations of Cd, As, Pb and Hg contaminants

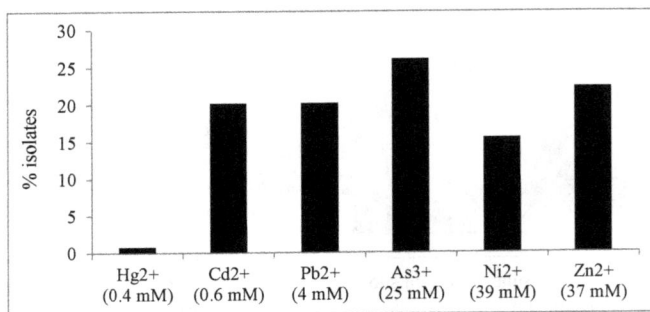

**Figure 3.** Percentage of actinomycetes showing tolerance to test concentration of different metal(loid) solutions in agar diffusion assay (primary screening).

reported in Bidor tailings (Ang et al. 2010) could have created the selective pressure for soil microbes, microorganisms within the environment may not be exposed to the total concentration of metal(loid)s but rather to the bioavailable fraction. Other factors, including pH and organic content can influence bioavailability (Saurav and Kannabiran 2009). Actinomycetes capable of growing at higher concentrations of Cd, As and Pb compared with Hg suggests that the former could have exerted a stronger selective pressure for the microbes in the tailings. Van Nostrand et al. (2007) used a concentration of > 1 mM $Ni^{2+}$ to benchmark resistance in bacteria. Others like Undabarrena et al. (2017) explained resistance when bacteria can tolerate a concentration of > 20 μM $Hg^{2+}$. In our study, the term tolerance is employed instead because according to Nies (2003), the ability to grow in the presence of metal toxicants is referred as resistance only when a specific mechanism has been identified.

Overall, 109 isolates (45.8% of the total isolates) possessed tolerance capability to either one or more metal(loid)s. Tolerance patterns of all isolates are presented in Table 3. The observed high percentage clearly indicated that metal tolerant phenotype can be dominant and widely distributed among actinomycetes in Bidor tailings. Among these isolates, 42 showed multi tolerance to three and more metals. Ten isolates showed the highest level of multi tolerance to five out of the six heavy metals. In the tetra-metal tolerant group, the highest pattern of tolerance (6.4%) was concurrently related to Pb, As, Ni and Zn. While in the tri-metal tolerant group, the highest pattern was related to As, Ni and Zn. Considering this pattern, the tolerance to these three metal(loid)s can be considerably greater than other metals (Table 3). According to Lal et al. (2013) and Nanda et al. (2019), multi-metal resistance in bacteria presents the ability to thrive in toxic environments with high level of heavy metals. The co-selection of genetic and biochemical pathways might be involved in multi-metal resistance and/or tolerance (Harrison et al. 2007). In a previous study, bacterial isolates from serpentine soils containing high contents of Co, Ni and Cr were compared with non-serpentine soils. The serpentine microflora was highly tolerant than non-serpentine ones and showed strong tolerance towards Ni, Co and Cr. The highly tolerant isolates also showed co-resistance to Cu and Zn, indicating that multi metal-resistance/tolerance was acquired for survival in such a stressed habitat (Pal et al. 2005).

In the secondary screening, 31 representative isolates from multi-metal group and from the di- and mono-metal tolerance group, were selected for determination of maximum tolerable concentration (MTC) to metals in broth cultures. Out of 17 multi-metal isolates screened, a high percentage (82.4% or 14 isolates) showed capability to tolerate very high concentration of at least one metal(loid) in the assay. On the other hand, only 2 out of the 14 isolates without multi-metal phenotype could tolerate and grow in such a high metal concentration (Table 4). This indicated importance of multi-metal tolerance for survival in the stressed environment. Our study observed some isolates of actinomycetes which can tolerate very high concentrations of Pb (43.2 mM), Ni (138.9 mM), Cd (21.8 mM) and As (182 mM), higher than

**Table 3.** Types of metal tolerance and tolerant patterns observed in a total of 109 isolates selected from primary screening.

| Types of metal tolerance (T) | No. of Isolates (% of total) | Isolates showing the tolerance pattern | | Metal tolerant patterns* | | | | | |
|---|---|---|---|---|---|---|---|---|---|
| | | No. (%) | % of total | As | Pb | Cd | Ni | Zn | Hg |
| Multi metal (Penta-T) | 10 (9.2) | 10 (100) | | + | + | + | + | + | − |
| Multi metal (Tetra-T) | 15 (13.8) | 7 (46.7) | 6.4 | + | + | − | + | + | − |
| | | 5 (33.3) | 4.6 | + | + | + | − | + | − |
| | | 1 (6.7) | 0.9 | + | − | + | + | + | − |
| | | 1 (6.7) | 0.9 | + | + | + | + | − | − |
| | | 1 (6.7) | 0.9 | + | − | + | + | + | − |
| Multi metal (Tri-T) | 17 (15.6) | 5 (29.4) | 4.6 | + | − | − | + | + | − |
| | | 2 (11.8) | 1.8 | + | + | − | − | + | − |
| | | 2 (11.8) | 1.8 | + | + | − | + | − | − |
| | | 2 (11.8) | 1.8 | + | − | + | − | + | − |
| | | 2 (11.8) | 1.8 | − | + | + | − | + | − |
| | | 1 (5.9) | 0.9 | + | + | + | − | − | − |
| | | 1 (5.9) | 0.9 | + | − | + | − | − | + |
| | | 1 (5.9) | 0.9 | − | − | + | + | + | − |
| | | 1 (5.9) | 0.9 | − | − | + | + | − | + |
| Di-T | 21 (19.3) | 4 (19) | 3.7 | + | + | − | − | − | − |
| | | 4 (19) | 3.7 | + | − | + | − | − | − |
| | | 4 (19) | 3.7 | − | − | + | − | + | − |
| | | 3 (9.7) | 2.8 | + | − | − | + | − | − |
| | | 3 (9.7) | 2.8 | + | − | − | − | + | − |
| | | 1 (4.8) | 0.9 | − | + | + | − | − | − |
| | | 1 (4.8) | 0.9 | − | + | − | + | − | − |
| | | 1 (4.8) | 0.9 | − | − | − | + | + | − |
| Mono-T | 46 (42.2) | 13 (28.3) | 11.9 | − | − | + | − | − | − |
| | | 11 (23.9) | 10.1 | − | + | − | − | − | − |
| | | 10 (21.7) | 9.2 | + | − | − | − | − | − |
| | | 9 (19.6) | 8.3 | − | − | − | − | + | − |
| | | 3 (6.5) | 2.8 | − | − | − | + | − | − |

* +: Tolerance to metal(loid) at the test concentration; −: growth inhibited.

some reported in previous studies: 4 mM Pb (Guo et al. 2009); 85.2 nM Ni (Van Nostrand et al. 2007); 130 mM Ni (Schmidt et al. 2009); 22 mM Cd (Lin et al. 2012); 60–90 mM As (Kermanshahi et al. 2007). Mercury, a highly toxic heavy metal, was tolerated at low concentration of 0.8 mM by one multi-metal isolate (Table 4). Low

**Table 4.** Maximum tolerable concentration (MTC) to metal(loid)s for 31 tolerant isolates (16 potential isolates in bold).

| Isolate codes | Metal MTC values (mM)* | | | | | Metal tolerance in primary screening (mM)** |
|---|---|---|---|---|---|---|
| | Pb²⁺ | Ni²⁺ | As³⁺ | Cd²⁺ | Hg²⁺ | |
| **TY046-078** | **43.2** | **138.9** | **182** | nt | nt | Pb²⁺(4); Ni²⁺(39); As³⁺(25); Zn²⁺(37) |
| **TY046-021** | **43.2** | **138.9** | **182** | **10.9** | nt | Pb²⁺(4); Ni²⁺(39); As³⁺(25); Cd²⁺(0.5); Zn²⁺(37) |
| **TY046-071** | 5.4 | **138.9** | 45.5 | 2.7 | nt | Pb²⁺(4); Ni²⁺(39); As³⁺(25); Cd²⁺(0.5); Zn²⁺(37) |
| **TY047-019** | 5.4 | **138.9** | 11.4 | nt | nt | Pb²⁺(4); Ni²⁺(39); As³⁺(25); Zn²⁺(37) |
| **TY046-027** | 5.4 | **138.9** | 11.4 | 2.7 | nt | Pb²⁺(4); Ni²⁺(39); As³⁺(25); Cd²⁺(0.5); Zn²⁺(37) |
| **TY046-073** | 2.7 | 17.4 | **182** | 2.7 | nt | Pb²⁺(4); Ni²⁺(39); As³⁺(25); Cd²⁺(0.5); Zn²⁺(37) |
| **TY046-016** | **43.2** | 69.5 | 11.4 | 2.7 | nt | Pb²⁺(4); Ni²⁺(39); As³⁺(25); Cd²⁺(0.5); Zn²⁺(37) |
| **TY046-018** | 5.4 | 17.4 | **91** | 2.7 | nt | Pb²⁺(4); Ni²⁺(39); As³⁺(25); Cd²⁺(0.5); Zn²⁺(37) |
| TY046-004 | 5.4 | 17.4 | 11.4 | 2.7 | nt | Pb²⁺(4); Ni²⁺(39); As³⁺(25); Cd²⁺(0.5); Zn²⁺(37) |
| **TY047-062** | 5.4 | 17.4 | 22.8 | **10.9** | nt | Pb²⁺(4); Ni²⁺(39); As³⁺(25); Cd²⁺(0.5); Zn²⁺(37) |
| TY046-017 | nt | 69.5 | nt | 2.7 | 0.8 | Ni²⁺(39); Hg²⁺(0.4); Cd²⁺(0.5) |
| **TY048-047** | nt | nt | 11.4 | **10.9** | – | As³⁺(25); Hg²⁺(0.4); Cd²⁺(0.5) |
| **TY049-057** | 5.4 | nt | 11.4 | **21.8** | nt | Pb²⁺(4); As³⁺(25); Cd²⁺(0.5) |
| TY046-070 | 5.4 | 34.7 | 22.8 | 2.7 | nt | Pb²⁺(4); Ni²⁺(39); As³⁺(25); Cd²⁺(0.5) |
| **TY045-023** | **43.2** | 34.7 | 22.8 | nt | nt | Pb²⁺(4); Ni²⁺(39); As³⁺(25); Zn²⁺(37) |
| **TY046-079** | **43.2** | nt | **182** | 2.7 | nt | Pb²⁺(4); As³⁺(25); Cd²⁺(0.5); Zn²⁺(37) |
| **TY047-009** | 5.4 | nt | **182** | nt | nt | Pb²⁺(4); As³⁺(25); Zn²⁺(37) |
| TY048-042 | nt | nt | nt | 2.7 | nt | Cd²⁺(0.5) |
| TY047-020 | nt | nt | nt | 2.7 | nt | Cd²⁺(0.5) |
| TY046-037 | 5.4 | nt | nt | 1.4 | nt | Pb²⁺(4); Cd²⁺(0.5) |
| TY047-024 | nt | 17.4 | 22.8 | nt | nt | As³⁺(25); Ni²⁺(39) |
| **TY048-049** | **43.2** | nt | nt | nt | nt | Pb²⁺(4) |
| **TY047-063** | nt | nt | **182** | nt | nt | As³⁺(25) |
| TY047-025 | nt | 34.7 | nt | nt | nt | Ni²⁺(39) |
| TY047-014 | 5.4 | nt | 11.4 | nt | nt | Pb²⁺(4); As³⁺(25) |
| TY047-023 | 5.4 | nt | nt | nt | nt | Pb²⁺(4) |
| TY047-027 | 5.4 | nt | nt | nt | nt | Pb²⁺(4) |
| TY048-038 | 5.4 | 17.4 | nt | nt | nt | Pb²⁺(4); Ni²⁺(39) |
| TY046-068 | nt | 17.4 | nt | nt | nt | Ni²⁺(39); Zn²⁺(37) |
| TY047-044 | nt | nt | 11.4 | nt | nt | As³⁺(25) |
| TY047-073 | 5.4 | nt | nt | nt | nt | Pb²⁺(4) |

*Metal MTC values supporting actinobacteria growth; **Test concentration showing bacterial tolerance.
–: no growth in all test concentration; nt: not tested.

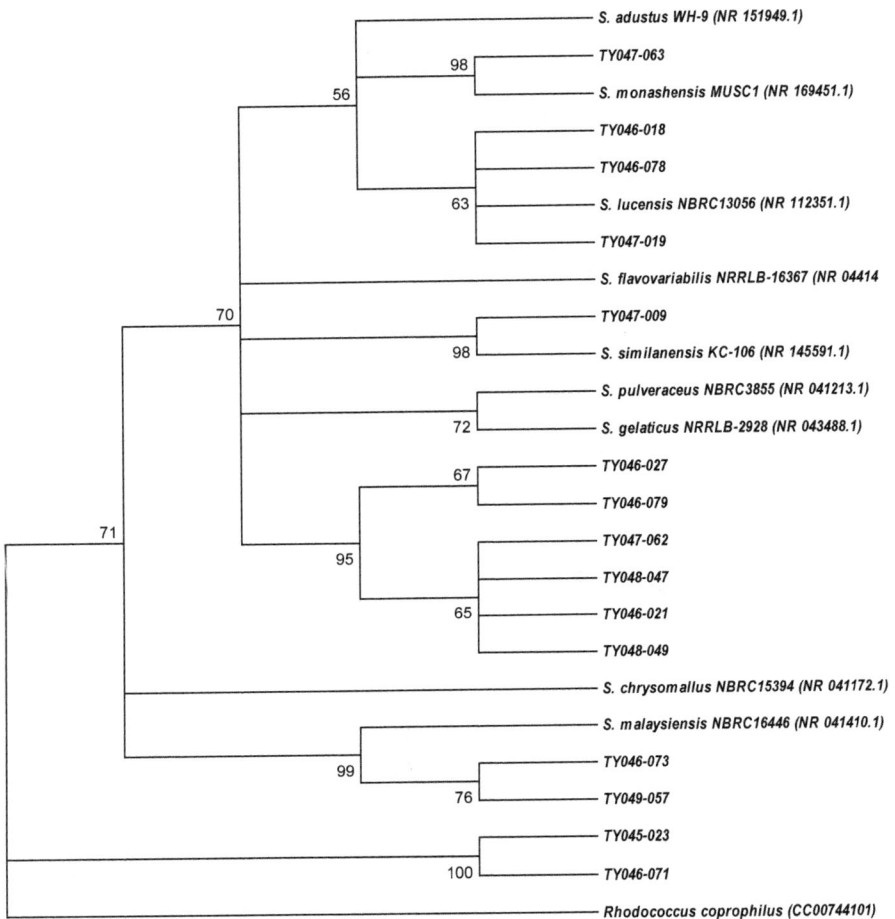

**Figure 4.** Phylogenetic tree (neighbor-joining method) based on partial 16S rRNA gene sequence showing relationship between heavy metal tolerant isolates with related members of the *Streptomyces* species. Numbers at the nodes indicate level of bootstrap support based on maximum likelihood analysis of 1000 resampled datasets.

Hg tolerance in actinomycetes has been reported previously (Amoroso et al. 1998, Ravel et al. 1998).

Analysis of 16S rRNA gene sequence was carried out for 16 metal(loid) tolerant isolates according to the methods described by Lili Sahira et al. (2015). The PCR products were sequenced at First Base Laboratories Sdn. Bhd. (Shah Alam, Selangor) and resulting 16S rRNA gene sequences were aligned by ClustalW program before constructing phylogenetic tree (neighbor joining method) using the MEGA 7.0 software. All isolates were identified as members of the *Streptomyces* genus except isolate TY046-016 which belonged to *Nocardia* genus (Figs. 4 and 5).

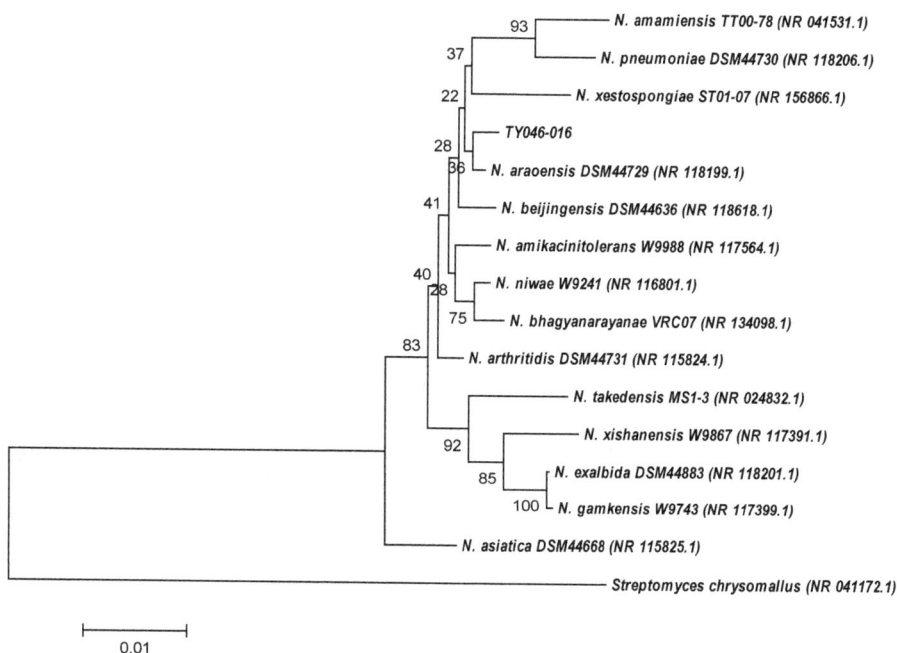

**Figure 5.** Phylogenetic tree (neighbor-joining method) based on partial 16S rRNA gene sequence showing relationship between heavy metal tolerant isolate TY046-016 with related members of the *Nocardia* species. Numbers at the nodes indicate level of bootstrap support based on maximum likelihood analysis of 1000 resampled datasets.

## 1.5 Cadmium Uptake in Streptomyces malaysiensis Strain TY049-057

A multi-metal tolerant actinomycetes, strain TY049-057, showed potential ability to tolerate high concentration of 21.8 mM Cd. The strain is phylogenetically related to *Streptomyces malaysiensis* NBRC 16446 (99.85% similarities) and based on comparison with literature (Al-Tai et al. 1999), it was identified as *S. malaysiensis* TY049-057. Strain TY049-057 was selected for metal uptake study since Cd is one of the most toxic amongst the metal(loid)s included in this study, besides Hg. With an exceptionally long biological half-life of more than 20 years and high mobility in soil-plant systems, Cd can exert serious detrimental effects on the ecosystems (Fashola et al. 2016). The ability to uptake metal(loid) ions by metal-tolerant actinomycetes has been well studied previously to look at the prospects of developing cheaper and more effective microorganism-based bioremediation techniques for contaminated industrial effluents (El Baz et al. 2015).

The method modified from Albarracin et al. (2005) was used in this study. Spore suspension of strain TY049-057 was inoculated in MM broth amended with 1 mM $CdCl_2$ and a Cd sensitive strain TY049-044 was included as a comparison. Cultures in MM broth without $Cd^{2+}$ were used as controls. The flask cultures were incubated in orbital shaker at 28°C and aliquots of the broth were collected at intervals between

2 to 10 days of growth. Cell viability was determined by calculation of average CFU (colony forming units) readings from triplicate flasks. Remaining cells were collected by centrifugation, rinsed and dried at 100°C before digesting with concentrated nitric acid to determine the $Cd^{2+}$ concentration in biomass by Inductively Coupled Plasma (ICP) analysis.

Presence of $Cd^{2+}$ showed remarkable effect in the time dependent growth profile of the metal sensitive strain. Growth of TY049-044 immediately reduced and ceased after 2 days of incubation (Fig. 6C). In strain TY049-057, presence of $Cd^{2+}$ during growth extended the exponential growth phase where time taken to achieve maximum cell mass was shifted from 2 days to about 4 days (Fig. 6A). The corresponding effects were seen clearly when there was no increase in $Cd^{2+}$ concentration in the biomass of TY049-044 because cell growth has already ceased (Fig. 6D). The time-course of metal uptake in strain TY049-057 demonstrated rapid uptake initially during the exponential growth phase (Fig. 6B). This is considered quite normal as biosorption is a spontaneous process and often occurs very rapidly (Lin et al. 2012). Growth of strain TY049-057 then slows down after the 4th day which was also reflected in the reduced uptake of $Cd^{2+}$. Cadmium uptake then enters a second phase with a maximum uptake on day 8 where cell growth was already declined. The uptake here was probably not only due to cell-surface binding, but most probably was also assisted by other mechanisms.

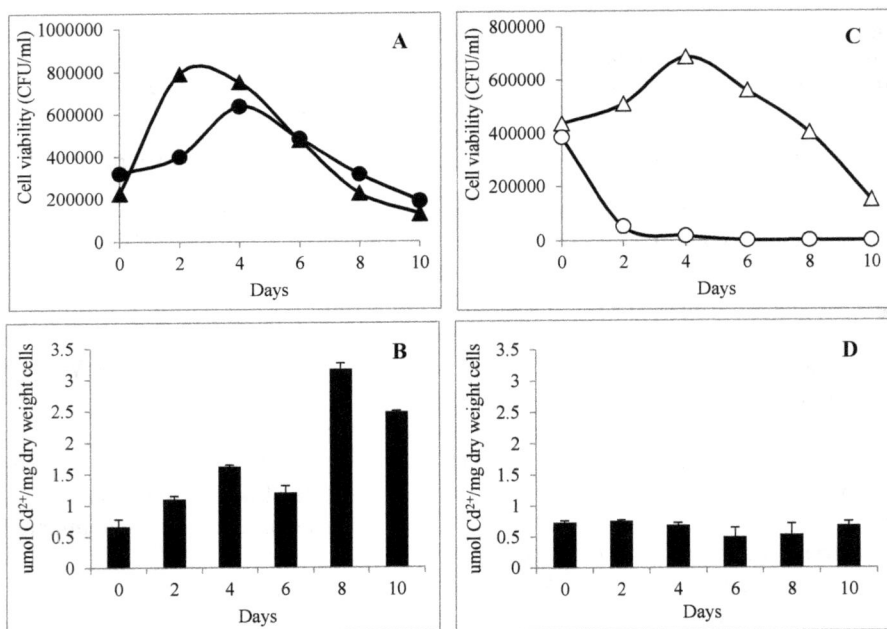

**Figure 6.** (A and B) Strain TY049-057 and (C and D) Strain TY049-044 grown in MM broth in the presence of 0 mM (▲, Δ) and 1 mM (●, ○) $CdCl_2$. (A and C: Cell growth/CFU; B and D: $Cd^{2+}$ concentration in biomass).

**Footnote:** Error bars are standard deviation of triplicate samples.

Lin et al. (2012) reported that when *S. zinciresistens* was grown in liquid media containing 2 mM $Cd^{2+}$, maximum metal adsorption in cell wall took place during the exponential growth phase. This was different from our results where maximum uptake occurred during declining growth phase of strain TY049-057. Production of secondary metabolites such as siderophores by *Streptomyces* is known to be stimulated by heavy metals (Lin et al. 2012). Cadmium was observed to stimulate production of siderophores in *S. tendae* F4 (Dimpka et al. 2008). In order to survive in an environment contaminated by metals, *Streptomyces* produce a wide range of metal ion chelators, such as siderophores and extracellular polymeric substances (EPS) which together play a role in uptake of toxic heavy metal (Timkova et al. 2018). Further studies are needed to explain this mechanism in *S. malaysiensis* strain TY049-057 to understand $Cd^{2+}$ uptake in this strain.

## 2. Conclusions and Future Perspectives

In view of studies from previous researches, investigation was carried out further to obtain a better knowledge on the impact of Bidor tin tailings to harbor heavy metal-resistant actinomycetes. The aim was to gain insight into types of multi-metal tolerance and levels observed in the isolated actinomycetes which could assist in survival and growth in this metalliferous environment. The current study is the first report of a *S. malaysiensis* strain from an abandoned tin mining area showing tolerance to toxic heavy metal cadmium and could be explored further for its potentials. Our results showed a high $Cd^{2+}$ uptake by *S. malaysiensis* strain TY049-057 with a maximum uptake of 358.5 mg $Cd^{2+}$/g dry cell (3.17 umol/mg) recorded after 8 days incubation in culture broth amended with 1 mM $CdCl_2$.

Many processes could be used in the bioremediation of contaminated wastewaters and soils. Microorganisms such as actinomycetes may be a good tool which can be used to treat metal contaminated wastes prior to discharge into the environment. Further studies on mechanisms of the uptake system and kinetic studies are needed in order to use the isolate/s from the current study in future as a bioremediation agent. The genome of *S. malaysiensis* strain TY049-057 will be interesting to be analyzed to understand the resistance mechanism/s of the strain towards the three-toxic metal(loid)s (Pb, As and Cd) which the strain showed tolerance.

## Acknowledgements

The authors gratefully thank the Ministry of Natural Resources, Environment and Climate Change for the RMK-12 Development project (24010704001) and Ministry of Science, Technology & Innovation for the Science Fund project (02-03-10-SF0108) grants. The authors would also like to thank the management of Bidor Research Station, Perak for permission to carry out sampling and Ms. Rodziah Hashim from the Geoinformation Programme, FRIM for providing the map images.

# References

Abbas, A. and Edwards, C. (1989). Effects of metals on a range of *Streptomyces* species. *Applied and Environmental Microbiology*, 55(8): 2030–2035.

Achal, V., Pan, X. and Zhang, D. (2011). Remediation of copper-contaminated soil by *Kocuria flava* CR1, based on microbially induced calcite precipitation. *Ecological Engineering*, 37(10): 1601–1605.

Ahmad, S. and Jones, D. (2013). Investigating the mining heritage significance for Kinta district, the industrial heritage legacy of Malaysia. *Procedia – Social and Behavioral Sciences*, 105: 445–457.

Ahmad, Z.Y., Jeyanny, V., Fakhri, M.I. and Wan Rasidah, K. (2018). Reversing soil degradation via phytoremediaton techniques in an ex-tin mine and gold mine in Peninsular Malaysia. pp. 686–691. *Proceedings of the Global Symposium of Soil Pollution*, 2–4 May 2018. Food and Agriculture Organization (FAO) Headquarters, Rome, Italy.

Albarracın, V.H., Alonso-Vega, P., Trujillo, M.E., Amoroso, M.J. and Abate, C.M. (2010). *Amycolatopsis tucumanensis* sp. nov., a copper-resistant actinobacterium isolated from polluted sediments. *International Journal of Systematic and Evolutionary Microbiology*, 60: 397–401.

Albarracin, V.H., Amoroso, M.J. and Abate, C.M. (2005). Isolation and characterisation of indigenous copper-resistant actinomycete strains. *Chemie der Erde - Geochemistry 65(Supplement 1)*, 145–156.

Albarracın, V.H., Winik, B., Kothe, E., Amoroso, M.J. and Abate, C.M. (2008). Copper bioaccumulation by the actinobacterium *Amycolatopsis* sp. AB0. *Journal of Basic Microbiology*, 48(5): 323–330.

Al-Kadeeb, S.A., Al-Rokban, A.H. and Awad, F. (2012). Characterization and identification of actinomycetes isolated from contaminated soil in Riyadh. *International Journal of Scientific & Engineering Research*, 3(8): 1–8.

Al-Tai, A., Kim, B., Kim, S.B., Manfio, G.P. and Goodfellow, M. (1999). *Streptomyces malaysiensis* sp. nov., a new streptomycete species with rugose, ornamented spores. *International Journal of Systematic Bacteriology*, 49: 1395–1402.

Alvarez, A., Saez, J.M., Costa, J.S.D., Colin, V.L., Fuentes, M.S., Cuozzo, S.A., Benimeli, C.S., Polti, M.A. and Amoroso, M.J. (2017). Actinobacteria: Current research and perspectives for bioremediation of pesticides and heavy metals. *Chemosphere*, 166: 41–62.

Amoroso, M.J., Castro, G.R., Carlino, F.J., Romero, N.C., Hill, R.T. and Oliver, G. (1998). Screening of heavy metal-tolerant actinomycetes isolated from the Sali River. *Journal of General & Applied Microbiology*, 44: 129–132.

Ang, L.H. and Ng, L.T. (2000). Trace element concentration in mango (*Mangifera indica* L.), seedless guava (*Psidium guajava* L.) and papaya (*Carica papaya* L.) grown on agricultural and ex-mining lands of Bidor, Perak. *Pertanika Journal of Tropical Agricultural Sciences*, 23(1): 15–22.

Ang, L.H., Ho, W.M. and Tang, L.K. (2014). A model of greened ex-tin mine as lowland biodiversity depository in Malaysia. *Journal of Wildlife and Parks*, 29: 61–67.

Ang, L.H., Tang, L.K., Ho, W.M., Hui, T.F. and Theseira, G.W. (2010). Phytoremedation of Cd and Pb by four tropical timber species grown on an ex-tin mine in Peninsular Malaysia. *International Journal of Environmental and Ecological Engineering*, 4(2): 70–74.

Bhakta, J.N., Munekage, Y., Ohnishi, K., Jana, B. and Balcazar, J.L. (2014). Isolation and characterization of cadmium- and arsenic-absorbing bacteria for bioremediation. *Water, Air & Soil Pollution*, 225: 2151–2161.

Castro, D.B.A., Pereira, L.B., eSilva, M.V.M., da Silva, B.P., Palermo, B.R.Z., Carlos, C., Belgini, D.R.B., Limache, E.E.C., Lacerda, G.V.J., Nery, M.B.P., Gomes, M.B., de Souza, S.S., da Silva, T.M., Rodrigues, V.D., Paulino, L.C., Vicentini, R., Ferraz, L.F.C. and Ottoboni, L.M.M. (2015). High quality draft genome sequence of *Kocuria marina* SO9-6, an actinobacterium isolated from a copper mine. *Genomics Data*, 5: 34–35.

Coelho, L.M., Rezende, H.C., Coelho, L.M., de Sousa, P.A.R., Melo, D.F.O. and Coelho, N.M.M. (2015). Bioremediation of polluted waters using microorganisms. *In: Advances in Bioremediation of Wastewater and Polluted Soil*, Naofumi Shiomi, IntechOpen, Doi: 10.5772/60770. Available from: https://www.intechopen.com/books/advances-in-bioremediation-of-wastewater-and-polluted-soil/bioremediation-of-polluted-waters-using-microorganisms.

de Quadros, P.D., Zhalnina, K., Davis-Richardson, A.G., Drew, J.C., Menezes, F.B., Camargo, F.A.O. and Triplett, E.W. (2016). Coal mining practices reduce the microbial biomass, richness and diversity of soil. *Applied Soil Ecology*, 98: 195–203.

Dimkpa, C.O., Merten, D., Svatos, A., Buchel, G. and Kothe, E. (2009). Siderophores mediate reduced and increased uptake of cadmium by *Streptomyces tendae* F4 and sunflower (*Helianthus annuus*), respectively. *Journal of Applied Microbiology*, 107: 1687–1696.

Dimkpa, C.O., Svatos, A., Merten, D., Buchel, G. and Kothe, E. (2008). Hydroxamate siderophores produced by *Streptomyces acidiscabies* E13 bind nickel and promote growth in cowpea (*Vigna unguiculata* L.) under nickel stress. *Canadian Journal of Microbiology*, 54: 163–172.

Divakar, G., Sameer, R.S. and Bapuji, M. (2018). Screening of multi-metal tolerant halophilic bacteria for heavy metal remediation. *International Journal of Current Microbiology and Applied Sciences*, 7(10): 2062–2076.

El Baz, S. (2017). Bioremediation of heavy metals by actinobacteria: Review. *American Journal of Innovative Research and Applied Sciences*, 5(5): 359–369.

El Baz, S., Baz, M., Barakate, M., Hassani, L., El Gharmali, A. and Imziln, B. (2015). Resistance to and accumulation of heavy metals by actinobacteria isolated from abandoned mining areas. *The Scientific World Journal* volume 2015, Article ID 761834. pp. 14. Doi: 10.1155/2015/761834.

El-Gendy, M.M.A.A. and El-Bondkly, A.M.A. (2016). Evaluation and enhancement of heavy metals bioremediation in aqueous solutions by *Nocardiopsis* sp. MORSY1948, and *Nocardia* sp. MORSY2014. *Brazilian Journal of Microbiology*, 47: 571–586.

Fashola, M.O., Ngole-Jeme, V.M. and Babalola, O.O. (2016). Heavy metal pollution from gold mines: Environmental effects and bacterial strategies for resistance. *International Journal of Environmental Research and Public Health*, 13: 1047–1066.

Feng, G., Xie, T., Wang, X., Bai, J., Tang, L., Zhao, H., Wei, W., Wang, M. and Zhao, Y. (2018). Metagenomic analysis of microbial community and function involved in Cd-contaminated soil. *BMC Microbiology*, 18: 11. Available at: https://doi.org/10.1186/s12866-018-1152-5.

Fernandes, C.C., Kishi, L.T., Lopes, E.M., Omori, W.P., de Souza, J.A.M., Alves, L.M.C. and de Macedo Lemos, E.G. (2018). Bacterial communities in mining soils and surrounding areas under regeneration process in a former ore mine. *Brazilian Journal of Microbiology*, 49: 489–502.

Gadd, G.M. (2010). Metals, minerals and microbes: Geomicrobiology and bioremediation. *Microbiology*, 156: 609–643.

Garrett, R.G. (2000). Natural sources of metals to the environment. *Human and Ecological Risk Assessment*, 6(6): 945–962.

Getha, K., Hatsu, M., Nur Fairuz, M.Y., Hema Thopla, G., Lili Sahira, H. and Muhd Syamil, A. (2014). Biosynthetic potential of phylogenetically and biochemically unique actinobacteria from Malaysian soil. pp. 152–157. *In*: Rahim Sudin, S., Lim, H.F., Huda Farhana, M.M. and Mahmudin, S. (eds.). *Proceedings of the Conference on Forestry and Forest Products Research 2013 (CFFPR): Forestry R&D: Meeting National and Global Needs*, 12–13 Nov 2013, Kuala Lumpur. ISBN 978-967-0622-26-2.

Guo, J.K., Lin, Y.B., Zhao, M.L., Sun, R., Wang, T.T., Tang, M. and Wei, G.H. (2009). *Streptomyces plumbiresistens* sp. nov., a lead-resistant actinomycete isolated from lead-polluted soil in north-west China. *International Journal of Systematic and Evolutionary Microbiology*, 59: 1326–1330.

Haferburg, G., Kloess, G., Schmitz, W. and Kothe, E. (2008). "Ni-struvite"—A new biomineral formed by a nickel resistant *Streptomyces acidiscabies*. *Chemosphere*, 72: 517–523.

Hamedi, J., Dehhaghi, M. and Mohammdipanah, F. (2015). Isolation of extremely heavy metal resistant strains of rare actinomycetes from high metal content soils in Iran. *International Journal of Environmental Research*, 9: 475–480.

Hema, T.G., Getha, K., Tan, G.Y.A., Lili Sahira, H., Muhd Syamil, A. and Nur Fairuz, M.Y. (2014). Actinobacteria isolates from tin tailings and forest soil for bioremediation of heavy metals. *Journal of Tropical Forest Science*, 26: 153–162.

Ho, W.M., Ang, L.H. and Lee, D.K. (2008). Assessment of Pb uptake, translocation and immobilization in kenaf (*Hibiscus cannabinus* L.) for phytoremediation of sand tailings. *Journal of Environmental Sciences*, 20: 1341–1347.

Igiri, B.E., Okoduwa, S.I.R., Idoko, G.O., Akabuogu, E.P., Adeyi, A.O. and Ejiogu, I.K. (2018). Toxicity and bioremediation of heavy metals contaminated ecosystem from tannery wastewater: A review. *Journal of Toxicology* Volume 2018, Article ID 2568038, 16 pages. Available at: https://doi.org/10.1155/2018/2568038.

Jiang, Y., Li, Q., Chen, X. and Jiang, C. (2016). Isolation and cultivation methods of actinobacteria. *In*: Dhanasekaran, D. and Jiang, Y. (eds.). Actinobacteria - Basics and Biotechnological Applications. IntechOpen, Doi: 10.5772/61457. Available from: https://www.intechopen.com/books/actinobacteria-basics-and-biotechnological-applications/isolation-and-cultivation-methods-of-actinobacteria.

Kapahi, M. and Sachdeva, S. (2019). Review: Bioremediation options for heavy metal pollution. *Journal of Health & Pollution*, 9(24): 1–20.

Karnachuk, O.V., Sasaki, K., Gerasimchuk, A.L., Sukhanova, O., Ivasenko, D.A., Kaksonen, A.H., Puhakka, J.A. and Tuovinen, O.H. (2008). Precipitation of Cu-sulfides by copper-tolerant *Desulfovibrio* isolates. *Geomicrobiology Journal*, 25(5): 219–227.

Kermanshahi, R.K., Ghazifard, A. and Tavakoli, A. (2007). Identification of bacteria resistant to heavy metals in the soils of Isfahan Province. *Iranian Journal of Science & Technology Transaction A*, 31(A1): 7–16.

Lakshmipathy, T.D., Prasad, A.S.A. and Kannabiran, K. (2010). Production of biosurfactant and heavy metal resistance activity of Streptomyces sp. VITDDK3—A novel halo tolerant actinomycetes isolated from saltpan soil. *Advances in Biological Research*, 4(2): 108–115.

Lal, D., Nayyar, N., Kohli, P. and Lal, R. (2013). *Cupriavidus metallidurans*: A modern alchemist. *Indian Journal of Microbiology*, 53(1): 114–115.

Lili Sahira, H., Getha, K., Muhd Syamil, A., Hema, T.G., Norhayati, I., Muhd Haffiz, J., Siti Syarifah, M.M. and Mohd Ilham, A. (2015). Taxonomic characterization and isolation of antitrypanosomal compound from *Streptomyces* sp. FACC-A032 isolated from Malaysian forest soil. *Malaysian Journal of Microbiology*, 11(2): 128–136.

Limcharoensuk, T., Sooksawat, N., Sumarnote, A., Awutpet, T., Kruatrachue, M., Pokethitiyook, P. and Auesukaree, C. (2015). Bioaccumulation and biosorption of $Cd^{2+}$ and $Zn^{2+}$ by bacteria isolated from a zinc mine in Thailand. *Ecotoxicology and Environmental Safety*, 122: 322–330.

Lin, Y., Wang, X., Wang, B., Mohamad, O. and Wei, G. (2012). Bioaccumulation characterisation of zinc and cadmium by *Streptomyces zinciresistens*, a novel actinomycete. *Ecotoxicology and Environmental Safety*, 77: 7–17.

Lin, Y.B., Wang, X.Y., Li, H.F., Wang, N.N., Wang, H.X., Tang, M. and Wei, G.H. (2011). *Streptomyces zinciresistens* sp. nov., a zinc-resistant actinomycete isolated from soil from a copper and zinc mine. *International Journal of Systematic and Evolutionary Microbiology*, 61: 616–620.

Morales, D.K., Ocampo, W. and Zambrano, M.M. (2007). Efficient removal of hexavalent chromium by a tolerant *Streptomyces* sp. affected by the toxic effect of metal exposure. *Journal of Applied Microbiology*, 103: 2704–2712.

Nafis, A., Raklami, A., Bechtaoui, N., El Khalloufi, F., El Alaoui, A., Glick, B.R., Hafidi, M., Kouisni, L., Ouhdouch, Y. and Hassani, L. (2019). Actinobacteria from extreme niches in Morocco and their plant growth-promoting potentials. *Diversity*, 11: 139. Doi:10.3390/d11080139.

Nagajyoti, P.C., Lee, K.D. and Sreekanth, T.V.M. (2010). Heavy metals, occurrence and toxicity for plants: A review. *Environmental Chemistry Letters*, 8: 199–216.

Nanda, M., Kumar, V. and Sharma, D.K. (2019). Multimetal tolerance mechanisms in bacteria: The resistance strategies acquired by bacteria that can be exploited to 'clean-up' heavy metal contaminants from water. *Aquatic Toxicology*, 212: 1–10.

Ng, S.H., Palombo, E.A. and Bhave, M. (2012). Identification of a copper-responsive promoter and development of a copper biosensor in the soil bacterium *Achromobacter* sp. AO22. *World Journal of Microbiology and Biotechnology*, 28: 2221–2228.

Nies, D.H. (1999). Microbial heavy-metal resistance. *Applied Microbiology & Biotechnology*, 51: 730–750.

Nies, D.H. (2003). Efflux-mediated heavy metal resistance in prokaryotes. *FEMS Microbiology Reviews*, 27: 313–339.

Nies, D.H. and Silver, S. (1995). Ion efflux systems involved in bacterial metal resistances. *Journal of Industrial Microbiology*, 14: 186–199.

Nouioui, I., Carro, L., García-López, M., Meier-Kolthoff, J.P., Woyke, T., Kyrpides, N.C., Pukall, R., Klenk, H.-P., Goodfellow, M. and Göker, M. (2018). Genome-based taxonomic classification of the Phylum *Actinobacteria*. *Frontiers in Microbiology*, 9: 2007. Doi: 10.3389/fmicb.2018.02007.

Oger, C., Mahillon, J. and Petit, F. (2003). Distribution and diversity of a cadmium resistance (*cadA*) determinant and occurrence of IS*257* insertion sequences in *Staphylococcal* bacteria isolated from a contaminated estuary (Seine, France). *FEMS Microbiology Ecology*, 43: 173–183.

Pal, A., Dutta, S., Mukherjee, P.K. and Paul, A.K. (2005). Occurrence of heavy metal resistance in microflora from serpentine soil of Andaman. *Journal of Basic Microbiology*, 45: 207–218.

Polti, M.A., Amoroso, M.J. and Abate, C.M. (2007). Chromium (VI) resistance and removal by actinomycete strains isolated from sediments. *Chemosphere*, 67(4): 660–667.

Polti, M.A., Amoroso, M.J. and Abate, C.M. (2010). Chromate reductase activity in *Streptomyces* sp. MC1. *Journal of General & Applied Microbiology*, 56: 11–18.

Polti, M.A., Garcia, R.O., Amoroso, M.J. and Abate, C.M. (2009). Bioremediation of chromium (VI) contaminated soil by *Streptomyces* sp. MC1. *Journal of Basic Microbiology*, 49(3): 285–292.

Poopal, A.C. and Laxman, R.S. (2009). Studies on biological reduction of chromate by *Streptomyces griseus. Journal of Hazardous Materials*, 169: 539–545.

Prithviraj, D., Divya, P., Patel, K., Aursang, R., Neelu, N., Balasaheb, K., Madhukar, K. and Abul, M. (2012). Biosorption of heavy metals by actinomycetes for treatment of industrial effluents. pp. 389–393. *Procedings of the UMT 11th International Annual Symposium on Sustainability Science and Management*, 9–11 July 2012, Terengganu, Malaysia. e-ISBN 978-967-5366-93-2.

Ramasamy, K., Kamaludeen, S. and Parwin, B. (2006). Bioremediation of metals, microbial processes and techniques. pp. 173–187. *In*: Singh, S.N. and Tripathi, R.D. (eds.). *Environmental Bioremediation Technologies*. Springer Publication, New York, USA.

Ravel, J., Amorosa, M.J., Colwell, R.R. and Hill, R.T. (1998). Mercury-resistant actinomycetes from the Chesapeake Bay. *FEMS Microbiology Letters*, 162: 177–184.

Remenar, M., Karelova, E., Harichova, J., Zamocky, M., Krcova, K. and Ferianc, P. (2014). Actinobacteria occurrence and their metabolic characteristics in the nickel-contaminated soil sample. *Biologia*, 69/11: 1453–1463.

Rho, J.Y. and Kim, J.H. (2002). Heavy metal biosorption and its significance to metal tolerance of Streptomycetes. *The Journal of Microbiology*, March 2002: 51–54.

Salam, N., Jiao, J.Y., Zhang, X.T. and Li, W.J. (2020). Update on the classification of higher ranks in the phylum *Actinobacteria. International Journal of Systematic and Evolutionary Microbiology*, 70(2): 1331–1355.

Saurav, K. and Kannabiran, K. (2009). Chromium heavy metal resistance activity of marine *Streptomyces* VITSVK5 spp. (GQ848482). *Pharmacologyonline*, 3: 603–613.

Schmidt, A., Haferburg, G., Sineriz, M., Merten, D., Buchel, G. and Kothe, E. (2005). Heavy metal resistance mechanisms in actinobacteria for survival in AMD contaminated soils. *Chemie der Erde*, 65(2005) S1: 131–144.

Schmidt, A., Haferburg, G., Schmidt, A., Lischke, U., Merten, D., Ghergel, F., Buchel, G. and Kothe, E. (2009). Heavy metal resistance to the extreme: *Streptomyces* strains from a former uranium mining area. *Chemie der Erde*, 69(2009) S2: 35–44.

Schmidt, A., Schmidt, A., Haferburg, G. and Kothe, E. (2007). Superoxide dismutases of heavy metal resistant *Streptomycetes. Journal of Basic Microbiology*, 47: 56–62.

Sedlakova-Kadukova, J., Kopcakova, A., Gresakova, L., Godany, A. and Pristas, P. (2019). Bioaccumulation and biosorption of zinc by a novel *Streptomyces* K11 strain isolated from highly alkaline aluminium brown mud disposal site. *Ecotoxicology and Environmental Safety*, 167: 204–211.

Sheoran, V., Sheoran, A.S. and Poonia, P. (2010). Soil reclamation of abandoned mine land by revegetation: A review. *International Journal of Soil, Sediment and Water*, 3(2): Article 13. Available at: https://scholarworks.umass.edu/intljssw/vol3/iss2/13.

Sineriz, M.L., Kothe, E. and Abate, C.M. (2009). Cadmium biosorption by *Streptomyces* sp. F4 isolated from former uranium mine. *Journal of Basic Microbiology*, 49 Suppl 1(S1): S55–62.

Silver, S. and Phung, L.T. (2005). A bacterial view of the periodic table: Genes and proteins for toxic inorganic ions. *Journal of Industrial Microbiology and Biotechnology*, 32: 587–605.

Soto, J., Ortiz, J., Herrera, H., Fuentes, A., Almonacid, L., Charles, T.C. and Arriagada, C. (2019). Enhanced arsenic tolerance in *Triticum aestivum* inoculated with arsenic-resistant and plant growth promoter microorganisms from a heavy metal-polluted soil. *Microorganisms*, 7(9): 348. Available at: https://doi.org/10.3390/microorganisms7090348.

Timkova, I., Kadukova, J.S. and Pristas, P. (2018). Review: Biosorption and bioaccumulation abilities of actinomycetes/streptomycetes isolated from metal contaminated sites. *Separations*, 5: 54. Doi:10.3390/separations5040054.

Undabarrena, A., Ugalde, J.A., Seeger, M. and Camara, B. (2017). Genomic data mining of the marine actinobacteria *Streptomyces* sp. H-KF8 unveils insights into multi-stress related genes and metabolic pathways involved in antimicrobial synthesis. *PeerJ*, 5: e2912. Doi:10.7717/peerj.2912.

Van Nostrand, J.D., Khijniak, T.V., Gentry, T.J., Novak, M.T., Sowder, A.G., Zhou, J.Z., Bertsch, P.M. and Morris, P.J. (2007). Isolation and characterization of four gram-positive nickel-tolerant microorganisms from contaminated sediments. *Microbial Ecology*, 53(4): 670–682.

Wakelin, S.A., Anand, R.R., Reith, F., Gregg, A.L., Noble, R.R.P., Goldfarb, K.C., Andersen, G.L., DeSantis, T.Z., Piceno, Y.M. and Brodie, E.L. (2011). Bacterial communities associated with a mineral weathering profile at a sulphidic mine tailings dump in arid Western Australia. *FEMS Microbial Ecology*, 79: 298–311.

Wuana, R.A. and Okieimen, F.E. (2011). Heavy metals in contaminated soils: A review of sources, chemistry, risks and best available strategies for remediation. *ISRN Ecology* Volume 2011, Article ID 402647. pp. 20. Doi:10.5402/2011/402647.

Yong, S.K. and Abdul-Talib, S. (2017). Case studies of successful mine site rehabilitation: Malaysia. p. 391. *In*: Bolan, N.S., Kirkham, M.B. and Ok, Y.S. (eds.). *Spoil to Soil: Mine Site Rehabilitation and Revegetation*. CRC Press. Available at: https://doi.org/10.1201/9781351247337.

Yuan, H.P., Zhang, J.H., Lu, Z.M., Min, H. and Wu, C. (2009). Studies on biosorption equilibrium and kinetics of $Cd^{2+}$ by *Streptomyces* sp. K33 and HL-12. *Journal of Hazardous Materials*, 164: 423–431.

Zibret, G., Gosar, M., Miler, M. and Alijagić, J. (2018). Impacts of mining and smelting activities on environment and landscape degradation - Slovenian case studies. *Land Degradation & Development*, 29: 4457–4470. https://doi.org/10.1002/ldr.3198.

# Chapter 8
# Extremotolerant *Rhodococcus* as an Important Resource for Environmental Biotechnology

*Irena B. Ivshina,* *Maria S. Kuyukina* and *Anastasiya V. Krivoruchko*

## 1. Introduction

Irrational consumer approach to the planet's resources, their unlimited exploitation, and the growing anthropogenic and technogenic pressure on natural ecosystems have led to a severe deterioration in the biosphere quality, unfavorable ecological situation, and an increased number of habitats, in which organisms are in extreme conditions of environmental pollution by various pollutants.

According to the World Health Organization (WHO), up to 500 thousand compounds, out of more than six million recognized chemicals, are in practical use. At the same time, about 200 thousand new compounds are synthesized annually for industrial activities, everyday life and treatment; of these, 40 thousand have harmful properties for humans, and 12 thousand are toxic. The biogeochemical cycles include an increasing list of foreign synthetic compounds unfamiliar to natural environments. Many xenobiotics that have come into common use, in various combinations with each other and environmental factors, can have dangerous biological effects. These include, in particular a large group of substances classified under a common term "pharmaceutical pollutant". Pharmaceutical pollutants represent a relatively recent new type of highly hazardous contaminants-environmental pollutants (antibiotics, hormonal and antidiabetic drugs, statins, antidepressants, etc.), which emerge because of the uncontrolled and unregulated flow of drugs and their metabolites into open ecosystems. Pharmaceutical pollutants, which are highly stable compounds

Institute of Ecology and Genetics of Microorganisms, Perm Federal Research Center, Russian Academy of Sciences, Perm State University, Perm, Russia.
* Corresponding author: ivshina@iegm.ru

with a diverse chemical nature and pronounced bioactivity, have been recognized as a new class of xenobiotics since the early 2000s. In terms of scale and ecological significance, the problem of drug-induced environmental pollution is now becoming truly planetary (aus der Beek et al. 2016, Kaushik and Thomas 2019, Patel et al. 2019, Tyumina et al. 2020).

The annually increasing planet's population, accelerated growth of industrial production and inappropriate disposal of toxic and recalcitrant wastes through landfilling have led to their accumulation in the world's oceans and soils. Currently, there is an urgent need for new effective methods and technologies of waste management in order to protect humans from the effects of potentially hazardous pollutants that cause oxidative stress and, as a consequence, contribute to the emerge of most non-infectious diseases in humans.

Environmental biotechnology (eco-biotechnology) allows solving problems of environmental protection and preventing the formation and accumulation of pollution. In recent years, the role of eco-friendly and relatively inexpensive biotechnological methods of pollution control using enzymatic activities of microorganisms, which contact xenobiotics and have to solve the problem of their detoxification, has steadily increased. For the rapidly developing applied eco-biotechnology, organic substances are of particular interest as raw materials to obtain various products by microbiological synthesis (Ceniceros et al. 2017, Busch et al. 2019, Cappelletti et al. 2020). Specificity and efficiency of enzymatic conversion reactions of most classes of organic compounds make them more advantageous compared to the chemical synthesis. Taken together, this prompted the expansion and intensification of research on the special features of microorganisms in contaminated environments (the so-called extremotolerant microorganisms or stress tolerators), which act as a primary response system to unfavorable or potentially hazardous environmental changes and trigger the adaptive reactions at the earlier stages.

Among these stress tolerators, indigenous actinomycetes of the genus *Rhodococcus* (Zopf 1891) emend Goodfellow et al. 1998 (phylum *Actinomycetota*, order *Mycobacteriales*, family *Nocardiaceae*) occupy a special place, standing out for the greatest variety of degradable xenobiotics and capable of complete mineralization of chemical pollutants via co-metabolism (exploiting alternative carbon sources), and forming valuable products (Jones and Goodfellow 2012, Busch et al. 2019, Cappelletti et al. 2020, Oren and Garrity 2021). Understanding the actual mechanisms of biological transformation of different types of pollutants is important in order to determine their ecological fates, evaluate associated risks, and neutralize them effectively.

Over the last 10–15 years, rhodococci due to their remarkably diverse properties and functions, and real benefits from their use have been the focus of several scientific reviews (de Carvalho and da Fonseca 2005b, Larkin et al. 2005, 2006, Martínková et al. 2009, Kim et al. 2018). Characterization of the current status of the genus *Rhodococcus*, its taxonomic history, various aspects of biochemical and genetic individuality of rhodococci are summed up in the Biology of *Rhodococcus* (Microbiology Monographs book series, https://www.springer.com/

gp/book/9783030114602) edited by Héctor M. Alvarez (second edition, 2019), an international scientific publication the first of its kind. It comprises monographic descriptions of one of the most complex groups of actinomycetes. The findings of these systematic reviews that examine rhodococci in space and time are of greater importance than those of individual original studies.

Rhodococci are relatively young eco-biotechnology agents. Compared to traditional methods of environmental biotechnology, the use of rhodococci provides more effective neutralization and utilization of ecotoxicants, allows retaining areas that would otherwise be occupied by landfills and waste storage facilities. Assessment of the real significance of *Rhodococcus* for applied purposes requires constant generalization of scientific information in order to determine the optimal efforts in fundamental and applied research of this group of microorganisms. However, despite the constantly increasing new scientific rhodococci-based developments, the number of unsolved problems is yet large; there are some fundamental issues that require further in-depth and holistic study. The nature of the regulating program and effective mechanisms of metabolic regulation have not been investigated yet in detail, and a number of questions arises about the adaptive responses of these stress tolerators throughout their developmental cycle under increasing anthropogenic influence. Also, there is no comprehensive understanding of functional genes important in eco-biotechnological processes, their differential activity and expression when exposed to xenobiotics.

Influenced by molecular biology, these studies are just being started due to the massive genome sequencing of non-pathogenic bacteria, *Rhodococcus* members among them, the novel genome-editing technology (CRISPR/Cas9), and the bioinformatics analysis enabling to compare the whole-genome data and individual genome sections (genes, operons) that encode degradative enzymes in bacteria (Larkin et al. 2010, Letek et al. 2010, DeLorenzo et al. 2018, Cappelletti et al. 2019b, Sangal et al. 2019, Garrido-Sanz et al. 2020, Liang et al. 2020). Addressing the above-mentioned issues is important for understanding structural and regulatory mechanisms of genetic processes involved in catabolism of anthropogenic pollutants by extremotolerant actinomycetes, and also for evaluating the possibility of durable biotechnological exploitation of this group of microorganisms, particularly for xenobiotic biodegradation and bioremediation of polluted ecosystems.

Taking into account the rate and volume of emerging data on novel genes confirming the catabolic versatility ("reliability") of rhodococci and on effective biodegrading strains with prospective application to detoxify anthropogenic substances, this review focuses on recent publications describing new aspects of waste management, cleaning and remediation of polluted environments using this biotechnology attractive actinomycete group. In the essential features, special attention is paid to the spectrum of possibilities and behavioral features of *Rhodococcus* as effective inactivators of relatively simple (e.g., heavy metals) xenobiotics and biodegraders of complex hydrocarbons, crude oil and petroleum products, and pharmaceutical pollutants as well as their influence on microbiocenoses and self-purification processes in soil and water ecosystems.

## 2. Rhodococci in Natural and Anthropogenic Environments

Ecological geography and array of possible habitats for *Rhodococcus* spp. are unprecedentedly broad and diverse—from nutrient-rich (human and animal organisms, plants) to oligotrophic (groundwater, snow, air) environments. The vast range of biological characteristics of rhodococci includes a peculiar composition of the lipophilic cell wall due to aliphatic chains of mycolic acids and high affinity for hydrophobic substrates; formation of capsule-like structures protecting against adverse environmental factors (e.g., osmotic stress and drying, etc.); morphological variability of colonies (dissociation) and associated variability of physiological and biochemical properties; cellular polymorphism and the presence of temporarily resting but potentially active cyst-like cells and low level of endogenous respiration (ensuring survival, for example, upon long starvation) in a three-stage developmental cycle. To be also listed are an increased production of carotenoid pigments and extracellular glycolipids; synthesis of protective and macroergic compounds; nitrogen fixation in the presence of hydrocarbons and oligonitrophilia; diauxotrophy, oligo- and psychrotrophy. Additionally, rhodococci are characterized by acido-, alkalo-, halo-, xero-, thermo- and osmotolerance; propensity for cell adhesion and surface colonization; and pathogenization of free-living forms. Taken together, the above mentioned is the result of adaptation mechanisms providing any possible forms of their existence in nature (Ivshina 2012, Zheng et al. 2013, Ivshina et al. 2015, Vereecke et al. 2020, Pátek et al. 2021).

Extremotolerant (psychroactive and thermotolerant, acid- and alkalitolerant, halo- and haloalkalitolerant) mycolic acid-containing nocardioform actinomycetes of the genus *Rhodococcus* are known mainly as typical dwellers of soil environments—from tropical to arctic soils. Rhodococci are isolated from extreme and unique environments, such as cold polar deserts, Antarctic and alpine soils, annually freezing and thawing tundra areas adjacent to glaciers, salt marshes, dry desert sand as well as mangrove sediments, snow, air, core, and a wide variety of anthropogenic habitats (Bej et al. 2000, Luz et al. 2004, Sheng et al. 2011, Urbano et al. 2013, Hassanshahian et al. 2013, Konishi et al. 2014, Mikolasch et al. 2015, Viggor et al. 2015, Acosta-González et al. 2016, Goordial et al. 2016, Röttig et al. 2016, Sinha et al. 2017, Auta et al. 2018, Habib et al. 2018, Catalogue of Strains 2023). They also colonize fresh surface, ground, mineral and stratal waters, and are found in marine coral reefs and bottom sediments of northern seas contaminated with petroleum products (Li et al. 2012, Urbano et al. 2013, Dastager et al. 2014, Ivshina et al. 2014, Hwang et al. 2015, Aggarwal et al. 2016, Táncsics et al. 2017, Ramaprasad et al. 2018). A rigid cell wall protects rhodococci from being eaten by protists. However, knowledge on distribution and seasonal dynamics of freshwater and marine populations of *Rhodococcus* is still scarce; there is little data on their physiological and ecological roles in bacterioplankton communities. Whether rhodococci are permanent inhabitants of aquatic ecosystems or introduced there from the soil is an unresolved question. There is only an assumption that rhodococci in their resting coccoid phase pass into fresh and sea waters (Anandan et al. 2016).

Rhodococci are found in the rhizosphere and phyllosphere of various plants (Zhao et al. 2012, Kämpfer et al. 2013, Ma et al. 2017, Silva et al. 2018).

Plant-associated rhodococci can not only stimulate their productivity (Hamedi and Mohammadipanah 2015) but also cause pathology. Phytopathogenic *R. corynebacterioides* and *"Rhodococcus fascians* assemblage" have been described as causative agents of Pistachio Bushy Top Syndrome (PBTS) (Stamler et al. 2015, Vereecke et al. 2020) and the leafy gall syndrome (Goethals et al. 2001, Depuydt et al. 2008). Currently, the pathogenic nature of *"Rhodococcus equi* (*Rhodococcus hoagii/Prescottella equi*)" for humans and animals have been validly documented (Kedlaya et al. 2001, Muscatello et al. 2007, Ng et al. 2013, Anastasi et al. 2016, MacArthur et al. 2017). However, the taxonomic status of individual representatives of pathogenic *Rhodococcus* (their invalid species are given in quotation marks) is to be clarified (Sangal et al. 2019). With many new *Rhodococcus* species described in the last decade, the prospects for their exploitation in open systems (soils, waters, treatment facilities) sets an urgent task of understanding the potential hazard when handling rhodococci in order to determine the risk level upon their introduction into natural environments.

Members of the genus *Rhodococcus* are highly tolerant to environmental stressors, such as extreme temperatures, salinity, water deficit, high/low pH, reactive oxygen species and radiation. Rhodococci are most often neutrophilic, whereas under particular conditions they are acidophilic or alkaliphilic. They are featured by acid tolerance, offering their relatively high content in the microbiota of forest soils. That they dwell in soils under coniferous cover is explained by their ability to withstand significant contents of mobile aluminum forms in highly acidic soils and to utilize abietinic acid available from the needle fall (Nguyen and Kim 2016). Among the most important abiotic factors that regulate the abundance and dynamics of *Rhodococcus* populations in natural environments, the organic composition of the substrate plays the leading role. Despite the detailed analyses of these factors separately, their cumulative effects on rhodococci with a variety of environmental pollutants are poorly understood yet.

*Rhodococcus* strains isolated from habitats with high anthropogenic loads are characterized by pronounced emulsifying and biodegradative abilities towards individual hydrocarbons and oil products, improved resistance to high concentrations (100–250 mM and more) of heavy metals (cadmium, zinc, nickel, copper, molybdenum, lead, chromium, vanadium, etc.) and organic solvents (from 20 to 80 vol.%). Additionally, they show stable activity under extreme acidity (pH 2.0–6.0) and salinity (2–6% NaCl) and maintain degradative ability at high (40°C and above) and low (4–15°C) temperatures (Ivshina 2012, Catalogue of Strains 2023) (Fig. 1).

The above properties determine the confinement of rhodococci to extreme specific ecosystems with harsh conditions that do not at all facilitate the active life of many microorganisms. Rhodococci occupy a permanent and dominant position in soil bacteriocenoses at oilfields under long and increased exposure to heavy metals and petroleum hydrocarbons. They are the key players in preventing the accumulation of hydrocarbon pollutants in the Earth's atmosphere as well as in the processes of natural restoration of oil-contaminated ecosystems (Goodfellow et al. 2004, de Carvalho et al. 2014, Ivshina et al. 2017, Cappelletti et al. 2019a).

*Rhodococcus ruber*[1] (Kruse 1896) Goodfellow and Alderson 1977[2]

IEGM[3] 231[4]

Co-identity: VKPM AC-1899

<- I.B. Ivshina, OEGM 29-1B-1[5]. Isolated from: water, spring[6], oil-extracting enterprise, Perm region, Russia[7]. Taxonomy/description: (245, 343)[9]. Shows positive result with *Rhodococcus ruber* primers in species-specific PCR (245)[9]; analysis of 16S rRNA gene sequence (GenBank accession number KF155234); whole genome was sequenced and deposited at DDBJ/EMBL/GenBank under accession numbers CCSD01000001 to CCSD01000115 (343)[9]. Properties: uses propane and *n*-butane as sole carbon source; produces biosurfactants when growing on *n*-alkanes ($C_{12}$–$C_{17}$) (248, 254, 327, 347, 348)[9]; degrades high-porous ceramic materials (237)[9]; degrades paracetamol (265)[9]; forms cholesterol oxidase; resistant to $Cd^{2+}$, $Mo^{6+}$, $Ni^{2+}$, $Pb^{2+}$, $VO^{2+}$, $VO_3^-$, $VO_4^{3-}$, accumulates molybdenum and nickel (286, 329)[9]; adheres to liquid hydrocarbons (*n*-hexadecane) (317, 320)[9]; bioremediation agent for oil-contaminated soil (324); uses *n*-hexadecane as sole carbon source (327)[9]; resistant to 1-butanol, ethanol (341, 349)[9]; produces glycolipid biosurfactant with immunomodulatory properties. (Medium 5 or 8, 11, 28°C)[8].

---

Medium: 5, 8, 11

Reference(s): 237, 245, 248, 253, 254, 257, 258, 261, 263, 264, 265, 271, 277, 286, 294, 317, 320, 324, 327, 329, 341, 343, 347, 348, 349

http://www.iegmcol.ru/strains/rhodoc/ruber/r_ruber231.html

**Figure 1.** Sample catalogue information on *Rhodococcus* strains maintained at the Regional Specialized Collection of Alkanotrophic Microorganisms (WFCC # 285, http://www.iegmcol.ru).

**Footnote:** [1] – valid genus/species name of bacteria; [2] – authors who described and re-described the species, year of validation; [3] – collection acronym; [4] – strain accession number in the collection; [5] – an individual or organization from where the strain was acquired; strain accession number upon acquisition; [6] – isolation substrate; [7] – geographical location of the strain isolation site; [8] – nutrient medium, cultivation temperature, conservation and storage methods; [9] – references to publications where this strain was used.

The internal, immanent property of rhodococci is to synthesize all cell components by using gaseous and liquid *n*-alkanes as a sole carbon source. This characteristic of the genus *Rhodococcus* can serve in a way as an additional distinctive feature. That rhodococci are capable of oxidative transformation and degradation of natural and anthropogenic hydrocarbons of different classes makes them least dependent on the ambient conditions, allows avoiding the fierce substrate competition in natural habitats and surviving in extreme hydrocarbon-polluted environments. Rhodococci therefore play a decisive role in maintaining the sustainability of the environment, actively participate in biogeochemical processes of the biosphere and significantly contribute to the hydrocarbon-free atmosphere of the Earth, thus contributing to the global ecosystem functioning.

From an ecological perspective, the capability of certain rhodococcal species (*R. rhodochrous* and/or *R. ruber*) to oxidize higher gaseous homologues of methane ($C_2$–$C_4$) is of practical interest. Studies of the distribution pattern of propane- and butane-oxidizing bacteria in groundwater and subsurface sediments of oil-bearing and non-oil-bearing regions revealed strong correlation between the species of populations of rhodococci of relatively high densities (from $10^3$ to $10^4$ cells/mL in water samples and from $10^5$ to $10^6$ cells/g in soil samples) and the boundary of the oil-bearing area. The prospects of rhodococci as bioindicators of hydrocarbon deposits have been identified (Ivshina et al. 1981, 1995). Propane- and butane-oxidizing rhodococci normally are not found in substrates where gaseous hydrocarbons are due to organic matter decomposition, namely in samples of silt, peat, vegetable, forest and garden soils, and lake water. This is apparently because higher gaseous methane homologues are found in negligible quantities in the composition of hydrocarbon gases of biological origin and make up one ten-thousandth of a percent in relation to methane. At the same time, hydrocarbon gases migrating from oil deposits are characterized by an increased content of heavy methane homologues, amounting to 17% or more. Consequently, in natural microbiocenoses, the bacteria assimilating higher gaseous hydrocarbons can mainly, if not exclusively, develop only due to gases diffusing from oil deposits.

In the area of oil fields, indicator bacteria form a powerful biological oxidizing filter that significantly influences the gaseous hydrocarbons diffusing from the underground reservoir to the Earth's surface. A bacterial filter usually develops to an extent that it completely intercepts the flow of gases from the generating layers. At the same time, the number of gas-oxidizing rhodococci is not subject to sharp seasonal fluctuations, and the main factor regulating the population size in these ecological niches is the constant flow rate of hydrocarbon gases.

Interestingly, subsoil sediments are characterized simultaneously by high percentage of two or more dominant *Rhodococcus* species, unlike underground waters of oil-bearing structures. Obviously, competition in poor trophic niches results in displacement of one or another species, whereas in nutrient-rich conditions, the competing species with marked ecological similarity can occur together and are not strong competitors in the soil (Williamson 1972). This probably explains the coexistence of several gas-oxidizing rhodococcal species in soil of oil-bearing regions, with one species usually found in underground (springs) waters above an oil field where hydrocarbon gases are the limiting carbon source. While mineral media containing liquid *n*-alkanes usually allow preferential isolation of other, non-gas-oxidizing, *Rhodococcus* species, i.e., *R. erythropolis*, *R. opacus*, *R. wratislaviensis*. It is quite probable that the process of displacing one species by another when their nutrient resources coincide in natural conditions can occur under the influence of various factors. The confinement of a particular *Rhodococcus* species to a certain ecological niche deserves a separate study.

Evaluating the ecological role of *Rhodococcus* in order to understand the peculiar mechanisms of interactions in natural microbiomes, leading to long-term coexistence of rhodococci in oligotrophic habitats, requires an in-depth *ex situ* study in laboratory microcosms in strictly controlled minimal synthetic media. Sustainable co-existence of competitive *Rhodococcus* species observed, e.g., in ground water, can probably be explained by the known ability of rhodococci to release metabolic byproducts into the environment (Mikolasch et al. 2015, Cappelletti et al. 2020) that can be used by other members of the bacterial community. Thus, rhodococci can "purposefully" influence the composition of the bacterial community and form their own "comfortable" environment. At the same time a carbon source, such as propane, *n*-butane ($C_3$–$C_4$) or liquid hydrocarbons, can have an important limiting effect and be a driving factor in the formation of a stable microbial community. In this research field, separate works have already emerged (Goldford et al. 2018, Mooshammer et al. 2021), confirming the fact that microbial communities can survive under conditions of prolonged hunger stress through cross-feeding.

The widespread occurrence of gas oxidizing *Rhodococcus* is proved by their detection in the air atmosphere of gas processing plants and in the snow cover over oil and gas deposits. Propane- and butane-oxidizing rhodococci make up to 5% of the total microbial population in the ambient air sampled at the gas processing plant territory, with their number varying within 10–30 cells/$m^3$ of the air volume.

Within the oilfield boundary, snow samples show high contents of heavy hydrocarbon gases and bacteria that oxidize them. That gas-oxidizing rhodococci are detected in snow is obviously explained by the mechanical movement of bacteria from the lower soil horizons to the upper ones influenced by the ascending gas flow during soil freezing in winter months. Snow cover is a good sorption substrate for bacteria and gaseous hydrocarbons. However, another explanation seems logical for the presence of rhodococci in snow and frozen soils that indicates high adaptability of these extremotolerants related to pronounced polymorphism of rhodococci in natural conditions. Today, there is no doubt that natural populations of organisms have a significant variety of phenotypes (morphs) specialized towards a particular spatial or temporal sub-niche and associated with certain environmental conditions. The detection of gas-oxidizing rhodococci during winter can be explained by a change in the bacterial morphs, and more specifically by an increase in the frequency of occurrence of a corresponding bacteriomorph in the population due to selective reproduction of individuals. In warm seasons, the soil is dominated by "summer" forms of bacteria, which suppress the development of "winter" ones, i,e., more psychroactive forms. Low temperatures favor the growth of specific "winter" forms of bacteria with energy metabolism modified for greater efficiency. Such a switch from one phenotype to another is apparently the organism's response to seasonal changes in the environment. In this regard, the adaptive implication of *Rhodococcus* polymorphism is that it reduces the probable population death and "smoothes out" variations in the number of individuals and increases the effective use of nutrient resources.

Rhodococci assimilating heavy gaseous hydrocarbons as the only carbon and energy source dwell not only over oil and gas fields but also thrive in leaks of combustible gases, in rocks and waters located in gas-polluted areas.

Typical polytrophs, rhodococci are an essential link in most food chains in soil and aquatic ecosystems. Given the ability of rhodococci to grow at low ambient concentrations of organic matter, they are part of the autochthonous microbiota. By virtue of their unique ecological features as well as trophic requirements and functional characteristics, rhodococci can be classified as dissipotrophs, which effectively use the scattered flow of monomers. In the microbial community, dissipotrophs are featured by high affinity for a carbon substrate; use of low substrate concentrations; relatively high growth rates in this concentration range to compete with "trivial" copiotrophs, while low specific maximum growth rates and adaptation to long-term survival under adverse conditions. In this regard, the concept of "dissipotrophs" is closely related to the concept of "oligotrophic microflora" (Yanagita et al. 1978).

Despite their broad physiological capabilities, rhodococci are at the end of the organic matter decomposition chain and, as the second-order decomposers, can actively occupy various ecological niches. Cyst-like forms present in the morphogenetic cycle of development ensure the rhodococcal survival in unfavorable environmental conditions and indicate their optimal adaptation to terrestrial existence. Rhodococci are slow-growers and efficiently degrade recalcitrant (to other bacteria and fungi) organic carbon sources. Adapted to limiting resource concentrations and are more competitive under aerobic oligotrophic conditions typical of natural ecosystems, they can most likely be assigned to a facultative form of moderate aerobic oligotrophs. However, oligotrophic rhodococci can also exhibit copiotrophic traits (Morrisey et al. 2016). Most likely, they are featured by a mixed competitive-stress tolerator-ruderal (C-S-R) life strategy (Ho et al. 2017).

A large-scale study of *Rhodococcus* species and their functional diversity in natural environments, the abundance dynamics and ecological characteristics gives an idea of rhodococci as a unique ecotrophic group exhibiting a variety of strategies for their survival under harsh conditions. Indeed, *Rhodococcus* representatives implement different adaptive strategies and their combinations, e.g., K- and L-strategies; elements of r- against the background of K-strategies. Since the r-, K-, and L-strategies of adaptation are interrelated, in nature there is practically no clear division of organisms into pronounced r-, K-types, or L-species (Whittaker 1975). *Rhodococcus* populations most often have more pronounced and obvious L-properties of K-strategists. These include cyclic development, extreme phenotypic variability, morphogenetic transitions and increased protection from adverse environmental factors; dominance in the later stages of succession; tendency to associate with hydrolytics and ability to function as dissipotrophs; slow growth in oligotrophic habitats in the absence of exogenous sources of phosphorus and nitrogen; resistance to starvation, drying, extreme concentrations of hydrogen ions, salts, mechanical and ultrasonic degradation, toxic agents and antiseptics.

Such a diverse "repertoire" of the *in-situ* behavior depending on the ecological niche parameters and environmental fluctuations, including different ecological strategies, broad responses combined with tolerance to adverse abiotic factors and high competitiveness for available substrates provide the ubiquitous distribution of rhodococci and their frequent predominance in extreme natural and technogenic environments. With the exceptional plasticity of their large genome (up to 10.4 Mbp), significant catabolic gene redundancy and a complex regulatory network, rhodococci have a superior ability to resist not only atmospheric influences and natural factor fluctuations but also global anthropogenic impacts that substantially transform the chemical composition of the environment (de Carvalho and da Fonseca 2005a,b, Larkin et al. 2006, 2010, de Carvalho et al. 2009, Cappelletti et al. 2019b, Guevara et al. 2019). The variety and redundancy of regulatory mechanisms in *Rhodococcus* are obviously the result of spontaneous rearrangements of the genome evolved under the changing environmental conditions that can explain and predict the emergence of new species, which are currently discovered and will be described.

Due to their perfect adaptability, rhodococci easily accommodate to disturbed environments where they "comfortably" and "expediently" dwell, playing an essential role in neutralizing the toxic anthropogenically-originated substances. Specificity of oxidoreductases involved in the oxygenation of complex hydrophobic and recalcitrant organic substrates, and also high lability of *Rhodococcus* metabolic systems that function in a wide range of temperatures and pH values make them suitable candidates to design industrial biocatalysts and to enhance bioremediation of polluted environments.

In this regard, new knowledge on the universal traits and peculiarities of *Rhodococcus* (specifically upon the induction of an oxygenase enzyme complex), and new data on their interaction with xenobiotic compounds are extremely essential.

They provide a scientific insight, more significant and fundamental than previously thought, in the role of this group of actinomycetes in the biosphere functioning, the removal or reduction of toxic components in the context of global climate destabilization. Finally, these studies create the prerequisites and opportunities for the development of new advanced biotechnologies for cleaning the environment, neutralizing, or effectively managing waste streams resulting from industrial production and consumption.

## 3. Mechanisms of *Rhodococcus* Resistance to Toxic Organic and Inorganic Pollutants

The most common organic environmental pollutants (oil and refined products) are highly toxic chemicals that act as solvents, accumulate in cell membranes, and destroy them, thus compromising the structural and functional integrity of cells. This is especially important for unicellular organisms, such as bacteria, which interact directly with the changing microenvironment and deleterious effects of solvents. A solvent tolerant phenotype is the result of several mechanisms, including changes in membrane properties and cell wall composition, molecular efflux pumps, altered energy metabolism, and modifications of cell size and shape (Sikkema et al. 1995). Most of these mechanisms are described for *Rhodococcus* in response to solvent and chaotropic impacts of hydrocarbon pollutants. Using combined confocal laser scanning and atomic force microscopy, diverse changes in the surface topography and relative surface area were revealed in *R. ruber* cells depending on the solvent hydrophobicity (Kuyukina et al. 2014, Korshunova et al. 2016) (Fig. 2). In particular, more hydrophobic and less toxic solvents (*n*-hexane, *n*-decane and cyclohexane) caused a gradual increase in the cell wall roughness measured at micro- and nanoscales and even a more pronounced increase in the surface-to-volume ratio, thus enhancing the cellular uptake of these hydrocarbons. While incubation with toluene (a more toxic and less hydrophobic aromatic hydrocarbon) resulted in the reduced relative surface area to decrease a cell contact with the toxic substrate, at the same time it was accompanied by an increase in the height of the surface micro-relief. Upon the exposure to alcohol chaotropic stressors, such as ethanol and butanol, *R. ruber* cells became smaller in size with increased surface roughness and looked shrunken under the dehydrating effect of solvents. Interestingly, the elastic modulus (Young's modulus) of *R. ruber* cells decreased by 1.8 to two times in the presence of all the solvents applied, indicating a decrease in the rigidness of bonds between the cell wall components. It can be assumed that the solvent exposure resulted in structural and mechanical rearrangements of bacterial cell walls: particularly, in the emergence of less rigid and flexible regions that can be caused by the enhanced synthesis of lipid components (Korshunova et al. 2016).

A change in the degree of saturation of cellular membrane fatty acids and isomerization of *cis* to *trans* unsaturated fatty acids are the mechanisms enabling bacteria to maintain the membrane fluidity status under the solvent impact. Gutiérrez et al. (1999) reported on an increased level of saturated fatty acids (57.0 to 63.8%) and the appearance of hexadecenoic acid (16:1ω6c) in *Rhodococcus* cells grown in the

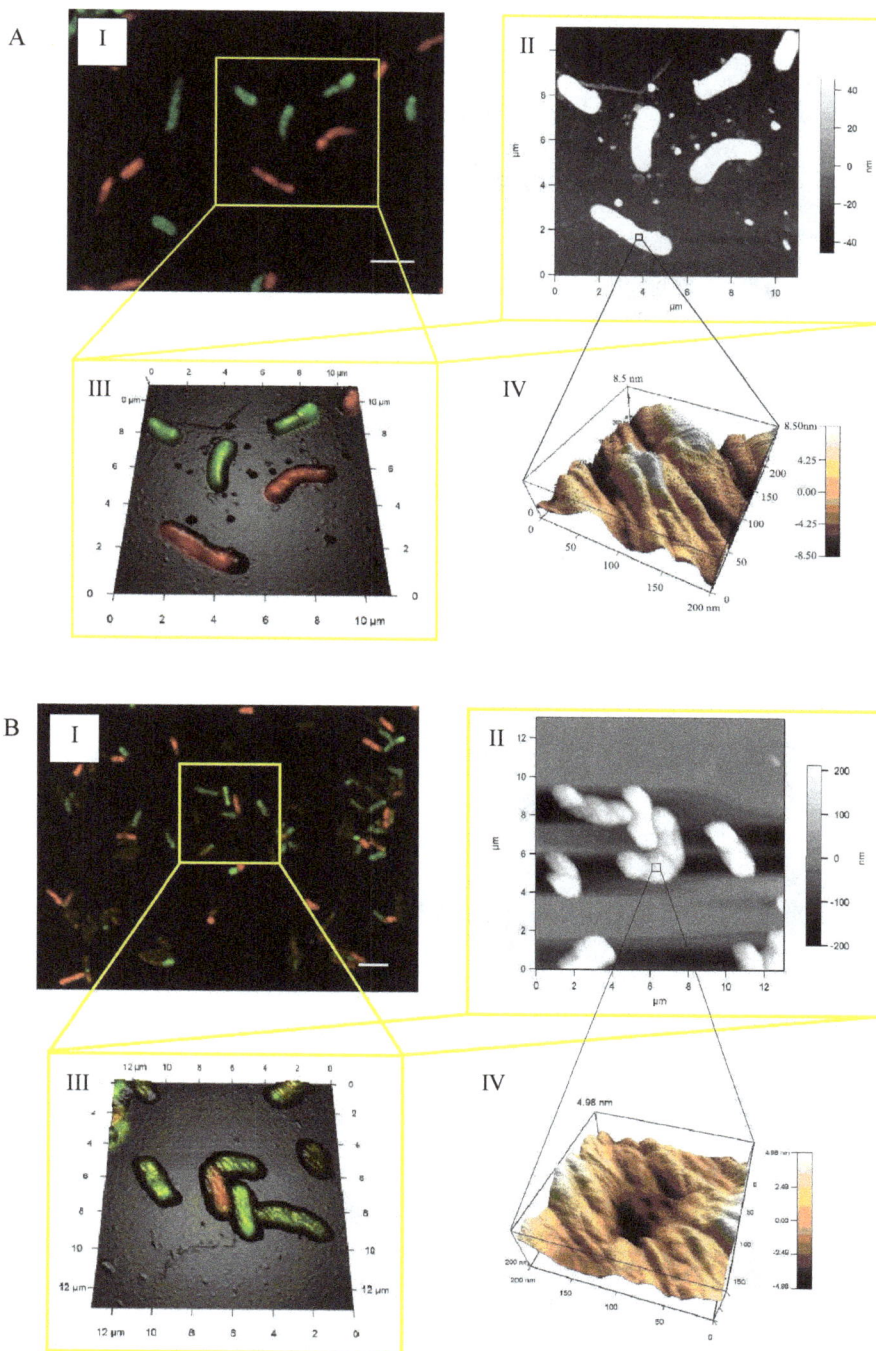

Figure 2 contd. ...

*...Figure 2 contd.*

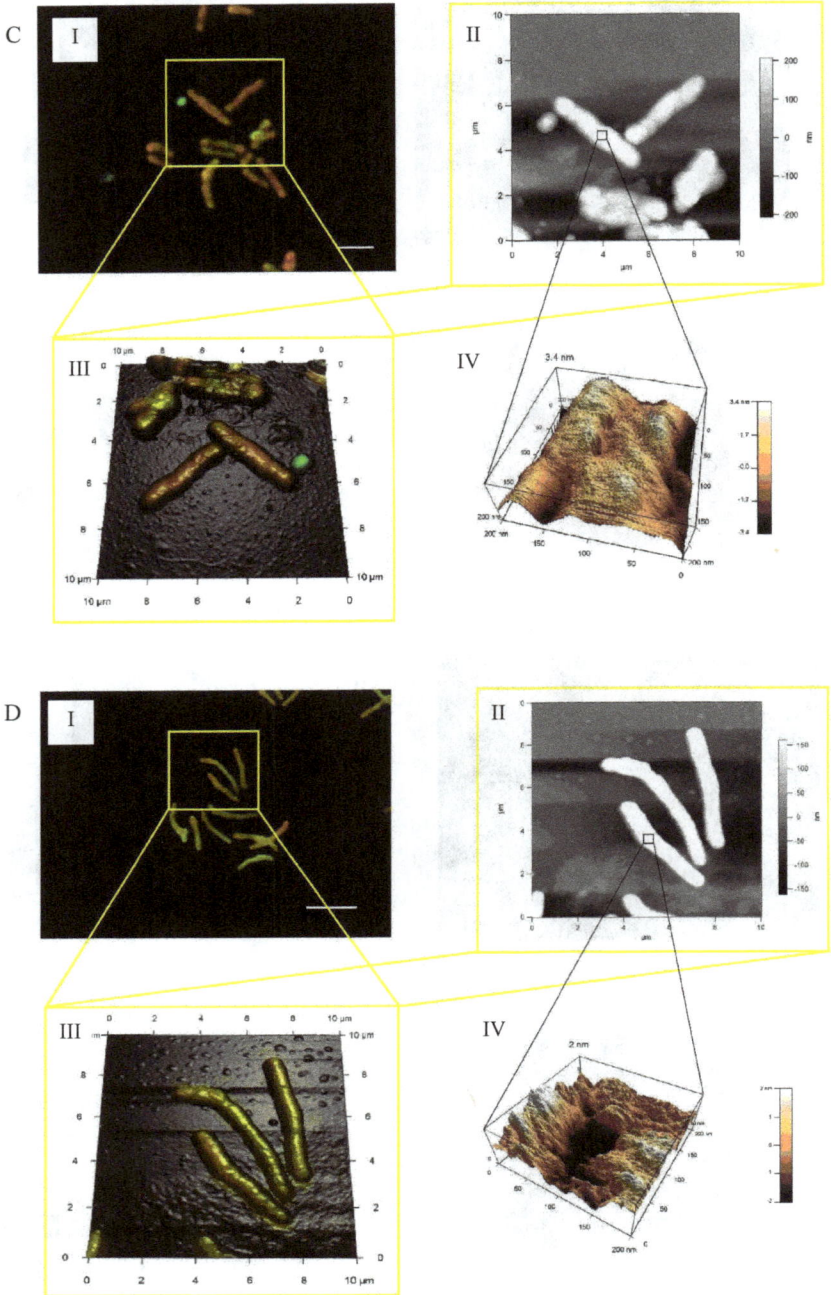

**Figure 2.** CLSM (I), AFM (II), combined CLSM/AFM (III) and high-resolution AFM (IV) of *R. ruber* IEGM 231 cells in the control (A) and exposed to *n*-decane (B), toluene (C) and ethanol (D). Scale bars represent 5 µm. Modified from (Kuyukina et al. 2014).

presence of benzene suggested as a possible mechanism to decrease the fluidity of the cell membrane. However, no changes in the ratio of *cis*/*trans* mono-unsaturated fatty acids were detected and all unsaturated fatty acids were *cis*-monoenoic. At the same time, Tsitko et al. (1999) determined that the contents of the branched (10-methyl) fatty acids increased from 3% to 15–34% of the total fatty acids along with a decrease from 50% to 20–35% of the proportion of unsaturated fatty acids when *R. opacus* cells were grown with benzene, phenol, 4-chlorophenol, chlorobenzene, or toluene as the sole source of carbon and energy in comparison with cells grown on fructose. Moreover, in the presence of phenol or chlorophenol, the *trans*-hexadecenoic acid content increased from 5% to 8–18%, presumably allowing its denser packing in the cytoplasmic membrane and protecting membrane against the fluidizing molecules of solvents. A physiological role of 10-methyl-branched fatty acids that occur in bacteria belonging to the genera *Dietzia*, *Gordonia*, *Nocardia*, *Mycobacterium*, *Rhodococcus*, and *Tsukamurella* has not been solved yet, while it has been found (Poger et al. 2014) that methyl branching reduces the lipid condensation, decreases the bilayer thickness, lowers the chain ordering, and promotes the formation of kinks at the branching point that leads to an increase in the membrane fluidity. Such diverse changes in the structure and properties of the cell membrane indicate a variety of instruments in *Rhodococcus* to mitigate the damaging effect of organic solvents. Furthermore, de Carvalho et al. (2005) revealed a dose-dependent increase in a saturation degree of cellular membrane lipids of *R. erythropolis* cells grown on *n*-pentane, cyclohexane, limonene, propanol and cyclohexanol, while methanol, ethanol and butanol caused a decrease in this sat/unsat ratio, thus suggesting different adaptive responses of rhodococcal cells related to the physicochemical properties of short-chain alcohols.

It should be noted that solvent tolerance mechanisms mentioned above for *Rhodococcus* include cell surface modifications that diminish solvent penetration into the cytoplasm. In addition, bacteria have an energy-dependent efflux system as an essential defense mechanism to extrude solvents that have entered the cell. There are five major families of efflux transporters: the adenosine triphosphate (ATP)-binding cassette (ABC) superfamily, the resistance-nodulation-division (RND) family, the small multidrug resistance (SMR) family, the major facilitator superfamily (MFS), and the multidrug and toxic compound extrusion (MATE) family. The majority of efflux pumps identified for solvents in bacteria belong to the RND family (a secondary transporter, which utilizes the proton gradient as a source of energy for solvent secretion), while ABC and MATE transporters were also reported (Kongpol et al. 2012). Our physiological experiments using efflux pump inhibitors suggested the involvement of proton- and sodium-dependent efflux systems in the organic solvent tolerance of *Rhodococcus* (Korshunova et al. 2016). Indeed, sodium orthovanadate-an inhibitor of ABC transporters—had no inhibitory effect on the growth of *R. ruber* IEGM 231 in the presence of *n*-hexane. This showed that the ATP-dependent pumps are most likely not involved in the removal of the excess of this solvent from the cells. However, phenyl-arginine β-naphthylamide (PAβN) and paroxetine almost completely (by 99%) inhibited the growth of this strain in the presence of the solvent that indicated the possible involvement of

RND, MFS and MATE secondary transporters in the formation of *Rhodococcus* resistance. Interestingly, a recent bioinformatic analysis of 33 published *Rhodococcus* genomes revealed the most abundant antibiotic resistance genes belonging to RND (*marR*, *acrR (tetR)*, *acrA*) and MFS (*emrB/qacA*, *mdtH*) efflux pump families (Cunningham et al. 2020). Also, the presence of the *req_39680* gene associated with a putative MATE family transporter was detected in *R. equi* isolates that correlated with the phenotypic expression of the ethidium bromide efflux mechanism (Gressler et al. 2014). Obviously, further research is needed to better understand the relationship between the efflux activities and solvent resistant phenotypes in the genus *Rhodococcus*, including the identification of key genes and their functional characterization using efflux inhibitors and creation of knockout mutants for these transport mechanisms.

In response to the action of toxic organic substrates, rhodococci exhibit the property of autoaggregation-a phenomenon that enables microorganisms belonging to the same strain to form multicellular aggregates (Lukic et al. 2014). Auto-aggregation has been demonstrated for different *Rhodococcus* species in the presence of hydrophilic solvents, such as butanol and dimethylformamide (de Carvalho and da Fonseca 2005b), and conversely, complex hydrophobic compounds, such as triterpenoid betulin (Tarasova et al. 2017), tricyclic diterpenoid dehydroabietic acid (Cheremnykh et al. 2018) and polycyclic non-steroidal anti-inflammatory drug diclofenac (Ivshina et al. 2019). Molecular mechanisms underlying *Rhodococcus* auto-aggregation are not fully understood and may include simple electrostatic and hydrophobic cell interactions mediated by cell-wall mycolic acids (long-chain α-branched and β-hydroxy fatty acids) and more specific biomolecules that are also involved in the adhesion of rhodococcal cells to biotic and abiotic surfaces, such as glycolipid biosurfactants and adhesive proteins (Ivshina et al. 2013 a, b, Krivoruchko et al. 2019).

To improve the tolerance of *R. opacus* to lignocellulose-derived inhibitors, the competent cells of strain MITXM-61 were treated by electroporation and subjected to the adaptive evolution at increasing sublethal concentrations of inhibitors (Kurosawa et al. 2015). It has been shown that laboratory adaptive evolution is an effective approach for improving the tolerance and fermentation performance even when the genetic constitution is unidentified. There are two methods of the evolutionary adaptation approach using (i) a single-batch cultivation at the pollutant (or its structural analogue) concentration lower than working ones and (ii) a short-term serial-transfer of the cell culture with a gradual increase of the pollutant concentration (Krivoruchko et al. 2019). For example, *R. rhodochrous* IEGM 608 cells were grown in the presence of 0.0007% isoquinoline prior to biodegradation of a toxic pharma pollutant drotaverine hydrochloride (Ivshina et al. 2012). Prieto et al. (2002) cultivated *R. erythropolis* UPV-1 cells at increased phenol concentrations (0.2 and then 0.4 g/L) to obtain an acclimated culture. Over 40 cultivation passages were carried out with increasing phenol concentrations (from 0.3 to 1.5 g/L) to obtain the adapted *R. opacus* PD630 mutants (Yoneda et al. 2016). Whole-genome sequencing and comparative transcriptomics identified the highly up-regulated degradation pathways and putative transporters for phenol in two adapted mutants, highlighting the important linkage between mechanisms of regulated phenol uptake, utilization, and evolved tolerance.

Although the ability of *Rhodococcus* spp. to degrade organic xenobiotics along with their physiological adaptations, such as modified cell wall and membrane compositions, are largely reported in the literature, much less is known about the rhodococcal capacity to tolerate toxic metals/metalloids. Since microorganisms evolved along with the planet geochemistry and inhabit all possible niches, many of them can utilize toxic metal(loid) compounds (As, Cd, Te, Se, etc.) as (1) energy source, (2) terminal electron acceptor during microbial respiration, and (3) cofactors in metalloproteins and enzymes (Presentato et al. 2019). For example, the ability of *R. aetherivorans* BCP1 cells to grow aerobically in the presence of the toxic oxyanion tellurite and to reduce it into elemental tellurium ($Te^0$) was reported (Presentato et al. 2016). It was shown that cells pre-exposed to $K_2TeO_3$ were able to reduce a greater amount of tellurite oxyanions at a faster rate than native cells and to produce intracellular rod-shaped nanostructures. Novel insights into pathways involved in Li response of *Rhodococcus* sp. A5wh were derived from the proteomic analysis (Belfiore et al. 2018), which identified 17 proteins participating in stress response, transcription, translations, and metabolism that were (over)expressed under Li stress. Similarly, the genomic analysis of *R. aetherivorans* BCP1 revealed the presence of three gene clusters responsible for organic and inorganic arsenic resistance (Firrincieli et al. 2019). A sequence similarity network (SSN) and phylogenetic analysis of these arsenate reductase genes indicated that two of them (ArsC2/3) are functionally related to thioredoxin (Trx)/thioredoxin reductase (TrxR)-dependent class and one (ArsC1) to the mycothiol (MSH)/mycoredoxin (Mrx)-dependent class. A targeted transcriptomic analysis performed by RT-qPCR indicated that the arsenate reductase genes as well as other genes included in the *ars* gene cluster (possible regulator gene *arsR*, and arsenite extrusion genes *arsA*, *acr3*, and *arsD*) were transcriptionally induced when BCP1 cells were exposed to As-(V) supplied at two different sublethal concentrations. The above examples and other numerous studies reviewed recently by Cappelletti et al. (2019b) illustrate the possibilities of "omics" technologies in elucidating diverse mechanisms of *Rhodococcus* resistance to heavy metal-associated and other stresses. Importantly, since hydrocarbon degradation and metal resistance genes are often located on plasmids, they can spread through horizontal gene transfer and be co-selected in the environments contaminated with oil and heavy metals. This explains the isolation of a large number of *Rhodococcus* strains as promising agents for bioremediation from oil-contaminated sites that are both resistant to metals and capable of degrading hydrocarbons (Ivshina et al. 1995, 2017).

A plethora of physiological and biochemical studies was aimed at elucidating the features of passive biosorption and active energy-dependent transport (efflux) of metal ions as mechanisms of *Rhodococcus* resistance to toxic metal(loids). It was shown that cell wall amphoteric molecules, such as polysaccharides, proteins, lipid groups, and mycolic acids, are responsible for intracellular sequestration of metal ions and oxyanions. Even dead biomass or isolated cell wall material can be used for metal biosorption with the efficiency greatly dependent on pH and the presence of competitor cations. Dobrowolski et al. (2017) revealed rapid (within 1–30 min) adsorption of Ni (II), Pb (II), Co (II), Cd (II), and Cr (VI) by an extracellular polymeric substance (EPS) derived from *R. opacus* and *R. rhodochrous*. A complex adsorption mechanism was suggested, including electrostatic attraction, surface complex formation and chemical interaction between metal ions and EPS functional groups (mainly hydroxyl, acetamide or amino groups). From our earlier study (Ivshina et al. 2013b), a non-specific resistance of *R. erythropolis* IEGM 186, IEGM 267, and IEGM 708; *R. fascians* IEGM 34 and IEGM 38; *R. rhodochrous* IEGM 647; and *R. ruber* IEGM 231 cultures to heavy metals correlated positively with their growth on hydrocarbons and biosurfactant production (estimated as emulsifying activity). Moreover, a trehalolipid biosurfactant isolated from *R. ruber* IEGM 231 cells was found to form complexes with $Ni^{2+}$ ions, resulting in a dose-dependent desorption of $Ni^{2+}$ from the ion-exchange resin. The molar ratio of the organic ligand (biosurfactant) to $Ni^{2+}$ was calculated to be 2.28 with a conditional stability constant of 9.15 for nickel, which is 2.5 times higher than that for nickel complexes with rhamnolipid from *Pseudomonas aeruginosa*.

Unlike biosorption, metal(loid) accumulation is an energy-consuming and metabolic-dependent process, as it requires an active uptake transporter across the cell membrane. Specific ion channels that transfer essential metals can facilitate the transport of toxic metals belonging to a closely related group of elements (Presentato et al. 2019). For example, accumulation of cesium by resistant *Rhodococcus* cells occurs via the potassium transport system and can be suppressed by the excess of $K^+$ or $Rb^+$ in the medium (Tomioka et al. 1994, Ivshina et al. 2002). It seems that at high metal concentrations, biosorption appears to be the primary sequestration mechanism in *Rhodococcus*, whereas intracellular accumulation followed by metabolic transformation prevails at lower metal concentrations. These findings are important for bioremediation of soil and water contaminated with radioactive cesium.

It can be assumed that to control the accumulation of toxic metals, rhodococci employ efflux pumps described above for solvent tolerance as a mechanism of cellular homeostasis in metal-rich environments. Indeed, an inducible efflux system was described in *R. opacus* as a resistance mechanism activated in the presence of Zn (II) or Cd (II) (Mirimanoff and Wilkinson 2000). Recent insights into metal homeostasis, including protein- and RNA-based sensors that interact directly with metals or metal-containing cofactors and trigger a stepwise transcriptional response to metal stress are reviewed by Chandrangsu et al. (2017). Such metalloregulatory systems have yet to be described in *Rhodococcus*, while a metal chaperone function was proposed for a new Fe-type nitrile hydratase activator protein from *R. equi* TG328-2, which was capable of accelerated hydrolyzing GTP to GDP upon binding divalent metal ions, such as Co (II) (Gumataotao et al. 2017).

Increasingly used in different industries, the engineered metal and metal oxide nanoparticles can be released into the environment, thus representing an emerging type of environmental pollutants along with pharmaceuticals and pesticides. It was found that *R. opacus* PD630 cells can resist strong oxidative stress caused by high concentrations (up to 1 mg/l) of $TiO_2$ nanoparticles (Sundararaghavan et al. 2020). The authors observed a dose-dependent increase in n$TiO_2$ uptake that correlated with the increased triacylglycerol synthesis and proposed this strain for the nanometal-contaminated water remediation and valuable metabolite production. Furthermore, stimulation effects of Co and Pd nanoparticles encapsulated in microporous silica crystallites were revealed on the biphenyl degradation by *R. erythropolis* T902.1 presumably due to activation of catechol 1,2-dioxygenase (a critical enzyme in the aromatic biodegradation pathway) (Wannoussa et al. 2015). In contrast, strong inhibitory effects were shown for $Cu^{2+}$ or $Ag^+$ ions, which were partly mitigated using corresponding Cu or Ag nanoparticles anchored in silica matrix as a result of decreased metal bioavailability and siderophore complexing. The authors suggested the advantages of metallic nanoparticles encapsulated in silica matrix to improve bacterial resistance to toxic metals and biodegradation activity towards recalcitrant organic pollutants.

# 4. Extremotolerant *Rhodococcus* for Environmental Remediation of Crude Oil-Contaminated Ecosystems

Upon oil exploitation and transportation, spills and leakages happen regularly. In the period between 1970 and 2017, more than 180 crude oil spill accidents were fixed only in marine ecosystems, with over 1 billion gallons of crude oil released and more than 3,500 km of coastline contaminated (Prabowo and Bae 2019). Crude oil-contaminated biotopes are detrimental and unfavorable for normal functioning of organisms that is related to direct and indirect impacts of hydrocarbons. Direct effects involve (1) influence of toxic compounds, such as mono- and polycyclic aromatic hydrocarbons, sulfur-containing components, and heavy metal salts, and (2) action of some fractions (light ones containing alkanes $C_5$–$C_9$ and monoaromatics) as solvents disintegrating cytoplasmic membranes and cell walls of eukaryotic and prokaryotic cells (Singer et al. 2000, de Carvalho and da Fonseca 2007, Brown et al. 2017). Soil co-contamination with crude oil and heavy metals is shown to be a common situation. Concentrations of heavy metal ions in crude oil-contaminated soils can come close to 0.6% (w/w) (Atagana 2011). An indirect influence of petroleum hydrocarbons is caused by their hydrophobicity. In soil, it results in changes of soil structure and physicochemical and biological characteristics, e.g., soil organic matter content, bulk density, porosity, permeability, soil respiration, nutrient transfer, and soil water repellency. Moreover, at accidents in exploitation sites, saline and hypersaline environments appear since the produced water contains increased (3–5%) concentrations of NaCl. As a result, osmotic, desiccation, starvation stresses and microaerobic conditions follow crude oil pollution in soils (Gao et al. 2014, Omran et al. 2020). In water ecosystems, crude oil can cover a significant area of water surface in the form of a thin impermeable film, which tends to expand; sedimentate at water body bottom, shores, corals, and parts of plants and animals; generate the formation of toxic soluble compounds; and change convection flows in site (Fate of Marine Oil Spills 2011). The situation becomes worse in fragile and sensitive ecosystems, such as extremely cold Antarctica sites where indigenous organisms are ideally accommodated to local conditions, and every allogenic agent can easily shift the ecological balance (Ruoppolo et al. 2013, Brown et al. 2017).

Biodegrading microorganisms have not only to survive under these adverse conditions but be able to effectively metabolize and degrade crude oil components. Members of the genus *Rhodococcus* realize a complex of strategies to maintain high hydrocarbon-oxidizing activities in crude-oil contaminated environments. They include (Fig. 3):

1. Diverse oxidative enzymes with overlapping substrate specificities: monooxygenases, dioxygenases, laccases, peroxydases, dehydrogenases (Classen et al. 2013, Ivshina et al. 2014, Busch et al. 2019). Among alkane monooxygenases alone, three superfamilies are distinguished, namely: (1) alkane hydroxylases (AlkB), (2) soluble di-iron monooxygenases (SDIMO), and (3) cytochromes P450 of the CYP153 (cytochrome P450) family. AlkB and CYP153-type monooxygenases are typically involved in oxidation of liquid alkanes $C_5$–$C_{16}$, and SDIMO act on short-chain hydrocarbons, such as gaseous and short-chain ($< C_8$) liquid alkanes (Panicker et al. 2010, Cappelletti et al. 2019a). Although substrate preferences are observed, these enzymes are involved in oxidation of a wide range of organic compounds. This is a molecular basis for catabolite reliability of *Rhodococcus*, their pantophagy and ability to utilize different types of crude oils and oil derivatives.

2. Different ways to assimilate petroleum hydrocarbons. Generally, rhodococci use two mechanisms: adhesion to large hydrocarbon droplets (it is associated with increased cell hydrophobicity due to changes in mycolic acid and phospholipid compositions) and surfactant-facilitated access through hydrocarbon emulsification, when cells encounter fine oil droplets and hydrocarbon molecules enter the cells. Molecules of biosurfactants can be localized and retained on the cell surface or excreted from cells. A direct contact of cells with hydrocarbons via high cell surface hydrophobicity is accompanied by aggregation of *Rhodococcus* cells in flocks, which controls the interfacial hydrocarbon entry mechanisms,

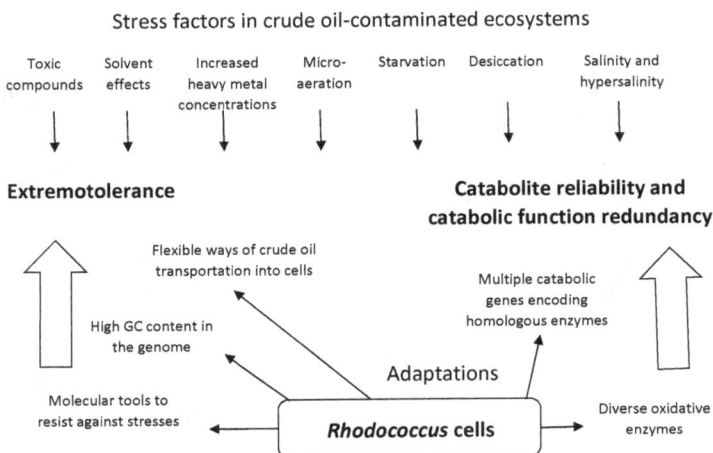

**Figure 3.** Adaptations of *Rhodococcus* bacteria make them perfectly fit for crude oil-contaminated ecosystem remediation.

and the ability of rhodococci to get into the oil phase. Additionally, involvement of exopolysaccharides in the adherence of rhodococcal cells to hydrocarbons is mentioned. Also, they can protect cells from toxic effects of volatile crude oil components. Apparently, *Rhodococcus* bacteria use the first mechanism towards various hydrocarbon substrates, including solid and dense ones (e.g., PAHs, long-chain alkanes, cuttings, and sludges), while the second mechanism is realized in the presence of liquid hydrocarbons (Bastiaens et al. 2000, Cappelletti et al. 2019a). Diversity of hydrocarbon assimilation mechanisms is the evidence of physiological plasticity and flexibility of *Rhodococcus* and capabilities of these bacteria to adequately respond to hydrocarbon substrates independently of their physical characteristics.

3. Advantageous combination of specific genome features. *Rhodococcus* genomes show relatively large (4 Mb to over 10 Mb) sizes, high GC content (61–71%), a high (4,668–9,589) number of coding sequences, presence of circular and linear plasmids harboring from 2 to 1,186 genes, and multiple catabolic genes encoding homologous enzymes. The last seems to be most important for survival and growth of *Rhodococcus* bacteria in contaminated sites. Catabolic gene redundancy is regarded as a basis for *Rhodococcus* catabolic versatility, catabolite reliability, functional robustness, competitiveness, and high adaptive potential. Large genome sizes, plasmids, and high number of genes further confirm the catabolic gene diversity and redundancy in *Rhodococcus*. As for the high GC content, it is a mechanism of DNA protection (Cappelletti et al. 2019b, Garrido-Sanz et al. 2020).

4. Typical stress tolerance mechanisms, including synthesis under stress conditions (high and low temperatures, high osmolarity, desiccation, starvation, reactive oxygen species, and solvents), heat shock proteins, DNA protection proteins, chaperones, universal stress proteins, ROS-neutralizing enzymes, compatible solutes, and exopolysaccharides; induction of appropriate sigma subunits of RNA polymerases; regulation of lipid and fatty acid composition of cell envelopes, and intensive cell aggregation (Bej et al. 2000, Iwabuchi et al. 2000, Alvarez et al. 2004, de Carvalho and da Fonseca 2007, LeBlanc et al. 2008, Puglisi et al. 2010). Furthermore, different heavy metal-associated genes, which encode heavy metal resistance proteins, heavy metal transporters, cation efflux enzymes, metal-binding proteins, mercuric reductases, and alkylmercuryl lyases were found in the genomes of *Rhodococcus* (Ivshina et al. 2014, Presentato et al. 2019). This shows that members of the genus *Rhodococcus* can potentially resist all types of stresses induced in crude oil-contaminated environments. Moreover, it is known that one stress signal frequently induces the resistance to other stresses. In *Rhodococcus*, it was shown that synthesis of many proteins involved in stress responses (e.g., chaperones, superoxide dismutase, catalase/ peroxidase) are induced in addition to catabolic proteins during the degradation of organic compounds (Cappelletti et al. 2019b).

To illustrate the advanced abilities of *Rhodococcus* bacteria as bioremediation agents for remediation of crude oil-contaminated ecosystems, following examples are provided. In soil microcosms, poly (vinyl alcohol) cryogel-immobilized *R. ruber* and *R. erythropolis* were shown to survive in oil-contaminated soil at contamination level of 5% (w/w) under drought conditions (no moistening), providing 45% removal of crude oil at ambient temperature in 14 months (Kuyukina et al. 2013). A piezotolerant strain *Rhodococcus* sp. PC20 isolated from the deep-sea sediments in 2013 at the depth of 1,100 m in the northern east of the Gulf of Mexico was able to degrade 77% of surrogate Light Louisiana Sweet crude oil in model water systems at contamination level of 1% (v/v), ambient temperature, and pressure 15 MPa in 96 h (Hackbusch et al. 2020). In field experiments, rhodococcal cells were introduced into a soil pile spiked with 0.4% (w/w) summer diesel fuel. The bulk agent as wheat straw, and mineral salt solution was added into the pile beforehand. After 6 weeks of bioremediation with regular irrigation, the biodegradation percentage of diesel fuel was 85% (Kis et al. 2017). In open soil ecosystems in temperate climate (Urals, Russia), rhodococci were introduced as a component of an oleophilic biopreparation, and its application provided a removal of 80–90% of crude oil at the initial contamination level of 30% (w/w) within 3 months (summer season) (Krivoruchko et al. 2019).

## 5. Preserving Viability and Stabilizing the Properties of *Rhodococcus* Biodegraders. Practice and Prospects of Long-Term Conservation-Reactivation

Prospective use of *Rhodococcus* in research and development (R&D) and their biotechnological exploitation raises the problem of efficiently preserving the laboratory cultures viable, non-contaminated and with unchanged genotypic and phenotypic characteristics for long periods. Successful implementation of scientific and practical tasks, e.g., potential use of rhodococci for bioremediation purposes, depends on the right choice of not only particular biodegrading strains but also an adequate method of their storage. Today, there is a return of close attention to the strains of microorganisms stored in collections that guarantee absolute safety and high quality of the biological objects maintained (Smith 2012, Prakash et al. 2013, Overmann 2015, Ivshina and Kuyukina 2018).

The existing practice of collection culture preservation has developed a variety of experimental approaches to reversible cessation of microbial cell activity. Various conservation methods for a wide spectrum of physiological groups of microorganisms are described. To date, among the methods that allow the vegetative cells of most microorganisms to be transferred to an anabiotic state with a minimum risk of genetic changes, the methods of choice are lyophilization (deep freezing and vacuum drying of biological objects, freeze-drying), cryopreservation (transfer of a biological object to a state of deep cold anabiosis at minus 196°C and minus 150°C in liquid nitrogen or its vapors, respectively) and low-temperature freezing (in freezers at minus 96°C and lower). The death rate of microbial cells is known to be 1,000 times less at these storage temperatures than at minus 10°C (Heckly 1978).

There is, however, no "gold" standard for *ex situ* preservation of microbial cultures. For each specific taxonomic group of microorganisms and often for individual strains of the same species, there is different cryo- and xerostability.

To preserve the valuable properties of each microbial group for a long time, an individual approach is required, including the process of preliminary preparation of the culture, selection of conservation methods and subsequent recovery of cells using specific protocols that is often associated with high material expenses (Ivshina et al. 1994, Suzuki 2017). The most important stages and components of the lyophilization process, low- and ultra-low temperature freezing of bacteria are reviewed (Tedeschi and De Paoli 2011, Prakash et al. 2013, Alonso 2016). Thus, for successful freeze-drying, rehydration and after-storage recovery of lyophilized cultures, in addition to the optimal mode of a freeze-drying process itself using modern freeze-drying equipment, there is a number of other additional requirements, e.g., a cryoprotectant, optimal cell concentration of the initial suspension, an appropriate rehydration agent, and a favorable recovery medium (Heckly 1978). It should also be taken into account that different species and even strains have a specific reaction to drying-rehydration. Drying allows obtaining a bacterial preparation in a convenient form to be delivered to users. One of the most available storage methods is low-temperature preservation, i.e., freezing and storing microbial cultures in specialized freezers or kelvinators, which maintain low (minus 85–96°C) temperatures for a long time. This method is less time-consuming and safer than cryopreservation in liquid nitrogen (Hubalek 2003).

Because *Rhodococcus* are relatively young biotechnology agents, the issues of their long-term *ex situ* storage are still at the stage of accumulating knowledge and working out individual preservation regimes. Considering the high scientific and potential commercial value of *Rhodococcus*, the practical task of maintaining viable, preferably "intact" (as to all possible genetic and phenotypic indicators) rhodococci consists of the individually selected conditions for effective long-term storage of each specific biotechnologically valuable culture without losing its priority properties. Taking into account the structural and physiological characteristics, a multi-year investigation of rhodococci maintained in the Regional Specialized Collection of Alkanotrophic Microorganisms (acronym IEGM, World Federation for Culture Collections # 285, http://www.iegmcol.ru) was performed, resulting in the effective conservation-reactivation protocols of stationary-phase cultures, including the selection of adequate cryoprotectants and appropriate conditions for cultivating the collection strains to ensure their long-term storage.

Among traditional methods of short-term storage aimed at reducing to minimum the cell functional activity (hypobiosis, "cells at rest"), methods of preserving live cultures of rhodococci on minimal agar, in water and 0.5% sodium chloride solution (along with subculturing) proved to be most appropriate ones. These essential in everyday practice methods seem attractive because they are most convenient and simple, do not require expensive bulky equipment, allowing relatively high (30–74%) survival rates and successful preservation of *Rhodococcus* metabolic activities for 2–7 years and often longer. Even after 10 years (the maximum observation period), individual representatives of *R. ruber* (e.g., strains IEGM 87, IEGM 443) retained a sufficient number of viable cells for guaranteed recovery of the culture. Relying on rhodococcal abilities to adhere and form the survival forms (cyst-like cells) under oligotrophic conditions, as well as their resistance to natural desiccation, a practice-proven beneficial technique to preserve *Rhodococcus* spp. with unchanged phenotypic properties is recommended. This technique uses membrane filters applied to the surface of Nutrient Agar or mineral agar with *n*-hexadecane (Catalogue of Strains 2023) and followed by removal of filters with grown cells and storage in sealed sterile test tubes at 4°C. In this case, the guaranteed period for immobilized non-dividing vegetative cells to retain their viability is between two and three years.

For long-term preservation, freeze-drying remains the most common method of stabilizing bacterial cultures. Rhodococci of different species survive lyophilization, not equally successful, depending on the conditions of freeze-drying and features of the strain. The revealed percentage of cell death is rather high in comparison with that of other taxonomic groups of saprotrophic forms. At the same time, one cannot speak about high sensitivity of particular *Rhodococcus* species to the lyophilization process, since the cultures demonstrated different levels of viability within the same species. For example, *R. ruber* IEGM 333 has the highest (39%) cell survival during the first year of storage, 100–150 times exceeding the viability of the least resistant to lyophilization *R. ruber* IEGM 71, IEGM 74 strains. Severe freeze-drying conditions necessitate the use of protective media that provide bioprotective and structure-forming effects. When selecting conditions for effective lyophilization-rehydration of strains, the maximum survival of rhodococci is observed when cell suspensions are mixed with a sucrose-gelatin medium using traditional components (g/l, sucrose – 100, gelatin – 15, agar – 10, distilled water – 1,000 mL) and rehydrated with 0.5% sodium chloride solution. Exposure of cells to a non-penetrating cryoprotectant obviously helps to minimize an osmotic pressure, which causes the cytoplasm shrinking (plasmolysis), and to prevent the intracellular ice formation.

An increase in the initial density of a cell suspension should lead to an increase in the percentage of lyophilization-survived cells. This effect is achieved by the release of "cold-protecting" substances from the lysed cells that act as cryoprotective agents for the surviving cells (Heckly 1978). In this regard, rhodococcal suspensions of various densities were tested, including the highest possible, i.e., $10^{11}$ cells/mL. It was found, however, that at the bacterial suspension concentration above $10^{10}$ cells/mL, the drying procedure leads to the massive cell death, which is caused by a possible failure in regulating the tight intercellular contacts and production of toxic metabolites (Uzunova-Doneva and Donev 2004–2005). For this case, the percentage of viable cells immediate after lyophilization was most often negligible (up to 0.00001%) of their initial number. Nevertheless, the initial concentration of $10^8$–$10^9$ cells/mL allowed maintaining sufficiently large numbers of *Rhodococcus* spp. cells in a viable state during the prolonged storage period. Cell viability analysis in lyophilized state during the first storage year showed that a high percentage of cell death upon freeze-drying does not entail the subsequent rapid degeneration of a bacterial suspension. Indeed, the rate of the cell death is significantly slowed down upon further storage compared to that during lyophilization; the most intensive cell death occurs in the first month. The "post-lyophilization effect" is observed, with the survival rate of rhodococci decreasing by 10.8 times on average. Half-year storage leads to a further gradual drop in the viable cell number. However, the lyophilized preparations still contain up to 44.8% of living cells by the end of the first year. A prolonged experimental period made it possible to evaluate the effectiveness of long-term storage of lyophilized *Rhodococcus* cultures (Table 1). The number of living cells retained after 8-year storage is still quite large, and at the same time, having prospects for a longer period of their preservation. Thus, for *R. ruber* strains, the logarithmic rate of viability loss (K) of lyophilized cells averages 0.24, and the feasible duration of storage determined using the accelerated storage test (Portner et al. 2007) as being 22.6 to 30.2 years. By the end of the observation period, only few strains demonstrated the viability level close to the critical one, indicating their urgent transfer on fresh media, followed by re-lyophilization.

In general, the dynamics of rhodococcal viability shows a steady reliable decrease in the number of living cells during the entire period of preservation. The average maximum survival rate (calculated relative to the number of cells determined before lyophilization) of freeze-dried *Rhodococcus* spp. cultures was 15.8% by the end of the fourth year of storage. The data obtained indicate that the death of the cell population that survived lyophilization occurs in two stages: stage I includes a rapid death of cells not adapted to freeze-drying, resulting from the damage caused by water crystallization and osmotic stress (Hubalek 2003) and stage II involves a slow death of adapted (surviving) cells even under standardized optimal storage conditions (in hermetically sealed glass ampoules at a temperature of 2–4°C) due to oxidative damage of proteins, nucleic acids, or lipids (Adams 2007). Despite this, the control of the bacterial cultures' identity after storage in a lyophilized state reveals that the main morphological characteristics, all patterns of substrate oxidation and degradative abilities inherent to intact *Rhodococcus* cells to be retained (Fig. 4).

To improve the survival rate of different *Rhodococcus* species upon their storage, we used cell populations that mainly realize their inherent processes of "self-preservation" upon transition to the stationary phase (when the biosynthesis of full-fledged cell structures is completed, and reserved substances accumulate). These "self-preservation" processes include the formation of capsule-like structures, cyst-like cells, and carotenoid pigments; synthesis of protective compounds, including trehalose, biosurfactants, and amino acids; cell aggregation and immobilization, which are a widespread phenomenon in nature that plays an essential role in the strategy of survival and maintaining the maximum catalytic functions. Strains belonging to the species *R. rhodochrous* and *R. ruber*, active producers of biosurfactants and red-orange pigments, retained high viability ($10^6$–$10^7$ cells/mL) by the end of the observation period. They most withstand long-term conservation in freeze-dried form. Carotenoids are known to contribute to strengthening the bacterial lipid bilayer. Like reinforcing rods, they penetrate the lipid bilayer sections of the cell membrane, changing its fluidity and facilitating structural stabilization, which leads to an increased membrane rigidity.

A relatively high (19.4–31.6%) survival rate of lyophilized *Rhodococcus* was obtained when they were pre-cultivated with *n*-alkanes. Endogenous respiration measurements of intact and freeze-dried *R. ruber* cultures pre-grown on different media indicate that organisms with a hydrocarbon metabolism are more resistant to lyophilization conditions and subsequent long-term storage, maintaining cellular respiration at a level twice as high than that of *R. ruber* grown on a rich nutrient

**Table 1.** Viability of lyophilized *Rhodococcus* spp. cultures upon 8-year storage.

| Species, strain | lg CFU/ml after | | | | K, year$^{-1}$ | Calculated storage time, yr |
|---|---|---|---|---|---|---|
| | 2 yrs | 4 yrs | 6 yrs | 8 yrs | | |
| *R. erythropolis* | | | | | | |
| IEGM[a] 7[T] | 7.26 | 6.81 | –[b] | – | 0.23 | 21.8 |
| IEGM 8 | – | 7.15 | 6.15 | – | 0.50 | 12.9 |
| IEGM 14 | 6.49 | – | 4.30 | – | 0.55 | 8.9 |
| IEGM 23 | 8.32 | 7.68 | – | – | 0.32 | 19.6 |
| *R. fascians* | | | | | | |
| IEGM 37 | 7.86 | 7.40 | – | – | 0.23 | 24.4 |
| IEGM 39 | 8.40 | 7.54 | – | – | 0.43 | 15.3 |
| IEGM 43 | – | 7.63 | – | 5.15 | 0.62 | 12.0 |
| IEGM 171 | 7.63 | – | 7.88[c] | – | | |
| *R. jostii* | | | | | | |
| IEGM 28 | – | 7.49 | – | 7.61[c] | | |
| IEGM 30 | – | 4.90 | – | 5.34[c] | | |
| IEGM 69 | 5.23 | – | 0.00 | 0.00 | | 5.9 |
| *R. opacus* | | | | | | |
| IEGM 56 | – | 6.85 | – | 2.30 | 1.14 | 7.6 |
| IEGM 59 | – | 6.72 | – | 2.56 | 1.04 | 7.9 |
| IEGM 61 | – | 1.60 | – | 0.00 | | |
| IEGM 458 | – | 6.72 | – | 5.60 | 0.28 | 18.4 |
| *R. rhdochrous* | | | | | | |
| IEGM 62[T] | 8.54[c] | 7.23 | – | – | 0.25 | 22.1 |
| *R. ruber* | | | | | | |
| IEGM 70[T] | – | 7.56 | – | – | 0.20 | 28.3 |
| IEGM 89 | 8.71 | 8.23 | – | – | 0.24 | 27.1 |
| IEGM 92 | 9.08 | – | – | 7.24 | 0.31 | 22.6 |
| IEGM 371 | – | 7.93 | 7.53 | – | 0.20 | 30.2 |

[a] Official acronym of the Regional Specialized Collection of Alkanotrophic Microorganisms (WFCC # 285, http://www.iegmcol.ru).

[b] No data.

[c] Deviation of experimental data from a theoretically calculated curve of the cell death upon increased storage period. The reasons for an increase in the number of viable cells during long-term storage can be various ranging from mechanical in the absence of an automated system for filling the ampoules with bacterial suspensions to purely technological. It is necessary to solve this problem, at least partially, in a methodical way: when testing the cell viability, it is obligatory to open at least three ampoules from each series of ampoules simultaneously placed for storage; to prepare an averaged sample and after a series of subsequent dilutions with inoculation on nutrient agar to determine the titer of cells that survived lyophilization.

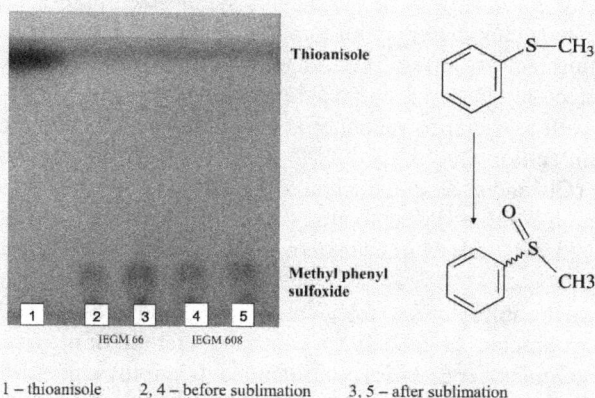

1 – thioanisole   2, 4 – before sublimation   3, 5 – after sublimation

**Figure 4.** Thin-layer chromatography of thioanizole biotransformation products using *Rhodococcus rhodochrous* cells after 4-year storage in lyophilized state.

medium (Ivshina et al. 1994). A high survival rate of lyophilized rhodococci during a long-term storage is primarily a representation of the revealed structural and physiological rearrangement of cells upon an induced alkanotrophy.

One of the features of *n*-alkane-grown rhodococci is to accumulate reserved nutrients in vegetative cells and deposit them in the form of water-insoluble granules as well as to change the fatty acid profile that correlates with the increased resistance of such cells to freezing-drying (Ivshina et al. 1982, 1994, Glazacheva et al. 1990). Thus, accumulation of poly-β-oxybutyrate granules contributes to their immobilization on the surface of cellular biopolymers interacting with the protective medium components via forming strong hydrogen bonds that stabilize cell structures, including the lipid bilayer and protein domains. The reserved substances are also used by cells at the rehydration-reactivation step to repair damage and thereby increase the survival rate of rhodococci exposed to stress.

*Rhodococcus* cell resistance to freezing-drying also correlated with the increased amounts of odd-numbered fatty acids. As compared to the fatty acid composition of cells grown on Nutrient Agar, the total odd-numbered fatty acids in propane- and *n*-hexadecane-grown cells increased two- and three-fold, respectively (Ivshina et al. 1994). Note, to ensure high survival of lyophilized rhodococci, it is advisable to pre-grow them on liquid *n*-alkanes rather than gaseous ones. It is known (Russell 1984, Benschoter and Ingram 1986) that a higher proportion of unsaturated fatty acids provides a more fluid state of membrane lipids and an optimal elasticity of the cell wall, which may be one of the mechanisms of bacterial cell resistance to low-temperature effects. Nevertheless, relatively high amounts of unsaturated fatty acids present in most strains of propane-oxidizing rhodococci significantly limited the use of the propane pre-growing procedure of cells. That is because unsaturated fatty acids are easily oxidized in dried cells, affecting drastically their survival after rehydration. The introduction of α-tocopherol acetate (1 mM) into a bacterial suspension prior lyophilizing helped to maintain the maximum number of viable cells with alkanotrophic metabolism during long-term storage.

Undoubtedly, *Rhodococcus* biosurfactants synthesized on *n*-alkanes and acting as intracellular protectors can play an important role in the metabolic resistance of cells to damaging factors during freeze-drying (Ivshina et al. 1998, Kuyukina et al. 2001). The protective effects of amino acids (alanine, glycine, asparagine) and their derivatives as well as vitamins (betaine) are also known. These are associated with their antioxidant activity and the ability to protect cells from free radical oxidation during storage (Cleland et al. 2004, Zhao and Zhang 2005, Adams 2007). However, these issues require further in-depth study to come closer to a detailed insight into complex mechanisms evolved in *Rhodococcus* to tolerate freezing and dehydration-rehydration.

Considering the ability of *Rhodococcus* to strongly adsorb onto cellulose (Ivshina et al. 2013a, Krivoruchko et al. 2019), a combined method of preserving cultures in the adsorbed state on paper disks is recommended. It employs immobilizing bacterial cells in the growth medium on paper disks, drying them at 28°C and then freezing and storing the disk-immobilized rhodococci in sterile cryotubes in a freezer at –85°C. Cultures can be recovered by placing the discs from a cryotube on the surface of the Nutrient Agar using sterile forceps, when 100%-disc fouling indicates the culture viability.

The viability dynamics of *R. erythropolis* and *R. ruber* strains stored without cryoprotective agents indicated a slight gradual decrease in cell numbers from $10^7$ to $10^6$ CFU/mL after seven-year storage (Table 2). Compared to cell viability before cryopreservation, the survival rate of these strains averaged 4.2% and 28.6%, respectively, ensuring complete culture recovery. Using the example of the *R. ruber* IEGM 333 strain, it was shown that an increased thawing rate (in a water bath at 37°C with shaking) provided an almost 20% increase in the number of viable cells. Employing cryopreservation (–85°C)—such an economic, in fact, near to an ideal technique—for *Rhodococcus* preservation has certain advantages (minimal labor efforts and material expenses, simplicity and safety) over lyophilization and offers reusability of the cryotube contents (upon ten cycles of cryopreservation-thawing, the number of viable cells decreased by only one or two orders of magnitude after 200 days after the start of the experiment) and acceleration of the dynamic control of culture viability during storage.

Other approaches applied in the IEGM Collection to cryopreserve *Rhodococcus* cell suspensions at lower ($4.3 \times 10^7$ cells/mL) concentration and with priority properties being maintained include the use of dimethyl sulfoxide and glycerol at optimal concentrations (5 and 10%, respectively), providing 45–94% viability of rhodococci by the end the first storage year at minus 85°C (with subsequent thawing at room temperature). Importantly, an increased resistance to such low temperature of cultures pre-grown on propane was registered compared with cells pre-grown on a nutrient-rich medium, resulting in the average survival rates of 76.6 and 56.5%, respectively. However, for certain *Rhodococcus* strains, cryopreservation without a protective agent was more effective. For example, after two years of storage, the survival rate of *R. ruber* IEGM 83 cells was 13% higher in distilled water than with the use of a cryoprotective agent. Thus, the IEGM Collection employs appropriate preservation techniques for each species and strain of *Rhodococcus* to ensure long-

**Table 2.** Viability dynamics of *R. erythropolis* and *R. ruber* collection strains after cryopreservation (−85°C) during 7-year storage (CFU/ml).

| Strain | Before cryopreservation | Storage time | | | | Survival, % |
|---|---|---|---|---|---|---|
| | | 6 mos | 1 yr | 4.5 yrs | 7 yrs | |
| *R. erythropolis* | | | | | | |
| IEGM[a] 74 | $5.5 \times 10^7$ | −[b] | $4.2 \times 10^7$ | $1.9 \times 10^7$ | $3.80 \times 10^6$ | 6.90 |
| IEGM 83 | $6.2 \times 10^7$ | $3.7 \times 10^6$ | − | $2.0 \times 10^6$ | $2.90 \times 10^6$ | 4.68 |
| IEGM 231 | $5.3 \times 10^7$ | $3.4 \times 10^7$ | $3.0 \times 10^7$ | $1.1 \times 10^7$ | $1.90 \times 10^6$ | 3.59 |
| IEGM 233 | $5.0 \times 10^7$ | $3.2 \times 10^7$ | $4.7 \times 10^7$ | $2.2 \times 10^7$ | $1.55 \times 10^6$ | 3.10 |
| IEGM 235 | $4.8 \times 10^7$ | $3.1 \times 10^7$ | $4.4 \times 10^7$ | $1.0 \times 10^7$ | $3.00 \times 10^6$ | 6.25 |
| IEGM 381 | $3.7 \times 10^7$ | $1.1 \times 10^7$ | $1.0 \times 10^7$ | $2.8 \times 10^6$ | $1.10 \times 10^6$ | 2.97 |
| IEGM 468 | $5.4 \times 10^7$ | $3.7 \times 10^7$ | $3.7 \times 10^7$ | $1.0 \times 10^7$ | $1.05 \times 10^6$ | 1.94 |
| *R. ruber* | | | | | | |
| IEGM 200 | $1.8 \times 10^7$ | $8.2 \times 10^6$ | $6.5 \times 10^6$ | − | $6.0 \times 10^6$ | 33.3 |
| IEGM 267 | $1.4 \times 10^7$ | $5.8 \times 10^6$ | $4.3 \times 10^6$ | − | $8.0 \times 10^5$ | 5.71 |
| IEGM 487 | $1.3 \times 10^7$ | $6.2 \times 10^6$ | $6.2 \times 10^6$ | − | $5.4 \times 10^6$ | 41.5 |
| IEGM 507 | $3.0 \times 10^7$ | $1.2 \times 10^7$ | $7.3 \times 10^6$ | − | $7.0 \times 10^6$ | 23.3 |
| IEGM 684 | $1.1 \times 10^7$ | $1.0 \times 10^7$ | $7.4 \times 10^6$ | − | $4.3 \times 10^6$ | 39.1 |

[a] Official acronym of the Regional Specialized Collection of Alkanotrophic Microorganisms (WFCC # 285, http://www.iegmcol.ru).

[b] No data.

term maintenance of their viability and essential biological characteristics. The most reliable methods of long-term storage are lyophilization or low temperature freezing of rhodococci with an induced alkanotrophic metabolism. The approximate time for *Rhodococcus* spp. to preserve viability in the lyophilized state is estimated at 5.9–43.9 years. Using the above-mentioned optimized preservation and reactivation protocols, the *Rhodococcus* pool of 665 environmental strains (stored at IEGM Collection) has been preserved for long-term storage. These strains isolated from soil, surface and groundwater, stratal water, sediment, snow, air and bore core samples obtained from sites with a history of pollution and in contrasting ecological-geographical regions can transform and completely decompose hazardous environmental pollutants (Catalogue of Strains 2023). According to the control testing for viability and authenticity of the lyophilized *Rhodococcus* cultures stored for 15 years (from 2005 to the present), of 64 reference strains tested, 73.4% were successfully recovered and retained cell integrity, main morphological and cultural properties, including practically valuable ones. The possibility of recovering the viability and authenticity of the IEGM collection strains after long-term storage indicates high physiological and reparative regeneration potential of *Rhodococcus*.

# 6. Conclusion

In the face of a severe decline in the pristine environmental quality, it is extremely important to find ways to prevent and eliminate (neutralize) anthropogenic (technogenic) pollution of natural ecosystems. The problem of cleaning up the biosphere from ecopollutants will remain highly relevant for a long time. Xenobiotics are a "time bomb". Once the new substances are synthesized and implemented into life, a prerequisite is created for their accumulation in open ecosystems. In this regard, the sizable majority of actinobacteriologists (actinomycetologists) have recently concentrated their efforts in that field of applied microbiology, which provides the search for *in situ* conditions, rational ways of biodegradation and effective biodegraders of new foreign compounds constantly entering the environment. Their harmful effects are enhanced by the presence of many other synergistic xenobiotics with varying degrees of degradability and toxicity.

The main reason that prompts more researchers to focus on polyextremotolerant *Rhodococcus*, dominating in anthropogenically disturbed biotopes, and adaptive mechanisms of their survival in unfavorable environmental conditions is their really feasible widespread application in modern ecobiotechnology fields.

Until recently, the study of *Rhodococcus* has been mostly of academic interest, but lately the interest in this group of actinomycetes becomes broader and more practical. Obviously, this is because rhodococci with their exceptional metabolic capabilities have almost no (like many other actinomycetes) "competitors" in the ability to decompose foreign organic compounds to inorganic products or low-molecular-weight organic fragments that can participate in the natural carbon cycle. To date, considerable knowledge has been accumulated on degradation of priority pollutants by rhodococci with their metabolic pathways understood, and *Rhodococcus* genome projects have been initiated (Busch et al. 2019, Garrido-Sanz et al. 2020). An increasing number of studies reported on relationship between the *Rhodococcus* ability to decompose recalcitrant xenobiotics, stress tolerance to their effects, and survival strategies of rhodococci under natural conditions of synergistic action of ecotoxicants and other harmful exogenous factors (Alvarez et al. 2004, LeBlanc et al. 2008, Fanget and Foley 2011, Corno et al. 2014, Su et al. 2015, Röttig et al. 2016, Zhang et al. 2017, Raymond-Bouchard et al. 2018, Firrincieli et al. 2019, Cappelletti et al. 2020, Hu et al. 2020, Sundararaghavan et al. 2020, Wang et al. 2020). Studying this fundamental relationship, understanding complex mechanisms of rhodococcal protection against environmental stresses, a thorough interpretation of the results of various stress reactions, providing cross-resistance of *Rhodococcus* to many chemical compounds can open new prospects of their biotechnological applications in the near future.

Territory-oriented studies of *Rhodococcus* diversity in biocenoses of anthropogenically-disturbed soils and aquatic ecosystems allowed accumulating the factual material represented by many pure non-pathogenic cultures and related characteristic descriptions. It was experimentally proved that hydrocarbon-oxidizing bacteriocenoses of soils in oil-producing regions are featured by stable abundance of *Rhodococcus* not subject to sharp seasonal fluctuations. Relative abundance of rhodococci of ecologically relevant species *R. erythropolis*, *R. globerulus*, *R. opacus*, *R. rhodochrous*, and *R. ruber* in soil heavily contaminated (up to 10 wt.%) with oil products makes up $10^4$–$10^5$ cells/g soil, confirming their leading role in processes of natural biodegradation of petroleum hydrocarbons. The isolated strains are characterized by useful degradation activities not only towards aliphatic and aromatic hydrocarbons, but also against other pollutants—heterocycles, oxygenated and halogenated compounds, nitroaromatics, organochlorine pesticides. Also, they are resistant to high concentrations of toxic metals and metalloids, active in a wide range of extreme temperatures and acidity, able to grow at high salt concentrations. These *Rhodococcus* strains hold much promise for biotechnological processes of toxic pollutant biodegradation and contaminated site bioremediation.

The most efficient environmental applications using *Rhodococcus* rely on proper selection and optimal *ex situ* storage of valuable strains. This review reports information on the appropriate methods for long-term storage of practically valuable strains of rhodococci. Biotechnology-focused microbial collections and their specialized databases ensure the safety and availability of stable viable cultures for basic and applied research. "*Microorganisms maintained in biological collections play an indispensable role in shaping a sustainable future*" (OECD 2004). The above mentioned is one of the necessary conditions enabling the dynamic development of bioeconomy sectors that implies improved energy efficiency, greening the industry, eliminating pollution, neutralization, and efficient waste management.

## Acknowledgements

The work was carried out within the framework of State Assignments (AAAA-A19-119112290008-4 and FSNF-2023-0004), the Russian Foundation for Basic Research grant (20-44-596001) and agreement of the Ministry of Science and Higher Education of the Russian Federation (075-15-2021-1051).

## References

Acosta-González, A., Martirani-von Abercron, S.-M., Rosselló-Móra, R., Wittich, R.M. and Marqués, S. (2016). The effect of oil spills on the bacterial diversity and catabolic function in coastal sediments: A case study on the Prestige oil spill. *Environmental Science and Pollution Research International*, 22(20): 15200–15214. Doi: 10.1007/s11356-015-4458-y.

Adams, J. (2007). The principles of freeze-drying. *Methods in Molecular Biology*, 368: 15–38. Doi: 10.1007/978-1-59745-362-2_2.

Aggarwal, R.K., Dawar, C., Phanindranath, R., Mutnuri, L. and Dayal, A.M. (2016). Draft genome sequence of a versatile hydrocarbon-degrading bacterium, *Rhodococcus pyridinivorans* strain KG-16, collected from oil fields in India. *Genome Announcements*, 4(1): e01704–15. Doi:10.1128/genomeA.01704-15.

Alonso, S. (2016). Novel preservation techniques for microbial cultures. pp. 7–33. *In*: Ojha, K.S. and Tiwari, B.K. (eds.). *Novel Food Fermentation Technologies*. Springer, Cham. DOI: 10.1007/978-3-319-42457-6_2.

Alvarez, H.M. (ed.). (2019). *Biology of Rhodococcus*. Second edition. Microbiology Monographs 16. Springer Nature. Switzerland AG. Doi: 10.1007/978-3-030-11461-9.

Alvarez, H.M., Silva, R.A., Cesari, A.C., Zamit, A.L., Peressutti, S.R., Reichelt, R. et al. (2004). Physiological and morphological responses of the soil bacterium *Rhodococcus opacus* strain PD630 to water stress. *FEMS Microbiology Ecology*, 50(2): 75–86. Doi: 10.1016/j.femsec.2004.06.002.

Anandan, R., Dhanasekaran, D. and Gopinath, P.M. (2016). An introduction to actinobacteria. pp. 3–37. *In*: Dhanasekaran, D. and Jiang, Y. (eds.). *Actinobacteria – Basics and Biotechnological Applications*. IntechOpen, London, UK. DOI: 10.5772/62329.

Anastasi, E., MacArthur, I., Scortti, M., Alvarez, S., Giguère, S. and Vázquez-Boland, J.A. (2016). Pangenome and phylogenomic analysis of the pathogenic actinobacterium *Rhodococcus equi*. *Genome Biology and Evolution*, 8(10): 3140–3148. Doi: 10.1093/gbe/evw222.

Atagana, H.I. (2011). Bioremediation of co-contamination of crude oil and heavy metals in soil by phytoremediation using *Chromolaena odorata* (L.) R.M. King & H.E. Robinson. *Water, Air, and Soil Pollution*, 215(1-4): 261–271. Doi: 10.1007/s11270-010-0476-z.

aus der Beek, T., Weber, F.A., Bergmann, A., Hickmann, S., Ebert, I., Hein, A. et al. (2016). Pharmaceuticals in the environment – Global occurrences and perspectives. *Environmental Toxicology and Chemistry*, 35: 823–835. https://doi.org/10.1002/etc.3339.

Auta, H.S., Emenike, C.U., Jayanthi, B. and Fauziah, S.H. (2018). Growth kinetics and biodeterioration of polypropylene microplastics by *Bacillus* sp. and *Rhodococcus* sp. isolated from mangrove sediment. *Marine Pollution Bulletin*, 127: 15–21. Doi: 10.1016/j.marpolbul.2017.11.036.

Bastiaens, L., Springael, D., Wattiau, P., Verachtert, H., Harms, H., DeWachter, R. et al. (2000). Isolation of adherent polycyclic aromatic hydrocarbon (PAH)-degrading bacteria using PAH-sorbing carriers. *Applied and Environmental Microbiology*, 66(5): 1834–1843. Doi: 10.1128/AEM.66.5.1834-1843.2000.

Bej, A.K., Saul, D. and Aislabie, J. (2000). Cold-tolerant alkane-degrading *Rhodococcus* species from Antarctica. *Polar Biology*, 23: 100–105. Doi: 10.1007/s003000050014.

Belfiore, C., Curia, M.V. and Farías, M.E. (2018). Characterization of *Rhodococcus* sp. A5$_{wh}$ isolated from a high-altitude Andean Lake to unravel the survival strategy under lithium stress. *Revista Argentina de Microbiología.*, 50(3): 311–322. Doi: 10.1016/j.ram.2017.07.005.

Benschoter, A.S. and Ingram, L.O. (1986). Thermal tolerance of *Zymomonas mobilis*: Temperature-induced changes in membrane composition. *Applied and Environmental Microbiology*, 51(6): 1278–1284. Doi: 10.1128/AEM.51.6.1278-1284.1986.

Brown, D.M., Bonte, M., Gill, R., Dawick, J. and Boogaard, P.J. (2017). Heavy hydrocarbon fate and transport in the environment. *Quarterly Journal of Engineering Geology and Hydrogeology*, 50(3): 333–346. Doi: 10.1144/qjegh2016-142.

Busch, H., Hagedoorn, P.-L. and Hanefeld, U. (2019). *Rhodococcus* as a versatile biocatalyst in organic synthesis. *International Journal of Molecular Sciences*, 20: 4787. (36 pp.). Doi: 10.3390/ijms20194787.

Cappelletti, M., Fedi, S. and Zannoni, D. (2019a). Degradation of alkanes in *Rhodococcus*. pp. 137–171. *In*: Alvarez, H.M. (ed.). *Biology of Rhodococcus*. Microbiology Monographs 16. Springer Nature. Switzerland AG. Doi: 10.1007/978-3-030-11461-9_6.

Cappelletti, M., Presentato, A., Piacenza, E., Firrincieli, A., Turner, R.J. and Zannon, D. (2020). Biotechnology of *Rhodococcus* for the production of valuable compounds. *Applied Microbiology and Biotechnology*, 104: 8567–8594. https://doi.org/10.1007/s00253-020-10861-z.

Cappelletti, M., Zampolii, J. and Zannoni, D. (2019b). Genomics of *Rhodococcus*. pp. 23–60. *In*: Alvarez, H.M. (ed.). *Biology of Rhodococcus*. Microbiology Monographs 16. Springer Nature. Switzerland AG. Doi: 10.1007/978-3-030-11461-9.

Catalogue of Strains of Regional Specialized Collection of Alkanotrophic Microorganisms [http://www.iegmcol/strains/index.html [10.08.2023].

Ceniceros, A., Dijkhuizen, L., Petrusma, M. and Medema, M.H. (2017). Genome-based exploration of the specialized metabolic capacities of the genus *Rhodococcus*. *BMC Genomics*, 18: 593. Doi: 10.1186/s12864-017-3966-1.

Chandrangsu, P., Rensing, C. and Helmann, J.D. (2017). Metal homeostasis and resistance in bacteria. *Nature Reviews Microbiology*, 15(6): 338–350. Doi: 10.1038/nrmicro.2017.15.

Cheremnykh, K.M., Luchnikova, N.A., Grishko, V.V. and Ivshina, I.B. (2018). Bioconversion of ecotoxic dehydroabietic acid using *Rhodococcus* actinobacteria. *Journal of Hazardous Materials*, 346: 103–112. Doi: 10.1016/j.jhazmat.2017.12.025.

Classen, T., Pietruszka, J. and Schuback, S.M. (2013). A new multicopper oxidase from Gram-positive bacterium *Rhodococcus erythropolis* with activity modulating methionine rich tail. *Protein Expression and Purification*, 89: 97–108. Doi: 10.1016/j.pep.2013.02.003.

Cleland, D., Krader, P., McCree, C., Tang, J. and Emerson, D. (2004). Glycine betaine as a cryoprotectant for prokaryotes. *Journal of Microbiological Methods*, 58(1): 31–38. Doi: 10.1016/j.mimet.2004.02.015.

Corno, G., Coci, M., Giardina, M., Plechuk, S., Campanile, F. and Stefani, S. (2014). Antibiotics promote aggregation within aquatic bacterial communities. *Frontiers in Microbiology.*, 5: 297. Doi: 10.3389/fmicb.2014.00297.

Cunningham, C.J., Kuyukina, M.S., Ivshina, I.B., Konev, A.I., Peshkur, T.A. and Knapp, C.W. (2020). Potential risks of antibiotic resistant bacteria and genes in bioremediation of petroleum hydrocarbon contaminated soils. *Environmental Science: Processes and Impacts*, 22: 1110–1124. Doi: 10.1039/C9EM00606K.

Dastager, S.G., Mawlankar, R., Tang, S.-K., Krishnamurthi, S., Ramana, V.V., Joseph, N. et al. (2014). *Rhodococcus enclensis* sp. nov., a novel member of the genus *Rhodococcus*. *International Journal of Systematic and Evolutionary Microbiology*, 64: 2693–2699. http://dx.doi.org/10.1099/ijs.0.061390-0.

de Carvalho, C.C.C.R. and da Fonseca, M.M.R. (2005a). Degradation of hydrocarbons and alcohols at different temperatures and salinities by *Rhodococcus erythropolis* DCL14. *FEMS Microbiology and Ecology*, 51(3): 389–399. Doi: 10.1016/j.femsec.2004.09.010.

de Carvalho, C.C.C.R. and da Fonseca, M.M.R. (2005b). The remarkable *Rhodococcus erythropolis*. *Applied Microbiology and Biotechnology*, 67(6): 715–726. Doi: 10.1007/s00253-005-1932-3.

de Carvalho, C.C.C.R. and da Fonseca, M.M.R. (2007). Preventing biofilm formation: Promoting cell separation with terpenes. *FEMS Microbiology and Ecology*, 61(3): 406–413. Doi: 10.1111/j.1574-6941.2007.00352.x.

de Carvalho, C.C.C.R., Costa, S.S., Fernandes, P., Couto, I. and Viveiros, M. (2014). Membrane transport systems and the biodegradation potential and pathogenicity of genus *Rhodococcus*. *Frontiers in Physiology*, 5: 133. Doi: 10.3389/fphys.2014.00133.

de Carvalho, C.C.C.R., Parreno-Marchante, B., Neumann, G., da Fonseca, M.M.R. and Heipieper, H.J. (2005). Adaptation of *Rhodococcus erythropolis* DCL14 to growth on *n*-alkanes, alcohols and terpenes. *Applied Microbiology and Biotechnology*, 67: 383–388. Doi: 10.1007/s00253-004-1750-z.

de Carvalho, C.C.C.R., Wick, L.Y. and Heipieper, H.J. (2009). Cell wall adaptations of planktonic and biofilm *Rhodococcus erythropolis* cells to growth on $C_5$ to $C_{16}$ n-alkane hydrocarbons. *Applied Microbiology and Biotechnology*, 82(2): 311–320. Doi: 10.1007/s00253-008-1809-3.

DeLorenzo, D.M., Rottinghaus, A.G., Henson, W.R. and Moon, T.S. (2018). Molecular toolkit for gene expression control and genome modification in *Rhodococcus opacus* PD630. *ACS Synthetic Biology*, 7(2): 727–738. Doi: 10.1021/acssynbio.7b00416.

Depuydt, S., Putnam, M., Holsters, M. and Vereecke, D. (2008). *Rhodococcus fascians*, an emerging threat for ornamental crops. pp. 480–489. *In*: Teixeira da Silva, J.A (ed.). *Floriculture, Ornamental, and Plant Biotechnology: Advances and Topical Issues*. First Edition. V. 5. Global Science Books, Ltd.

Dobrowolski, R., Szcześ, A., Czemierska, M. and Jarosz-Wikołazka, A. (2017). Studies of cadmium (II), lead (II), nickel (II), cobalt (II) and chromium (VI) sorption on extracellular polymeric substances produced by *Rhodococcus opacus* and *Rhodococcus rhodochrous*. *Bioresource Technology*, 225: 113–120. Doi: 10.1016/j.biortech.2016.11.040.

Fanget, N.V. and Foley, S. (2011). Starvation/stationary-phase survival of *Rhodococcus* 1454 *erythropolis* $SQ_1$: A physiological and genetic analysis. *Archives in Microbiology*, 193(1): 1–13. Doi: 10.1007/s00203-010-0638-9.

Fate of Marine Oil Spills. (2011). ITOPF Tech. *Inf. Pap.*, 2: 1–12. https://www.itopf.org/knowledge-resources/documents-guides/document/tip-02-fate-of-marine-oil-spills.

Firrincieli, A., Presentato, A., Favoino, G., Marabottini, R., Allevato, E., Stazi, S.-R. et al. (2019). Identification of resistance genes and response to arsenic in *Rhodococcus aetherivorans* BCP1. *Frontiers in Microbiology*, 110: 888. Doi: 10.3389/fmicb.2019.00888.

Gao, Y., Guo, S., Wang, J., Li, D., Wang, H. and Zeng, D. (2014). Effects of different remediation treatments on crude oil contaminated saline soil. *Chemosphere*, 117: 486–493. Doi: 10.1016/j.chemosphere.2014.08.070.

Garrido-Sanz, D., Redondo-Nieto, M., Martín, M. and Rivilla, R. (2020). Comparative genomics of the *Rhodococcus* genus shows wide distribution of biodegradation traits. *Microorganisms*, 8(5): 774. Doi: 10.3390/microorganisms8050774.

Glazacheva, L.E., Ivshina, I.B. and Oborin, A.A. (1990). Changes in the fine structure of *Rhodococcus rhodochrous* and *Rhodococcus ruber* cells assimilating propane and *n*-butane. *Microbiology*, 59(2): 301–306.

Goethals, K., Vereecke, D., Jaziri, M., Van, M.M. and Holsters, M. (2001). Leafy gall formation by *Rhodococcus fascians*. *Annual Review of Phytopathology*, 39: 27–52. Doi: https://en.wikipedia.org/wiki/Doi_(identifier): 10.1146/annurev.phyto.39.1.27.

Goldford, J., Lu, N., Bajić, D., Estrela, S., Tikhonov, M., Sanchez-Gorostiaga, A. et al. (2018). Emergent simplicity in microbial community assembly. *Science*, 361(6401): 469–474. Doi: 10.1126/science.aat1168.

Goodfellow, M., Jones, A.L., Maldonado, L.A. and Salanitro, J. (2004). *Rhodococcus aetherivorans* sp. nov., a new species that contains methyl t-butyl ether-degrading actinomycetes. *Systematic and Applied Microbiology*, 27(1): 61–65. Doi: 10.1078/0723-2020-00254.

Goordial, J., Raymond-Bouchard, I., Zolotarov, Y., de Bethencourt, L., Ronholm, J., Shapiro, N. et al. (2016). Cold adaptive traits revealed by comparative genomic analysis of the eurypsychrophile *Rhodococcus* sp. JG3 isolated from high elevation McMurdo Dry Valley permafrost, Antarctica. *FEMS Microbiology Ecology*, 92(2): fiv154. https://doi.org/10.1093/femsec/fiv154.

Gressler, L.T., de Vargas, A.C., da Costa, M.M., Pötter, L., da Silveira, B.P., Sangioni, L.A. et al. (2014). Genotypic and phenotypic detection of efflux pump in *Rhodococcus equi*. *Brazilian Journal of Microbiology*, 45(2): 661–665. Doi: https://en.wikipedia.org/wiki/Doi_(identifier):10.1590/s1517-83822014000200040.

Guevara, G., Lopez, M.C., Alonso, S., Perera, J. and Navarro-Llorens, J.M. (2019). New insights into the genome of *Rhodococcus ruber* strain Chol-4. *BMC Genomics*, 20: 332. Doi: 10.1186/s12864-019-5677-2.

Gumataotao, N., Lankathilaka, K.P.W., Bennett, B. and Holz, R.C. (2017). The iron-type nitrile hydratase activator protein is a GTPase. *Biochemical Journal*, 474(2): 247–258. Doi: 10.1042/bcj20160884.

Gutiérrez, J., Nichols, P. and Couperwhite, I. (1999). Changes in whole cell-derived fatty acids induced by benzene and occurrence of the unusual 16:1ω6c in *Rhodococcus* sp. 33. *FEMS Microbiology Letters*, 176(1): 213–218. Doi: 10.1111/j.1574-6968.1999.tb13664.x.

Habib, S., Ahmad, S.A., Johari, W.L.W., Shukor, M.Y.A., Alias, S.A., Khalil, K.A. et al. (2018). Evaluation of conventional and response surface level optimisation of *n*-dodecane (*n*-C$_{12}$) mineralisation by psychrotolerant strains isolated from pristine soil at Southern Victoria Island, Antarctica. *Microbial Cell Factories*, 17: 44. Doi: 10.1186/s12934-018-0889-8.

Hackbusch, S., Noirungsee, N., Viamonte, J., Sun, X., Bubenheim, P., Kostka, J.E. et al. (2020). Influence of pressure and dispersant on oil biodegradation by a newly isolated *Rhodococcus* strain from deep-sea sediments of the Gulf of Mexico. *Marine Pollution Bulletin*, 150: 110683. Doi: 10.1016/j.marpolbul.2019.110683.

Hamedi, J. and Mohammadipanah, F. (2015). Biotechnological application and taxonomical distribution of plant growth promoting actinobacteria. *Journal of Industrial Microbiology and Biotechnology*, 42(2): 157–171. Doi: 10.1007/s10295-014-1537-x.

Hassanshahian, M., Ahmadinejad, M., Tebyanian, H. and Kariminik, A. (2013). Isolation and characterization of alkane degrading bacteria from petroleum reservoir wastewater in Iran (Kerman and Tehran provenances). *Marine Pollution Bulletin*, 73(1): 300–305. Doi: 10.1016/j.marpolbul.2013. 05.002.

Heckly, R.J. (1978). Preservation of microorganisms. *Advances in Applied Microbiology*, 24: 1–53. Doi: 10.1016/s0065-2164(08)70635-x.

Ho, A., Di Lonardo, D.P. and Bodelier, P.L.E. (2017). Revisiting life strategy concepts in environmental microbial ecology. *FEMS Microbiology Ecology*, 93: fix006. Doi: 10.1093/femssec/fix006.

Hu, X., Li, D., Qiao, Y., Song, Q., Guan, Z., Qiu, K. et al. (2020). Salt tolerance mechanism of a hydrocarbon-degrading strain: Salt tolerance mediated by accumulated betaine in cells. *Journal of Hazardous Materials*, 392: 122326. Doi: 10.1016/j.jhazmat.2020.122326.

Hubalek, Z. (2003). Protectants used in the cryopreservation of microorganisms. *Cryobiology*, 46(3): 205–229. Doi: 10.1016/s0011-2240(03)00046-4.

Hwang, C.Y., Lee, I., Cho, Y., Lee, Y.M., Baek, K., Jung, Y.J. et al. (2015). *Rhodococcus aerolatus* sp. nov., isolated from subarctic rainwater. *International Journal of Systematic and Evolutionary Microbiology*, 65(2): 465–471. Doi: 10.1099/ijs.0.070086-0.

Ivshina, I.B. (2012). Current situation and challenges of specialized microbial resource centres in Russia. *Microbiology*, 81(5): 509–516. Doi: 10.1134/S0026261712050098.

Ivshina, I.B. and Kuyukina, M.S. (2018). Specialized microbial resource centers: A driving force of the growing bioeconomy. pp. 111–140. *In*: Sharma, S.K. and Varma, A. (eds.). *Microbial Resource Conservation, Soil Biology*. Vol. 54. (451 pp.). Springer International Publishing AG, part of Springer Nature, 2018. https://doi.org/10.1007/978-3-319-96971-8_4.

Ivshina, I.B., Berdichevskaya, M.V., Zvereva, L.V., Rybalka, L.V. and Elovikova, E.A. (1995). Phenotypic characterization of alkanoprophic rhodococci from various ecosystems. *Microbiology*, 64(4): 507–513.

Ivshina, I.B., Kamenskikh, T.N., Kuyukina, M.S., Richkova, M.I., Shadrin, O.A., Rybalka, L.V. et al. (1994). Methods of *Rhodococcus* spp. cultures conservation and their usage in practice of supporting alkanotrophic *Rhodococcus* specialized fund. Expedient for culture conservation practice. *Microbiology*, 63(1): 118–128.

Ivshina, I.B., Kuyukina, M.S. and Krivoruchko, A.V. (2017). Hydrocarbon-oxidizing bacteria and their potential in eco-biotechnology and bioremediation. pp. 121–148. *In*: Kurtböke, I. (ed.). *Microbial Resources: From Functional Existence in Nature to Industrial Applications*. Elsevier, New York. Doi:10.1016/B978-0-12-804765-1.00006-0.

Ivshina, I.B., Kuyukina, M.S. and Kostina, L.V. (2013b). Adaptive mechanisms of nonspecific resistance to heavy metal ions in alkanotrophic actinobacteria. *Russian Journal of Ecology*, 44(1): 123–130. Doi: 10.1134/S1067413613020082.

Ivshina, I.B., Kuyukina, M.S., Krivoruchko, A.V., Barbe, V. and Fischer, C. (2014). Draft genome sequence of propane- and butane-oxidizing actinobacterium *Rhodococcus ruber* IEGM 231. *Genome Announcements*, 2(6): e01297-14. Doi: 10.1128/genomeA.01297-14.

Ivshina, I.B., Kuyukina, M.S., Krivoruchko, A.V., Plekhov, O.A., Naimark, O.B., Podorozhko, E.A. et al. (2013a). Biosurfactant-enhanced immobilization of hydrocarbon-oxidizing *Rhodococcus ruber* on sawdust. *Applied Microbiology and Biotechnology*, 97: 5315–5327. Doi: 10.1007/s00253-013-4869-y.

Ivshina, I.B., Kuyukina, M.S., Philp, J.C. and Christofi, N. (1998). Oil desorption from mineral and organic materials using biosurfactant complexes produced by *Rhodococcus* species. *World Journal of Microbiology and Biotechnology*, 14: 711–717. Doi: 10.1023/A:1008885309221.

Ivshina, I.B., Mukhutdinova, A.N., Tyumina, H.A., Vikhareva, H.V., Suzina, N.E. and El'-Registan, G.I. (2015). Drotaverine hydrochloride degradation using cyst-like cells of *Rhodococcus ruber*. *Current Microbiology*, 70(3): 307–314. Doi: 10.1007/s00284-014-0718-1.

Ivshina, I.B., Nesterenko, O.A., Glazacheva, L.E. and Shekhovtsev, V.P. (1982). Facultative gas assimilating *Rhodococcus rhodochrous* studied by electron microscopy. *Microbiology*, 51(3): 477–481.

Ivshina, I.B., Oborin, A.A., Nesterenko, O.A. and Kasumova, S.A. (1981). *Rhodococcus* bacteria in ground water from oil fields in the Perm Cisural region. *Microbiology*, 50(4): 531–538.

Ivshina, I.B., Peshkur, T.A. and Korobov, V.P. (2002). Efficient uptake of cesium Ions by *Rhodococcus* cells. *Microbiology*, 71(3): 357–361. Doi: 10.1023/A:1015875216095.

Ivshina, I.B., Tyumina, E.A., Kuzmina, M.V. and Vikhareva, E.V. (2019). Features of diclofenac biodegradation by *Rhodococcus ruber* IEGM 346. *Scientific Reports*, 9(1): 9159. Doi: 10.1038/s41598-019-45732-9.

Ivshina, I.B., Vikhareva, E.V., Richkova, M.I., Mukhutdinova, A.N. and Karpenko, J.N. (2012). Biodegradation of drotaverine hydrochloride by free and immobilized cells of *Rhodococcus rhodochrous* IEGM 608. *World Journal of Microbiology and Biotechnology*, 28: 2997–3006. Doi: 10.1007/s11274-012-1110-6.

Iwabuchi, N., Sunairi, M., Anzai, H., Nakajima, M. and Harayama, S. (2000). Relationships between colony morphotypes and oil tolerance in *Rhodococcus rhodochrous*. *Applied and Environmental Microbiology*, 66(11): 5073–5077. Doi: 10.1128/AEM.66.11.5073-5077.2000.

Jones, A.L. and Goodfellow, M. (2012). Genus IV. *Rhodococcus* (Zopf 1891) emend. Goodfellow, Alderson and Chun 1998a. pp. 437–464. *In*: Goodfellow, M., Kämpfer, P., Busse, H.-J., Trujillo, M.E., Suzuki, K., Ludwig, W., and Whitman, W.B. (eds.). *Bergey's Manual of Systematic Bacteriology*, 2nd edn, vol. 5, The Actinobacteria, part A, Edited by New York: Springer.

Kämpfer, P., Wellner, S., Lohse, K., Lodders, N. and Martin, K. (2013). *Rhodococcus cerastii* sp. nov. and *Rhodococcus trifolii* sp. nov., two novel species isolated from leaf surfaces. *International Journal of Systematic and Evolutionary Microbiology*, 63(3): 1024–1029. Doi:10.1099/ijs.0.044958-0.

Kaushik, G. and Thomas, M.A. (2019). The potential association of psychoactive pharmaceuticals in the environment with human neurological disorders. *Sustainable Chemistry and Pharmacy*, 13: 100148. https://doi.org/10.1016/j.scp.2019.100148.

Kim, D., Choi, K.Y., Yoo, M., Zylstra, G.J. and Kim, E.J. (2018). Biotechnological potential of *Rhodococcus* biodegradative pathways. *Journal of Microbiology and Biotechnology*, 28(7): 1037–1051. Doi: 10.4014/jmb.1712.12017.

Kis, Á.E., Laczi, K., Zsíros, S., Kós, P., Tengölics, R., Bounedjoum, N. et al. (2017). Characterization of the *Rhodococcus* sp. MK1 strain and its pilot application for bioremediation of diesel oil-contaminated soil. *Acta Microbiologica et Immunologica Hungarica*, 64(4): 463–482. Doi: 10.1556/030.64.2017.037.

Kedlaya, I., Ing, M.B. and Wong, S.S. (2001). *Rhodococcus equi* infections in immunocompetent hosts: case report and review. *Clinical Infectious Diseases*, 32(3): 39–46. Doi: 10.1086/318520.

Kongpol, A., Kato, J., Tajima, T. and Vangnai, A.S. (2012). Characterization of acetonitrile-tolerant marine bacterium *Exiguobacterium* sp. SBH81 and its tolerance mechanism. *Microbes and Environments*, 27(1): 30–35. Doi: 10.1264/jsme2.ME11228.

Konishi, M., Nishi, S., Fukuoka, T., Kitamoto, D., Watsuji, T.-O., Nagano, Y. et al. (2014). Deep-sea *Rhodococcus* sp. BS-15, lacking the phytopathogenic *fas* genes, produces a novel glucotriose lipid biosurfactant. *Marine Biotechnology*, 16(4): 484–493. Doi: 10.1007/s10126-014-9568-x.

Korshunova, I.O., Pistsova, O.N., Kuyukina, M.S. and Ivshina, I.B. (2016). The effect of organic solvents on the viability and morphofunctional properties of *Rhodococcus*. *Applied Biochemistry and Microbiology*, 52(1): 43–50. Doi: 10.1134/S0003683816010075.

Krivoruchko, A., Kuyukina, M. and Ivshina, I. (2019). Advanced *Rhodococcus* biocatalysts for environmental biotechnologies. *Catalysts*, 9(3): 236. 19 pp. Doi: 10.3390/catal9030236.

Kurosawa, K., Radek, A., Plassmeier, J. and Sinskey, A. (2015). Improved glycerol utilization by a triacylglycerol producing *Rhodococcus opacus* strain for renewable fuels. *Biotechnology for Biofuels and Bioproducts*, 8: 31. Doi: 10.1186/s13068-015-0209-z.

Kuyukina, M.S., Ivshina, I.B., Kamenskikh, T.N., Bulicheva, M.V. and Stukova, G.I. (2013). Survival of cryogel-immobilized *Rhodococcus* cells in crude oil-contaminated soil and their impact on biodegradation efficiency. *International Biodeterioration and Biodegradation*, 84: 118–125. Doi: 10.1016/j.ibiod.2012.05.035.

Kuyukina, M.S., Ivshina, I.B., Korshunova, I.O. and Rubtsova, E.V. (2014). Assessment of bacterial resistance to organic solvents using a combined confocal laser scanning and atomic force microscopy (CLSM/AFM). *Journal of Microbiological Methods*, 107: 23–29. Doi: 10.1016/j.mimet.2014.08.020.

Kuyukina, M.S., Ivshina, I.B., Philp, J.C., Christofi, N., Dunbar, S.A. and Ritchkova, M.I. (2001). Recovery of *Rhodococcus* biosurfactants using methyl tertiary-butyl ether extraction. *Journal of Microbiological Methods*, 46: 149–156. Doi: 10.1016/s0167-7012(01)00259-7.

Larkin, M.J., Kulakov, L.A. and Allen, C.C.R. (2005). Biodegradation and *Rhodococcus* –masters of catabolic versatility. *Current Opinion in Biotechnology*, 16(3): 282–290. Doi: 10.1016/j.copbio.2005.04.007.

Larkin, M.J., Kulakov, L.A. and Allen, C.C.R. (2006). Biodegradation by members of the genus *Rhodococcus*: biochemistry, physiology, and genetic adaptation. *Advances in Applied Microbiology*, 59: 1–29. Doi: 10.1016/S0065-2164(06)59001-X.

Larkin, M.J., Kulakov, L.A. and Allen, C.C.R. (2010). Genomes and plasmids in *Rhodococcus*. pp. 73–90. In: Alvares, H.M. (ed.). *Biology of Rhodococcus*. Microbiology Monographs. Springer-Verlag, Berlin, Heidelberg. Doi: 10.1007/978-3-642-12937-7_3.

LeBlanc, J.C., Gonçalves, E.R. and Mohn, W.W. (2008). Global response to desiccation stress in the soil actinomyete *Rhodococcus jostii* RHA1. *Advances in Applied Microbiology*, 74(9): 2627–2636. Doi: 10.1128/AEM.02711-07.

Letek, M., González, P., Macarthur, I., Rodríguez, H., Freeman, T.C., ValeroRello, A. et al. (2010). The genome of a pathogenic *Rhodococcus*: Cooptive virulence underpinned by key gene acquisitions. *PLOS Genetics*, 6: e1001145. Doi: 10.1371/journal.pgen.1001145.

Li, J., Zhao, G.-Z., Long, L.-J., Wang, F.-Z., Tian, X.-P., Zhang, S. et al. (2012). *Rhodococcus nanhaiensis* sp. nov., an actinobacterium isolated from marine sediment. *International Journal of Systematic and Evolutionary Microbiology*, 62: 2517–2521. Doi: 10.1099/ijs.0.038067-0.

Liang, Y., Jiao, S., Wang, M., Yu, H. and Shen, Z. (2020). A CRISPR/Cas9-based genome editing system for *Rhodococcus ruber* TH. *Metabolic Engineering*, 57: 13–22. https://doi.org/10.1016/j.ymben.2019.10.003.

Lukic, J., Strahinic, I., Milenkovic, M., Nikolic, M., Tolinacki, M., Kojic, M. et al. (2014). Aggregation factor as an inhibitor of bacterial binding to gut mucosa. *Microbial Ecology*, 68: 633–644. Doi: 10.1007/s00248-014-0426-1.

Luz, A.P., Pellizari, V.H., Whyte, L.G. and Greer, C.W. (2004). A survey of indigenous microbial hydrocarbon degradation genes in soils from Antarctica and Brazil. *Canadian Journal of Microbiology*, 50(5): 323–333. Doi: 10.1139/w04-008.

Ma, J., Zhang, L., Wang, G., Zhang, S., Zhang, X., Wang, Y. et al. (2017). *Rhodococcus gannanensis* sp. nov., a novel endophytic actinobacterium isolated from root of sunflower (*Helianthus annuus* L.). *Antonie van Leeuwenhoek*, 110(9): 1113–1120. Doi: 10.1007/s10482-017-0884-9.

MacArthur, I., Anastasi, E., Alvarez, S., Scortti, M. and Vázquez-Boland, J.A. (2017). Comparative genomics of *Rhodococcus equi* virulence plasmids indicates host-driven evolution of the *vap* pathogenicity island. *Genome Biology and Evolution*, 9(5): 1241–1247. Doi: 10.1093/gbe/evx057.

Martínková, L., Uhnáková, B., Pátek, M., Nesvera, J. and Kren, V. (2009). Biodegradation potential of the genus *Rhodococcus*. *Environment International*, 35(1): 162–177. Doi: 10.1016/j.envint.2008.07.018.

Mikolasch, A., Omirbekova, A., Schumann, P., Reinhard, A., Sheikhany, H., Berzhanova, R. et al. (2015). Enrichment of aliphatic, alicyclic and aromatic acids by oil-degrading bacteria isolated from the rhizosphere of plants growing in oil-contaminated soil from Kazakhstan. *Applied Microbiology and Biotechnology*, 99(9): 4071–4084. Doi: 10.1007/s00253-014-6320-4.

Mirimanoff, N. and Wilkinson, K.J. (2000). Regulation of Zn accumulation by a freshwater Gram-positive bacterium (*Rhodococcus opacus*). *Environmental Science and Technology*, 34: 616–622. Doi: 10.1021/es990744g.

Mooshammer, M., Kitzinger, K., Schintlmeister, A., Ahmerkamp, S., Nielsen, J.L., Nielsen, P.H. et al. (2021). Flow-through stable isotope probing (Flow-SIP) minimizes cross-feeding in complex microbial communities. *The ISME Journal*, 15(1): 348–353. https://doi.org/10.1038/s41396-020-00761-5.

Morrisey, E.M., Mau, R.L., Schwartz, E., Dijkstra, P., Gestel, N.V., Lin, C.M. et al. (2016). Phylogenetic organization of bacterial activity. *The ISME Journal*, 10(9): 2336–2340. Doi: 10.1038/ismej.2016.28.

Muscatello, G., Leadon, D.P., Klay, M., Ocampo-Sosa, A., Lewis, D.A., Fogarty, U. et al. (2007). *Rhodococcus equi* infection in foals: the science of 'rattles'. *Equine Veterinary Journal*, 39(5): 470–478. Doi:10.2746/042516407x209217.

Ng, S., King, C.S., Hang, J., Clifford, R.J., Lesho, E.P., Kuschner, R.A. et al. (2013). Severe cavitary pneumonia caused by a non-*equi Rhodococcus* species in an immunocompetent patient. *Respiratory Care*, 58(4): 47–50. Doi: 10.4187/respcare.02017.

Nguyen, T.M. and Kim, J. (2016). *Rhodococcus pedocola* sp. nov. and *Rhodococcus humicola* sp. nov., two antibiotic-producing actinomycetes isolated from soil. *International Journal of Systematic and Evolutionary Microbiology*, 66(6): 2362–2369. Doi: 10.1099/ijsem.0.001039.

OECD. (2004). Guidance for the Operation of Biological Resource Centres (BRCs) [Biological Research Centres' is erroneously printed on the cover of this document], OECD http://www.oecd.org/dataoecd/60/42/23547743.pdf.

Omran, S.E., Shorafa, M., Zolfaghari, A.A. and Toolarood, A.A.S. (2020). The effect of biochar on severity of soil water repellency of crude oil-contaminated soil. *Environmental Science and Pollution Research*, 27(4): 6022–6032. Doi: 10.1007/s11356-019-07246-9.

Oren, A. and Garrity, G.M. (2021). Valid publication of the names of forty-two phyla of prokaryotes. *International Journal of Systematic and Evolutionary Microbiology*, 71(10): 005056.

Overmann, J. (2015). Significance and future role of microbial resource centers. *Systematic and Applied Microbiology*, 38(4): 258–265. Doi: 10.1016/j.syapm.2015.02.008.

Panicker, G., Mojib, N., Aislabie, J. and Bej, A.K. (2010). Detection, expression and quantitation of the biodegradative genes in Antarctic microorganisms using PCR. *Antonie van Leeuwenhoek*, 97: 275–287. Doi: 10.1007/s10482-009-9408-6.

Pátek, M., Grulich, M. and Nešvera, J. (2021). Stress response in *Rhodococcus*. *Biotechnology Advances*, 107698. Doi: 10.1016/j.biotechadv.2021.107698.

Patel, M., Kumar, R., Kishor, K., Mlsna, T., Pittman, C.U. and Mohan, D. (2019). Pharmaceuticals of emerging concern in aquatic systems: Chemistry, occurrence, effects, and removal methods. *Chemical Reviews*, 119(6): 3510–3673. https://doi.org/10.1021/acs.chemrev.8b00299.

Poger, D., Caron, B. and Mark, A.E. (2014). Effect of methyl-branched fatty acids on the structure of lipid bilayers. *The Journal of Physical Chemistry B: Biophysics, Biomaterials, Liquids, and Soft Matter*, 118(48): 13838–13848. Doi: 10.1021/jp503910r.

Portner, D.C., Leuschner, R.G.K. and Murray, B.S. (2007). Optimising the viability during storage of freeze-dried cell preparations of *Campylobacter jejuni*. *Cryobiology*, 54(3): 265–270. Doi: 10.1016/j.cryobiol.2007.03.002.

Prabowo, A.R. and Bae, D.M. (2019). Environmental risk of maritime territory subjected to accidental phenomena: Correlation of oil spill and ship grounding in the Exxon Valdez's case. *Results in Engineering*, 4: 100035. https://doi.org/10.1016/j.rineng.2019.100035.

Prakash, O., Nimonkar, Y. and Shouche, Y.S. (2013). Practice and prospects of microbial preservation. *FEMS Microbiology Letters*, 339(1): 1–9. Doi: 10.1111/1574-6968.12034.

Presentato, A., Piacenza, E., Anikovskiy, M., Cappelletti, M., Zannoni, D. and Turner, R.J. (2016). *Rhodococcus aetherivorans* BCP1 as cell factory for the production of intracellular tellurium nanorods under aerobic conditions. *Microbial Cell Factories*, 15: 204. Doi: https://doi.org/10.1186/s12934-016-0602-8.

Presentato, A., Piacenza, E., Cappelletti, M. and Turner, R.J. (2019). Interaction of *Rhodococcus* with metals and biotechnological applications. pp. 333–357. *In*: Alvarez, H.M. (ed.). *Biology of Rhodococcus*. Microbiology Monographs 16. Springer Nature, Switzerland AG. Doi: 10.1007/978-3-030-11461-9_12.

Prieto, M.B., Hidalgo, A., Rodríguez-Fernández, C., Serra, J.L. and Llama, M.J. (2002). Biodegradation of phenol in synthetic and industrial wastewater by *Rhodococcus erythropolis* UPV-1 immobilized in an air-stirred reactor with clarifier. *Applied Microbiology and Biotechnology*, 58(6): 853–859. Doi: 10.1007/s00253-002-0963-2.

Puglisi, E., Cahill, M.J., Lessard, P.A., Capri, E., Sinskey, A.J., Archer, J.A.C. et al. (2010). Transcriptional response of *Rhodococcus aetherivorans* I24 to polychlorinated biphenyl-contaminated sediments. *Microbial Ecology*, 60: 505–515. Doi: 10.1007/s00248-010-9650-5.

Ramaprasad, E.V.V., Mahidhara, G., Sasikala, C. and Ramana, C.V. (2018). *Rhodococcus electrodiphilus* sp. nov., a marine electro active actinobacterium isolated from coral reef. *International Journal of Systematic and Evolutionary Microbiology*, 68(8): 2644–2649. Doi: 10.1099/ijsem.0.002895.

Raymond-Bouchard, I., Tremblay, J., Altshuler, I., Greer, C.W. and Whyte, L.G. (2018). Comparative transcriptomics of cold growth and adaptive features of a eury- and steno-psychrophile. *Frontiers in Microbiology*, 9: 1565. Doi: 10.3389/fmicb.2018.01565.

Röttig, A., Hauschild, P., Madkour, M.H., Al-Ansari, A.M., Almakishah, N.H. and Steinbüchel, A. (2016). Analysis and optimization of triacylglycerol synthesis in novel oleaginous *Rhodococcus* and *Streptomyces* strains isolated from desert soil. *Journal of Biotechnology*, 225: 48–56. Doi: 10.1016/j.jbiotec.2016.03.040.

Ruoppolo, V., Woehler, E.J., Morgan, K. and Clumpner, C.J. (2013). Wildlife and oil in the Antarctic: A recipe for cold disaster. *Polar Record*, 49(249): 97–109. Doi: 10.1017/S0032247411000763.

Russel, N.J. (1984). Mechanism of thermal adaptation in bacteria: Blueprints for survival. *Trends in Biochemical Sciences*, 9(3): 108–112. Doi: 10.1016/0968-0004(84)90106-3.

Sangal, V., Goodfellow, M., Jones, A.L., Seviour, R.J. and Sutcliffe, I.C. (2019). Sutcliffe refined systematics of the genus *Rhodococcus* based on whole genome analyses. pp. 1–21. *In*: Alvarez, H.M. (ed.). *Biology of Rhodococcus*. Microbiology Monographs 16. Springer Nature. Switzerland AG. DOI: 10.1007/978-3-030-11461-9_1.

Sheng, H.M., Gao, H.S., Xue, L.G., Ding, S., Song, C.L., Feng, H.Y. et al. (2011). Analysis of the composition and characteristics of culturable endophytic bacteria within subnival plants of the Tianshan Mountains, northwestern China. *Current Microbiology*, 62(3): 923–932. Doi: 10.1007/s00284-010-9800-5.

Sikkema, J., de Bont, J.A.M. and Poolman, B. (1995). Mechanisms of membrane toxicity of hydrocarbons. *Microbiology Reviews*, 59(2): 201–222. Doi:10.1128/MMBR.59.2.201-222.1995.

Silva, L.J., Souza, D.T., Genuario, D.B., Hoyos, H.A.V., Santos, S.N., Rosa, L.H. et al. (2018). *Rhodococcus psychrotolerans* sp. nov., isolated from rhizosphere of *Deschampsia antarctica*. *Antonie van Leeuwenhoek*, 111(4): 629–636. Doi: 10.1007/s10482-017-0983-7.

Singer, M.M., Aurand, D., Bragin, G.E., Clark, J.R., Coelho, G.M. and Sowby, M.L. (2000). Standardization of the preparation and quantitation of water-accommodated fractions of petroleum for toxicity testing. *Marine Pollution Bulletin*, 40(11): 1007–1016. Doi: 10.1016/S0025-326X(00)00045-X.

Sinha, R.K., Krishnan, K.P., Hatha, A.A., Rahiman, M., Thresyamma, D.D. and Kerkar, S. (2017). Diversity of retrievable heterotrophic bacteria in Kongsfjorden, an Arctic fjord. *Brazilian Journal of Microbiology*, 48(1): 51–61. Doi: 10.1016/j.bjm.2016.09.011.

Smith, D. (2012). Culture collections. *Advances in Applied Microbiology*, 79: 73–118. Doi: 10.1016/B978-0-12-394318-7.00004-8.

Stamler, R.A., Kilcrease, J., Kallsen, C., Fichtner, E.J., Cooke, P., Heerema, R.J. et al. (2015). First report of *Rhodococcus* isolates causing Pistachio Bushy Top Syndrome on 'UCB-1' rootstock in California and Arizona. *Plant Disease*, 99(11): 1468–1476. Doi: 10.1094/PDIS-12-14-1340-RE.

Su, X., Sun, F., Wang, Y., Hashmi, M.Z., Guo, L., Ding, L. et al. (2015). Identification, characterization and molecular analysis of the viable but nonculturable *Rhodococcus biphenylivorans*. *Scientific Reports*, 5: 18590. Doi: 10.1038/srep18590.

Sundararaghavan, A., Mukherjee, A. and Suraishkumar, G.K. (2020). Investigating the potential use of an oleaginous bacterium, *Rhodococcus opacus* PD630, for nano-TiO$_2$ remediation. *Environmental Science and Pollution Research*, 27(22): 27394–27406. Doi: 10.1007/s11356-019-06388-0.

Suzuki, K. (2017). An overview of biological resource center-maintenance of microbial resources and their management. pp. 257–274. *In*: Kurtböke, I. (ed.). *Microbial Resources: From Functional Existence in Nature to Industrial Applications*. Elsevier, New York. ISBN 978-0-12-804765-1. Doi:10.1016/B978-0-12-804765-1.00013-8.

Táncsics, A., Máthe, I., Benedek, T., Toth, E.M., Atasayar, E., Sproer, C. et al. (2017). *Rhodococcus sovatensis* sp. nov., an actinomycete isolated from the hypersaline and heliothermal Lake Ursu. *International Journal of Systematic and Evolutionary Microbiology*, 67(2): 190–196. Doi: 10.1099/ijsem.0.001514.

Tarasova, E., Grishko, V. and Ivshina, I. (2017). Cell adaptations of *Rhodococcus rhodochrous* IEGM 66 to betulin biotransformation. *Process Biochemistry*, 52: 1–9. Doi: 10.1016/j.procbio.2016.10.003.

Tedeschi, R. and De Paoli, P. (2011). Collection and preservation of frozen microorganisms. pp. 313–326. *In*: Dillner, J. (ed.). *Methods in Biobanking*. Methods in Molecular Biology. 675. Springer+Business Media, LLC. Doi: 10.1007/978-1-59745-423-0_18.

Tomioka, N., Uchiyama, H. and Yagi, O. (1994). Cesium accumulation and growth characteristics of *Rhodococcus erythropolis* CS98 and *Rhodococcus* sp. strain CS402. *Applied and Environmental Microbiology*, 60(7): 2227–2231. Doi:10.1128/AEM.60.7.2227-2231.1994.

Tsitko, I.V., Zaitsev, G.M., Lobanok, A.G. and Salkinoja-Saloneni, M.S. (1999). Effect of aromatic compounds on cellular fatty acid composition of *Rhodococcus opacus*. *Applied and Environmental Microbiology*, 65(2): 853–855. Doi: 10.1128/AEM.65.2.853-855.1999.

Tyumina, E.A., Bazhutin, G.A., Cartagena Gómez, A.d.P. and Ivshina, I.B. (2020). Nonsteroidal anti-inflammatory drugs as emerging contaminants. *Microbiology*, 89(2): 148–163. Doi: 10.1134/S0026261720020125.

Urbano, S.B., Albarracin, V.H., Ordonez, O.F., Farias, M.E. and Alvarez, H.M. (2013). Lipid storage in high-altitude Andean Lakes extremophiles and its mobilization under stress conditions in *Rhodococcus* sp. A5, a UV-resistant actinobacterium. *Extremophiles*, 17(2): 217–227. Doi: 10.1007/s00792-012-0508-2.

Uzunova-Doneva, T. and Donev, T. (2004–2005). Anabiosis and conservation of microorganisms. *Journal of Culture Collections*, 4: 17–28.

Vereecke, D., Zhang, Y., Francis, I.M., Lambert, P.Q., Venneman, J., Stamler, R.A. et al. (2020). Functional genomics insights into the pathogenicity, habitat fitness, and mechanisms modifying plant development of *Rhodococcus* sp. PBTS1 and PBTS2. *Frontiers in Microbiology*, 11: 14. Doi: 10.3389/fmicb.2020.00014.

Viggor, S., Jõesaar, M., Vedler, E., Kiiker, R., Pärnpuu, L. and Heinaru, A. (2015). Occurrence of diverse alkane hydroxylase *alkB* genes in indigenous oil-degrading bacteria of Baltic Sea surface water. *Marine Pollution Bulletin*, 101(2): 507–516. Doi: 10.1016/j.marpolbul.2015.10.064.

Wang, C., Chen, Y., Zhou, H., Li, X. and Tan, Z. (2020). Adaptation mechanisms of *Rhodococcus* sp. CNS16 under different temperature gradients: Physiological and transcriptome. *Chemosphere*, 238: 124571. Doi: 10.1016/j.chemosphere.2019.124571.

Wannoussa, W., Hiligsmann, S., Tasseroul, L., Masy, T., Lambert, S.D., Heinrichs, B. et al. (2015). Effect of metal ions and metal nanoparticles encapsulated in porous silica on biphenyl biodegradation by *Rhodococcus erythropolis* T902.1. *Journal of Sol-Gel Science and Technology*, 75: 235–245. Doi: 10.1007/s10971-015-3694-4.

Whittaker, R.H. (1975). *Communities and Ecosystems*/2nd ed. N.Y.; London: MacMillan Publ. Co., Inc., 387p.

Williamson, M. (1972). *The Analysis of Biological Population*. Special Topics in Biology Series. E. Arnold, London, Crane Russak, New York.

Yanagita, T., Ichikama, T., Tsuji, T., Kamata, Y., Ito, K. and Sasaki, M. (1978). Two trophic groups of bacteria, oligotrophs and eutrophs: Their distributions in fresh and sea water areas in the central northern Japan. *The Journal of General and Applied Microbiology*, 24(1): 59–88. Doi: 10.2323/jgam.24.59.

Yoneda, A., Henson, W.R., Goldner, N.K., Park, K.J., Forsberg, K.J., Kim, S.J. et al. (2016). Comparative transcriptomics elucidates adaptive phenol tolerance and utilization in lipid accumulating *Rhodococcus opacus* PD630. *Nucleic Acids Research*, 44(5): 2240–2254. Doi: 10.1093/nar/gkw055.

Zhang, C., Yang, L., Ding, Y., Wang, Y., Lan, L., Ma, Q. et al. (2017). Bacterial lipid droplets bind to DNA via an intermediary protein that enhances survival under stress. *Nature Communications*, 8: 15979. Doi: 10.1038/ncomms15979.

Zhao, G. and Zhang, G. (2005). Effect of protective agents, freezing temperature, rehydration media on viability of malolactic bacteria subjected to freeze-drying. *Journal of Applied Microbiology*, 99(2): 333–338. Doi: 10.1111/j.1365-2672.2005.02587.x.

Zhao, G.-Z., Li, J., Zhu, W.-Y., Tian, S.-Z., Zhao, L.-X., Yang, L.-L. et al. (2012). *Rhodococcus artemisiae* sp. nov., an endophytic actinobacterium isolated from the pharmaceutical plant *Artemisia annua* L. *International Journal of Systematic and Evolutionary Microbiology*, 62: 900–905. Doi: 10.1099/ijs.0.031930-0.

Zheng, Y.-T., Toyofuku, M., Nomura, N. and Shigeto, S. (2013). Correlation of carotenoid accumulation with aggregation and biofilm development in *Rhodococcus* sp. SD-74. *Analytical Chemistry*, 85: 7295–7301. Doi: 10.1021/ac401188f.

# Chapter 9

# Dereplication of the Termite Gut-associated Actinomycete Metabolome as a Source of Bioactive Secondary Metabolites

*Christian A. Romero,*[1,2,4] *D. İpek Kurtböke*[3,*] and *Ronald J. Quinn*[1]

## 1. Introduction

Natural products (NPs) are composed of relatively few chemical elements-carbon, hydrogen, oxygen and an occasional nitrogen, sulphur or halogen (Thornburg et al. 2018). They encompass compounds with various molecular sizes, chemotypes, and structural features which varies from the simplest salicylic acid to one of the most complex vancomycin (Guo 2017). The molecular complexity of NPs which is driven by the continual evolution of the biosynthetic genes that synthesize compounds that interact with the unique three-dimensional structures of nucleic acids and proteins, has provided unparalleled opportunity for the identification of novel scaffold structures that has served as a source of inspiration for drug discovery and development (Quinn et al. 2008, Thornburg et al. 2018). NP-based drug discovery leading to robust and viable lead candidates is an intricate, interdisciplinary pursuit of chemistry, pharmacology, and clinical sciences (Butler 2004, Lahlou 2013). Despite the significant number of NP-derived drugs being ranked in the top 35 worldwide in 2000, 2001, and 2002, most of the big pharmaceutical industries have terminated or significantly scaled down their operations (Lahlou 2013). This decline was due in part to the costs behind high rates of rediscovery in the late stages of the isolation process and for the development of powerful new approaches such as automated

[1] Griffith Institute for Drug Discovery, Griffith University, QLD 4111, Brisbane, Australia.
[2] Escuela Superior Politécnica del Litoral, Facultad de Ciencias de la Vida, Centro de Investigaciones Biotecnológicas del Ecuador, CIBE, Campus Gustavo Galindo Km 30.5 Vía Perimetral, P.O. Box 09-01-5863, Guayaquil, Ecuador.
[3] School of Science, Technology and Engineering, University of the Sunshine Coast, Maroochydore BC, QLD 4558, Australia.
[4] Current Address: Universidad Bolivariana del Ecuador UBE, Facultad de Enfermería, Km 5.5 vía Durán-Yaguachi, Durán, Ecuador.
* Corresponding author: ikurtbok@usc.edu.au

separation techniques, high-throughput screening, and combinatorial chemistry (Yang et al. 2013, Guo 2017). High-throughput screening (HTS) has evolved since its introduction during the early 1990s (Volochnyuk et al. 2019). Currently, HTS includes fragment-based lead discovery (FBLD), encoded library technologies, and phenotypic approaches to form a comprehensive screening toolbox to successfully identify new lead molecules (Erlanson et al. 2016, Goodnow Jr. et al. 2017, Moffat et al. 2017). Therefore, over recent years, drug companies have been predominantly relying upon the screening of large synthetic compound libraries over NP libraries as the screening of the latter is generally more problematic as they contain complex mixtures of mostly uncharacterised compounds, including some molecules exhibiting undesirable properties which could interfere with the biological effect of the lead candidate (Butler 2004, Butler et al. 2014, Volochnyuk et al. 2019).

Nevertheless, the major advantage of the screening of NPs in biological assays is their inherently large structural diversity which is unsurpassable by any synthetic compound (Khanna and Ranganathan 2009, Lahlou 2013). NPs are generally biosynthesized in an enantiomerically pure form, thus comprise larger numbers of chiral centres and have increased steric complexity than either synthetic drugs or combinatorial libraries (Batista et al. 2018, Lahlou 2013). Furthermore, it has been shown that 83% of the core ring scaffolds present in NPs were absent from both commercially available molecules and screening libraries (Pascolutti and Quinn 2014). To capture NP-like characteristics, libraries may have to be generated following four main approaches: (i) target-oriented synthesis (TOS) (Schreiber 2000, Burke and Lalic 2002); (ii) diversity-oriented synthesis (DOS) (Schreiber 2000, Tan 2005); (iii) biology-oriented synthesis (BIOS) (Noren-Muller et al. 2006, van Hattum and Waldmann 2014); and (iv) functional-oriented synthesis (FOS) (Pascolutti and Quinn 2014, Wender et al. 2007). The unique and vast chemical diversity of NPs has been optimised through evolutionary selection to bind to multiple, unrelated classes of protein receptors as high affinity ligand (Nicolaou et al. 2000, Rosén et al. 2009). This means that these compounds are not only biologically active but also likely to be substrates for one or more of the many transporter systems that can deliver them to their intracellular site of action (Harvey et al. 2015). Consequently, by including molecules with a NP-product-like scaffold into a screening library, the number of hit rates can be increased (Hert et al. 2009). With the emergence of novel high-content phenotypic cell-based screening systems, the need to rapidly identify effective, novel chemical structures and bioactive lead molecules has become a vital necessity (Baker et al. 2007, Cragg and Newman 2013). Thus, government, industry and academic research groups have incorporated chromatographic separation techniques including solid phase extraction (SPE), counter-current chromatography (CCR), high performance liquid chromatography (HPLC), or supercritical fluid chromatography (SFC) to partially purify components of natural product extracts prior to perform biological assays (Pauli et al. 2008, Camp et al. 2012, Thornburg et al. 2018).

Only a small fraction of the world's biodiversity has been evaluated for biological activity, it can thus be assumed that NPs will continue to be a major source of lead molecules for clinical development (Miller et al. 2009, Cragg and Newman 2013). In an extensive review performed by Newman and Cragg (2020) on the new drugs introduced between 1981 and 2019, it was estimated that from the 1,881 new chemical entities reported (NCEs), up to 23.5% were either directly derived from a NP or were inspired by a NP. A further 25.7% of these entities were synthetic compounds based on a NP pharmacophore. Thus, from the 1,881 NCEs, 930 (49.2%) were classified either as an unmodified NP or synthetic molecule modelled on a NP scaffold (Newman and Cragg 2016, 2020). Between 2018 and 2019, around 59 NPs and NP-derived compounds were being either evaluated in clinical trials or under registration process. Twenty-two of these compounds were investigated as potential oncology treatments, twelve as anti-infectives, three for multiple sclerosis, three for diabetes and one as an antiglaucoma agent (Graul et al. 2018, 2019, Newman and Cragg 2020).

## 1.1 Microbial Resources

### 1.1.1 Microbial Metabolites as Pharmacological Agents

Microorganisms have proven to be promising candidates for the isolation of novel scaffolds exhibiting a wide array of biological effects and pharmaceutically relevant activities (Harvey et al. 2015). The large capacity exhibited by microorganisms to synthesize specialized compounds and perform biocatalytic transformations, is the result of evolutionary adaptations that have enabled them to devote significant portion of their genomes to the production of secondary metabolites some of which have been useful to mankind (Schmidt 2008, Hug et al. 2020). Although these biomolecules are not essential for the survival of the microorganism, under appropriate environmental conditions, they might improve their ability to grow, disperse and induce competition or predation behaviour (Vining 2007, Perez et al. 2016).

Since the discovery of penicillin from *Penicillium notatum* by Alexander Fleming in 1928, drug discovery significantly shifted from plants to microorganisms as a source of natural products (Pham et al. 2019). Microbial secondary metabolites are privileged structures that specifically interact with biological targets, they have been extensively utilized in agriculture, food industry, scientific research, and medicine (Hug et al. 2020). Over the last 90 years, more than 34,000 bioactive microbial secondary metabolites displaying a remarkable and diverse array of biological activities have been isolated from marine, terrestrial and insect-associated microorganisms (Berdy 2012, Kato et al. 2012, Beemelmanns et al. 2016, Mevers et al. 2017, Wiese and Imhoff 2019). Microorganisms have yielded some of the most economically relevant leads for the pharmaceutical industry, including antibacterial agents, such as penicillin G (**1**) (sourced from *Penicillium chrysogenum*); cephalosporin C (**2**) (sourced from *Cephalosporium acremonium*); tetracycline (**3**), aminoglycosides, and other polyketides of many structural types (sourced from different *Streptomyces* species); cholesterol lowering agents, such as mevastatin (**4**) (sourced from *Penicillum brevicompactum*) and lovastatin (**5**) (sourced from *Aspergillus terreus*); immunosuppressive agents, such as rapamycin (sirolimus) (**6**) (sourced from *Streptomyces hygroscopicus*) and ciclosporin A (**7**) (sourced from *Tolypocladium inflatum*); as well as anthelmintics and antiparasitic drugs such as invermectin (**8**) (sourced from *Streptomyces avermitilis*) (Fig. 1) (Buss et al. 2003, Baker et al. 2007). A significant number of chemotherapeutic agents

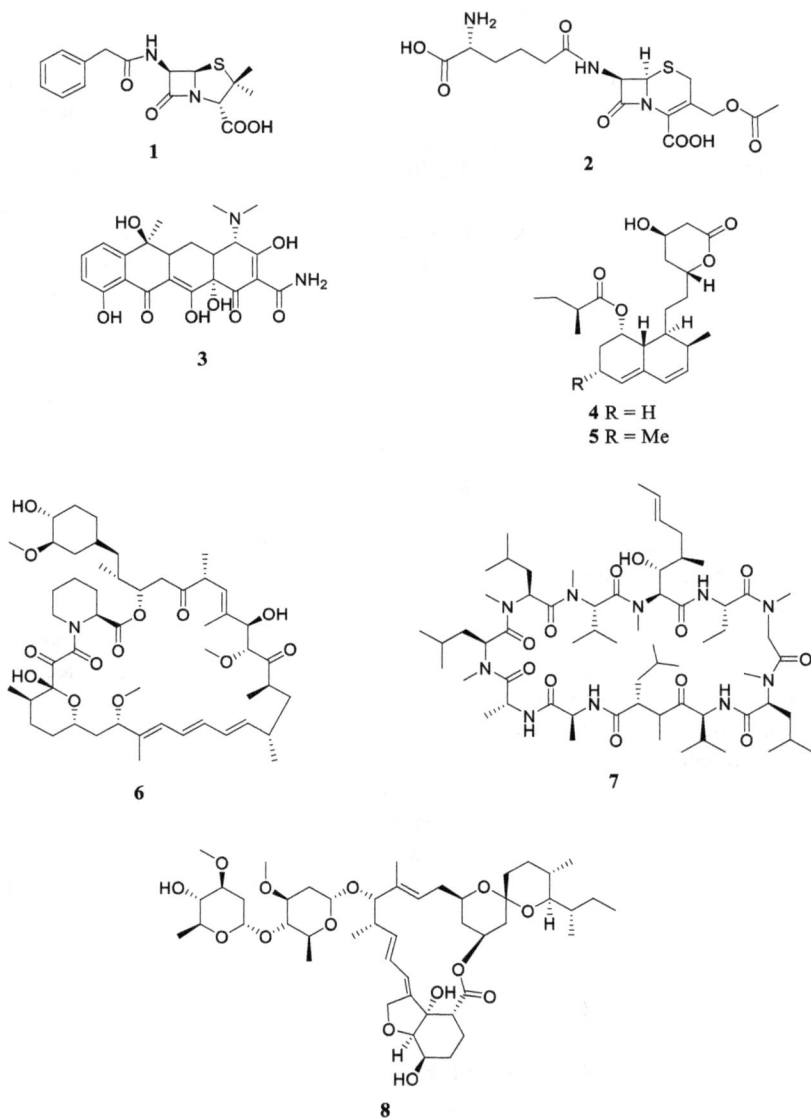

**Figure 1.** Chemical structures of economically relevant microbial metabolites.

isolated from microorganisms have been used to treat bacterial infections and have greatly contributed to the improvement of human health during the past century (Sarker et al. 2006).

## 1.2 Significance of Actinomycetes

Phylum Actinomycetota (Nouioui et al. 2018) includes the former order of *Actinomycetales* which is currently included within the recently proposed Class

*Actinomycetia* (https://lpsn.dsmz.de/class/actinomycetia) (Salam et al. 2020) and commonly known as "Actinomycetes". This common name will be used in this chapter. They are the most widely distributed group of bacteria in nature forming a large part of the microbial population of soil and aquatic ecosystems such as rivers, lakes, streams, marine environments, and salt marshes (Goodfellow and Williams 1983, Terkina et al. 2006, Barka et al. 2016). The most abundant and frequently isolated species are from the genus *Streptomyces* which is ubiquitous in soil (Barka et al. 2016). The next most common actinomycete genera are in descending order, *Micromonospora* (up to $\times 10^4$–$10^5$ colony forming units/g of dry soil), *Actinoplanes*, *Actinomadura* and *Nocardia* (Janssen 2006, Ranjani et al. 2016). Actinomycetes are prolific producers of a variety of bioactive secondary metabolites with diverse chemical structures and biological activities (Bérdy 2005, 2012, Barka et al. 2016). These small molecules are biosynthesised during the aerial hyphae formation from the substrate mycelium and often hold complex structures which result from long enzymatic pathways (Martin et al. 2005). Consequently, it seems that the genes involved in secondary metabolite production may be subjected to some of the regulatory mechanisms that control aerial mycelium formation (Ruiz et al. 2010). Under laboratory conditions, the biosynthesis of these metabolites is believed to be triggered by fermentation-dependent events such as the depletion of nutrients, the biosynthesis of an inducer or a decrease in growth rate. In response to these conditions, actinomycetes generate signals which trigger a cascade of regulatory events resulting in chemical differentiation (secondary metabolism) and morphological differentiation (morphogenesis) (Martin et al. 2005, Ruiz et al. 2010, Traxler and Kolter 2015).

The actinomycetes, particularly the ones from the genus *Streptomyces*, are the richest source of microbial small molecules (mainly antibiotics) and lead compounds as they have provided approximately 12,000 of all described bioactive metabolites (Berdy 2012, Ranjani et al. 2016, Takahashi and Nakashima, 2018). *Streptomyces* species are the largest producers, accounting for approximately 80% of the total amount (Bérdy 2012, Kurtböke 2012a). They have yielded many clinically essential antimicrobial compounds, including streptomycin (**9**) (sourced from *Streptomyces griseus*), actinomycin (**10**) (sourced from *Streptomyces costaricanus*), and streptothricin (**11**) (sourced from *Streptomyces griseus*), (Barka et al. 2016, Takahashi and Nakashima 2018). The remaining 20% have been isolated from rare actinomycete genera, including *Salinispora, Actinoplanes, Micromonospora, Actinomadura,* and *Streptoverticillium* (Kurtböke 2012b,c, Tiwari and Gupta 2012). Diverse and unique compounds exhibiting high biological activity and low toxicity have been identified from rare actinomycetes (Subramani and Aalbersberg 2013). The discovery of the aminoglycoside gentamicin (**12**) (sourced from the actinomycete *Micromonospora purpurea*) in 1963 (Weinstein et al. 1963), an antibiotic that inhibits bacterial protein synthesis, greatly increased the interest in rare actinomycetes (Lazzarini et al. 2000, Kurtböke 2012a, Tiwari and Gupta, 2012). *Micromonospora* is the second most important producer of bioactive compounds (Eccleston et al. 2008) (more than 740 bioactive secondary metabolites have been described) after *Streptomyces* (Wagman and Weinstein 1980). Further commercially relevant antibiotics from rare actinomycetes include rifamycin SV (**13**) (sourced from *Amycolatopsis rifamycinica*);

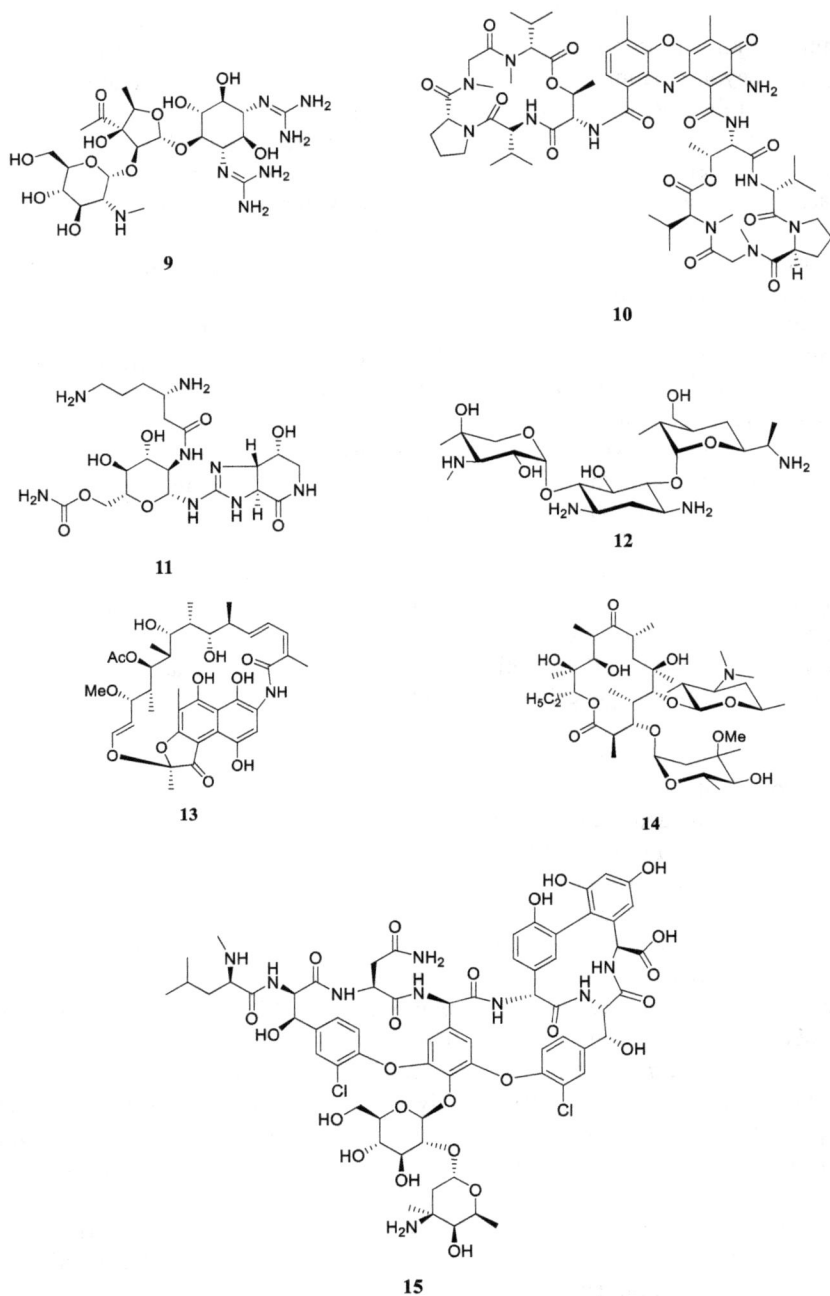

**Figure 2.** Examples of bioactive metabolites synthesised by actinomycetes.

erythromycin (**14**) (sourced from *Saccharopolyspora erythrea*) and vancomycin (**15**) (sourced from *Amycolatopsis orientalis*) (Genilloud 2017) (Fig. 2).

## 1.3 Mining for Novel Sources of Actinomycete Diversity

During the last decades intensive screening programs were carried out worldwide to access the actinomycete biodiversity (Genilloud et al. 2011, Genilloud 2017). Large numbers of samples from a wide range of geographical locations and habitats were processed and millions of strains were isolated and screened in industrial laboratories and research centres (Genilloud et al. 2011, Zahir et al. 2019). Consequently, the rate of discovering commercially relevant bioactive small molecules from common actinomycete sources has decreased as this practice frequently reverts to the re-isolation of known compounds (Dias et al. 2012, Katz and Baltz 2016) New approaches have been developed to address the problem of rediscovery of microbial compounds (Kurtböke 2012a, Tiwari and Gupta 2012). One of these strategies involves the isolation, characterisation, and screening of novel/rare actinomycete taxa sourced from unique and underexplored environments (Hayakawa 2008, Kurtböke 2012a,b, Brinkmann et al. 2017). Novel actinomycete strains producing new structurally diverse bioactive natural products have been discovered from desert biomes, marine ecosystems, deep-sea sediments, and insect associated actinomycetes (Poulsen et al. 2011, Carr et al. 2012, Subramani and Aalbersberg 2013 Kurtböke et al. 2015, Romero et al. 2015, Beemelmanns et al. 2016). Marine environments have been investigated for the isolation and characterisation of sediment actinomycetes collected from geographically diverse areas using taxon-specific isolation procedures. Sediment samples comprising novel actinomycete genera, including *Demequina*, *Iamia*, *Marinactinospora*, *Marisediminicola*, *Paraoerskovia*, *Verrucosispora* were collected from the Canary Basin, the Japan Trench, the Norwegian fjords and the Challenger Deep of the Mariana Trench in the western Pacific Ocean (Riedlinger et al. 2004, Fiedler et al. 2005, Hohmann et al. 2009, Abdel-Mageed et al. 2010, Goodfellow and Fiedler 2010). A series of unique polycyclic polyketide synthase type 1-antibiotics, namely, abyssomicins B (**16**) and C (**17**) and *atrop*-abyssomicins C (**18**), D (**19**), G (**20**) and H (**21**) (Fig. 3) were isolated from *Verrucosispora maris* using a combination of a targeted assay and HPLC-DAD monitoring. *Atrop*-abyssomicin C (**13**) exhibited antibiotic activity against two multi-drug resistance clinical isolates of *Staphylococcus aureus* (N315 and Mu50). The minimum inhibitory concentration (MIC) values of a*trop*-abyssomicin C against *S. aureus* N315 and *S. aureus* Mu50 were in the range of 4 µg/mL and 13 µg/mL, respectively (Riedlinger et al. 2004, Fiedler et al. 2005, Hohmann et al. 2009, Abdel-Mageed et al. 2010, Goodfellow and Fiedler 2010).

Lately, attention has been paid in studying actinomycetes associated to eusocial insects such as termites, beewolves and beetles, which have hardly been exploited (Beemelmanns et al. 2016, Kurtböke and French 2007, Kurtböke et al. 2015). Examples of novel small molecules discovered from these sources include, microtermolides A (**22**) and B (**23**) isolated from a *Streptomyces* sp. strain associated with fungus-growing termites (Carr et al. 2012); sceliphrolactam (**24**) (Dong-Chang et al. 2011) a previously unreported 26-membered polyene macrocyclic lactam displaying antifungal activity against amphotericin B-resistant *Candida albicans* (MIC = 4 µg/mL); and mycangimycin (**25**) (Oh et al. 2009), a polyene peroxide with pronounce antifungal activity against the antagonistic

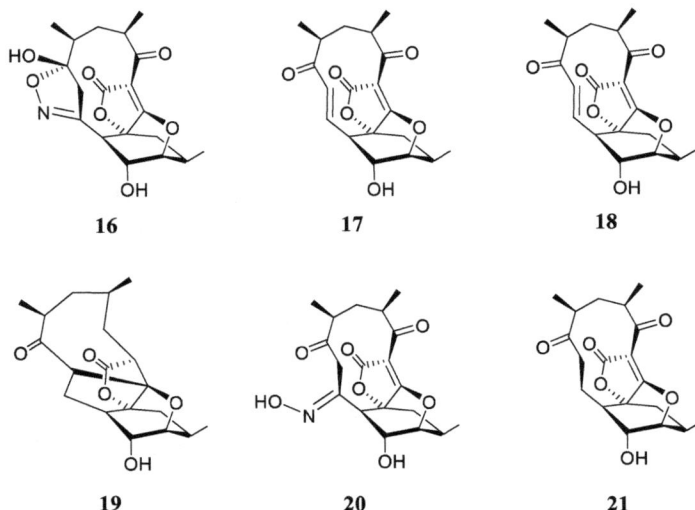

**Figure 3.** Unique polycyclic polyketide synthase type 1-antibiotics isolated from the marine actinomycete *Verrucosispora maris.*

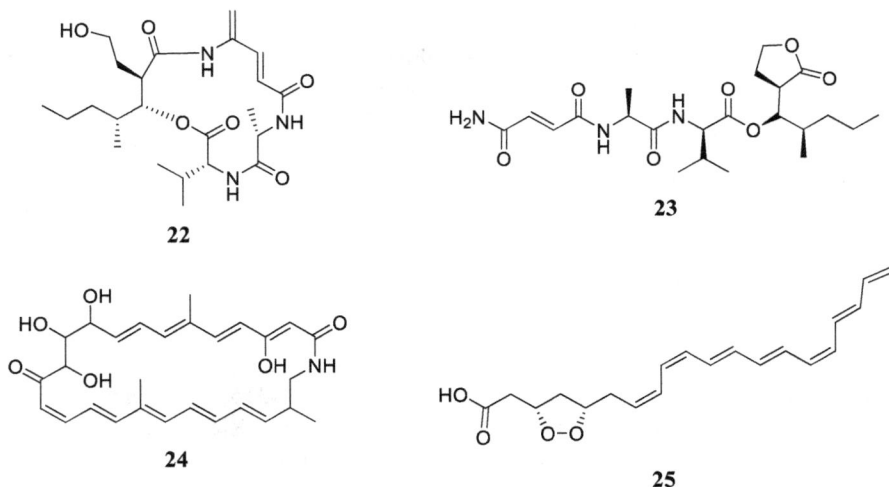

**Figure 4.** Examples of antibiotics isolated from insect-associated actinomycetes.

ascomycetes, *Ophiostoma minus* (MIC = 1.2 µg/mL), *Saccharomyces cerevisiae* (MIC = 0.4 µg/mL) and *Candida albicans* ATCC 10231 (MIC = 0.2 µg/mL), produced from a pine beetle-associated fungus *Dendroctonus frontalis* (Fig. 4).

## 1.4 Dereplication of Bioactive Natural Products

Dereplication plays a crucial role in natural products discovery, it allows to increase the number of new/novel natural products that can be identified (Hou et al. 2012). Furthermore, it provides rapid and competent identification and quantification of known secondary metabolites present in fractionated or unfractionated crude extracts, thus avoids time-consuming isolation procedures (Poulsen et al. 2011, Abdelmohsen et al. 2014, Carnevale-Neto et al. 2016). Dereplication approaches can vary, but typically combining chromatographic and spectroscopic methods (i.e., LC-UV, LC-MS, LC-MS/MS, and GC-MS) with database searching enables the comparison of known metabolites (Patel et al. 2010).

Some of the most comprehensive databases include:

> Chapman & Hall's Dictionary of Natural Products (DNP) containing over 313,000 natural products
  http://dnp.chemnetbase.com/faces/chemical/ChemicalSearch.xhtml

> AntiBase with more than 48,000 entries
  https://sciencesolutions.wiley.com/solutions/technique/lc-ms/antibase-the-natural-compound-identifier/

> MarinLit comprising more than 31,000 marine natural products
  http://pubs.rsc.org/marinlit?_ga=2.144261308.820213707.1607020116-750125451.1607020116

Several authors have proposed different early stage dereplication strategies to identify novel bioactive small molecules from a variety of microbial strains. Carnevale-Neto et al. (2016) developed a new GC-MS-based protocol for the rapid identification of plant metabolites using a combination of two different methods, the Ratio Analysis of Mass Spectrometry deconvolution tool (RAMSY) together with an Automated Mass Spectral Deconvolution and Identification System software (AMDIS) to provide an improved spectral identification workflow. The authors suggested that the incorporation and automation of RAMSY jointly with AMDIS may improve compound identification in mass spectra of complex biological mixtures, such as plants extracts. Bioactivity-directed fractionation methodologies have been largely used for the isolation of promising drug candidates such as platensimycin and platencin (Saleem et al. 2011, Allahverdiyev et al. 2013, Manallack et al. 2013). Hou et al. (2012) developed a method to support drug discovery efforts and evaluate rapidly and efficiently marine-derived bacterial natural products using a LC-MS-PCA (principal component analysis) based metabolomic approach. This method was effective at prioritising strains and significantly increased the efficiency to discover new natural products compared with traditional LC-MS trace analyses. Similarly, (Carr et al. 2012) have reported an early stage dereplication approach to rapidly identify novel compounds from eusocial insect-associated actinomycetes using HPLC-HRMS based metabolomics. This approach relied on careful processing of bacterial extracts employing PCA of pre-processed samples to promptly identify unique actinomycete producers from similar ecological niches. Using this strategy, two new compounds synthesised by hybrid nonribosomal-polyketide (NRPS-PKS) pathways, namely, microtermolides A (**22**) and B (**23**) were identified (Carr et al. 2012). Although LC-MS-based dereplication approaches offer the major advantage of detecting the accurate mass of the analytes (usually in the fentomolar-attomolar range) present in the extracts or fractions. It can also be problematic for some compound classes, especially if they have molecular masses lower than

300 Da, as they may generate a mix of fragment ion adducts and dimeric and double charged ions hence, complicating the task of identifying the elemental composition of the desired compounds (Nielsen and Larsen 2015). $^1$H NMR spectroscopy on the other hand is a quantitative, non-selective and non-destructive technique that allows the rapid, high-throughput and automated analysis of all molecules containing hydrogen nuclei including compounds that are less tractable to LC-MS analysis, such as sugars, amines, volatile ketones, and relatively nonreactive compounds (Mercier et al. 2011). Metabolomics-type NMR spectroscopy has increasingly been used to analyse the small molecule composition of tissue or biofluid samples to determine changes in the organism's metabolic status because of disease, genetic manipulation, or environmental stress (Griffin 2003). The incorporation of NMR-based metabolomics to explore the microbial drug-like metabolome in a broader sense than just dereplicating the known secondary metabolites of complex mixtures or fractions is not very common in natural products-based literature (Harvey et al. 2015). Recently, Lang et al. (2008) described a HPLC-NMR-ESMS/UV based dereplication methodology for the rapid identification of known compounds from fungal and bacterial extracts. Having access to $^1$H NMR data at the initial steps of the dereplication proved to be highly discriminating for the recognition of a wide range of known compounds, as the structural information of the small molecules comprising the extracts could be obtained and interpreted in a relatively short period of time (Lang et al. 2008). Moreover, to improve natural product research competitiveness, more innovative and productive strategies are needed to rapidly identify novel lead structures from natural sources (Shu 1998, Cragg and Newman 2013, Yang et al. 2013). A strategy using the partition coefficient (log P < 5) to front-load NP extracts with lead-and drug-like molecules, facilitated the drug discovery process of samples derived from Australian plants, fungi and marine invertebrates and resulted in the generation of 20,000 natural product extracts and 112,000 LLE fractions (Camp et al. 2012, Grkovic et al. 2014, Nature Bank 2020). This optimised method allowed the isolation of NP occupying mid-polarity physicochemical space, an essential property for oral bioavailability and cell permeability (Camp et al. 2012, Basu et al. 2013). Furthermore, an innovative $^1$H NMR metabolic fingerprinting approach to uncover and reveal the unique spectral patterns of the drug-like natural product metabolome of the Australian marine sponge from the Poecilosclerida family was developed. This methodology enabled the identification of four new natural products and one novel compound, named iotrochotazine A (Grkovic et al. 2014).

## 2. Discovery of New Compounds from Termite-gut Associated Actinomycetes: Australian Examples

New structurally diverse small-molecules with a variety of biological activities have been identified from novel *Streptomyces* species and other rare actinomycetes isolated from underexplored environments such as desert biomes, marine ecosystems, deep-sea sediments and insect-associated actinomycetes (Bérdy 2005, 2012, Oh et al. 2009, Dong-Chang et al. 2011, Carr et al. 2012, Kurtböke 2012a, Romero et al. 2015). In order to access to the unique components of the drug-like natural product metabolome of termite gut-associated actinomycetes, a unique $^1$H NMR metabolic fingerprinting approach was used (Grkovic et al. 2014, Romero et al. 2015). The methodology consisted of growing twenty-one actinomycete strains, previously isolated from the gut of the wood-feeding termite *Coptotermes lacteus* (Froggatt) (Kurtböke and French 2007, 2008), in four different solid culture conditions, oatmeal agar (OMA), lupin flour agar (LFA), rye flour agar (RFA) and glucose yeast extract agar (GYES) to determine how the variation of media components could induce the production of new microbial natural products (Fig. 5). The actinomycete cultures were incubated at 28°C for 15 days and then extracted overnight with ethyl acetate. The EtOAc extracts were dried down under reduced pressure to yield between 10 to 15 mg of each culture condition and were subsequently subjected to a metabolic fingerprinting approach to generate lead-like enhanced (LLE) fractions containing components with desirable physicochemical properties (Fig. 6). The filter used to maximize the recovery of the desired molecules was the partition coefficient (log P < 5) (Quinn et al. 2008, Camp et al. 2012, Grkovic et al. 2014, Romero et al. 2015).

An exhaustive examination of 420 LLE fractions for the occurrence of unique chemical profiles (i.e., non-repetitive or unique NMR resonances and distinctive ESIMS ion peaks) was performed manually. Based on this analysis, five strains, namely, *Streptomyces* spp. (USC-590, USC-592, USC-593, USC-597) and a *Microbispora* sp. (USC-6900) showing unique chemotypes were selected to be grown in 40–60 Petri dishes (100 × 15 mm) containing RFA, OMA, or GYES solid media (Table 1). NMR-guided isolation of the metabolites from the *Streptomyces* spp. (USC-590, USC-592) (Fig. 7), (USC-593, USC-597) and the *Microbispora* sp.

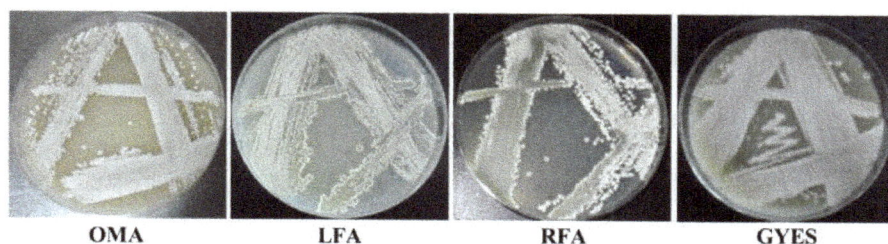

| OMA | LFA | RFA | GYES |

**Figure 5.** Colony morphologies of *Streptomyces* sp. USC 592 in four different solid culture conditions. (adapted from Romero et al. 2015).

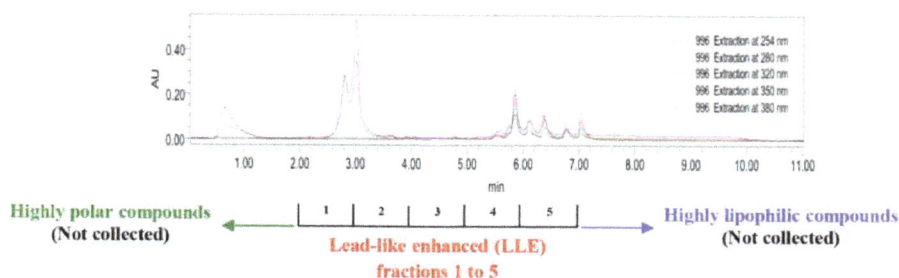

**Figure 6.** HPLC chromatogram depicting the drug-like/lead-like region containing the desired constituents of one of the selected crude extracts (adapted from Camp et al. 2012).

**Table 1.** Colony characteristics and chemical profiles of the selected actinomycete strains (adapted from Romero 2016).

| Actinomycete species | AM[a] | SM[b] | DP[c] | RT (micampn)[d] | ESIMS [e] | NMR Fingerprints[f] |
|---|---|---|---|---|---|---|
| *Streptomyces* sp. (USC 590) | White | Cherry | Cherry red | 7.0 | 529.12 | **LLE 5.** 13.38, 12.53, 7.84, 7.81, 7.76, 7.41, 6.87, 4.72, 4.56, 1.43. |
| *Streptomyces* sp. (USC 592) | Lime | Yellow | Light yellow | 3.0 | 393.14 | **LLE 1.** 12.25, 7.99, 7.89, 6.80, 5.01, 4.46, 3.98. **LLE 4.** 8.54, 4.26, 4.13, 2.62. |
| | | | | 5.8 | 401.61 | |
| *Streptomyces* sp. (USC 593) | Yellow | Orange | Orange | 5.2 5.6 | 408.15 369.14 | **LLE 4.** 7.66, 7.60, 7.48, 7.55, 7.42, 7.33, 7.25, 7.06, 7.02, 6.87, 6.12, 5.68, 3.73, 3.65, 3.63, 2.39, 2.28. |
| *Streptomyces* sp. (USC 597) | Beige light pink | Dark brown | Light brown | 2.9 | 266.18 | **LLE 1.** 8.26, 7.90, 7.16, 2.37, 2.11, 2.01, 1.46, 1.00, 0.96. **LLE 2.** 8.28, 8.21, 7.24, 5.91, 3.70, 3.58, 1.77, 0.96. |
| | | | | 3.1 | 252.17 | |
| *Microbispora* sp. (USC 6900) | White | Cherry | Light brown | 4.9 | 225.5 | **LLE 3.** 10.26, 8.93, 8.44, 8.31, 8.09, 8.05. |

[a] Aerial mycelium. [b] Substrate mycelium. [c] Diffusible pigment. [d] Retention times. [e] Positive ionization mode [M+H]⁺. [f] Unusual/interesting resonances in ppm. All samples were acquired in DMSO-$d_6$ at 600 MHz.

(USC-6900) led to the identification of six new drug-like natural products, namely, actinoglycosidines A (**26**) and B (**27**), actinopolymorphol D (**28**), and niveamycins A (**29**), B (**30**) and C (**31**); together with eight co-occurring known compounds, namely, BE-54017-derivative 4 (**32**), BE-54017 (**33**), 2-amino-6-methoxy-9H-pyrrolo[2,3-d] pyrimidine-7carbonitrile (**34**) and WS-5995 A (**35**) and B (**36**), 3H-Pyrrolo[2,3-d] pyrimidine-5-carboxylic acid, 2-amino-4,7-dihydro-4-oxo-, methyl ester (**37**) phenazine-1-carboxamide (**38**) and TMC-66 (**39**).

## 3. Determination of the Absolute Configuration of the New Natural Products

Determination of the absolute configuration (AC) of natural products often poses a challenging problem in structure elucidation (Nugroho and Morita 2014). To address this, several methods such as X-ray crystallography, chiroptical spectroscopy and NMR anisotropy methods, each having its own limitations, have been developed over the last years (Freedman et al. 2003, Nugroho and Morita 2014). The AC of chiral natural products comprising chromophores has been most commonly elucidated using chiroptical methods, including electronic circular dichroism (ECD), vibrational circular dichroism (VCD) and Raman optical activity (ROA) (Bruhn et al. 2013, Nugroho and Morita 2014).

**Indolotryptoline core**

**Indolocarbazole core**

Indole

Indole

Tryptoline

Carbazole

LLE fraction 4

Pure compound

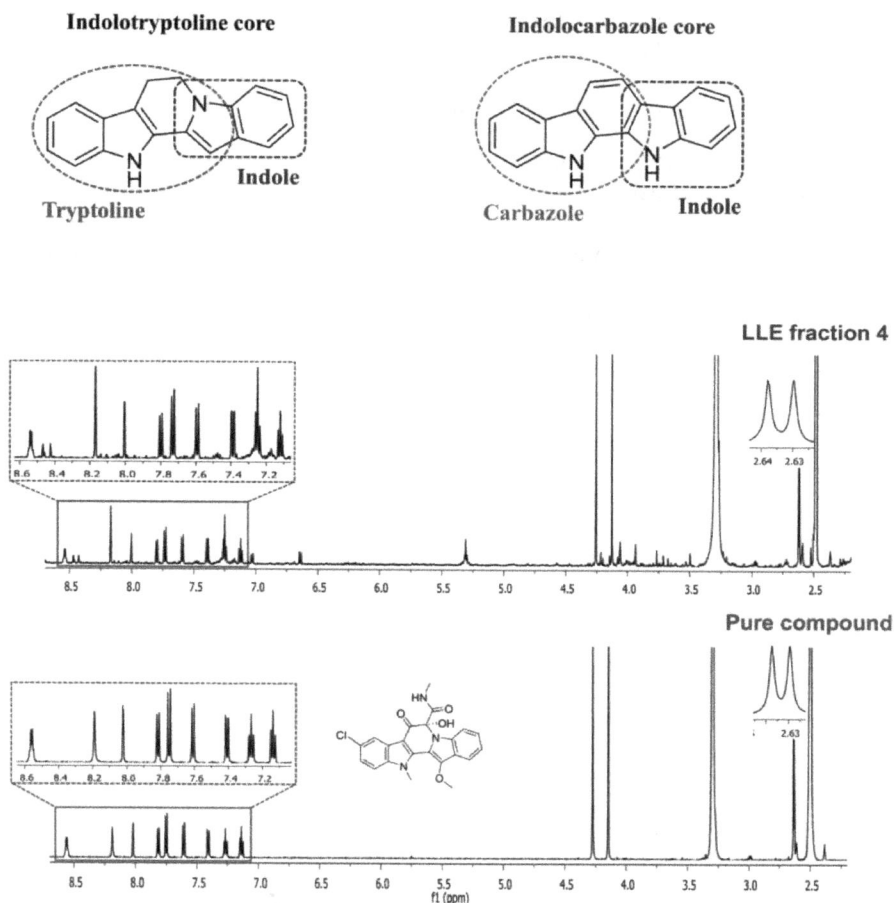

**Figure 7.** $^1$H NMR fingerprint spectra of *Streptomyces* sp. (USC-592).
**Footnote:** Top spectrum exhibits the distinctive proton resonances of LLE fraction 4. The bottom spectrum displays the proton NMR resonances of compound 29, after comparison of its spectral values with those of the literature led to the rapid identification of 29 as the known natural product BE-54017-derivative 4 (adapted from Romero 2016).

Among these approaches, ECD has been most widely used over the past decade (Kong and Wang 2013). ECD measures the differential response of a chiral molecule to the modulation of UV/Vis radiation between left- and right-circularly polarised states (Ling-Yi and Wang 2013). In general, AC determination using ECD compares the spectrum of new compounds against analogous molecules having a known AC. However, recently an alternative non-empirical method involving ECD calculations of time-dependent density functional theory (TDDFT) has become a rapid and reliable way to establish the AC of chiral compounds (Lin et al. 2014, Nugroho and Morita 2014). ECD calculations usually include two steps, a conformational search to obtain the candidate conformers and their subsequent optimisation using TDDFT (Nugroho and Morita 2014). The accuracy of TDDFT calculations depends mainly on the basis set and functional use for the calculations, thus, the larger the basis set is, the more accurate the results will be.

## 3.1 Absolute Configuration of Niveamycin B

The absolute configuration of the side chain chiral carbon at C-9 of niveamycin B (33) was determined using ECD calculations. The ECD spectra of the most stable conformers for 33 were calculated at the B3LYP/6-31+G(d,p)//CAM-B3LYP/SVP (Fig. 8) and B3LYP/6-311 + G(d,p)//B3LYP/6-311 + G(2d,p) (Fig. 8) level

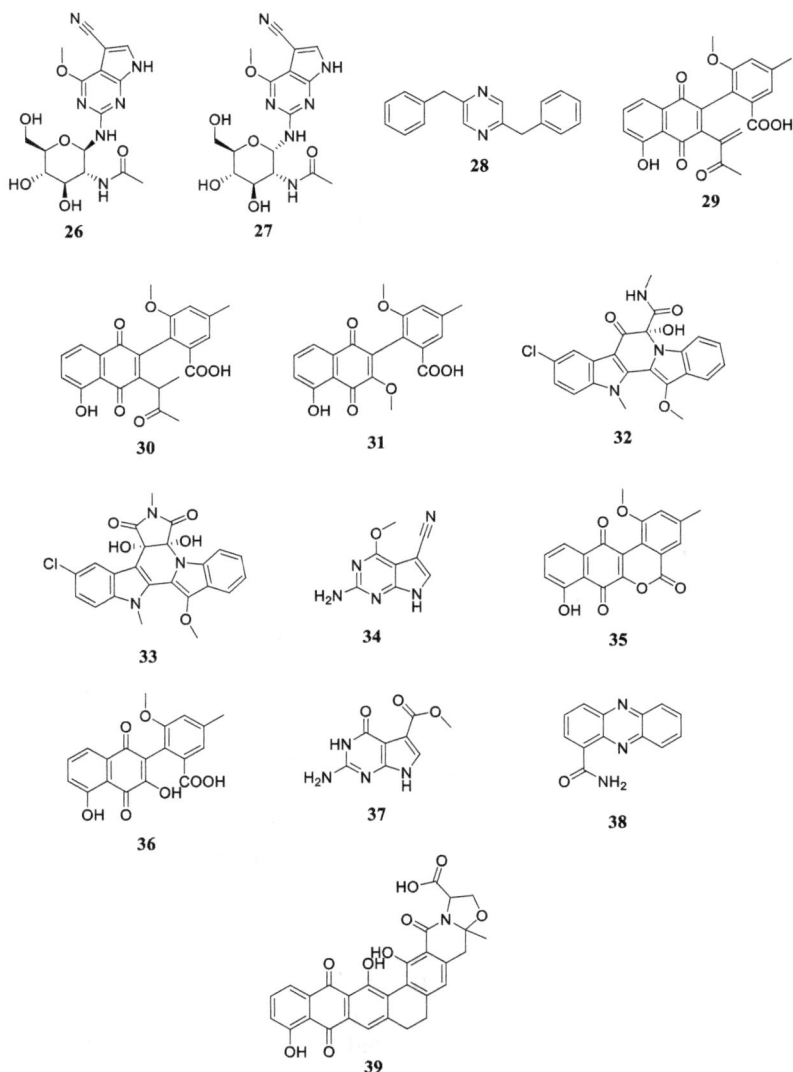

**Figure 8.** Natural products isolated from termite-gut associated actinomycete strains (adapted from Romero et al. 2015).

on six stable conformers. The ECD spectra for the six conformers were Boltzman-averaged to obtain the ECD spectrum of the isomers. Although the two calculation levels agreed well with that of the experimental and led to the conclusion that the absolute configuration at C-9 was *S*, the B3LYP/6-311 + G(d,p)// B3LYP/6-311 + G(2d,p) computed ECD spectrum provide a more accurate result as it matched the experimental ECD spectrum better (Romero 2016).

## 4. Conclusions

The continual evolution of microbial biosynthetic gene clusters has led to the synthesis of complex bioactive NPs with novel scaffold structures and large number of chiral centers which are likely to penetrate cell membranes and bind to multiple, unrelated classes of protein receptors as high affinity ligands. As only a small fraction of the world's biodiversity has been evaluated for biological activity, it can be assumed that NPs will continue to be a major source of therapeutic agents and potential drug leads. Although the rate of discovering commercially relevant novel bioactive small molecules from common actinomycete sources has decreased, innovative dereplication strategies involving the isolation, characterisation, and screening of *Streptomyces* and rare actinomycete species sourced from underexplored environments will improve the prospects. New structurally diverse small-molecules with a variety of biological activities have already been identified from novel *Streptomyces* species and other rare actinomycetes isolated from insect-associated actinomycetes using an early stage dereplication strategy. The approach relied on careful processing of bacterial extracts employing LC-MS-PCA to promptly identify actinomycetes producing new bioactive metabolites. Similarly, the developed in-house $^1$H NMR metabolic fingerprinting approach described in this chapter facilitated access to unique components of the drug-like natural product metabolome of termite gut-associated actinomycetes. It allowed the analysis by high resolution $^1$H NMR spectroscopy of HPLC fractions containing constituents with drug-like properties. The effectiveness of the technique was demonstrated by the isolation and elucidation of six new natural products, namely, actinoglycosidines A (**26**) and B (**27**), actinopolymorphol D (**28**), niveamycin A (**32**), B (**33**) and C (**34**). Advances in the field thus will again bring the focus onto most bioactive bacterial cluster, the actinomycetes for realization of many future discoveries

## Acknowledgements

C.A. Romero gratefully acknowledges the Escuela Superior Politécnica del Litoral (ESPOL), Centro de Investigaciones Biotecnológicas del Ecuador (CIBE) and Secretaría Nacional de Educación Superior Ciencia y Tecnología (SENESCYT) for the PhD scholarship provided.

# References

Abdel-Mageed, W.M., Milne, B.F., Wagner, M., Schumacher, M., Sandor, P., Pathom-aree, W., Goodfellow, M., Bull, A.T., Horikoshi, K., Ebel, R., Diederich, M., Fiedler, H.P. and Jaspars, M. (2010). Dermacozines, a new phenazine family from deep-sea dermacocci isolated from a Mariana Trench sediment. *Organic & Biomolecular Chemistry*, 8(10): 2352–2362. doi:10.1039/c001445a.

Abdelmohsen, U.R., Cheng, C., Viegelmann, C., Zhang, T., Grkovic, T., Ahmed, S., Quinn, R.J., Hentschel, U. and Edrada-Ebel, R. (2014). Dereplication strategies for targeted isolation of new antitrypanosomal actinosporins A and B from a marine sponge associated-*Actinokineospora* sp. EG49. *Marine Drugs*, 12(3): 1220–1244. doi:10.3390/md12031220.

Allahverdiyev, A.M., Bagirova, M., Abamor, E.S., Ates, S.C., Koc, R.C., Miraloglu, M., Elcicek, S., Yaman, S. and Unal, G. (2013). The use of platensimycin and platencin to fight antibiotic resistance. *Infection and Drug Resistance*, 6: 99–114. doi:10.2147/IDR.S25076.

Baker, D.D., Chu, M., Oza, U. and Rajgarhia, V. (2007). The value of natural products to future pharmaceutical discovery. *Natural Product Reports*, 24(6): 1225–1244. doi:10.1039/b602241n.

Barka, E.A., Vatsa, P., Sanchez, L., Gaveau-Vaillant, N., Jacquard, C., Meier-Kolthoff, J.P., Klenk, H.P., Clement, C., Ouhdouch, Y. and van Wezel, G.P. (2016). Taxonomy, physiology, and natural products of actinobacteria. *Microbiology and Molecular Biology Reviews*, 80(1): 1–43. doi:10.1128/MMBR.00019-15.

Basu, S., Hu, M., Gao, S. and Deb, A. (2013). Oral bioavailability challenges of natural products used in cancer chemoprevention. *Progress in Chemistry*, 25(9): 1553–1574.

Batista, A.N.L., dos Santos Jr, F.M., Batista Jr, J.M. and Cass, Q.B. (2018). Enantiomeric mixtures in natural product chemistry: Separation and absolute configuration assignment. *Molecules*, 23(2): 492. doi:10.3390/molecules23020492.

Beemelmanns, C., Guo, H., Rischer, M. and Poulsen, M. (2016). Natural products from microbes associated with insects. *Beilstein Journal of Organic Chemistry*, 12: 314–327. doi:10.3762/bjoc.12.34.

Berdy, J. (2012). Thoughts and facts about antibiotics: Where we are now and where we are heading. *Journal of Antibiotics*, 65(8): 385–395. doi:10.1038/ja.2012.27.

Bérdy, J. (2005). Bioactive microbial metabolites. *Journal of Antibiotics*, 58: 1–26.

Brinkmann, C.M., Marker, A. and Kurtböke, D.I. (2017). An overview on marine sponge-symbiotic bacteria as unexhausted sources for natural product discovery. *Diversity*, 9(4): 40. doi:10.3390/d9040040.

Bruhn, T., Schaumloffel, A., Hemberger, Y. and Bringmann, G. (2013). SpecDis: Quantifying the comparison of calculated and experimental electronic circular dichroism spectra. *Chirality*, 25(4): 243–249. doi:10.1002/chir.22138.

Burke, M.D. and Lalic, G. (2002). Teaching target-oriented and diversity-oriented organic synthesis at Harvard University. *Chemistry & Biology*, 9: 535–541.

Buss, A.D., Cox, B. and Waigh, R.D. (2003). Natural products as lead for new pharmaceuticals *Burger's Medicinal Chemistry and Drug Discovery* (pp. 847–900): John Wiley & Sons. Inc.

Butler, M.S. (2004). The role of natural product chemistry in drug discovery. *Journal of Natural Products*, 67: 2141–2153.

Butler, M.S., Robertson, A.A. and Cooper, M.A. (2014). Natural product and natural product derived drugs in clinical trials. *Natural Product Reports*, 31(11): 1612–1661. doi:10.1039/c4np00064a.

Camp, D., Davis, R.A., Campitelli, M., Ebdon, J. and Quinn, R.J. (2012). Drug-like properties: Guiding principles for the design of natural product libraries. *Journal of Natural Products*, 75(1): 72–81. doi:10.1021/np200687v.

Carnevale-Neto, F., Pilon, A.C., Selegato, D.M., Freire, R.T., Gu, H., Raftery, D., Lopes, N.P. and Castro-Gamboa, I. (2016). Dereplication of natural products using GC-TOF mass spectrometry: Improved metabolite identification by spectral deconvolution ratio analysis. *Frontiers in Molecular Biosciences*, 3: 59. doi:10.3389/fmolb.2016.00059.

Carr, G., Poulsen, M., Klassen, J.L., Hou, Y., Wyche, T.P., Bugni, T.S., Currie, C.R. and Clardy, J. (2012). Microtermolides A and B from termite-associated *Streptomyces* sp. and structural revision of vinylamycin. *Organic Letters*, 14: 2822–2825.

Cragg, G.M. and Newman, D.J. (2013). Natural products: A continuing source of novel drug leads. *Biochimica et Biophysica Acta*, 1830(6): 3670–3695. doi:10.1016/j.bbagen.2013.02.008.

Dias, D.A., Urban, S. and Roessner, U. (2012). A historical overview of natural products in drug discovery. *Metabolites*, 2(2): 303–336. doi:10.3390/metabo2020303.

Dong-Chang, O., Poulsen, M., Currie, C.R. and Clardy, J. (2011). Sceliphrolactam, a polyene macrocyclic lactam from a wasp-associated *Streptomyces* sp. *Organic Letters*, 13: 752–755.

Eccleston, G.P., Brooks, P.R. and Kurtboke, D.İ. (2008). The occurrence of bioactive micromonosporae in aquatic habitats of the Sunshine Coast in Australia. *Marine Drugs*, 6(2): 243–261. doi:10.3390/md20080012.

Erlanson, D.A., Fesik, S.W., Hubbard, R.E., Jahnke, W. and Jhoti, H. (2016). Twenty years on: The impact of fragments on drug discovery. *Nature Reviews Drug Discovery*, 15(9): 605–619. doi:10.1038/nrd.2016.109.

Fiedler, H.P., Bruntner, C., Bull, A.T., Ward, A.C., Goodfellow, M., Potterat, O., Puder, C. and Mihm, G. (2005). Marine actinomycetes as a source of novel secondary metabolites. *Antonie van Leeuwenhoek*, 87(1): 37–42. doi:10.1007/s10482-004-6538-8.

Freedman, T.B., Cao, X., Dukor, R.K. and Nafie, L.A. (2003). Absolute configuration determination of chiral molecules in the solution state using vibrational circular dichroism. *Chirality*, 15: 743–758.

Genilloud, O. (2017). Actinomycetes: Still a source of novel antibiotics. *Natural Product Reports*, 34(10): 1203–1232. doi:10.1039/c7np00026j.

Genilloud, O., Gonzalez, I., Salazar, O., Martin, J., Tormo, J.R. and Vicente, F. (2011). Current approaches to exploit actinomycetes as a source of novel natural products. *Journal of Industrial Microbiology and Biotechnology*, 38(3): 375–389. doi:10.1007/s10295-010-0882-7.

Goodfellow, M. and Fiedler, H.P. (2010). A guide to successful bioprospecting: Informed by actinobacterial systematics. *Antonie van Leeuwenhoek*, 98(2): 119–142. doi:10.1007/s10482-010-9460-2.

Goodfellow, M. and Williams, S.T. (1983). Ecology of Actinomycetes. *Annual Review of Microbiology*, 37: 189–216.

Goodnow Jr, R.A., Dumelin, C.E. and Keefe, A.D. (2017). DNA-encoded chemistry: Enabling the deeper sampling of chemical space. *Nature Reviews Drug Discovery*, 16(2): 131–147. doi:10.1038/nrd.2016.213.

Graul, A.I., Pina, P., Cruces, E. and Stringer, M. (2019). The year's new drugs and biologics 2018: Part I. *Drugs Today (Barc)*, 55(1): 35–87. doi:10.1358/dot.2019.55.1.2959663.

Graul, A.I., Pina, P. and Stringer, M. (2018). The year's new drugs and biologics 2017: Part I. *Drugs Today (Barc)*, 54(1): 35–84. doi:10.1358/dot.2018.54.1.2766396.

Griffin, J. (2003). Metabonomics: NMR spectroscopy and pattern recognition analysis of body fluids and tissues for characterisation of xenobiotic toxicity and disease diagnosis. *Current Opinion in Chemical Biology*, 7(5): 648–654. doi:10.1016/j.cbpa.2003.08.008.

Grkovic, T., Pouwer, R.H., Vial, M.L., Gambini, L., Noel, A., Hooper, J.N., Wood, S.A., Mellick, G.D. and Quinn, R.J. (2014). NMR fingerprints of the drug-like natural-product space identify iotrochotazine A: A chemical probe to study Parkinson's disease. *Angewandte Chemie-International Edition*, 53(24): 6070–6074. doi:10.1002/anie.201402239.

Guo, Z. (2017). The modification of natural products for medical use. *Acta Pharmaceutica Sinica B*, 7(2): 119–136.

Harvey, A.L., Edrada-Ebel, R. and Quinn, R.J. (2015). The re-emergence of natural products for drug discovery in the genomics era. *Nature Reviews Drug Discovery*, 14(2): 111–129. doi:10.1038/nrd4510.

Hayakawa, M. (2008). Studies on the isolation and distribution of rare actinomycetes in soil. *Actinomycetologica*, 22: 12–19.

Hert, J., Irwin, J.J., Laggner, C., Keiser, M.J. and Shoichet, B.K. (2009). Quantifying biogenic bias in screening libraries. *Nature Chemical Biology*, 5(7): 479–483. doi:10.1038/nchembio.180.

Hohmann, C., Schneider, K., Bruntner, C., Irran, E., Nicholson, G., Bull, A.T., Jones, A.L., Brown, R., Stach, J.E., Goodfellow, M., Beil, W., Kramer, M., Imhoff, J.F., Sussmuth, R.D. and Fiedler, H.P. (2009). Caboxamycin, a new antibiotic of the benzoxazole family produced by the deep-sea strain *Streptomyces* sp. NTK 937. *Journal of Antibiotics*, 62(2): 99–104. doi:10.1038/ja.2008.24.

Hou, Y., Braun, D.R., Michel, C.R., Klassen, J.L., Adnani, N., Wyche, T.P. and Bugni, T.S. (2012). Microbial strain prioritization using metabolomics tools for the discovery of natural products. *Analytical Chemistry*, 84(10): 4277–4283. doi:10.1021/ac202623g.

Hug, J.J., Krug, D. and Müller, R. (2020). Bacteria as genetically programmable producers of bioactive natural products. *Nature Reviews Chemistry*, 4(4): 172–193. doi:10.1038/s41570-020-0176-1.

Janssen, P.H. (2006). Identifying the dominant soil bacterial taxa in libraries of 16S rRNA and 16S rRNA genes. *Applied and Environmental Microbiology*, 72(3): 1719–1728. doi:10.1128/AEM.72.3.1719-1728.2006.

Kato, N., Takahashi, S., Nogawa, T., Saito, T. and Osada, H. (2012). Construction of a microbial natural product library for chemical biology studies. *Current Opinion in Chemical Biology*, 16: 101–108. doi:10.1016/j.cbpa.2012.02.016.

Katz, L. and Baltz, R.H. (2016). Natural product discovery: Past, present, and future. *Journal of Industrial Microbiology and Biotechnology*, 43(2-3): 155–176. doi:10.1007/s10295-015-1723-5.

Khanna, V. and Ranganathan, S. (2009). Physiochemical property space distribution among human metabolites, drugs and toxins. *BMC Bioinformatics*, 10 Suppl 15: S10. doi:10.1186/1471-2105-10-S15-S10.

Kong, L.Y. and Wang, P. (2013). Determination of the absolute configuration of natural products. *Chinese Journal of Natural Medicines*, 11(3): 193–198.

Kurtböke, D.İ. (2012a). Biodiscovery from rare actinomycetes: An eco-taxonomical perspective. *Applied Microbiology and Biotechnology*, 93(5): 1843–1852. doi:10.1007/s00253-012-3898-2.

Kurtböke, D.İ. (2012b). A phage-guided route to discovery of bioactive rare actinomycetes. *In*: Kurtböke, D.I. (ed.). *Bacteriophages*: InTech.

Kurtböke, D.İ. and French, J.R. (2007). Use of phage battery to investigate the actinofloral layers of termite gut microflora. *Journal of Applied Microbiology*, 103(3): 722–734. doi:10.1111/j.1365-2672.2007.03308.x.

Kurtböke, D.İ., French, J.R., Hayes, R.A. and Quinn, R.J. (2015). Eco-taxonomic insights into actinomycete symbionts of termites for discovery of novel bioactive compounds. pp. 111–135. *In*: Mukherjee, J. (ed.). *Biotechnological Applications of Biodiversity* (Vol. 147). New York: Springer Heidelberg.

Kurtböke, D.İ. and French, J.R.J. (2008). Actinobacterial resources from termite guts for regional bioindustries. *Microbiology Australia*, 29: 42–44.

Lahlou, M. (2013). The success of natural products in drug discovery. *Pharmacy and Pharmacology*, 04(03): 17–31. doi:10.4236/pp.2013.43A003.

Lang, G., Mayhudin, N.A., Mitova, M.I., Sun, L., van der Sar, S., Blunt, J.W., Cole, A.L.J., Ellis, G., Laatsch, H. and Munro, M.H. (2008). Evolving trends in the dereplication of natural product extracts: new methodology for rapid, small-scale investigation of natural product extracts. *Journal of Natural Products*, 71: 1595–1599.

Lazzarini, A., Cavaletti, L., Toppo, G. and Marinelli, F. (2000). Rare genera of actinomycetes as potential producers of new antibiotics. *Antonie van Leeuwenhoek*, 78: 399–405.

Lin, L.L., Wang, L. and Si, Y. (2014). Electronic circular dichroism behavior of chiral Pht. *Acta Pharmaceutica Sinica B*, 2014(4): 167–171.

Ling-Yi, K. and Wang, P. (2013). Determination of the absolute configuration of natural products. *Chin. Journal of Natural Medicines*, 11: 193–198.

Manallack, D.T., Crosby, I.T., Khakham, Y. and Capuano, B. (2013). Platensimycin: A promising antimicrobial targeting fatty acid synthesis. *Current Medicinal Chemistry*, 15: 705–710.

Martin, J.F., Casqueiro, J. and Liras, P. (2005). Secretion systems for secondary metabolites: How producer cells send out messages of intercellular communication. *Current Opinion in Microbiology*, 8(3): 282–293. doi:10.1016/j.mib.2005.04.009.

Mercier, P., Lewis, M.J., Chang, D., Baker, D. and Wishart, D.S. (2011). Towards automatic metabolomic profiling of high-resolution one-dimensional proton NMR spectra. *Journal of Biomolecular NMR*, 49(3-4): 307–323. doi:10.1007/s10858-011-9480-x.

Mevers, E., Chouvenc, T., Su, N.Y. and Clardy, J. (2017). Chemical interaction among termite-associated microbes. *Journal of Chemical Ecology*, 43(11-12): 1078–1085. doi:10.1007/s10886-017-0900-6.

Miller, J.R., Dunham, S., Mochalkin, I., Banotai, C., Bowman, M., Buist, S., Dunkle, B., Hanna, D., Harwood, H.J., Huband, M.D., Karnovsky, A., Kuhn, M., Limberakis, C., Liu, J.Y., Mehrens, S., Mueller, W.T., Narasimhan, L., Ogden, A., Ohren, J., Prasad, J.V., Shelly, J.A., Skerlos, L., Sulavik, M., Thomas, V.H., VanderRoest, S., Wang, L., Wang, Z., Whitton, A., Zhu, T. and Stover, C.K. (2009). A class of selective antibacterials derived from a protein kinase inhibitor pharmacophore. *Proceedings of the National Academy of Sciences of the United States of America*, 106(6): 1737–1742. doi:10.1073/pnas.0811275106.

Moffat, J.G., Vincent, F., Lee, J.A., Eder, J. and Prunotto, M. (2017). Opportunities and challenges in phenotypic drug discovery: an industry perspective. *Nature Reviews Drug Discovery*, 16: 531–543.

Nature Bank. Griffith Institute for Drug Discovery. (2020). Retrieved from https://www.griffith.edu.au/institute-drug-discovery/unique-resources/naturebank.

Newman, D.J. and Cragg, G.M. (2016). Natural products as sources of new drugs from 1981 to 2014. *Journal of Natural Products*, 79(3): 629–661. doi:10.1021/acs.jnatprod.5b01055.

Newman, D.J. and Cragg, G.M. (2020). Natural products as sources of new drugs over the nearly four decades from 01/1981 to 09/2019. *Journal of Natural Products*, 83(3): 770–803. doi:10.1021/acs.jnatprod.9b01285.

Nicolaou, K.C., Pfefferkorn, J.A., Roecker, A.J., Cao, G.-Q., Barluenga, S. and Mitchell, H.J. ( 2000). Natural product-like combinatorial libraries based on privileged structures. 1. General principles and solid-phase synthesis of Benzopyrans. *Journal of the American Chemical Society*, 122: 9939–9953.

Nielsen, K.F. and Larsen, T.O. (2015). The importance of mass spectrometric dereplication in fungal secondary metabolite analysis. *Frontiers in Microbiology*, 6(71): 1–15. doi:10.3389/fmicb.2015.00071.

Noren-Muller, A., Reis-Correa Jr, I., Prinz, H., Rosenbaum, C., Saxena, K., Schwalbe, H.J., Vestweber, D., Cagna, G., Schunk, S., Schwarz, O., Schiewe, H. and Waldmann, H. (2006). Discovery of protein phosphatase inhibitor classes by biology-oriented synthesis. *Proceedings of the National Academy of Sciences of the United States of America*, 103(28): 10606–10611. doi:10.1073/pnas.0601490103.

Nouioui, I., Carro, L., Garcia-Lopez, M., Meier-Kolthoff, J.P., Woyke, T., Kyrpides, N.C., Pukall, R., Klenk, H.P., Goodfellow, M. and Goker, M. (2018). Genome-based taxonomic classification of the phylum actinobacteria. *Frontiers in Microbiology*, 9: 2007. doi:10.3389/fmicb.2018.02007.

Nugroho, A.E. and Morita, H. (2014). Circular dichroism calculation for natural products. *Journal of Natural Medicines*, 68(1): 1–10. doi:10.1007/s11418-013-0768-x.

Oh, D.-C., Scott, J.J., Currie, C.R. and Clardy, J. (2009). Mycangimycin, a polyene peroxide from a mutualist Streptomyces sp. *Organic Letters*, 11(3): 633–636.

Pascolutti, M. and Quinn, R.J. (2014). Natural products as lead structures: Chemical transformations to create lead-like libraries. *Drug Discovery Today*, 19(3): 215–221. doi:10.1016/j.drudis.2013.10.013.

Patel, K.N., Patel, J.K., Patel, M.P., Rajput, G.C. and Patel, H.A. (2010). Introduction to hyphenated techniques and their applications in pharmacy. *Pharmaceutical Methods*, 1(1): 2–13. doi:10.4103/2229-4708.72222.

Pauli, G.F., Pro, S.M. and Friesen, J.B. (2008). Countercurrent Separation of Natural Products. *Journal of Natural Products*, 71: 1489–1508.

Perez, J., Moraleda-Munoz, A., Marcos-Torres, F.J. and Munoz-Dorado, J. (2016). Bacterial predation: 75 years and counting! *Environmental Microbiology*, 18(3): 766–779. doi:10.1111/1462-2920.13171.

Pham, J.V., Yilma, M.A., Feliz, A., Majid, M.T., Maffetone, N., Walker, J.R., Kim, E., Cho, H.J., Reynolds, J.M., Song, M.C., Park, S.R. and Yoon, Y.J. (2019). A review of the microbial production of bioactive natural products and biologics. *Frontiers in Microbiology*, 10: 1404. doi:10.3389/fmicb.2019.01404.

Poulsen, M., Oh, D.C., Clardy, J. and Currie, C.R. (2011). Chemical analyses of wasp-associated Streptomyces bacteria reveal a prolific potential for natural products discovery. *PLoS One*, 6(2): e16763. doi:10.1371/journal.pone.0016763.

Quinn, R.J., Carroll, A.R., Pham, N.B., Baron, P., Palframan, M.E., Suraweera, L., Pierens, G.K. and Muresan, S. (2008). Developing a drug-like natural product library. *Journal of Natural Products*, 71: 464–468.

Ranjani, A., Dhanasekaran, D. and Gopinath, P.M. (2016). An introduction to Actinobacteria. pp. 1–37. *In*: Dhanasekaran, D. and Y. Jiang (eds.). *Actinobacteria-Basics and Biotechnological Applications*. InTech Publisher.

Riedlinger, J., Reicke, A., Zähner, H., Krismer, B., Bull, A.T., Maldonado, L.A., Ward, A.C., Goodfellow, M., Bister, B., Bischoff, D., Süssmuth, R.D. and Fiedler, H.-P. (2004). Abyssomicins, inhibitors of the para-aminobenzoic acid pathway produced by the marine *Verrucosispora* strain AB-18-032. *Journal of Antibiotics*, 57(4): 271–279.

Romero, C.A. (2016). *Dereplication of the Actinomycete Metabolome as a Source of Bioactive Secondary Metabolites.* (PhD Doctorate), Griffith University, Brisbane, Australia.

Romero, C.A., Grkovic, T., Han, J., Zhang, L., French, J.R.J., Kurtböke, D.İ. and Quinn, R.J. (2015). NMR fingerprints, an integrated approach to uncover the unique components of the drug-like natural product metabolome of termite gut-associated *Streptomyces* species. *RSC Advances*, 5(126): 104524–104534. doi:10.1039/c5ra17553d.

Rosén, J., Gottfries, J., Muresan, S., Backlund, A. and Oprea, T.I. (2009). Novel chemical space exploration via natural products. *Journal of Medicinal Chemistry*, 52: 1953–1962.

Ruiz, B., Chavez, A., Forero, A., Garcia-Huante, Y., Romero, A., Sanchez, M., Rocha, D., Sanchez, B., Rodriguez-Sanoja, R., Sanchez, S. and Langley, E. (2010). Production of microbial secondary metabolites: regulation by the carbon source. *Critical Reviews in Microbiology*, 36(2): 146–167. doi:10.3109/10408410903489576.

Salam, N., Jiao, J.Y., Zhang, X.T. and Li, W.J. (2020). Update on the classification of higher ranks in the phylum Actinobacteria. *International Journal of Systematic and Evolutionary Microbiology*, 70(2): 1331–1355. doi:https://doi.org/10.1099/ijsem.0.003920.

Saleem, M., Hussain, H., Ahmed, I., van Ree, T. and Krohn, K. (2011). Platensimycin and its relatives: A recent story in the struggle to develop new naturally derived antibiotics. *Natural Product Reports*, 28(9): 1534–1579. doi:10.1039/c1np00010a.

Sarker, S.D., Latif, Z. and Gray, A.I. (2006). Natural product isolation. An overview. pp. 1–26. *In*: Sarker, S.D., Latif, Z. and Gray, A.I. (eds.). *Natural Products Isolation* (Second edition). Totowa, New Jersey: Human Press.

Schmidt, E.W. (2008). Trading molecules and tracking targets in symbiotic interactions. *Nature Chemical Biology*, 4(8): 466–473. doi:10.1038/nchembio.101.

Schreiber, S.L. (2000). Target-oriented and diversity-oriented organic synthesis in drug discovery. *Science*, 287: 1964–1969.

Shu, Y.Z. (1998). Recent natural products based drug development: A pharmaceutical industry perspective. *Journal of Natural Products*, 61: 1053–1071.

Subramani, R. and Aalbersberg, W. (2013). Culturable rare Actinomycetes: Diversity, isolation and marine natural product discovery. *Applied Microbiology and Biotechnology*, 97(21): 9291–9321. doi:10.1007/s00253-013-5229-7.

Takahashi, Y. and Nakashima, T. (2018). Actinomycetes, an inexhaustible source of naturally occurring antibiotics. *Antibiotics*, 7(2): 45. doi:10.3390/antibiotics7020045.

Tan, D.S. (2005). Diversity-oriented synthesis: Exploring the intersections between chemistry and biology. *Nature Chemical Biology*, 1: 74–84.

Terkina, I.A., Parfenova, V.V. and Ahn, T.S. (2006). Antagonistic activity of actinomycetes of Lake Baikal. *Applied Biochemistry and Microbiology*, 42(2): 173–176. doi:10.1134/s0003683806020104.

Thornburg, C.C., Britt, J.R., Evans, J.R., Akee, R.K., Whitt, J.A., Trinh, S.K., Harris, M.J., Thompson, J.R., Ewing, T.L., Shipley, S.M., Grothaus, P.G., Newman, D.J., Schneider, J.P., Grkovic, T. and O'Keefe, B.R. (2018). NCI program for natural product discovery: A publicly-accessible library of natural product fractions for high-throughput screening. *ACS Chemical Biology*, 13(9): 2484–2497. doi:10.1021/acschembio.8b00389.

Tiwari, K. and Gupta, R.K. (2012). Rare actinomycetes: A potential storehouse for novel antibiotics. *Critical Reviews in Biotechnology*, 32(2): 108–132. doi:10.3109/07388551.2011.562482.

Traxler, M.F. and Kolter, R. (2015). Natural products in soil microbe interactions and evolution. *Natural Product Reports*, 32(7): 956–970. doi:10.1039/c5np00013k.

van Hattum, H. and Waldmann, H. (2014). Biology-oriented synthesis: Harnessing the power of evolution. *Journal of the American Chemical Society*, 136(34): 11853–11859. doi:10.1021/ja505861d.

Vining, L.C. (2007, September). Roles of secondary metabolites from microbes. pp. 184–198. *In*: Chadwick, D.J. and Whelan, J. (eds.). *Ciba Foundation Symposium* 171-Secondary Metabolites: their Function and Evolution. Chichester, UK: John Wiley & Sons, Ltd..

Volochnyuk, D.M., Ryabukhin, S.V., Moroz, Y.S., Savych, O., Chuprina, A., Horvath, D., Zabolotna, Y., Varnek, A. and Judd, D.B. (2019). Evolution of commercially available compounds for HTS. *Drug Discovery Today*, 24(2): 390–402. doi:10.1016/j.drudis.2018.10.016.

Wagman, G.H. and Weinstein, M.J. (1980). Antibiotics from *Micromonospora*. *Annual Review of Microbiology*, 34: 537–557.

Weinstein, M.J., Luedemann, G.M., Oden, E.M. and Wagman, G.H. (1963). Gentamicin, a new antibiotic complex fom *Micromonospora*. *Antimicrobial Agents and Chemotherapy*, 161: 1.

Wender, P.A., Verma, V.A., Paxton, T.J. and Pillow, T.H. (2007). Function-oriented synthesis, step economy, and drug design. *Accounts of Chemical Research*, 41(1): 40–49.

Wiese, J. and Imhoff, J.F. (2019). Marine bacteria and fungi as promising source for new antibiotics. *Drug Development Research*, 80(1): 24–27. doi:10.1002/ddr.21482.

Yang, J.Y., Sanchez, L.M., Rath, C.M., Liu, X., Boudreau, P.D., Bruns, N., Glukhov, E., Wodtke, A., de Felicio, R., Fenner, A., Wong, W.R., Linington, R.G., Zhang, L., Debonsi, H.M., Gerwick, W.H. and Dorrestein, P.C. (2013). Molecular networking as a dereplication strategy. *Journal of Natural Products*, 76(9): 1686–1699. doi:10.1021/np400413s.

Zahir, T., Camacho, R., Vitale, R., Ruckebusch, C., Hofkens, J., Fauvart, M. and Michiels, J. (2019). High-throughput time-resolved morphology screening in bacteria reveals phenotypic responses to antibiotics. *Communications Biology*, 2: 269. doi:10.1038/s42003-019-0480-9.

# Chapter 10

# Removal of Termite-associated Antifungal Streptomycete Defence Barrier using Streptophages for Successful Implementation of Biological Control Fungi

*Fenton V. Case, John R.J. French* and *D. İpek Kurtböke**

## 1. Introduction

Termites (Infra-order: Isoptera) are eusocial arthropods that have colonised all continents besides Antarctica and consist of more than 2,800 species (Brune 2014, Verma et al. 2009). The termite lineage is classified under the cockroach order *Blattodea* with the sister cockroach group *Cryptocercus* (Rosengaus et al. 2013, Kurtböke et al. 2014, Evans and Iqbal 2015). Approximately 280 genera of termites are recognised, 84% of which belong to the Termitidae family (Engel et al. 2009). Although notoriously known for the damage they cause to timber infrastructure due to their wood (lignocellulose) eating abilities, termites contribute significantly to most of the earth's ecosystems by recycling plant biomass and by aerating soil via tunnelling, which increases water absorbency and improves soil fertility (McGarry et al. 2000, Bignell 2019). The composition and distribution of termite pest species varies greatly by continents and the ecosystems and only about 80–120 species are considered of economic importance due to the damage they cause to timber structures, forestry, and the agricultural industry (Chouvenc et al. 2011). In Australia, the main termite species causing timber damage belongs to the genera *Coptotermes, Macrotermes, Nasutitermes, Porotermes* and *Heterotermes* (Lee et al. 2007). *Mastotermes darwiniensis* is the most destructive termite in Australia and is generally found north of the Tropic of Capricorn. *Coptotermes* species, however, are accountable for the greatest economic loss than all other Australian termite pest

School of Science, Technology and Engineering, University of the Sunshine Coast, Maroochydore DC, Queensland 4558, Australia.
* Corresponding author: ikurtbok@usc.edu.au

species due to their extensive coverage across the continent as well as in the other parts of the world (Ahmed and French 2008, Evans 2021).

## 1.1 Lignocellulose Digestion

All termites feed on lignocellulose in different stages of its decomposition. This includes lignocellulose in wood, dry grass, plant litter, herbivore dung or organic matter in the later stages of humification (Brune 2014). Lignocellulose is comprised of cellulose, hemicellulose and lignin, which occur in a polymer complex in the cell walls and middle lamelle of plants. This complex is covalently bound, forming a durable matrix which hinders enzymatic degradation (Jørgensen et al. 2007). Enzymatic hydrolysis of cellulose to glucose is caused by a mixture of enzymes that are collectively referred to as cellulases. Termites utilize three major classes of these enzymes; endoglucanase, β-glucosidase and exoglucanases (Ni and Tokuda 2013). Both the termite host and cellulolytic symbionts of the hindgut produce these enzymes with the exception of exoglucanases, which are exclusive to the symbionts (Tokuda and Watanabe 2007).

The hindgut contains $\times 10^6$–$10^8$ microorganisms comprising of over 300 species from all three domains of life; eukarya (fungi and protists), bacteria and archaea (Brune and Freiedich 2000, Hongoh et al. 2005, Hongoh 2010). The ability of the lower termites to utilize lignocellulose as a nutritional substrate is mediated by a tripartitite symbiosis of the termite host, the cellulolytic protist symbiont/s and the bacterial hindgut community (Noda et al. 2007). Cleveland (1924) was the first to prove that this intimate symbiosis was necessary for the insects' survival showing that after symbiont removal the host would continue to eat but would die from starvation. Higher termites whereas differ as they partake in a bipartite symbiosis, generally lacking protists and harbouring only bacteria and archaea, they are however, as equally dependent on their microbial symbionts for survival (Brune 2014).

Before the lignocellulose enters the hindgut, the termites' mandibles and muscular gizzard cause the mechanical breakdown or communition of the material into smaller particles which increases the surface area, generating a greater susceptibility to enzymatic hydrolysis (Watanabe and Tokuda 2010). As it passes through the foregut the host cellulases (endoglucanases and β-glucosidases) produced in salivary glands are mixed, releasing glucose which is resorbed via the epithelium. The remaining partially digested particles continue into the hindgut where it is immediately phagocytised by the larger protists, which hydrolyse the remaining cellulose/hemicellulose, with the combination of exoglucanases, endoglucanases and β-glucosidases from protist and bacterial symbiont origins. Smaller protist species lack the ability to phagocytise the wood particles and probably acquire soluble substrates from the hindgut fluid (Ohkuma 2008, Ohkuma and Brune 2011). The microbial fermentation products, generally short chain fatty acids, are resorbed by the host whilst the lignin, which was unable to be digested, is excreted in the faeces (Noda et al. 2007).

Most hindgut protists, if not all, harbour prokaryotes as either endosymbionts that exclusively live within the cytoplasm of the protist cell or ectosymbionts that adhere to the cell surface. Common protist-associated bacterial species include endosymbiotic methanogens (Inoue et al. 2008, Ohkuma 2008), ectosymbiotic spirochetes (Inoue et al. 2008), endo- and ectosymbiotic *Bacteroidales* (Noda et al. 2009), and endosymbiotic bacteria belonging to the candidate phylum Termite Group 1 (TG1) (Ohkuma et al. 2007). Some of the bacterial and archaea community of the hindgut also adhere to the hindgut wall, or if motile, can be free living within the gut. The composition of the bacterial community varies between the termite species as does the protozoa community (Bignell et al. 2010), which is suggested to be linked to the host diet and physiological gut conditions, i.e., pH and oxygen gradients (Brune et al. 1995). The composition of the microbial community is highly conserved within the species (Noda et al. 2009), which is due to proctodeal trophallaxis, the exchange of gut fluids from anus to mouth between nest mates, allowing transmission of not only food and water but also transmission and colonisation of the microbiota in the hindgut in each generation (Nalepa et al. 2001).

The foregut and midgut are small unlike the large hindgut paunch which harbours most microbial symbionts. Higher termites however, due to the loss of gut protozoa, have acquired other means to digest cellulose. Most notably, members of the sub family Macrotermitinae maintain external fungus gardens in the nest (*Termitomyces* sp.) to predigest plant litter before consumption (Bignell et al. 2010). This variation in the nature of their diet, their primary cellulolytic symbionts and the microbial community which colonize the compartments of the alimentary tracts in the different termite lineages, has driven anatomical and physiological adaptations of the gut to maximise the degradation of their nutritional source. These include elongation of sections to increase the time wood particles spend in the compartment, and an increased pH value which has been suggested to promote autoxidative processes which cleave lignin-cellulose complexes (Kappler and Brune 2002). The anterior of the hindgut in soil feeding termite species has the highest pH value that has been recorded in biological systems, with an abundance of alkaline tolerant Firmicutes colonising this niche (Brune and Kühl 1996).

## *1.2 Termite Control Methods*

Although termites are an important decomposer of woody and plant biomass and thus are an essential part of the natural ecosystem (Jouquet et al. 2014), they can also cause economic losses by damaging human made infrastructure such as buildings, bridges, dams and even roads; or by damaging crops, plantations and forestry (Khan and Ahmad 2018). Therefore, effective methods to prevent damage and/or control termite populations are required, and examples are listed below.

### *1.2.1 Chemical*

The most commonly used termite control method has been the use of chemical termiticides including the ones relied heavily upon the application of persistent organochlorines such as Chlordane, Aldrin and Heptachlor. Additionally, these chemicals were known to have lipophilic properties which bio-magnify to cause toxic effects to humans, aquatic animals, birds, mosquitoes, rodents, domestic household animals and beneficial insects such as pollinating bees (Chandler et al. 2004). The United States Environmental Protection Agency (EPA) banned the use of Chlorpyrifos since 2001 after studies linked its usage to lower neurodevelopment, hyperactivity disorders and increased risk of brain tumours in children (Pogoda and Preston-Martin 1997, Rauh et al. 2006).

Chemical termiticides are mixed into the soil surrounding the structure to be protected to create a soil barrier, however, detrimental environmental effects from their overuse led Australia also to ban organochlorine use in 1995 (Ahmed and French 2008). The termiticides like Chlorpyrifos, Permethrin, Bifenthrin, Deltamethrin, Imidacloprid, Fipronil and/or Triflumuron have a broad spectrum of insecticide activity, mainly disrupting the central nervous system by blocking chloride ion channels resulting in hyper-excitation and central nervous system toxicity (Tingle et al. 2003). Additionally, these chemicals have some lipophilic properties which biomagnify to cause toxic effects to humans, aquatic animals, birds, mosquitoes, rodents, domestic household animals (i.e., cats and dogs), and beneficial insects such as pollinating bees (Grace 1994, Chandler et al. 2004).

Chemical termite control methods also comprise trap-treat-release (TTR) techniques or dusting, fumigation, baits and wood preservatives. Dusting involves the application of a slow acting termiticide such as Fipronil to the termite cuticle of a captured member of a colony then releasing said member back into the colony allowing transmission of the chemical during allogrooming (social grooming) (Green III et al. 2013). Mortality has been shown to reach 1000 untreated individuals per dusted termite in laboratory trials, however, under field conditions mortality is reduced to 50–100 per termite.

Chemical fumigation treatment applies a toxic volatile within an enclosed space, killing termites within infested timber. Active ingredients in various fumigants include; carbon dioxide (asphyxiant), phosphine, ethanedinitrile, sulfuryl fluoride (Profume Gas Fumigant) and methyl bromide (Derrick et al. 1990). Chemical treatment of wood with a preservative to prevent termite and also fungal attack is a common and effective method. Wood treated with high concentrations of butylene oxide, boric acid, triethylamine, disodium octaborate tetrahydrate, chromated copper arsenate (CCA), or copper in compounds such as copper borate, copper boronaphthanate and N-naphthaloylhydroxylamine, have inhibited termite damage (Schultz and Nicholas 2002, Yamaguchi 2003, Kartal et al. 2004).

Chemical baiting uses the active ingredients; hexaflumuron, flufenoxuron, chlorfluazuron, diflubenzuron (benzoylurea compounds) or sulfluramid, which are added to a suitable cellulose substrate in a baiting container. Chemicals used in baiting act as metabolic inhibitors or insect growth regulators by inhibiting synthesis of chitin (Gautam and Henderson 2014), an essential component of the exoskeleton. Like dusting, these chemicals need to have a delayed effect so the foraging termite can take the chemical back to the nest allowing the spread and eventual collapse of the colony.

Additionally juvenile hormone analogues (JHAs or juvenoids) can be used with the baiting technique. JHAs induce the differentiation of workers and nymphs into pre-soldier/soldier caste. Therefore, without the abundance of workers to forage and feed the excessive soldier caste, the colony collapses after several days (Saiki et al. 2014). Hrdý et al. (2001) reported high mortality rates in *Reticulitermes flavipes* and *Coptotermes formosanus* when treated with the carbamate derivative juvenoid,

2-(4-hydroxybenzyl)-1-cyclohexane (W-328), which induced differentiation of soldiers in 41–82% of individuals (50 ppm). Mortality rates seem to be dependent on the termite species which is likely due to the natural proportion of the soldier caste maintained in individual species (Vargo and Husseneder 2009). For example, in *Reticulitermes santonensis,* 0.3–1% (Hrdý et al. 2001) of the colony population are soldiers, whilst a 5–10% soldier population is found in *Coptotermes* species (Haverty 1977). Therefore, higher natural soldier proportions leads to faster rates of colony mortality.

### 1.2.2 Physical

Physical barriers are a popular method of impeding subterranean termite attack. These barriers are materials that are positioned under and surround the structure to exclude termites because they are too small for the termite to pass between and yet too large to be moved, thus preventing access and subsequent damage to the site. Types of physical barriers include concrete slabs (Lenz et al. 1997), graded particles between 1.4–2.8 mm diameter (depending on termite species) such as; granite, sand, cinder (Keefer et al. 2013) and glass (Ahmed and French 2011), solid sheeting such as stainless steel, marine grade aluminium and plastic (Su et al. 2004) and woven high grade stainless steel meshing (Grace et al. 1996). Stainless steel meshing works on a similar principle as it is too small for the termite to pass through, yet too thick to bite through. The meshing must be laid underneath the foundation but also be exposed above the ground around the perimeter of the structure. Despite this, termites can bridge or breach barrier systems (Ahmed and French 2008).

Physical control methods also include heating, electrocution, freezing and microwaving. Heat treatment of buildings is performed by enclosing the structure and circulating propane gas at a temperature of 45°–50°C for 1 hour (Hansen et al. 2011). Similarly, freezing utilises an enclosed space into which liquid nitrogen is pumped to decrease the ambient temperature to –50°C, which is reported to be higher than the critical thermal minimum required to kill the western drywood termite *Incisitermes minor* and the Formosan subterranean termite *Coptotermes formosanus* Shiraki (–21.3°C and –13.9°C respectively) (Rust et al. 1997). Electrical treatment (Electrogun®) involves the application of an electric current to termite infested timber. The electric shock is of low current (0.5 amp), high voltage (90,000 V) and high frequency (60,000 cycles) and is passed through the wood and the termite galleries within, resulting in mortality rates of 81–99% four weeks post application (Lewis and Haverty 1996). Microwave treatment with ≥ 500 watt (W) for ≥ 20 seconds has shown mortality rates to exceed 84%, due to temperatures reaching above 100°C, however unequal heat distribution allows some termites to escape treatment (Lewis et al. 2000). Heating, freezing, electrocuting and microwaving are not suitable for the eradication of termite pests in most housing as these methods apply great stress to already weakened building structures and may cause fracturing in wood, plastics and glass (Lewis et al. 2000, Verma et al. 2009). Therefore, alternative control measures must be implemented to minimize the damage these chemical and physical control methods cause to buildings, environment as well as constituting public and animal health risk.

## 1.2.3 Biological

Biological control of termites using fungi, bacteria and nematodes has been the focus of much research in the last 50 years (Chouvenc et al. 2011) and although not considered as classical biological control, plant-derived products (botanicals) have also received a great deal of interest (Verma et al. 2009). Additionally, the use of viruses and predators (mainly ants) for termite control has been investigated, however, information and feasibility studies are still lacking (Culliney and Grace 2000) and to date no commercially-sustained product has been established (Chouvenc et al. 2011) in the past and the researchers have attributed these findings to the complex chemical, behavioural and immunological defence mechanisms termites employ to impede biological control agents from infecting the entire colony (Myles 2002, Chouvenc et al. 2009, 2010, 2013, Yanagawa et al. 2011, Rosengaus et al. 2013).

Chouvenc and Su (2012) noted that the "inability of the fungal pathogen *M. anisopliae* to complete its life cycle within a *Coptotermes formosanus* (Isoptera: Rhinotermitidae) group was mainly the result of cannibalism and the burial behavior of the nest mates, even when termite mortality reached up to 75%. Because a subterranean termite colony, as a superorganism, can prevent epizootics of *M. anisopliae*, the traditional concepts of epizootiology may not apply to this social insect when exposed to fungal pathogens, or other pathogen for which termites have evolved behavioral and physiological means of disrupting their life cycle".

### 1.2.3.1 Botanical

Different parts and extracts of a vast number of plants, such as essential oils, leafs, fruits, resins and roots, contain bioactive components that can be utilised for termite control. The bioactive components possessing insecticidal, repellency or feeding deterrent activity has been examined, however, a great deal is yet to be discovered (Regnault-Roger et al. 2012). In most cases these plant extracts act by significantly reducing the microbial population of the hindgut whilst also inhibiting some protist species such as *Pseudotrichonympha grassii* koidzumi, which is the primary cellulolytic symbiont of Formosan subterranean termites (Raina et al. 2012). Botanical control is a promising alternative to chemical termiticides as 100% termite mortality can be achieved with concentration as low as 1ppm (neem extract) (Doolittle et al. 2007) and Nix et al. (2006) has shown that incorporating the essential oils (vetiver oil) into a mulch treatment significantly decreases tunnelling activity and wood consumption of the *Coptotermes formosanus* (Shiraki).

### 1.2.3.2 Nematode

Four families of nematode, Mermithidae, Allantonematidae, Steinernematidae and Heterorhabditidae, are obligate insect parasites; however, most research has focused on species from the latter two families. These nematodes harbour bacterial symbionts, *Photorhabdus* sp. and *Xenorhabdus* sp., which are released into the insect hemocoel during the infective juvenile stage (free-living in soil). Upon entering the hemocoel, these bacteria rapidly multiply to cause septicaemia and subsequent death of the insect host (Kaya and Gaugler 1993). Experiments with nematode control of termites appear to be dependent on the species of nematode and termite used.

*H. indica, H. bacteriophora, S. carpocapsae* and *S. riobrave* were effective against *C. formosanus* and *R. flavipes* in laboratory assays, with the exception of *S. riobrave* which had no effect against *R. flavipes* (Wang et al. 2002). Higher rates of infection were also reported in *N. costalis* when exposed in laboratory assays, yet field studies have been unsuccessful due to reports of repellency (Epsky and Capinera 1988). Additionally, Mankowski et al. (2005) have shown that nematodes are removed during allogrooming. These results suggest nematodes have a limited use for termite control as termite species have developed the ability to detect and defend against nematode infection (Culliney and Grace 2000, Wang et al. 2002).

### 1.2.3.3 Bacterial

Studies report high mortality rates, 95–100%, achieved with the use of soluble endotoxins, inclusion bodies and/or spores from commercial and local isolates of *Bacillus thuringiensis*, which is a commonly used microbial control agent of pest moths, butterflies, flies and beetles (Lepidoptera, Diptera and Coleoptera), against *Reticulitermes flavipes, Reticulitermes hesperus, Microcerotermes championi, Heterotermes indicola* and *Bifiditermes beesoni* (Smythe and Coppel 1965, Khan et al. 1977a, 1985). *Serratia marcescens* has also been found to produce 100% mortality in laboratory colonies of *M. championi, H. indicola* and *Coptotermes heimi* 7–13 days following exposure (Khan et al. 1977b). Somewhat lower mortality from exposure to *Pseudomonas aeruginosa* ranging from 25–52% in seven days to 84–100% 25 days post exposure was attained in the same termite species (Khan et al. 1992). *B. thuringiensis, S. marcescens* and *P. aeruginosa* have been unsuccessful under field conditions, which is partly due to *B. thuringiensis'* poor survival in soil conditions (Culliney and Grace 2000). Three species of hydrogen cyanide producing rhizobacteria, *Aeromonsa caviae, Rhizobium radiobacter* and *Alcaligenes latus*, were reported to kill *Odontotermes obesus* under laboratory conditions (Devi et al. 2007), however, no repellency, transmission or field evaluations were discussed. More recent studies were summarized by Jabeen et al. (2018) and Ahmad et al. (2021).

### 1.2.3.4 Fungal

Approximately 20 fungal species have shown pathogenicity in laboratory testing, however, the majority of research on fungal control of termites has been focused on the use of two entomopathogenic fungi, *Metarhizium anisopliae* (52%) and *Beauvaria bassiana* (26%) (Hyphomycetes) (Chouvenc et al. 2011). These two species of fungi have a broad spectrum of insecticidal activity and are highly virulent, having adapted several mechanisms to overcome their insect hosts. Within the first 24 hours of the conidia (fungal spore) attaching to the insect host, the hyphae rapidly expands over the cuticle gathering nutrients, followed by the formation of germ tubes and appressorium, combined with the production of hydrolytic enzymes such as chitinases, proteases, lipases and esterases which degrade the host's cuticle allowing access to the hemocoel (Avulova and Rosengaus 2011). Once inside the insects' circulatory system the fungi begins the secretion of mycotoxins (destruxins) which act as immunosuppressants (Turnbull et al. 2004), reducing phagocytic and encapsulation activity, and also opening calcium channels that have cytotoxic effects in several immune cells and epithelial cells (Vey 2002). High concentration of intracellular calcium leads to lethargy and eventual paralysis (Vey 2002). Additionally, hyphal bodies, which are the main reproductive stage of fungus replication, produces a collagenous protective coat which masks the β-1,3 glucan antigen which primarily elicits cellular responses in the host (Wang and Leger 2006). Several studies have reported significant reduction of termite populations in laboratory and/or field experiments for both *M. anisopliae* and *B. bassiana* against *Nasutitermes exitiosus*, *C. formosanus*, *C. acinaciformis*, *R. flavipes*, *R. santonensis* and *M. obesi* (Hänel and Watson 1983, Rath and Tidbury 1996, Wright et al. 2005, Singha et al. 2011), although mortality seems to be dependent on the strain of fungus, termite species, concentration of conidia suspension and the mode of delivery used. Hänel and Watson (1983) found that *M. anisopliae* persisted in the mound of the *N. exitiosus* termite for up to 15 weeks, despite this termite mortality declined progressively. Additionally, Chouvenc et al. (2008) were unsuccessful in initiating an epizootic in colonies of *C. formosanus* and *R. flavipes* even though conidial concentrations were sufficient in causing high mortality within 3–7 days in laboratory experiments. These results were attributed to the removal of conidia from the treated mound via grooming, antifungal agents in the soil and also via the repellence effect induced by the conidia causing termites to avoid contaminated sites. TTR and baiting techniques incorporating the fungal conidia has been shown to diminish the repellency effect and establish lethal concentrations of conidia per individual, however, the lack of transmission of the pathogens between colony members is an important element inhibiting a maintained infection (Delate et al. 1995, Staples and Milner 2000, Sun et al. 2003, Wang and Powell 2004). Therefore, direct inoculation of large volumes of high concentration conidia into the mound has shown to provide the greatest success in field trials (Culliney and Grace 2000), however, this proves to be problematic for subterranean species.

## 2. Termite Defense Strategies

The use of fungal pathogens for biological control of termites is based upon the same principles as classical biological control, that is; the etiological agent is virulent, self-replicating, easily transmittable, has the capacity to survive in the hosts conditions, and the host is susceptible (Hajek and Delalibera Jr 2010). Termites however, have adapted several strategies to impede the dispersion of a fungal biological control agent from infecting the entire colony. Examples are listed below:

### 2.1 Antimicrobial Compounds in Soil

The conidia must be able to survive in the hosts' habitat long enough to come in contact with the host cuticle. However, antifungal metabolites produced by soil microorganisms that may be opportunistic, commensal or synergistic in the termite nest inhibit fungal growth (Chouvenc et al. 2013). Antifungal metabolites are also produced and excreted in faecal matter used for building nests (Rosengaus et al. 1998, 2013) or secreted by the frontal or sternal glands by soldiers of some species

(Fuller 2007, Rosengaus et al. 2000, 2004). Antifungal metabolites Termicin and GNBP-2 have been identified in these glands (Bulmer and Crozier 2004).

## 2.2 Antifungal Activity in the Gut

Conidia that are ingested are inhibited by exposure to antifungal metabolites in the alimentary tract and dispersed in the faecal material. A small fraction of the conidia processed this way may retain the potential to germinate, yet the overall reduction of the pathogen load helps to prevent an epizootic episode (Chouvenc et al. 2011).

## 2.3 Behavioural Avoidance

To cause infection the conidia must first come in contact with the cuticle of the termite. However, several studies show that the presence of conidia causes termites to avoid contaminated sites, with the effect amplified by high concentrations (Myles 2002, Rosengaus et al. 1999, Staples and Milner 2000). Chouvenc et al. (2008) showed that by collecting, inoculating and releasing termites (TTR technique) back into the colony, bypasses the avoidance strategy however, the success was hindered by the effects of other defence strategies. Additionally, cellulose baiting techniques have been successful in reducing the repellency effect caused by *M. anisopliae* conidia (Wang and Powell 2004).

## 2.4 Grooming Behaviour

Reserved to social insects, termites practice routine social grooming (allogrooming) to remove foreign matter from their cuticles (Bignell et al. 2010). Originally this social interaction was believed to benefit the transmission of a biological control agent throughout the population, however, termites are able to detect conidia using their antennae (Yanagawa et al. 2009) and ingest them, subsequently inhibiting germination as the conidia passes through the alimentary tract. Although transmission of the etiological agent occurs during allogrooming, the overall reduction of pathogenic load diminishes the risk of disease (Yanagawa and Shimizu 2006).

## 2.5 Necrophagy and Corpse Avoidance

If termites of the colony die from fungal infection the remaining nest mates may dispose of the corpse by cannibalising the deceased so that the spores are inhibited as they pass through the alimentary tract, or if there are a large proportion of deceased termites the healthy population will seal them off from the rest of the nest (Chouvenc et al. 2008). Additionally, should a member of the colony be infected the uninfected nest mates have been shown to attack, kill and bury the infected member (Yanagawa et al. 2011).

## 2.6 Cellular and Humoral Immunity

In case the conidia manage to reach the cuticle and begin germination, the termite elicits an immune response to a fungal antigen, i.e., β-1,3-Glucan (Avulova and Rosengaus 2011). Cellular encapsulation of the fungal hyphae has been shown to increase termite survival (Chouvenc et al. 2009). However, the authors discuss that encapsulation is a last line of defence in case grooming has not removed all conidia, reporting in another study that 96% of cases of encapsulation occurred in the invaginated regions of the cuticle which were inaccessible to groomers (Chouvenc and Su 2010).

Several researchers have suggested that as these defence mechanisms can prevent or reduce the risk of infection, then inhibiting these mechanisms could make the termite more susceptible (Bulmer et al. 2009, Chouvenc et al. 2011). Some studies have used imidacloprid to reduce the grooming activity to allow initiation of a fungal epizootic, however, this method is limited by the large volumes of imidacloprid that would be required which negates the concept of biological controls reducing chemical pesticide use. Additionally, behavioural avoidance has been countered by the use of cellulose baits, whilst technology to circumvent corpse avoidance and cannibalisation, and also cellular and humoral immunity remain undiscovered. The identification of the microbial flora in the termite nest which prevents fungal infection has had little attention, although symbionts in other insect species have been shown to prevent infections of other parasitic or entomopathogenic fungi.

## 3. Actinomycetes Providing Symbiont Mediated Protection in Insects

Microbial symbionts are known to function as a key source of nutrients for successful growth and reproduction of their arthropod host. In the case of the termite the microbiota are responsible for the fixing of nitrogen, the degradation of cellulose, and the production of methane and acetone, which is exchanged for a constant supply of metabolic substrates and a temperature stable niche in which to grow (Brownlie and Johnson 2009, Schwitalla et al. 2020). Recent studies have shown that microbial symbionts also provide protection for the host and their nutritional resource from pathogen infection. In particular, actinomycetes have been found to prevent the colonisation of entomopathogenic fungi in termite, ant, beetle and wasp species (Table 1) by producing potent antimicrobials. Kaltenpoth (2009) suggests the ability of actinomycetes to utilise a vast range of carbon and nitrogen sources and the diversity of diffusible and volatile secondary metabolites able to be produced would predispose this phylum to engage in defensive/protective symbioses (symbiont mediated protection), and also implies that these symbioses are a widespread theme in the insect community.

The fungus growing termites cultivate fungi (*Termitomyces* sp.) which they use as a nitrogen rich nutritional source (Makonde et al. 2013). Fungi of the sub-genus *Pseudoxylaria* threaten the colony health via substrate competition with *Termitomyces*, however, a variety of actinomycetes (*Streptomyces, Kitasatospora, Micromonospora* and *Actinomadura*) produce antibiotics which inhibit this growth. Visser et al. (2012) investigating actinoflora of *Macrotermes natalensis, Microtermes* sp. and *Odontotermes* sp. nests, observed a lack of selective antifungal activity, reporting that the termite cultivar was more susceptible than the antagonist. Therefore, the authors deduced that the theme of actinomycetes functioning as

**Table 1.** Example cases of actinomycete symbiont mediated protection (SMP) from fungal infections in insect species.

| Host Insect | Bacterial symbiont | Enemy | Mechanism |
|---|---|---|---|
| Attine ants | *Pseudonocardia* sp. | *Escovopsis* sp. | Antifungal compound (dentigerumycin, nystatin P1, candicidin, antimycins) |
| | *Streptomyces* sp. | | |
| Beewolves | *Candidatus Streptomyces philanthi* | Fungi and bacteria | Antimicrobial compounds (sceliphrolactam) |
| Southern Pine Beetles | *Streptomyces* sp. | Competing fungi | Antifungal compound (mycangimycin) |
| Termites (lower) | *Streptomyces* sp. | Fungi (*Metarhizium anisopliae*) | Antifungal compound (unknown) |

Adapted from (Kaltenpoth 2009, Seipke et al. 2012, Chouvenc et al. 2013).

defensive symbionts in insect communities is unclear in regard to termites. However, this conclusion was reached without acknowledging the potential of actinomycetes as defensive symbionts in wood and humus-feeding termites. *Streptomyces* species have also been isolated from the nest, gut and cuticle of various termite species (Hongoh et al. 2005, Kurtböke and French 2007, Makonde et al. 2013). Chouvenc et al. (2013) report a *Streptomyces* species isolated from the carton material of the *Coptotermes formosanus* mound with antimicrobial activity against *Metarhizium anisopliae* and *Beauvaria bassiana* able to inhibit growth by 69%. They again question whether these strains coevolved with the host or are recruited from free living strains that have been obtained as workers forage. The same 16S ribotype with similar antimicrobial activity was found in all five nests sampled, which was also identical to many isolates from soil samples, supporting that these symbionts originated from free living strains.

The majority of literature suggests that actinomycetes provide symbiont mediated protection in insects from a range of bacterial and fungal pathogens (Kaltenpoth 2009, Seipke et al. 2012). The origin for most of these symbioses is at this time unclear, however, the acquisition and turnover of free living strains in the nest from the insect interaction with the soil seems likely (Menegatti et al. 2021).

Only two antifungal metabolites from termite associated *Streptomyces* symbiont origins, resistomycin and tetracenomycin D have been identified (Zhang et al. 2013). However, detection of antifungal metabolites from *Streptomyces* symbionts in some wasp, beetle and ant species were reported (Kaltenpoth 2009, Kaltenpoth et al. 2010). Examples include, sceliphrolactam from a *Streptomyces* species associated with mud dauber wasps (Oh et al. 2011, Poulsen et al. 2011), four endophenazine compounds A–D from endosymbiont *Streptomyces* species in four different arthropods (Gebhardt et al. 2002) and mycangimycin from a *Streptomyces* species associated with the fungus growing pine beetle *Dendroctonus frontalis* (Oh et al. 2009) (for further examples please see Chapter 9). Continued profiling of the actinomycete community in the termite nest, gut and the surrounding soil and subsequent screening of their antimicrobial activity will help to elucidate the mode of acquisition by the host as well as establishing evidence for the presence or absence of similar types of symbiosis in other termite species.

### 3.1 Breaching Symbiont-mediated Protection Mechanisms

Termites employ several strategies to prevent the infection of the colony by *M. anisopliae* and *B. bassiana*, these include; antifungal activity in the gut, nest and soil, behavioural avoidance of the conidia, allogrooming to remove conidia from nestmates cuticles, necrophagy and corpse avoidance, and cellular and humoral immunity. However, they lack complex adaptive immune system of a vertebrate, instead relying upon physical barriers and innate defences to prevent infection (Ahmed and French 2008). Chouvenc et al. (2013) has shown that *Streptomyces* species maintained in the nest material of the *Coptotermes formosanus* termite, inhibits the growth of *M. anisopliae,* providing a significant survival benefit to the termite, and has been shown to be present in other insects such as some ant, beetle and wasp species.

The literature shows that defensive mechanisms prevent epizootics, therefore removing them could increase termite susceptibility to fungal infections (Bulmer et al. 2009, Chouvenc et al. 2011). *Streptomyces* spp. have been the producers of around 60–70% of all known antibiotics (Challis and Hopwood 2003) including antifungal ones. Accordingly, the removal of these species from the termite environment could increase its susceptibility to *M. anisopliae*. They can easily be removed from insect systems with antibiotics such as the gentamicin (Mattoso et al. 2012), however, the broad spectrum of activity of antibiotics would displace many other organisms as well as facilitate the development of antibiotic resistance among environmental species.

### 3.2 Use of Streptophages to Break Down Termite Antifungal Streptomycetes Barrier

Arguably, the most interesting defensive mechanism is the symbiosis with various antimicrobial compound producing microorganisms. *Streptomyces* species isolated from a carton material from *Coptotermes formosanus* mound displayed antagonistic activity against *Metarhizium anisopliae* and *Beauvaria bassiana* and were able to inhibit growth of these fungi by 69% (Chouvenc et al. 2013). Information related to streptomycetes engaging in protective symbiosis with termites is emerging (Chouvenc and Su 2010, Chouvenc et al. 2018, Zhou et al. 2021) however, it is not known if this defence mechanism is somehow damaged would it result in the weakening termite immunity. A novel alternative method might therefore be the use of bacteriophages, as natural antagonists, which are superior to antibiotics due to a more specific activity spectrum (Kurtböke 2017). In the light of the given information a target-directed approach was used for removal of streptomycetes from termite surfaces to increase their susceptibility to infection by entomopathogenic fungi and increasing the chances of effective biological control.

In a laboratory test firstly, streptomycete isolates from the hindgut of *Coptotermes lacteus* termites located at the Sunshine Coast region (Kurtböke and French 2007) (Fig. 1) were selected to test their antifungal activity and possible involvement in the prevention of the host termite species from fungal infestation. From these isolates the most antifungal activity possessing ones (USC-6102, 6118, 6119 and 6153) (Fig. 1) were selected to test their interaction with the biological

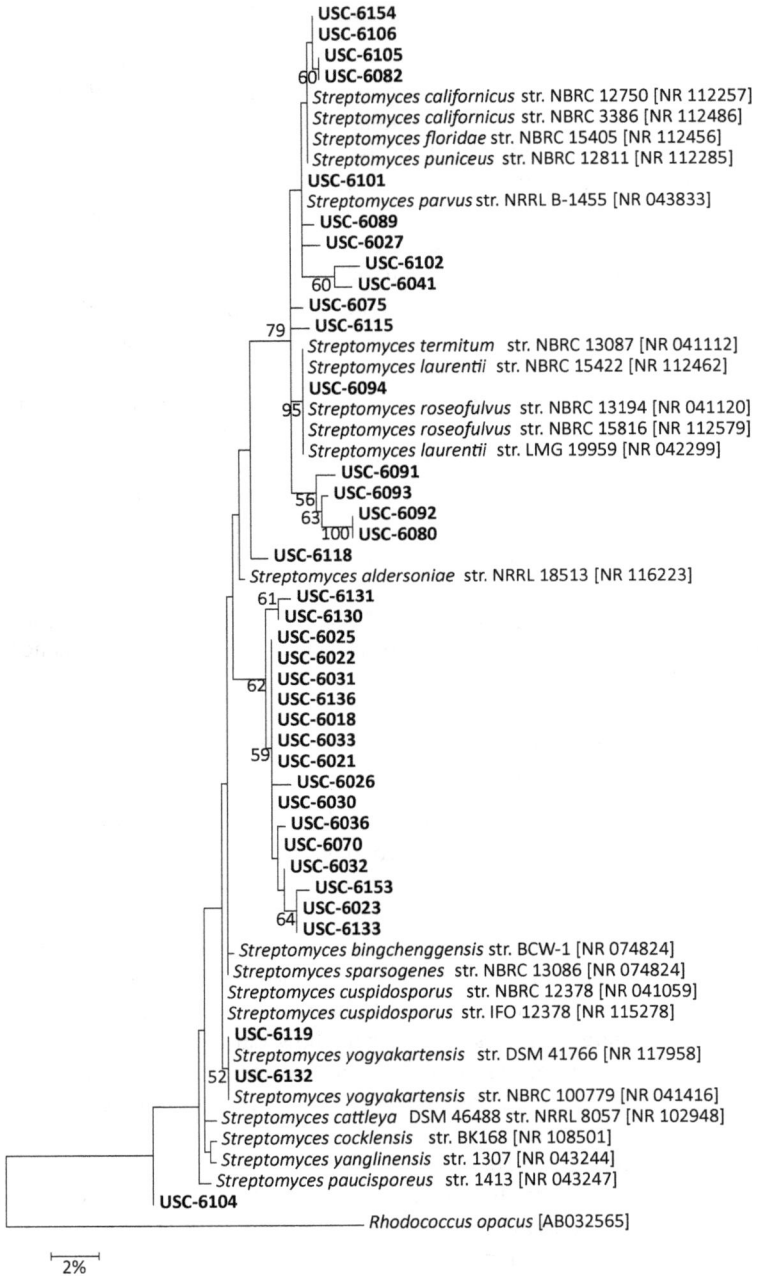

**Figure 1.** Phylogenetic tree of the 16S rRNA gene sequences of the isolates in relation to their closest relatives.

**Footnote:** Bootstrap values ≥ 50% are indicated at nodes. The scale bar represents 0.02% sequence divergence. GenBank accession numbers of reference sequences are presented. *Rhodococcus opacus* was used as an outgroup sequence to root the tree.

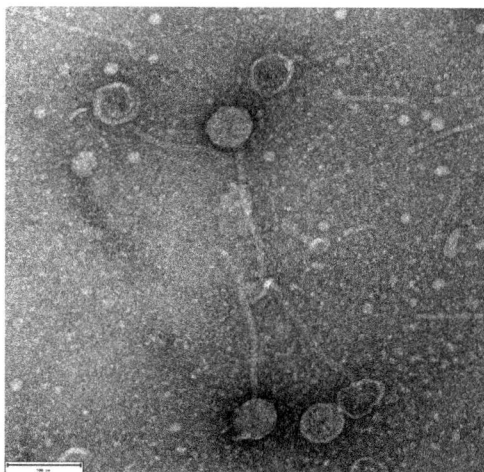

**Figure 2.** TEM micrograph of the streptophage cocktail displaying typical siphoviridae morphology.
**Footnote:** Bar indicate 100 um.

control agent in the presence and absence of the streptophages using a sterile termite model. Five polyvalent streptophages (Fig. 2) were selected from the University of the Sunshine Coast's Microbial Library (Kurtböke 2004a, b) for this purpose. They were previously isolated using host streptomycete type-strains; *Streptomyces albidoflavus* (ACM 4011), *Streptomyces hydroscopicus* (ACM 4209), *Streptomyces griseus* (ACM 474); *Streptomyces rimosus* (ACM 4341), *Streptomyces antibioticus* (ACM 4036) (Kurtböke, 2004, a, b, 2017).

Metarhizium anisopliae* (DAR 34191) and *Beauvaria bassiana* (DAR 34194) were used as biological control agents.

**Termite Treatment**

Surface sterilized (3% NaClO), antibiotic (cycloheximide (100 ppm), nystatin (100 ppm), pimaricin (100 ppm)) and microwave (30 seconds in a 1000 W) treated (Vela et al. 1976) non-live Sunshine Coast regional *Coptotermes lacteus* termites (obtained using a baiting technique described by French 1991) were used. Sterilized termites were divided into groups of 20 individuals containing 18 workers + 2 soldiers. Controls and the treatment groups used in this experiment are shown in Table 2. Each treatment group was performed in triplicates (n = 264; total termites = 5,280). Groups receiving streptomycetes were submerged in a suspension composed of a mixture of selected isolates ($\times 10^7$ cfu/ml) for 30 minutes before being placed into petri dishes and incubated at 28°C. Following the streptomycete treatment, groups that received streptophages were inoculated with 1ml of the streptophage suspension ($\times 10^7$ pfu/ml; host-bacteriophage ratio of 1:1,000) on a 48 hour schedule relevant to the time of bacterial wash starting at 0 hours to 14 days. Whilst groups receiving *M. anisopliae* were inoculated with 1ml of conidial suspension ($\times 10^7$ cfu/ml) on a 24 hour schedule relevant to the time of streptophage inoculation starting at 0 hours to 7 days. After *M. anisopliae* inoculation these groups were

**Table 2.** Treatment groups applied onto sterile termites used in the test.

| Treatment | (+) Streptomycete, (+) streptophage, (+) *M. anisopliae* |
|---|---|
| Streptomycete control | (+) Streptomycete, (–) streptophage, (–) *M. anisopliae* |
| Biological control agent control | (–) Streptomycete, (–) streptophage, (+) *M. anisopliae* |
| Streptophage control | (+) Streptomycete, (+) streptophage, (–) *M. anisopliae* |
| Control | (–) Streptomycete, (–) streptophage, (–) *M. anisopliae* |

+: presence of treatment –: absence of treatment.

transferred to a 26°C incubator for 14 days to aid fungal growth. Following incubation, the termites were rinsed (1 ml of sterile distilled water) and vortexed. Fungal concentration per termite in each treatment group was determined via colony count performed on Potato Dextrose Agar (OXOID, Australia), supplemented with 50ppm streptomycin sulphate to prevent bacterial growth. Streptomycete concentration was determined on Oatmeal Agar (OXOID, Australia), supplemented with 50 ppm cycloheximide to prevent fungal growth.

## 4. Conclusions

Following the streptophage applications the concentrations of the streptomycetes per termite were significantly reduced ($p \leq 0.01$). As a result within a week of the post fungal inoculations, *M. anisopliae* concentrations per termite were restored to positive control levels ($p \leq 0.05$). The results indicate that the application of streptophages can successfully remove antifungal streptomycetes from termite surfaces resulting in increased termite susceptibility to biocontrol fungal infection. These preliminary findings might be of importance to the pest control industry that is interested in the implementation of effective and safer control measures. However, large scale laboratory and field trials are required to confirm the effectivity and reproducibility of the technique under field conditions. Chouvenc et al. (2011) expressed a concern about the "unrealistic optimism" of biological controls in the termite industry based on data from "bioassays of poor biological relevancy". *In vitro* conditions used in experiments, including the optimal growth temperatures of the *M. anisopliae* and *Streptomyces* isolates, naturally do not reflect the conditions of the termite environment. Further studies using termite nest conditions with the addition of a living termite model to investigate the bacteriophage-*Streptomyces*-*M. anisopliae* interactions in the presence of all behavioural defence strategies termites employ are needed before concluding if the integration of the two methods can be a successful termite control strategy. Additionally, the mode of inoculation, i.e., inundative or baiting, and its effects on bacteriophage efficiency, and also the effects to non-pest termite species in field trials, will require investigation. Moreover, Ashfield-Crook et al. (2018, 2021) indicated that the use of polyvalent streptophages might create non-target effects in the environment including removal of benefical soil streptoflora which in turn might result in plant pathogenic fungal flourishment.

# Acknowledgements

Authors thank Agricultural Scientific Collections Trust located at the Orange Agricultural Institute (Orange, NSW) for supplying the biological control fungi used in the study. Authors thank Dr Ken Wasmund, University of Portsmouth, UK for the construction of the phylogenetic tree. Authors also thank the Central Analytical Research Facility (CARF) at the Queensland University of Technology (QUT), a *Microscopy Australia* linked laboratory for the electron micrograph.

# References

Ahmad, F., Fouad, H., Liang, S.Y., Hu, Y. and Mo, J.C. (2021). Termites and Chinese agricultural system: Applications and advances in integrated termite management and chemical control. *Insect Science*, 28(1): 2–20.

Ahmed, B.M. (Shiday) and French, J.R.J. (2008). An overview of termite control methods in Australia and their link to aspects of termite biology and ecology. *Pakistan Entomologist*, 30(2): 101–18.

Ahmed, B.M. (Shiday) and French, J.R.J. (2011). Termite foraging behaviour and biological response against sintered glass screening as a potential graded physical barrier. *African Journal of Environmental Science and Technology*, 5(4): 327–336.

Ashfield-Crook, N.R., Woodward, Z., Soust, M. and Kurtböke, D.İ. (2018). Assessment of the detrimental impact of polyvalent streptophages intended to be used as biological control agents on beneficial soil streptoflora. *Current Microbiology*, 75(12): 1589–1601.

Ashfield-Crook, N.R., Woodward, Z., Soust, M. and Kurtböke, D.İ. (2021). Bioactive streptomycetes from isolation to applications: A Tasmanian potato farm example. In *The Plant Microbiome* (pp. 219–249). Humana, New York, NY.

Avulova, S. and Rosengaus, R. (2011). Losing the battle against fungal infection: Suppression of termite immune defenses during mycosis. *Journal of Insect Physiology*, 57(7): 966–71.

Bignell, D., Roisin, Y. and Lo, N. (eds.). (2010). *Biology of Termites: A Modern Synthesis*, Springer.

Bignell, D.E. (2019). Termite ecology in the first two decades of the 21st century: A review of reviews. *Insects*, 10(3): 60.

Brownlie, J. and Johnson, K. (2009). Symbiont-mediated protection in insect hosts. *Trends Microbiolology*, 17(8): 348–54.

Brune, A. (2014). Symbiotic digestion of lignocellulose in termite guts. *Nature Reviews Microbiology*, 12(3): 168–80.

Brune, A. and Freiedich, M. (2000). Microecology of the termite gut: Structure and function on a microscale. *Current Opinion in Microbiology*, 3(3): 263–9.

Brune, A. and Kühl, M. (1996). pH profiles of the extremely alkaline hindguts of soil-feeding termites (Isoptera: Termitidae) determined with microelectrodes. *Journal of Insect Physiology*, 42(11): 1121–7.

Brune, A., Emerson, D. and Breznak, J. (1995). The Termite gut microflora as an oxygen sink: Microelectrode determination of oxygen and pH gradients in guts of lower and higher termites. *Applied and Environmental Microbiology*, 61(7): 2681–7.

Bulmer, M. and Crozier, R. (2004). Duplication and diversifying selection among termite antifungal peptides. *Molecular Biology and Evolution*, 21(12): 2256–64.

Bulmer, M., Bachelet, I., Raman, R., Rosengaus, R., Sasisekharan, R. and Wogan, G. (2009). Targeting an antimicrobial effector function in insect immunity as a pest control strategy. *Proceedings of the National Academy of Sciences of the United States of America*, 106(31): 12652–7.

Chandler, G., Cary, T., Bejarano, A., Pender, J. and Ferry, J. (2004). Population consequences of fipronil and degradates to copepods at field concentrations: an integration of life cycle testing with Leslie matrix population modeling. *Environmental Science & Technology*, 38(23): 6407–14.

Challis, G.L. and Hopwood, D.A. (2003). Synergy and contingency as driving forces for the evolution of multiple secondary metabolite production by *Streptomyces* species. *Proceedings of the National Academy of Sciences*, 100(2): 14555–61.

Chouvenc, T. and Su, N. (2010). Apparent synergy among defense mechanisms in subterranean termites (Rhinotermitidae) against epizootic events: Limits and potential for biological control. *Journal of Economic Entomology*, 103(4): 1327–37.

Chouvenc, T., Su, N. and Elliott, M. (2008). Interaction between the Subterranean Termite *Reticulitermes flavipes* (Isoptera: Rhinotermitidae) and the entomopathogenic fungus *Metarhizium anisopliae* in foraging arenas. *Journal of Economic Entomology*, 101(3): 885–93.

Chouvenc, T., Su, N. and Robert, A. (2009). Cellular encapsulation in the eastern subterranean termite, *Reticulitermes flavipes* (Isoptera), against infection by the entomopathogenic fungus *Metarhizium anisopliae*. *Journal of Invertebrate Pathology*, 101(3): 234–41.

Chouvenc, T., Su, N. and Grace, J. (2011). Fifty years of attempted biological control of termites – Analysis of a failure. *Biological Control*, 59(2): 69–82.

Chouvenc, T., Efstathion, C., Elliott, M. and Su, N. (2013). Extended disease resistance emerging from the faecal nest of a subterranean termite. *Proceedings of the Royal Society B: Biological Sciences*, 280(1770): 20131885.

Chouvenc, T., Elliott, M.L., Šobotník, J., Efstathion, C.A. and Su, N.Y. (2018). The termite fecal nest: A framework for the opportunistic acquisition of beneficial soil *Streptomyces* (*Actinomycetales*: Streptomycetaceae). *Environmental Entomology*, 47(6): 1431–1439.

Cleveland, L. (1924). The physiological and symbiotic relationships between the intestinal protozoa of termites and their host, with special reference to *Reticulitermes flavipes* Kollar. *Biological Bulletin*, 46(5): 203–27.

Culliney, T. and Grace, J. (2000). Prospects for the biological control of subterranean termites (Isoptera: Rhinotermitidae), with special reference to *Coptotermes formosanus*. *Bulletin of Entomological Research*, 90(1): 9–21.

Delate, K., Grace, J. and Tome, C. (1995). Potential use of pathogenic fungi in baits to control the Formosan subterranean termite (Isopt., Rhinotermitidae). *Journal of Applied Entomology*, 119(1-5): 429–33.

Derrick, M., Burgess, H., Baker, M. and Binnie, N. (1990). Sulfuryl fluoride (Vikane): A review of its use as a fumigant. *Journal of the American Institute for Conservation*, 29(1): 77–90.

Devi, K., Seth, N., Kothamasi, S. and Kothamasi, D. (2007). Hydrogen cyanide-producing rhizobacteria kill subterranean termite *Odontotermes obesus* (rambur) by cyanide poisoning under *in vitro* conditions. *Current Microbiology*, 54(1): 74–8.

Doolittle, M., Raina, A., Lax, A. and Boopathy, R. (2007). Effect of natural products on gut microbes in Formosan subterranean termite, *Coptotermes formosanus*. *International Biodeterioration & Biodegradation*, 59(1): 69–71.

Engel, M., Grimaldi, D. and Krishna, K. (2009). Termites (Isoptera): their phylogeny, classification, and rise to ecological dominance. *American Museum Novitates*, 3650(1): 1–27.

Epsky, N. and Capinera, J. (1988). Efficacy of the entomogenous nematode *Steinernema feltiae* against a subterranean termite, *Reticulitermes tibialis* (Isoptera: Rhinotermitidae). *Journal of Economic Entomology*, 81(5): 1313–7.

Evans, T.A. (2021). Predicting ecological impacts of invasive termites. *Current Opinion in Insect Science*, 46: 88–94.

Evans, T.A. and Iqbal, N. (2015). Termite (order *Blattodea*, infraorder *Isoptera*) baiting 20 years after commercial release. *Pest Management Science*, 71(7): 897–906.

French, J.R.J. (1991). Baits and foraging behavior of Australian species of *Coptotermes*. *Sociobiology*, 19(1): 171–86.

Fuller, C. (2007). Fungistatic activity of freshly killed termite, *Nasutitermes acajutlae*, soldiers in the Caribbean. *Journal of Insect Science*, 7(14): 1–8.

Gautam, B. and Henderson, G. (2014). Comparative evaluation of three chitin synthesis inhibitor termite baits using multiple bioassay designs. *Sociobiology*, 61(1): 82–7.

Gebhardt, K., Schimana, J., Krastel, P., Dettner, K., Rheinheimer, J., Zeeck, A. and Fiedler, H.-P. (2002). Endophenazines AD, new phenazine antibiotics from the arthropod associated endosymbiont *Streptomyces anulatus*. I. Taxonomy, fermentation, isolation and biological activities. *The Journal of Antibiotics*, 55(9): 794–800.

Grace, J. (1994). Protocol for testing effects of microbial pest control agents on nontarget subterranean termites (Isoptera: Rhinotermitidae). *Journal of Economic Entomology*, 87(2): 269–74.

Grace, J., Yates, III, J. and Tome, C. (1996). Termite-resistant construction: Use of a stainless-steel mesh to exclude *Coptotermes formosanus*. *Sociobiology*, 28(3): 365–72.

Green, III, F., Arango, R., Esenther, G., Rojas, M. and Morales-Ramos, J. (2013). Synergy of diflubenzuron baiting and NHA dusting on mortality of Reticulitermes flavipes. In *Series: Conference Proceedings*.

Hajek, A. and Delalibera Jr, I. (2010). Fungal pathogens as classical biological control agents against arthropods. *Biological Control*, 55(1): 147–58.

Hänel, H. and Watson, J. (1983). Preliminary field tests on the use of *Metarhizium anisopliae* for the control of *Nasutitermes exitiosus* (Hill)(Isoptera: Termitidae). *Bulletin of Entomological Research*, 73(2): 305–13.

Hansen, J., Johnson, J. and Winter, D. (2011). History and use of heat in pest control: A review. *International Journal of Pest Management*, 57(4): 267–89.

Haverty, M. (1977). The proportion of soldiers in termite colonies: A list and a bibliography (Isoptera). *Sociobiology (USA)*, 2(3): 199–216.

Hongoh, Y. (2010). Diversity and genomes of uncultured microbial symbionts in the termite gut. *Bioscience, Biotechnology, and Biochemistry*, 74(6): 1145–51.

Hongoh, Y., Deevong, P., Inoue, T., Moriya, S., Trakulnaleamsai, S., Ohkuma, M., Vongkaluang, C., Noparatnaraporn, N. and Kudo, T. (2005). Intra- and interspecific comparisons of bacterial diversity and community structure support coevolution of gut microbiota and termite host. *Appl. Environ. Microbiol.*, 71(11): 6590–9.

Hrdý, I., Kuldová, J. and Wimmer, Z. (2001). A juvenile hormone analogue with potential for termite control: laboratory tests with *Reticulitermes santonensis, Reticulitermes flaviceps* and *Coptotermes formosanus* (Isopt., Rhinotermitidae). *Journal of Applied Entomology*, 125(7): 403–11.

Inoue, J., Noda, S., Hongoh, Y., Ui, S. and Ohkuma, M. (2008). Identification of endosymbiotic methanogen and ectosymbiotic spirochetes of gut protists of the termite *Coptotermes formosanus*. *Microbes and Environments*, 23(1): 94–7.

Jabeen, F., Hussain, A., Manzoor, M., Younis, T., Rasul, A. and Qazi, J.I. (2018). Potential of bacterial chitinolytic, *Stenotrophomonas maltophilia*, in biological control of termites. *Egyptian Journal of Biological Pest Control*, 28(1): 1–10.

Jørgensen, H., Kristensen, J. and Felby, C. (2007). Enzymatic conversion of lignocellulose into fermentable sugars: Challenges and opportunities. *Biofuels, Bioproducts and Biorefining*, 1(2): 119–34.

Jouquet, P., Blanchart, E. and Capowiez, Y. (2014). Utilization of earthworms and termites for the restoration of ecosystem functioning. *Applied Soil Ecology*, 73(1): 34–40.

Kaltenpoth, M. (2009). Actinobacteria as mutualists: General healthcare for insects? *Trends in Microbiology*, 17(12): 529–35.

Kaltenpoth, M., Schmitt, T., Polidori, C., Koedam, K. and Strohm, E. (2010). Symbiotic streptomycetes in antennal glands of the South American digger wasp genus *Trachypus* (Hymenoptera, Crabronidae). *Physiological Entomology*, 35(2): 196–200.

Kappler, A. and Brune, A. (2002). Dynamics of redox potential and changes in redox state of iron and humic acids during gut passage in soil-feeding termites (*Cubitermes* spp.). *Soil Biology and Biochemistry*, 34(2): 221–7.

Kartal, S., Yoshimura, T. and Imamura, Y. (2004). Decay and termite resistance of boron-treated and chemically modified wood by *in situ* co-polymerization of allyl glycidyl ether (AGE) with methyl methacrylate (MMA). *International Biodeterioration & Biodegradation*, 53(2): 111–7.

Kaya, H. and Gaugler, R. (1993). Entomopathogenic nematodes. *Annual Review of Entomology*, 38(1): 181–206.

Keefer, T., Zollinger, D. and Gold, R. (2013). Evaluation of aggregate particles as a physical barrier to prevent subterranean termite incursion into structures. *Southwestern Entomologist*, 38(3): 447–64.

Khan, A. and Ahmad, W. (2018). *Termites and Sustainable Management*, Volume 1-Biology, social behaviour and economic importance. Part of the sustainability in plant and crop protection book series. Springer Nature, Switzerland.

Khan, K., Fazal, Q. and Jafri, R. (1977a). Pathogenicity of locally discovered *Bacillus thuringiensis* strain to the termites: *Heterotermes indicola* (Wassman) and *Microcerotermes championi* (Shyder). *Pakistan Journal of Scientific Research*, 29(1): 12–3.

Khan, K., Fazal, Q., Jafri, R. and Ahmad, M. (1977b). Susceptibility of various species of termites to a pathogen, *Serratia marcescens*. *Pakistan Journal of Scientific Research*, 29(1): 46–7.

Khan, K., Jafri, R. and Ahmad, M. (1985). The pathogenicity and development of *Bacillus thuringiensis* in termites. *Pakistan Journal of Zoology*, 30(1): 117–9.

Khan, K., Jafri, R., Ahmad, M. and Khan, K. (1992). The pathogenicity of *Pseudomonas aeruginosa* against termites. *Pakistan Journal of Zoology*, 24(3): 243–5.

Kurtböke, D.İ. (2017). Ecology and habitat distribution of actinobacteria. Chapter 6. pp. 123–149. *In:* Wink, J., Mohammadipanah, F. and Hamedi, J. (eds.). *Biology and Biotechnology of Actinobacteria*. Springer, Cham. https://doi.org/10.1007/978-3-319-60339-1_6.

Kurtböke, D.İ. (2004a). Uniqueness of the Smart State's microbial diversity: From an actinomycete collection to biodiscovery at the University of the Sunshine Coast. *Microbiology Australia*, 25(2): 36–38.

Kurtböke, D.İ. (2004b). Actino-Rush on the Sunshine Coast: Prospects for bioprospecting. pp. 319–322. *In:* Watanabe, M.M., Suzuki, K. and Seki, T. (eds.). *In the Proceedings of the 10th International Congress on Culture Collections*. Tsukuba, Japan, 2004.

Kurtböke, D.İ. and French, J. (2007). Use of phage battery to investigate the actinofloral layers of termite gut microflora. *Journal of Applied Microbiology*, 103(3): 722–34.

Kurtböke, D.İ., French, J.R.J., Hayes, R.A. and Quinn, R.J. (2014). Eco-taxonomic insights into actinomycete symbionts of termites for discovery of novel bioactive compounds. *In:* Mukherjee, J. (ed.). *Biotechnological Applications of Biodiversity*. Advances in Biochemical Engineering/Biotechnology, vol 147. Springer, Berlin, Heidelberg. https://doi.org/10.1007/10_2014_270.

Lee, C., Vongkaluang, C. and Lenz, M. (2007). Challenges to subterranean termite management of multi-genera faunas in Southeast Asia and Australia. *Sociobiology*, 50(1): 213–22.

Lenz, M., Schafer, B., Runko, S. and Glossop, L. (1997). The concrete slab as part of a termite barrier system: Response of Australian subterranean termites to cracks of different width in concrete. *Sociobiology*, 30(1): 103–18.

Lewis, V. and Haverty, M. (1996). Evaluation of six techniques for control of the western drywood termite (Isoptera: Kalotermitidae) in structures. *Journal of Economic Entomology*, 89(4): 922–34.

Lewis, V., Power, A. and Haverty, M. (2000). Laboratory evaluation of microwaves for control of the western drywood termite. *Forest Products Journal*, 50(5): 79–87.

Makonde, H., Boga, H., Osiemo, Z., Mwirichia, R., Stielow, J., Goker, M. and Klenk, H. (2013). Diversity of *Termitomyces* associated with fungus-farming termites assessed by cultural and culture-independent methods. *PLoS One*, 8(2): e56464.

Mankowski, M.E., Kaya, H.K., Kenneth Grace, J. and Sipes, B. (2005). Differential susceptibility of subterranean termite castes to entomopathogenic nematodes. *Biocontrol Science and Technology*, 15(4): 367–77.

Mattoso, T., Moreira, D. and Samuels, R. (2012). Symbiotic bacteria on the cuticle of the leaf-cutting ant *Acromyrmex subterraneus subterraneus* protect workers from attack by entomopathogenic fungi. *Biology Letters*, 8(3): 461–4.

McGarry, D., Bridge, B. and Radford, B. (2000). Contrasting soil physical properties after zero and traditional tillage of an alluvial soil in the semi-arid subtropics. *Soil and Tillage Research*, 53(2): 105–15.

Menegatti, C., Fukuda, T.T. and Pupo, M.T. (2021). Chemical ecology in insect-microbe interactions in the neotropics. *Planta Medica*, 87(01/02): 38–48.

Myles, T. (2002). Alarm, aggregation, and defense by Reticulitermes flavipes in response to a naturally occurring isolate of *Metarhizium anisopliae*. *Sociobiology*, 40(2): 243–55.

Nalepa, C., Bignell, D. and Bandi, C. (2001). Detritivory, coprophagy, and the evolution of digestive mutualisms in Dictyoptera. *Insectes Sociaux*, 48(3): 194–201.

Ni, J. and Tokuda, G. (2013). Lignocellulose-degrading enzymes from termites and their symbiotic microbiota. *Biotechnology Advances*, 31(6): 838–50.

Nix, K., Henderson, G., Zhu, B. and Laine, R. (2006). Evaluation of vetiver grass root growth, oil distribution, and repellency against Formosan subterranean termites. *HortScience*, 41(1): 167–71.

Noda, S., Kitade, O., Inoue, T., Kawai, M., Kanuka, M., Hiroshima, K., Hongoh, Y., Constantino, R., Uys, V. and Zhong, J. (2007). Cospeciation in the triplex symbiosis of termite gut protists (*Pseudotrichonympha* spp.), their hosts, and their bacterial endosymbionts. *Molecular Ecology*, 16(6): 1257–66.

Noda, S., Hongoh, Y., Sato, T. and Ohkuma, M. (2009). Complex coevolutionary history of symbiotic *Bacteroidales* bacteria of various protists in the gut of termites. *BMC Evol. Biol.*, 9: 158.

Oh, D., Poulsen, M., Currie, C. and Clardy, J. (2011). Sceliphrolactam, a polyene macrocyclic lactam from a wasp-associated *Streptomyces* sp. *Organic Letters*, 13(4): 752–5.

Oh, D., Scott, J., Currie, C. and Clardy, J. (2009). Mycangimycin, a polyene peroxide from a mutualist *Streptomyces* sp. *Organic Letters*, 11(3): 633–6.

Ohkuma, M. (2008). Symbioses of flagellates and prokaryotes in the gut of lower termites. *Trends in Microbiology*, 16(7): 345–52.

Ohkuma, M. and Brune, A. (2011). Diversity, structure, and evolution of the termite gut microbial community. *In Biology of Termites: A Modern Synthesis*, Springer, pp. 413–38.

Ohkuma, M., Sato, T., Noda, S., Ui, S., Kudo, T. and Hongoh, Y. (2007). The candidate phylum 'Termite Group 1' of bacteria: Phylogenetic diversity, distribution, and endosymbiont members of various gut flagellated protists. *FEMS Microbiology Ecology*, 60(3): 467–76.

Pogoda, J. and Preston-Martin, S. (1997). Household pesticides and risk of pediatric brain tumors. *Environmental Health Perspectives*, 105(11): 1214.

Poulsen, M., Oh, D., Clardy, J. and Currie, C. (2011). Chemical analyses of wasp-associated *Streptomyces* bacteria reveal a prolific potential for natural products discovery. *PLoS One*, 6(2): 16763.

Raina, A., Bedoukian, R., Florane, C. and Lax, A. (2012). Potential of natural products and their derivatives to control Formosan subterranean termites (Isoptera: Rhinotermitidae). *Journal of Economic Entomology*, 105(5): 1746–50.

Rath, A. & Tidbury, C. (1996). Susceptibility of *Coptotermes acinaciformis* (Isoptera: Rhinotermitidae) and *Nasutitermes exitiosus* (Isoptera: Termitidae) to two commercial isolates of *Metarhizium anisopliae*. *Sociobiology (USA)*, 28(1): 67–72.

Rauh, V., Garfinkel, R., Perera, F., Andrews, H., Hoepner, L., Barr, D., Whitehead, R., Tang, D. and Whyatt, R. (2006). Impact of prenatal chlorpyrifos exposure on neurodevelopment in the first 3 years of life among inner-city children. *Pediatrics*, 118(6): 1845–59.

Regnault-Roger, C., Vincent, C. and Arnason, J. (2012). Essential oils in insect control: Low-risk products in a high-stakes world. *Annual Review of Entomology*, 57: 405–24.

Rosengaus, R., Guldin, M. and Traniello, J. (1998). Inhibitory effect of termite fecal pellets on fungal spore germination. *Journal of Chemical Ecology*, 24(10): 1697–706.

Rosengaus, R., Jordan, C., Lefebvre, M. and Traniello, J. (1999). Pathogen alarm behavior in a termite: A new form of communication in social insects. *Naturwissenschaften*, 86(11): 544–8.

Rosengaus, R., Lefebvre, M. and Traniello, J. (2000). Inhibition of fungal spore germination by *Nasutitermes*: Evidence for a possible antiseptic role of soldier defensive secretions. *Journal of Chemical Ecology*, 26(1): 21–39.

Rosengaus, R., Mead, K., Du Comb, W., Benson, R. and Godoy, V. (2013). Nest sanitation through defecation: Antifungal properties of wood cockroach feces. *Naturwissenschaften*, 100(11): 1051–9.

Rosengaus, R., Traniello, J., Lefebvre, M. and Maxmen, A. (2004). Fungistatic activity of the sternal gland secretion of the dampwood termite *Zootermopsis angusticollis*. *Insectes Sociaux*, 51(3): 259–64.

Rust, M., Paine, E. and Reierson, D. (1997). Evaluation of freezing to control wood-destroying insects (Isoptera, Coleoptera). *Journal of Economic Entomology*, 90(5): 1215–21.

Saiki, R., Yaguchi, H., Hashimoto, Y., Kawamura, S. and Maekawa, K. (2014). Reproductive soldier like individuals induced by juvenile hormone analog treatment in *Zootermopsis nevadensis* (Isoptera, Archotermopsidae). *Zoological Science*, 31(9): 573–81.

Schultz, T. and Nicholas, D. (2002). Development of environmentally-benign wood preservatives based on the combination of organic biocides with antioxidants and metal chelators. *Phytochemistry*, 61(5): 555–60.

Schwitalla, J.W., Benndorf, R., Martin, K., Vollmers, J., Kaster, A.K., de Beer, Z.W., Poulsen, M. and Beemelmanns, C. (2020). *Streptomyces smaragdinus* sp. nov., isolated from the gut of the fungus growing-termite *Macrotermes natalensis*. *International Journal of Systematic and Evolutionary Microbiology*, 70(11): 5806.

Seipke, R.F., Kaltenpoth, M. and Hutchings, M.I. (2012). *Streptomyces* as symbionts: An emerging and widespread theme? *FEMS Microbiology Reviews*, 36(4): 862–876.

Singha, D., Singha, B. and Dutta, B. (2011). Potential of *Metarhizium anisopliae* and *Beauveria bassiana* in the control of tea termite *Microtermes obesi* Holmgren *in vitro* and under field conditions. *Journal of Pest Science*, 84(1): 69–75.

Smythe, R. & Coppel, H. (1965). The susceptibility of *Reticulitermes flavipes* (Kollar) and other termite species to an experimental preparation of *Bacillus thuringiensis* Berliner. *Journal of Invertebrate Pathology*, 7(4): 423–6.

Staples, J. and Milner, R. (2000). A laboratory evaluation of the repellency of *Metarhizium anisopliae* conidia to *Coptotermes lacteus* (Isoptera: Rhinotermitidae). *Sociobiology*, 36(1): 133–48.

Su, N., Ban, P. and Scheffrahn, R. (2004). Polyethylene barrier impregnated with lambda-cyhalothrin for exclusion of subterranean termites (Isoptera: Rhinotermitidae) from structures. *Journal of Economic Entomology*, 97(2): 570–4.

Sun, J., Fuxa, J. and Henderson, G. (2003). Effects of virulence, sporulation, and temperature on *Metarhizium anisopliae* and *Beauveria bassiana* laboratory transmission in *Coptotermes formosanus*. *Journal of Invertebrate Pathology*, 84(1): 38–46.

Tingle, C., Rother, J., Dewhurst, C., Lauer, S. and King, W. (2003). Fipronil: Environmental fate, ecotoxicology, and human health concerns. *In Reviews of Environmental Contamination and Toxicology*, Springer, pp. 1–66.

Tokuda, G. and Watanabe, H. (2007). Hidden cellulases in termites: Revision of an old hypothesis. *Biology Letters*, 3(3): 336–9.

Turnbull, M., Martin, S. and Webb, B. (2004). Quantitative analysis of hemocyte morphological abnormalities associated with *Campoletis sonorensis* parasitization. *Journal of Insect Science*, 4(11): 15–30.

Vargo, E. and Hussender, C. (2009). Biology of subterranean termites: insights from molecular studies of *Reticulitermes* and *Coptotermes*. *Annual Review of Entomology*, 54(1): 379–403.

Vela, G., Wu, J. and Smith, D. (1976). Effect of 2450 MHz microwave radiation on some soil microorganisms *in situ*. *Soil Science*, 121(1): 44–51.

Verma, M., Sharma, S. and Prasad, R. (2009). Biological alternatives for termite control: A review. *International Biodeterioration & Biodegradation*, 63(8): 959–72.

Vey, A. (2002). Effects of the peptide mycotoxin destruxin E on insect haemocytes and on dynamics and efficiency of the multicellular immune reaction. *Journal of Invertebrate Pathology*, 80(3): 177–87.

Visser, A., Nobre, T., Currie, C., Aanen, D. and Poulsen, M. (2012). Exploring the potential for Actinobacteria as defensive symbionts in fungus-growing termites. *Microbial Ecology*, 63(4): 975–85.

Wang, C. and Leger, R.S. (2006). A collagenous protective coat enables *Metarhizium anisopliae* to evade insect immune responses. *Proceedings of the National Academy of Sciences*, 103(17): 6647–52.

Wang, C. and Powell, J. (2004). Cellulose bait improves the effectiveness of *Metarhizium anisopliae* as a microbial control of termites (Isoptera: Rhinotermitidae). *Biological Control*, 30(2): 523–9.

Wang, C., Powell, J. and Nguyen, K. (2002). Laboratory evaluations of four entomopathogenic nematodes for control of subterranean termites (Isoptera: Rhinotermitidae). *Environmental Entomology*, 31(2): 381–7.

Watanabe, H. and Tokuda, G. (2010). Cellulolytic systems in insects. *Annual Review of Entomology*, 55(1): 609–32.

Wright, M., Raina, A. and Lax, A. (2005). A strain of the fungus *Metarhizium anisopliae* for controlling subterranean termites. *Journal of Economic Entomology*, 98(5): 1451–8.

Yamaguchi, H. (2003). Silicic acid: Boric acid complexes as wood preservatives. *Wood Science and Technology*, 37(3): 287–97.

Yanagawa, A. and Shimizu, S. (2006). Resistance of the termite, *Coptotermes formosanus* Shiraki to *Metarhizium anisopliae* due to grooming. *Biological Control*, 52(1): 75–85.

Yanagawa, A., Yokohari, F. and Shimizu, S. (2009). The role of antennae in removing entomopathogenic fungi from cuticle of the termite, *Coptotermes formosanus*. *Journal of Insect Science*, 9(6): 1–9.

Yanagawa, A., Fujiwara-Tsujii, N., Akino, T., Yoshimura, T., Yanagawa, T. and Shimizu, S. (2011). Behavioral changes in the termite, *Coptotermes formosanus* (Isoptera), inoculated with six fungal isolates. *Journal of Invertebrate Pathology*, 107(2): 100–6.

Zhang, Y.L., Li, S., Jiang, D.H., Kong, L.C., Zhang, P.H. and Xu, J.D. (2013). Antifungal activities of metabolites produced by a termite-associated *Streptomyces canus* BYB02. *Journal of Agricultural and Food Chemistry*, 61(7): 1521–1524.

Zhou, L.F., Wu, J., Li, S., Li, Q., Jin, L.P., Yin, C.P. and Zhang, Y.L. (2021). Antibacterial potential of termite-associated *Streptomyces* spp. *ACS Omega*, 6(6): 4329–4334.

# Chapter 11

# Diversity and Biotechnological Potential of Actinomycetes in Arid Lands of Mongolia

*Tsetseg Baljinova*

## 1. Introduction

The UN streamline documents have for the first time emphasized the key role of the green economy for world development towards sustainability (UNDP 2012). The green economy includes amongst its essential parts the bioeconomy. The Organization for Economic Cooperation and Development (OECD) estimated that by 2030 most countries would develop their bioeconomy "...where biotechnology contributes to a significant share of economic output" (OECD 2009). Indeed, biotechnology is "...recognized universally as one of the key enabling technologies for the 21st century" (Bull et al. 2000). The OECD has also underlined that biotechnology "...offers technological solutions for many of the health and resource-based challenges facing the world" and stressed that the bioeconomy should be based on achievements of the biological sciences, notably on genes and cell processes in living organisms. Consequently, these documents have reinforced the importance of new knowledge and sound scientific data on biological diversity of living organisms and genetic resources as a main biological resource for the development of biotechnology.

A recent estimation of biotechnological development has revealed that in 2019 the market size of biotechnology was US$ 295 billion and in the next five years "...will achieve a steady increase". It is also interesting that between 2015 and 2020 the worldwide growth rate of biotechnology was 1.3% (Martin et al. 2021). Consequently, it seems likely that demands for new knowledge and development in the field of biotechnology, in particular microbial biotechnology, will grow. As it was stated by Demain (2007), "...microbiology is a major participant in modern global industry".

Laboratory of Microbiology, Institute of Biology, Mongolian Academy of Sciences, Ulaanbaatar-13330, Mongolia.
Email: tsetseg110@gmail.com

Amongst microorganisms, actinomycetes are the most prolific producers of primary and secondary metabolites that are important for the development of biotechnology, especially for the production of pharmaceuticals (Bèrdy 2005, Kurtböke 2010, Barka et al. 2016, Genilloud 2017, Mukhtar et al. 2017, Takahashi and Nakashima 2018, Al-shaibani et al. 2021). In addition, problems caused by antibiotic resistance have promoted further interest in actinomycetes as the most useful microorganisms for the biosynthesis of novel antibiotics for the treatment of infectious diseases. Consequently, knowledge on actinomycetes occurring in biotopes of unexplored countries is of considerable importance. Recently, extreme environments such as arid lands, deep sea sediments, saline soils and salt lakes have been recognized as the most favourable ecosystems for the isolation of bioactive actinomycetes (Okoro et al. 2009, Bull 2011, Meklat et al. 2011, Subramani and Aalbersberg 2012, Mohammadipanah and Wink 2015, Arasu et al. 2016, Carro et al. 2018, Genilloud 2018, Goodfellow et al. 2018, Corral et al. 2020, Xie and Pathom-aree 2021). In addition, studies on endophytic actinomycetes show that they are a prolific source of bioactive and chemically novel compounds with huge medicinal and agricultural potential (Igarashi et al. 2002, Janso and Carter 2010, Golinska et al. 2015, Matsumoto and Takahashi 2017).

Mongolia is rich in extreme environments due to its geographic location between the taiga of Southern Siberia and the Gobi Desert. Indeed, 74% of Mongolia consists of arid lands (drylands) including steppe, semi-desert and desert regions (Table 1). The country is also rich in saline soils and salt lakes (Sevastiyanov et al. 2014). Amongst salt lakes, soda lakes are of particular interest as a source of relic microbial communities which are considered to be the origin of terrestrial microbial biodiversity (Zavarzin 1993). DasSarma (2006) noted that saline environments could be used as "models for studying of mechanisms of survival away from our planet".

Studies on actinomycetes isolated from extreme environments in Mongolia have shown that they are a source of taxonomically novel and biotechnologically important microorganisms (Tsetseg 1973, 1974, 2003, 2010, 2012, 2018, Dugarjav 1977, Puntsag et al. 1977, Skalon 1977, Onkhor et al. 1984, Rachev et al. 1987, Davaadorj et al. 1993, Fedorova et al. 1993, Vasilchenko et al. 1996, Enkh-Amgalan et al. 1998, 1999, Tsetseg et al. 1999, Ara et al. 2010a, b, 2011a, b, 2012a, b, 2013). In particular, actinomycetes are widely distributed in desert soils (Onkhor et al. 1983, Tsetseg et al. 1994, Daram and Tsetseg 2001) and rhizospheres of desert and dry steppe plants (Tsetseg 1973, 2008, Norovsuren et al. 2007a). They have also been found to be abundant in saline desert soils and in takyr-like soils where they accounted for 26.4% and 12.1% of microbial communities, respectively, indicating that they are an integral part of the microbial flora (Tsetseg 2003). Halophilic and haloalkaliphilic streptomycetes which grow in the presence of 5% NaCl and from pH 8–9 have been isolated from desert-steppe soils (Norovsuren et al. 2007b). Similarly, halotolerant and moderately haloalkaliphilic actinomycetes belonging to the genera *Isoptericola*, *Nesterenkonia*, *Nocardiopsis* and *Streptomyces* which grow on media supplemented with 10% NaCl have been isolated from steppe

**Table 1.** Characteristics of the arid lands of Mongolia*.

| Zone | Subzone | Country's territory (%) | Average temperature°C | | Annual precipitation (mm) | Index of aridity ($R_0$/Lr) | Humus, % (0–5 cm depth) |
|------|---------|------|----------|------|------------|---------|---------|
| | | | January | July | | | |
| Steppe | Moderately dry steppe with chernozem-like and dark-chestnut soils | 19.2 | –22.3 | 18.8 | 250–300 | 1.7–3.0 | 4.0–6.5 |
| | Dry steppe with chestnut and light chestnut soils | 24.0 | –19.8 | 21.6 | 150–250 | 5.0–7.0 | 1.5–3.4 |
| Semi-desert | Desertified steppe (desert-steppe) with brown soils | 11.22 | –19.0 | 23.0 | 120–150 | 7.5 | 0.4–1.0 |
| | Steppified desert with paleo-brown soils | 8.2 | –18.7 | 23.5 | 100–105 | 9.4 | 0.3–0.6 |
| Desert | True desert with gray-brown soils | 7.87 | –18.2 | 24.2 | 75–100 | 13.0 | 0.09–0.15 |
| | Hyper-arid desert with gray-brown hyper-arid soils | 3.54 | –17.0 | 28.0 | 20–50 | > 20.0 | ? |

* Compiled from Bespalov (1951), Umarov and Yakunin (1974), Zhirnov et al. (2005) and Gunin et al. (2007). ($R_0$ – amount of radiation balance for moistened surface; Lr – latent heat of evaporation).

and semi-desert saline soils (Tsetseg et al. 2011, Ara et al. 2013). It has also been shown that thermotolerant and thermophilic actinomycetes assigned to the genera *Actinomadura, Micromonospora, Streptomyces* and *Streptosporangium* are more prevalent than their mesophilic counterparts in desert soils of Mongolia (Kurapova et al. 2012). It can be concluded that Mongolia's natural environments are a valuable source of taxonomically novel and biotechnologically important microorganisms with metabolisms adapted to extreme environments.

The aim of this chapter is to describe the diversity and biotechnological potential of actinomycetes present in the natural environments of Mongolia based on knowledge generated over the past 20 years. This evaluation of actinomycete diversity rests on 16S rRNA gene sequence data. In turn, the biotechnological potential of selected actinomycetes was based on whole genome sequences, biosynthetic genes and phenotypic data. Particular attention is given to the ability of the actinomycetes to produce extracellular enzymes and amino acids, to their plant growth promotion activities, antimicrobial and anti-quorum sensing properties and to their ability to act as biological control agents. Special attention is paid to actinomycetes from steppe and semi-desert saline soils compared with those present in adjacent soils. In addition, a short description of natural environments found in Mongolia is given.

## 2. Geographic Location, Climate and Natural Environments of Mongolia

Mongolia is a land-locked country in the heart of Asia that lies between the Russian Federation and the People's Republic of China. Geographically it runs from north to south between N52°06′ and N41°32′, and from west to east between E87°47′ and E119°54′ (Murzaev 1952). Administratively, Mongolia consists of 21 provinces (aimags) and is the seventh largest country in Asia (Batjargal and Enkhbat 1998) covering an area of 1564 thousand square kilometers. According to the geographic regionalization scheme, Mongolia consists of four landscape-geographic regions: the Altai mountains, the Khangai-Khentei mountains, the Eastern Mongolian steppe, and the Gobi Desert (Tsegmid 1962).

Mongolia represents the critical transition zone in Central Asia where the great Siberian taiga forest, the Central Asian steppe, the high Altai mountains and the Gobi Desert converge (Batima et al. 2004) giving rise to three different natural belts and three zones, namely the alpine, mountain taiga and mountain forest-steppe belts, and the steppe, semi-desert and desert zones (Gunin et al. 2007). It is interesting that the border of the semi-desert zone in some places reaches N50.5° being the most northern in Eurasia (Murzaev 1952).

Zhirnov et al. (2005) stated that 74% of Mongolia's territory is occupied by arid ecosystems (drylands) which include the steppe, semi-desert and desert zones, i.e., areas with low precipitation, high solar radiation and soils poor in organic matter (Table 1). According to the classification of McGinnies et al. (1968), extreme arid ecosystems have less than 60–100 mm mean annual precipitation, arid regions – from 60–100 to 50–250 mm and semi-arid areas – from 150–250 to 250–500 mm.

Mongolia has a continental climate, namely long and cold winters, low precipitation and large annual, seasonal, monthly and diurnal fluctuations in air temperature. Extreme temperatures range from –50°–56°C in winter to +35°–+41°C in summer. The mean annual precipitation ranges from 38.4 mm in desert regions to 389.3 mm in the forest-steppe area (Vostokova et al. 1995) which means that the forest-steppe area can be considered as arid.

About 85% of the country lies at an altitude of 1000 m above sea level. There are around 190 glaciers, a permafrost area that covers approximately 63% of the country's territory and expands in some places up to N47°, and numerous hot and cold mineral springs (Murzaev 1952, Tumurbaatar and Mijiddorj 2006). About 10.5% of the country is covered by saline soils (Bespalov 1951). According to Egorov (1993), soil salinization embraces 2.5% of the Altai Mountain region, 4.0% of the Khangai-Khentei region and 30.5% of the combined territories of the Gobi Desert and Eastern-Mongolian steppe regions.

It can be concluded that due to its geographic location and climate, Mongolia contains extreme environments ranging from the cold high mountain taiga of Southern Siberia to the hyper-arid region of the Gobi Desert. The harsh ecological conditions in these ecosystems are a challenge for animal, plant and microbial life. It is generally assumed that novel microorganisms have become adapted to harsh conditions, such as water shortage, salinity, high levels of radiation, low content of organic matter and large annual, seasonal, monthly and diurnal fluctuations in air and soil temperatures. The resulting unique microbiota is of particular interest to biotechnologists, ecologists and evolutionary biologists and is an invaluable source of new scientific knowledge that promotes our understanding of the role of microorganisms in extreme biotopes.

## 3. Taxonomic Diversity of Actinomycetes Found in Mongolia

Chemotaxonomic studies of actinomycetes isolated from steppe and forest-steppe soils revealed the presence of actinomycetes belonging to the genera *Actinomadura*, *Micromonospora*, *Saccharothrix* and *Streptomyces* (Tsetseg et al. 1997). Subsequent partial sequencing of 16S rRNA genes of 12 strains, including ones from the genera cited above, showed a high probability that some of them represent new taxa (Enkh-Amgalan et al. 1998). Their 16S rDNA similarities to actinomycetes classified in the genera *Actinobispora*, *Actinomadura*, *Glycomyces*, *Kibdelosporangium*, *Micromonospora*, *Microtetraspora*, *Saccharothrix* and *Streptomyces* varied between 96-99%. Isolates showing similarities of 96%, 97% and 98% to members of the genera *Actinobispora*, *Saccharothrix* and *Kibdelosporangium*, respectively, were isolated from hyper-arid desert soils (Enkh-Amgalan et al. 1999). The genus *Actinobispora* is now a synonym of the genus *Pseudonocardia* (Huang et al. 2002).

In a large-scale study almost complete sequences of 16S rRNA genes of 2443 actinomycetes isolated from soil and lichens of 11 Mongolian provinces was carried out under the auspices of the Mongolia-Japan Joint Research Project (MJP) "Taxonomic and ecological studies of microorganisms in Mongolia and their utilization" (2006–2017). These organisms were assigned to 66 validly published genera that belonged to 25 families and 10 orders of the classes *Actinobacteria* and *Thermoleophilia* (Tsetseg and Ando 2013, Tsetseg et al. 2016a, Tsetseg 2018). The number of genera identified accounted for about 30% of those described in the second edition of Bergey's Manual of Systematic Bacteriology (Goodfellow et al. 2012).

Phylogenetic analysis of the actinomycetes indicated that many of them belonged to taxonomically novel taxa, as exemplified by *Luteipulveratus* gen. nov. of the family *Dermacoccaceae*, which accomodated *Luteipulveratus mongoliensis* sp. nov. (Ara et al. 2010a). Subsequently, *Actinoplanes toevensis* sp. nov. and *A. tereljensis* sp. nov. (Ara et al. 2010b), *Pseudonocardia mongoliensis* sp. nov. and *P. khuvsgulensis* sp. nov. (Ara et al. 2011a), *Actinophytocola burenkhanensis* sp. nov. (Ara et al. 2011a), *Herbidospora mongoliensis* sp. nov. (Ara et al. 2012a) and *Cryprosporangium mongoliense* sp. nov. (Ara et al. 2012b) of the families *Cryptosporangiaceae*, *Micromonosporaceae*, *Pseudonocardiaceae* and *Streptosporangiaceae* were validly published.

New species of actinomycetes were also found in traditional fermented milk products and in plant tissues. *Bifidobacterium mongoliense* sp. nov. was isolated from airag, traditional fermented mare's milk (Watanabe et al. 2009). Partial 16S rRNA gene sequencing of 123 actinomycetes isolated from surface-sterilized plants growing in the desert zone of Mongolia showed that they belonged to the genera *Actinocatenispora*, *Geodermatophilus*, *Kribbella*, *Microlunatus*, *Micromonospora*, *Nocardia*, *Nocardioides*, *Promicromonospora*, *Pseudonocardia*, *Saccharothrix*, *Streptomyces* and *Streptosporangium* (Oyunbileg et al. 2021a). The endophytic new species, *Actinocatenispora comari* was described for a strain isolated from surface-sterilized aerial parts of the medicinal plant *Comarum salesowianum*, which grows in the mountainous steppe belt of the Gobi Desert (Oyunbileg et al. 2021b).

A haloalkaliphilic hydrolytic actinomycete isolated from saline alkaline soil in the north-eastern Mongolia was proposed as a new monospecific genus, *Natronoglycomyces*, belonging to the family *Glycomycetaceae* (Sorokin et al. 2021). The genome of the type-strain *Natronoglycomyces albus* consisted of a chromosome of 3.94 Mbp and two plasmids of 59.8 and 14.8 Kbp. The strain was shown to have multiple glycosidase-encoding genes. Furthermore, a culture-independent study carried out to compare bacterial communities in the meromictic Lake Oigon (western part of Mongolia) and Lakes Shira and Shunet (Siberian Russia) revealed that actinomycetes belonging to the genus *Nitriliruptor* (Sorokin et al. 2009) were the predominant group in Lake Oigon where they accounted for ~ 17.7% of the relative abundance whereas in Lakes Shira and Shunet they only formed ~ 1.0% and ~ 5.6%, respectively (Baatar et al. 2016).

Table 2 shows the extent of actinomycete diversity found in Mongolian biomes. In total, 75 genera belonging to 31 families and 18 orders of the class *Actinomycetia sensu* Salam et al. (2020) were recognized. Other taxa included the genus *Patulibacter* of the family *Patulibacteraceae* of the order *Solirubrobacterales* of the class *Thermoleophilia* and the genus *Nitriliruptor* of the family *Nitriliruptoraceae* of the order *Nitriliruptorales* of the class *Nitriliruptoria*. It should be noted that actinomycetes are now classified in the phylum *Actinomycetota* (Oren and Garrity 2021).

Mongolian isolates included members of the genera *Actinomadura*, *Actinoplanes*, *Amycolatopsis*, *Micromonospora*, *Nocardia*, *Saccharopolyspora*, *Saccharothrix*, *Streptomyces* and *Streptosporangium*, all of which are prolific source of antibiotics (Lazzarini et al. 2000, Bérdy 2005, Goodfellow et al. 2012, Takahashi and Nakashima 2018). The genera *Actinoplanes*, *Cryptosporangium*, *Kribbella*, *Micromonospora*, *Nocardia*, *Nonomuraea*, *Rhodococcus*, *Streptomyces* and *Streptosporangium* were found to be the dominant genera in Mongolia (Tsetseg et al. 2016a). It is also interesting that amycolate actinomycetes, i.e., actinomycetes that do not form mycelia, such as those belonging to the families *Cellulomonadaceae*, *Microbacteriaceae* and *Micrococcaceae*, were also isolated. It can, therefore, be concluded that actinomycetes from Mongolian biomes represent a potentially rich source of novel metabolites for biodiscovery.

It can be seen from Table 3 that the number of the actinomycete genera found in Mongolian biomes was comparable with those found in other Asian countries,

**Table 2.** Taxonomic diversity of actinomycetes isolated from Mongolian habitats*.

| Class | Order | Family and number of genera (in brackets) |
|---|---|---|
| *Actinomycetia* | *Actinocatenisporales* | *Actinocatenisporaceae* (1) |
| | *Bifidobacteriales* | *Bifidobacteriaceae* (1) |
| | *Cellulomonadales* | *Cellulomonadaceae* (1), *Actinotaleaceae* (1), *Jonesiaceae* (2), *Oerskoviaceae* (2), *Promicromonosporaceae* (4) |
| | *Cryptosporangiales* | *Cryptosporangiaceae* (1) |
| | *Dermabacterales* | *Dermabacteraceae* (1) |
| | *Dermatophilales* | *Dermacoccaceae* (1), *Intrasporangiaceae* (2) |
| | *Geodermatophilales* | *Geodermatophilaceae* (1) |
| | *Glycomycetales* | *Glycomycetaceae* (2) |
| | *Jatrophihabitantales* | *Jatrophihabitantaceae* (1) |
| | *Kineosporiales* | *Kineosporiaceae* (2) |
| | *Microbacteriales* | *Microbacteriaceae* (9) |
| | *Micrococcales* | *Micrococcaceae* (6) |
| | *Micromonosporales* | *Micromonosporaceae* (8) |
| | *Mycobacteriales* | *Corynebacteriaceae* (1), *Dietziaceae* (1), *Gordoniaceae* (1), *Mycobacteriaceae* (1), *Nocardiaceae* (2) |
| | *Propionibacteriales* | *Kribbellaceae* (1), *Nocardioidaceae* (2), *Propionibacteriaceae* (1) |
| | *Pseudonocardiales* | *Pseudonocardiaceae* (8) |
| | *Streptomycetales* | *Streptomycetaceae* (2) |
| | *Streptosporangiales* | *Streptosporangiaceae* (5), *Nocardiopsaceae* (1), *Thermomonosporaceae* (3) |
| *Nitriliruptoria* | *Nitriliruptorales* | *Nitriliruptoraceae* (1) |
| *Thermoleophilia* | *Solirubrobacterales* | *Patulibacteraceae* (1) |

* Classifications of Nouioui et al. (2018), Gupta (2019) and Salam et al. (2020).

notably, mega-diversity countries such as China, Indonesia and Malaysia. In contrast to Mongolia, these countries have fewer extreme biomes. It is interesting that in the arid lands of Asia, besides of Mongolia, the richest actinomycete diversity was found in the Taklamakan desert, China, as isolates were assigned to 55 validly published genera, one new genus and 19 new species (Liu et al. 2021).

The highest tally of actinomycete genera in the tropical countries has been found in Indonesia and Vietnam, namely 64 and 53 genera, respectively. Similarly, 59 genera of actinomycetes were found in the subtropical region of Japan. Muramatsu et al. (2003) noted that while similar numbers of actinomycete genera were found in a tropical area of Malaysia and in a temperate region of Japan there was little taxonomic overlap. This result underpins the view that actinomycete diversity of all countries is of interest, the more so as the same species isolated from separate

**Table 3.** Comparison of actinomycete genera found in Mongolia with those detected in subtropical and tropical regions of selected Asian countries*.

| Country | Sampling areas | Number | | | References |
|---------|----------------|--------|---|---|------------|
| | | Samples | Isolates identified | Genera | |
| Mongolia | High mountain forest, forest-steppe, steppe, semi-desert and desert regions (13 provinces) | 201 | 2448 | 77 | This chapter |
| China | Primeval subtropical ever-green broadleaf forest, primeval tropical rain forest, secondary ever-green broadleaf forest, primeval alpine taiga, primeval forest in Southwestern China (6 locations) | 815 | 1998 | 33 | Jiang et al. (2019) |
| Indonesia | Tropical rainforests (13 locations) | 903 | 3193 | 64 | Widyastuti and Ando (2010) |
| Japan | Rishiri Island, cool-temperate region | 74 | 668 | 40 | Hayakawa et al. (2010) |
| | Iriomoto Island, subtropical region | 95 | 566 | 59 | Hayakawa et al. (2010) |
| | Natural forests or forested areas, Okutama, Tokyo, temperate region | ? | 981 | 22 | Muramatsu et al. (2003) |
| Malaysia | Natural forests or forested areas, Selangor, tropical region | ? | 790 | 23 | Muramatsu et al. (2003) |
| Singapore | Tropical rainforests | 350 | 5000 | 36 | Wang et al. (1999) |
| Vietnam | Tropical and subtropical regions (5 locations) | 202 | 1882 | 53 | Hop et al. (2011) |

* Based on analyses of 16S rRNA gene sequences.

environments can display different metabolic properties (Jensen et al. 2007, Schlatter and Kinkel 2014).

# 4. Biotechnological Potential of Actinomycetes from Mongolian Habitats

## 4.1 Antimicrobial and Anti-quorum Sensing Activities of Endophytic Actinomycetes

The global spread of multi-drug resistant pathogens led the World Health Organization (WHO)-member countries to endorse the Global Action Plan on Antimicrobial Resistance (WHO 2015). This plan was a challenge for researchers to develop strategies to combat the resistance of pathogens to antibiotics. It had previously been shown that quorum sensing (QS) systems of pathogenic bacteria controlled the production of virulence factors, including biofilm formation (De Kievit and Iglewski 2000, Zhu et al. 2002). It was then hypothesized that compounds inhibiting QS might have the potential to treat bacterial infectious diseases (Molina et al. 2003, Otto 2004). Subsequently, it was shown that the use of antibiotics in combination with quorum sensing inhibitors (QSIs) increased the susceptibility of pathogenic bacteria to antibiotics (Brackman et al. 2011).

Recently, intensive searches for QSIs and new antimicrobials have targeted endophytic actinomycetes (Chankhamhaengdecha et al. 2013, Polkade et al. 2016, Singh and Dubey 2018, Chen et al. 2019). In Mongolia there are 3191 species, subspecies and varieties of vascular plants belonging to over 684 genera and 108 families (Urgamal et al. 2019) that represent a large "storehouse" for the isolation and screening of endophytic actinomycetes for antimicrobial and QSI activity. Oyunbileg et al. (2021a) isolated over 800 actinomycetes from 53 species of surface-sterilized plants collected in the Gobi Desert region and found that nearly 12% of them showed anti-quorum sensing activity and nearly 24% were active against *Micrococcus luteus* ATCC 9341 and *Chromobacterium violaceum* CV026. Strain 2556, isolated from *Astragalus variabilis*, inhibited the growth of both test-organisms and produced a phenolic complex. In contrast, strain 3050, isolated from *Oxytropis grubovii*, not only inhibited the growth of *Micrococcus luteus* ATCC 9341 but showed QSI activity and synthesized flavonoids.

Flavonoids and phenolic compounds synthesized by plants have been shown to have many useful attributes, including anti-inflammatory, anti-allergic, anti-artherogenic, antihypertensive, antithrombotic, antirheumatic, antioxidant, anticancer and antimicrobial properties (Taechowisan et al. 2014, 2017). These compounds were found to be synthesized by endophytic streptomycetes isolated from the roots of *Boesenbergia rotunda*, which grows in Thailand. Six flavonoids, isolated from *Streptomyces* sp. BT01, showed a high antimicrobial activity against Gram-positive bacteria (Taechowisan et al. 2014). Two biologically active biphenyls: 3'-hydroxy-5-methoxy-3,4-methylenedioxybiphenyl and 3'-hydroxy-5,5'-dimethoxy-3,4-methylenedioxybiphenyl, isolated from the culture medium of *Streptomyces* sp. BO-07, were not only active against Gram-positive bacteria but exhibited antioxidant and anticancer properties (Taechowisan et al. 2017). Consequently, endophytic actinomycetes isolated from Mongolian plants are relevant in the search for antimicrobial and QSI activities.

### 4.2 Antimicrobial Activities of Mongolian Soil Actinomycetes

The antimicrobial activities of 1030 actinomycetes isolated from forest-steppe, steppe, semi-desert, true desert and hyper-arid desert zones and subzones of Mongolia were screened against five test-organisms using the agar diffusion method. It can be seen from Fig. 1 that actinomycetes from each of these areas showed a specific pattern of antimicrobial activity. Strains isolated from the forest-steppe and semi-desert zones and from the hyper-arid subzone were particularly active against *S. aureus* whereas those from the steppe zone were especially active against *B. subtilis*. Similarly, most of the isolates showing activity against *E. coli* were from the forest-steppe area and the true-desert subzone. In contrast, none of strains from

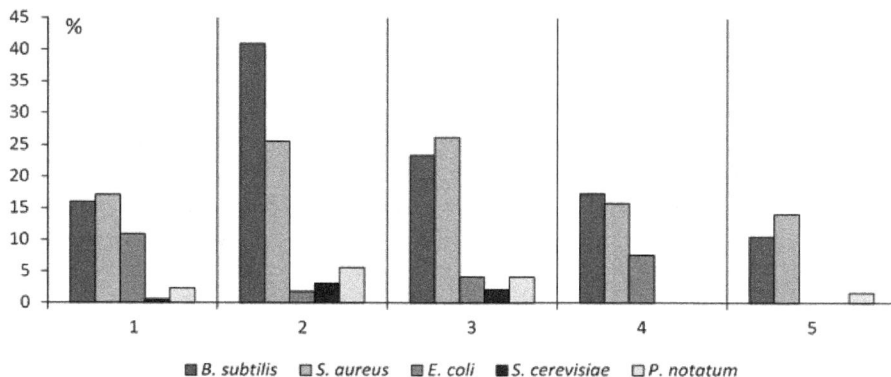

Figure 1. Percentage of actinomycetes isolated from soils collected from different natural zones of Mongolia showing antimicrobial activity.
1-forest-steppe, 2-steppe, 3-semi-desert, 4-true desert, 5-hyper-arid desert.

the true desert subzone were active against *S. cerevisiae* or *P. notatum* and none of those from the hyper-arid subzone inhibited either *E. coli* or *S. cerevisiae*.

Twenty-three strains of desert actinomycetes cultivated in a liquid medium containing soybean meal were examined for their ability to inhibit the growth of 17 test-organisms including nine strains of bacteria, four strains of yeasts and four strains of microscopic fungi (Tsetseg 2010). Antimicrobial activity was tested in methanol extracts of mycelia and filtrates of culture fluids. Most strains were active against microscopic fungi (52.2%), the corresponding percentages for the Gram-positive bacteria, *E. coli* and yeasts were 34.8%, 4.4% and 8.7%, respectively. Amongst them strain 605 that inhibited the growth of *Candida albicans* 562, *Candida parapsilosis*, *Candida utilis*, *Torula utilis* and the plant pathogenic fungi *Botrytis cinerea* and *Alternaria tenuis*, but not that of *Penicillium chrysogenum*, this strain was identified as *Streptomyces levoris* and found to produce an aromatic polyene antibiotic (Tsetseg 2010). It is evident from the studies outlined above that the expression and spectrum of antimicrobial activities depend on cultivation methods, media composition and the test-organisms used. Moreover, very often strains that do not show antimicrobial activity when using the agar diffusion method are found to have genomes rich in biosynthetic gene clusters predicted to encode for novel specialized metabolites.

## 4.3 Whole Genome Mining of Actinomycetes Isolated from Mongolia

Recent advances in molecular biology and bioinformatics have made it possible to estimate the biosynthetic potential of organisms by mining their whole-genome sequences. These developments helped to overcome shortcomings of conventional natural product discovery programs that often left 80–90% of microbial biosynthetic potential undiscovered (Nett et al. 2009). Baltz (2017) emphasized that preference in biodiscovery campaigns should be given to "gifted" microbes with the largest genomes as they have the greatest potential to synthesize novel secondary metabolites which are also known as specialized metabolites. Indeed, he considered that actinomycetes with genome sizes above 8 Mb and with more than 30 biosynthetic gene clusters to be "highly gifted microbes", and because of this, they should retain their position as the most prolific source of bioactive secondary metabolites. Several researchers (Nett et al. 2009, Sayed et al. 2019) also noted that actinomycetes with moderate genome sizes of 5.0–7.9 Mb and 19–20 bioclusters had the ability to produce novel specialized metabolites.

Until recently, members of the genera *Luteipulveratus* and *Herbidospora* were not included in antimicrobial screens designed to find new secondary metabolites. However, whole genome sequencing of Mongolian isolate *L. mongoliensis* NBRC 105296[T] (Ara et al. 2010a) revealed that the strain had the largest genome among species assigned to this genus (Juboi et al. 2015). The genome size of the *L. mongoliensis* strain, which was isolated from an arid Mongolian soil, was 20% larger than that of *Luteipulveratus halotolerans* C296001[T], an isolate from a tropical rain forest soil in Sarawak, Malaysia (Juboi et al. 2015). The genome sizes of the *Luteipulveratus* strains were 5.4 and 4.5 Mb, respectively, the corresponding numbers of BGCs were 30 and 23. The genome of the *L. mongoliensis* strain contained bioclusters predicted to encode for a lantipeptide, a type II polyketide (II-PKS), a terpene, two siderophores, ectoine and two bacteriocins. However, the genomes of neither of *Luteipulveratus* strains contained biosynthetic gene clusters encoding for non-ribosomal peptide synthetases (NRPS) or type 1 polyketide synthases (I-PKS) (Juboi et al. 2015).

When the genome of *Herbidospora mongoliensis* NBRC 105882[T] was compared with those of the other type strains of *Herbidospora* species, it was found to have the largest genome (9.1 Mb) (Table 4). The genomes of the *Herbidospora* strains contained 15 to 18 biosynthetic gene clusters predicted to encode for 32 NRPS, PKS/NRPS hybrids and type-I PKS metabolites. Interestingly, 15 of the bioclusters were strain specific. Similarly, the genome of the *H. mongoliensis* strain contained 17 gene clusters, four of which were strain-specific, namely a PKS/NRPS gene cluster and three specific PKS gene clusters (Komaki et al. 2015). It is evident from these data that representative *Herbidospora* strains have the genetic capacity to produce known and unknown secondary metabolites and hence can be considered as gifted *sensu* Baltz (2017).

**Table 4.** Genome sizes and biosynthetic potential of the type strains of *Herbidospora** species.

| Species | Country and source of isolates** | Genome size (Mb) | Number of | | |
|---|---|---|---|---|---|
| | | | NRPS clusters | PKS/NRPS hybrids | PKS-I clusters |
| *H. cretaceae* NBRC 15474[T] | Japan, soil | 8.3 | 9 | 1 | 5 |
| *H. daliensis* NBRC 106372[T] | Taiwan, sediment | 8.5 | 10 | 1 | 4 |
| *H. yilanensis* NBRC 106371[T] | Taiwan, sediment | 7.9 | 9 | 2 | 4 |
| *H. mongoliensis* NBRC 105882[T] | Mongolia, soil | **9.1** | 9 | 2 | 6 |
| *H. sakaeratensis* NBRC 102641[T] | Thailand, soil | 8.4 | 12 | 2 | 4 |

* Modified from Komaki et al. (2015). **Data from Kudo et al. (1993), Tseng et al. (2010), Boondaeng et al. (2011) and Ara et al. (2012a).

It is, however, important to understand that the success of genome mining studies depends on the quality of the whole genome sequences of the strains under study (Baltz 2021) and the completeness of natural product databases. As stated by Hoskisson and Seipke (2020), high-throughput chemical analyses should "catch up with genomics".

### 4.4  Biosynthetic Potential of Actinomycetes of the Genus Actinoplanes Isolated from Mongolian Soils

The versatile biosynthetic potential of actinomycetes is a valuable source of novel secondary metabolites for agriculture and medicine. More than 50% of secondary metabolites are synthesized by assembling acetate building blocks into polyketides using PKSs, and by assembling amino acids to peptides using NRPSs (Nett et al. 2009, Fisch 2013). The clinically important antibiotic teicoplanin from *Actinoplanes teichomyceticus* (Somma et al. 1984) and friulimicin from *Actinoplanes friuliensis* (Aretz et al. 2000) are, for example, the products of NRPS gene clusters. Indeed, *Actinoplanes* strains are a prolific source of specialized metabolites including antibiotics (Lazzarini et al. 2000), glucose isomerases (Gong et al. 1980) and the alpha-glucosidase inhibitor acarbose, a potent drug used worldwide in the treatment of type 2 diabetes (Schwientek et al. 2012).

Members of the genus *Actinoplanes* are widely distributed across Mongolian habitats (Tsetseg et al. 2016a). In studies designed to evaluate the biosynthetic potential of 23 *Actinoplanes* strains isolated from Mongolian habitats it was found that they had the capacity to produce NRPS (74%), type-I PKS (74%) and type-II PKS (61%) compounds (Enkh-Amgalan et al. 2012) while 65% of them showed antimicrobial activity. Bioclusters predicted to express for all three metabolites were found in 11 strains, for PKS-I and NRPS moieties in six strains while coding for PKS-II and NRPS compounds were present in three strains. Five out of eight strains which gave negative results in the antimicrobial screens were shown to contain NRPS and/ or PKS genes whereas only NRPS encoding genes were found in three of the strains. Diverse PKS-I and NRPS encoding genes were found amongst the isolates though this was not the case with PKS-II genes. Interestingly, most of the strains contained an unusual PKS-I gene, PKS-NRPS hybrid genes and iterative genes that did not correspond to any known secondary metabolite. These data show that the genomes of *Actinoplanes* strains isolated from Mongolian habitats have the capacity to produce novel secondary metabolites. Further details can be found in the publication by Enkh-Amgalan et al. (2012). In general, good correlation was found between the results of the antimicrobial assays and the ability of isolates to produce secondary metabolites.

### 4.5  Enzymes Production

Microbial diversity continues to be a valuable resource in the search for enzymes needed in various branches of industry, agriculture, and medicine (Adrio and Demain 2014, Mukhtar et al. 2017). They have also found applications in the treatment of cardiovascular diseases and cancer (Broome 1981, Peng et al. 2005). According to the WHO, 17 million people die from cardiovascular diseases every year, a figure that is expected to rise to 23.3 million by 2030 (Mathers and Loncar 2006, Raju and Divakar 2014). *Nocardia* and *Streptomyces* strains have been found to produce proteases that can lyse fibrin in blood clots which cause cardiovascular disease (Landau and Egorov 1971, 1996, Peng et al. 2005, Ju et al. 2012, Silva et al. 2015, Tsetseg et al. 2016b). Microbial L-asparaginase is used to treat cancers, such as acute lymphoblastic leukemia and lymphosarcoma (Broome 1981, Verma et al. 2007). This enzyme is produced by actinomycetes belonging to the genera *Amycolatopsis*, *Corynebacterium*, *Mycobacterium*, *Nocardia*, *Pseudonocardia* and *Streptomyces* (Khamna et al. 2009, Deepthi and Devamma 2012, Varalakshmi 2013, Shrivastava et al. 2016). Khamna et al. (2009) reported that 6.7% out of 445 actinomycetes isolated from the rhizosphere of Thai medicinal plants showed such activity.

Nineteen actinomycetes isolated from Mongolian soils were screened for the production of L-asparaginase and fibrinolytic proteases. Nine of these strains belonged to the genus *Amycolatopsis*, five to the genus *Streptomyces*, three to the genus *Nocardia* and single strains to the genera *Actinomadura* and *Glycomyces*. L-asparaginase production was determined by a rapid plate assay using asparagine-dextrose salts agar (ADS) supplemented with phenol red (0.009% final concentration) whereas fibrinolytic activity was estimated by measuring the diameter of clear zones formed on plasminogen-free fibrin plates after incubation at 37°C for 6, 12 and 24 hours. After 48 hours, L-asparaginase activity was detected in 14 isolates, including eight *Amycolatopsis* strains, all of the *Streptomyces* strains and in the single *Glycomyces* strain. In contrast, none of the *Actinomadura* or *Nocardia* strains produced L-asparaginase. In turn, fibrinolytic activity was detected in an *Actinomadura* strain after 6 hours incubation. However, after 24 hours nine strains gave positive results, including three *Amycolatopsis*, three *Streptomyces* and single strains belonging to the genera *Actinomadura*, *Glycomyces* and *Nocardia*. Low fibrinolytic activity was shown by the *Amycolatopsis* strains (clear zones were between 7–12.5 mm). The *Actinomadura* strain that did not produce L-asparaginase showed the highest fibrinolytic activity, namely a zone of 17 mm after 24 hours. These results are encouraging as 73.7% of the strains produced L-asparaginase and 47.4% exhibited fibrinolytic activity. Furthermore, nearly half of 108 actinomycetes isolated from forest-steppe soils produced L-asparaginase (Daram and Tsetseg 2014).

Just over half of the *Kribbella* strains isolated from Mongolian soils produced amylases, 4.5% proteases and 2.3% - cellulases. Strains closely related to *Kribbella antibiotica*, *Kribbella ginsengisoli*, *Kribbella italica* and *Kribbella koreensis* showed the highest amylolytic activity and a strain closely related to *Kribbella qitaiheensis* had the highest proteolytic activity (Tsetseg and Badamgarav 2021). These initial results are encouraging as they show that actinomycetes from Mongolian soils have an ability to produce enzymes of agricultural and medical value.

### 4.6 Amino Acid Production

The production of amino acids is second only to that of antibiotics (excluding ethanol) in the vast biotechnological fermentation industry, which had a global market of US$ 4.5 billion in 2004 (Leuchtenberger et al. 2005). Amino acids are widely used as animal feed additives, flavor enhancers and sweeteners in the food industry, in cosmetics and in medicine and healthcare (Leuchtenberger et al. 2005, Demain 2007, Ivanov et al. 2013). Since the discovery of *Corynebacterium glutamicum* (previously *Micrococcus glutamicus*) as a source of glutamic acid (Kinoshita et al. 1957) actinomycetes have been the most important amino acid-producing microorganisms.

In a study of 384 actinomycetes isolated from 13 different soil types from desert, steppe, and forest-steppe zones of Mongolia 73.8% of the desert, 78.5% of the steppe and 34.4% of the forest-steppe actinomycetes produced free extracellular amino acids (Tsetseg 2012). The free amino acid-producing actinomycetes were the most abundant in the gray-brown saline and hyper-arid sandy desert soils. In India, 35–77% of the actinomycetes studied excreted 1–2 amino acids (Krishna et al. 1971, cited by Pavlovicha 1978) whereas in the Ukraine the tested actinomycetes produced 3–6 amino acids, amongst which alanine and glutamic acid prevailed (Andreyuk et al. 1968). In contrast, the actinomycetes from Mongolian habitats synthesized 8–12 free amino acids. Most of these strains excreted alanine (86.7–94.1%), aspartic acid (94.1–100%), glycine (80–82.4%) with 46.7–76.5% and 40–64.7% of them synthesizing glutamic acid and lysine, respectively. They also excreted essential amino acids such as histidine, methionine and valine. These data show that Mongolian soil actinomycetes are a promising source for amino acids of potential value.

Nomadic cattle breeding is a main supplier of raw materials and food for Mongolians. However, 70% of pastures where domestic animals feed and graze have been degraded to varying levels (Batjargal and Enkhbat 1998). Consequently, restoration of pastures has become one of the most important issues in the economic development of Mongolia. In this respect, it is important to determine whether small amounts of bioactive compounds, such as amino acids released by actinomycetes into rhizospheres of plants and soils can enhance the yield of pasture plants. Krasilnikov (1958) discovered that alanine and phenylethylamine (product of decarboxylation of phenylalanine) in small concentrations increased the growth of peas while Näsholm et al. (2000) found that plants could absorb amino acids through their roots.

## 4.7 Biological Control Agents

Microbial pathogens that cause plant diseases are a threat to global food security. Each year an estimated 10–16% of the global harvest is lost to plant diseases (Strange and Scott 2005, Oerke 2006). In financial terms, losses caused by phytopathogens cost US$220 billion (Chakraborty and Newton 2011). Measures to control plant pathogens are now concentrated on discovery of microbial biological control agents as they offer an ecologically friendly alternative to the use of hazardous chemicals.

After wheat, potatoes are the most important food crop in Mongolia. Itgel (1995) found that in Mongolia 11.5–35% of the potato yield was infected with common scab, a disease that is known to be widespread in dry conditions (Liu et al. 2004). The spread of potato scab in the country is exacerbated by climate change, notably increased temperature, reduced precipitation and desertification. In Mongolia, *Streptomyces scabiei* (Itgel 1995) and *Streptomyces turgidiscabies*, (Tsetseg and Nyamsuren 2013) have been identified as potato scab pathogens.

Worldwide several *Streptomyces* species are known to cause potato scab, namely *Streptomyces acidiscabies* (Lambert and Loria 1989a), *Streptomyces europaeiscabiei*, *Streptomyces stelliscabiei* and *Streptomyces reticuliscabiei* (Bouchek-Mechiche et al. 2000), *Streptomyces luridiscabiei*, *Streptomyces puniciscabiei* and *Streptomyces niveiscabiei* (Park et al. 2003), as well as *Streptomyces scabiei* (Lambert and Loria 1989b) and *S. turgidiscabies* (Miyajima et al. 1998).

It is clearly important that measures are taken to control potato scab in Mongolia. Tsetseg et al. (2008) found that 66 out of 78 actinomycetes (84.6%) isolated from potato fields in the forest-steppe zone suppressed the growth of *S. turgidiscabies*. Furthermore, 20 out of 49 actinomycetes (40%) isolated from the semi-desert zone inhibited the growth of *S. scabiei* and *S. turgidiscabies*. Amongst them, 17 streptomycetes showed activity against *S. turgidiscabies* and 14 – against *S. scabiei* while a *Cryptosporangium* isolate suppressed the growth of *S. scabiei*, and a *Nocardia* strain was shown to be antagonistic to *S. scabiei* and *S. turgidiscabies*. Finally, when 19 endophytic streptomycetes isolated from leaves of *Hordeum vulgare* L. (barley) were screened against fungal pathogens (Norovsuren and Filippova 2020) around 63% of them inhibited the growth of *Aspergillus niger* and 75% that of *Fusarium* sp.

### *4.8 Plant Growth Promotion Activity*

Dry climate, frequent droughts, a short vegetation season and low soil fertility significantly decrease crop productivity in Mongolia. There is, therefore, a need to find indigenous microorganisms that can address this problem. Actinomycetes are known to have an important role in the promotion of plant growth as they produce a range of bioactive compounds, such as amino acids, antibiotics, auxins, phytohormones, toxins and vitamins (Krasilnikov 1958, Yadav and Yadav 2019, Orozco-Mosqueda et al. 2021). It has, for instance, been shown that *Arthrobacter arilaitensis* and *Streptomyces pseudovenezuelae* are effective as biofertilizers as they have been shown to increase maize biomass under drought stress due to their ability to secrete high levels of indole-acetic acid (IAA) and 1-aminocyclopropane-1-carboxylic acid (ACC) (Chukwuneme et al. 2020).

In a pilot experiment, nine actinomycetes isolated from potato fields in Mongolia were tested for antimicrobial activity against *S. turgidiscabies* and for their ability to promote the growth of wheat seeds (Table 5). Six strains stimulated root growth and three – shoot growth. Amongst them, strains 5 and 32 showed the highest root growth promotion, namely 29.7% and 37.7%, respectively, whereas strain 59 enhanced shoot growth by 23.3% compared with the control. It can be seen from Table 5 that plant growth promotion did not correlate with the diameters of inhibition zones. It is also interesting that the same strains not only promoted plant growth but also inhibited that of *S. turgidiscabies*.

It has also been shown that a strain of *Glycomyces lechevalierae* isolated from potato scab lesions stimulated the growth of radish seedlings by 30%, and that a *Streptomyces* strain, which showed 99.9% 16S rRNA gene sequence similarity to *Streptomyces canus* and *Streptomyces ciscaucasicus*, stimulated the growth of radish seedlings by 62% (Tsetseg and Nyamsuren 2013). It seems likely that actinomycetes

**Table 5.** Effect of actinomycetes showing antagonistic activity on enhancing the germination of wheat seeds.

| Isolate No. | Antagonistic activity (diameter of inhibition zone, mm) | | Effect on wheat seeds (%) | |
|---|---|---|---|---|
| | *S. turgidiscabies* P11 | *S. turgidiscabies* A33 | Root | Sprout |
| 1 | 20 | 25 | +14.8 | −5.5 |
| 5 | 24 | 27 | **+29.7** | +7.6 |
| 8 | 27 | 26 | +0.7 | −19.4 |
| 19 | 21 | **31** | +8.8 | −2.8 |
| 32 | **31** | 29 | **+37.7** | +2.3 |
| 37 | 29 | 25 | −11.6 | −12.3 |
| 41 | **28** | **31** | +3.5 | −13.8 |
| 59 | **28** | 27 | −7.2 | **+23.3** |
| 74 | 20 | **28** | +5.5 | −23.3 |

isolated from Mongolian soils and plants will prove to be of practical importance in enhancing the growth of agricultural plants in Mongolia.

## 4.9 Heavy Metal Resistance

Mining is important in Mongolia given large deposits of coal, copper, gold, iron, molybdenum and uranium (Bulgamaa 2010). At present, the mining industry supplies 15–20% of the country's GDP and accounts for about 50% of all exports. The Government actively encourages mineral exploration and extraction and seeks to attract new investment to develop mining technologies and modern mine management practices. Consequently, mining is expected to expand and support economic growth through metal exports. However, the mining industry is also a source of significant negative impacts on environmental sustainability, notably pollution by hazardous heavy metals, which may also cause risks to human health. Heavy metal pollution can be mitigated through bioremediation processes, including ones that involve actinomycetes. In Argentina, Iran and Morocco heavy metal resistance was found in actinomycetes belonging to the genera *Amycolatopsis*, *Nonomuraea*, *Promicromonospora*, *Saccharothrix*, *Streptomyces* and *Streptosporangium* (Costa et al. 2012, El Baz et al. 2015, Hamedi et al. 2015). It has been shown that changes in the metabolism of *Amycolatopsis tukumanensis* were responsible for its resistance to copper, these adaptations comprise what has been described as the copper resistome (Costa et al. 2012).

Oyunbileg and Tsetseg (2002) tested 81 actinomycetes isolated from soils of the Erdenet copper mine, the Zaamar gold mine and the Shariin Gol coal mine areas for their ability to grow in the presence of various concentrations of Cd, Co, Cu, Ni and Pb. They found that 46.9% of the isolates were resistant to Ni, 44.4% to Cu, 37.0% to PB, 13.5% to Co and 6.2% to Cd in concentrations of 10 µM. Six of the strains were resistant to 15 µM of these metals. These results indicate that actinomycetes indigenous to Mongolia may prove to be effective agents of bioremediation with respect to heavy metals.

In general, it can be concluded that extreme biomes found in Mongolia are a rich source of actinomycete diversity which can be developed for agricultural, environmental and medicinal biotechnology. Less attention has been given to the extent of actinomycete diversity in saline soils and lakes even though the areas concerned cover territory four times the size of Switzerland. The exploration of actinomycete diversity in saline environments of the steppe and semi-desert regions from a biotechnological perspective are considered below.

## 5.  Diversity and Biotechnological Potential of Actinomycetes in Saline and Adjacent Soils in Arid Lands of Mongolia

Characterization of actinomycete diversity in saline and adjacent soils was based on the isolation of strains from samples collected in the steppe zone (Dornod province) and the semi-desert (desert-steppe) zone (Uvs province, which is situated at the edge of the northern part of the semi-desert zone in the Great Lakes Basin). These sampling locations are in the western and eastern parts of Mongolia approximately 2,000 km from each other. Geographically, the sampling areas are in the Eastern Mongolian steppe and the Gobi Desert regions. These areas are rich in saline lakes and saline soils. Indeed, in these regions soil salinization accounts for 30.5% of the combined territories (Egorov 1993).

There are about 1500 permanent and ephemeral lakes with salinities up to 320 g/l in Dornod province (Egorov 1993). Sampling there was carried out in dry steppe far from saline lakes, and from muds and shores of four ephemeral saline lakes: Ikh Khotont Lake, Khulstain Lake, Galuut Tsagaan Lake and an unnamed saline lake. Samples from Uvs province were collected from the shores and adjacent areas near the largest in Mongolia saline lakes, such as Lake Uvs and Lake Khyargas, which have salinity values around of 18.8 g/l and 8.5 g/l, respectively (Sevastyanov et al. 2014). In addition, two samples were collected from Lake Khokh Ereg and one from the unnamed saline lake. In total, ten samples were collected from the sampling sites in each province. Samples composed of salt or mud and soils with visible salt incrustations were considered to be saline, their pH varied in the steppe zone between 9.3 and 10.3 and in the semi-desert zone between 8.0 and 9.4. The corresponding pH of the adjacent soils were 6.6–9.3 and 6.5–9.1.

### *5.1  Diversity of Actinomycetes Isolated from the Saline and Adjacent Soils*

The 65 actinomycetes isolated from the steppe samples consisted of 17 strains from the saline and 48 from the adjacent soils. Similarly, the strains isolated from the semi-desert samples consisted of 33 from the saline samples and 81 from the adjacent soils. Taxonomically the steppe actinomycetes were assigned to nine genera classified into eight families and seven orders whereas the semi- desert isolates belonged to 19 genera classified into 12 families and eight orders. The total taxonomic diversity is shown in Table 6.

The 16S rRNA gene sequence similarities of the isolates when compared to the type strains of related validly described species varied between 96.0–100%. Isolates

**Table 6.** Taxonomic diversity and numbers of actinomycetes isolated from the steppe and semi-desert soils in Mongolia*.

| Order | Family | Genus | Similarity (%) | Steppe | | Semi-desert | |
|---|---|---|---|---|---|---|---|
| | | | | S | A | S | A |
| *Cellulomonadales* | *Cellulomonadaceae* | *Cellulomonas* | **98.3** | | | 1 | |
| | *Actinotaleaceae* | *Actinotalea* | **98.3–98.8** | | | 6 | |
| | *Oerskoviaceae* | *Oerskovia* | 99.9 | | | | 1 |
| | | *Sediminihabitans* | **98.3** | | | 3 | 1 |
| | *Promicromonosporaceae* | *Isoptericola* | **98.8–99.5** | 2 | | 1 | |
| *Mycobacteriales* | *Dietziaceae* | *Dietzia* | 99.9 | | | 2 | |
| | *Nocardiaceae* | *Nocardia* | 99.4–99.7 | | 1 | 1 | |
| | | *Rhodococcus* | **96.0**–100 | | | 1 | 12 |
| *Microbacteriales* | *Microbacteriaceae* | *Microbacterium* | **98.6**–100 | | | | 6 |
| | | *Agrococcus* | 100 | | | 1 | |
| *Micrococcales* | *Micrococcaceae* | *Micrococcus* | 99.9 | | | 1 | |
| | | *Arthrobacter* | 99.9 | | | | 1 |
| | | *Nesterenkonia* | 99.9 | 1 | | | |
| | | *Pseudarthrobacter* | 99.6 | | | 1 | |
| *Micromonosporales* | *Micromonosporaceae* | *Micromonospora* | 99.6–99.8 | 1 | 2 | | 1 |
| | | *Actinoplanes* | 99.4–99.5 | | 2 | 1 | |
| *Pseudonocardiales* | *Pseudonocardiaceae* | *Saccharopolyspora* | 99.6 | 1 | | | |
| *Streptomycetales* | *Streptomycetaceae* | *Streptomyces* | **98.8**–100 | 4 | 42 | 11 | 52 |
| | | *Kitasatospora* | 99.7–99.9 | | | | 7 |
| *Streptosporangiales* | *Streptosporangiaceae* | *Nonomuraea* | **98.8**–100 | | 1 | 1 | |
| | *Nocardiopsaceae* | *Nocardiopsis* | 99.7–100 | 8 | | 2 | |

\* Based on classifications of Nouioui et al. (2018), Gupta (2019) and Salam et al. (2020) (S – saline soils, A –adjacent soils; in bold – potentially novel taxa).

showing > 99.0% sequence similarity with their nearest neighbours were considered to be potentially novel species, as recommended by Meier-Kolthoff et al. (2013). Table 6 shows some novel *Actinotalea, Cellulomonas, Isoptericola, Microbacterium, Nonomuraea, Rhodococcus, Sediminihabitans* and *Streptomyces* species.

Isolates from the steppe zone were assigned to the genera *Actinoplanes, Isoptericola, Micromonospora, Nesterenkonia, Nocardia, Nocardiopsis, Nonomuraea, Saccharopolyspora* and *Streptomyces*, six of these genera were isolated from saline soils and five from the adjacent soils (Table 6). Interestingly, the genera, *Isoptericola, Nesterenkonia, Nocardiopsis* and *Saccharopolyspora* were only isolated from the saline soils whereas those assigned to the genera *Actinoplanes, Nocardia,* and *Nonomuraea* were restricted to the adjacent soils. In contrast, *Micromonospora* and *Streptomyces* strains were recovered from both environments. Nearly 90% of the isolates from the adjacent soils were streptomycetes, the corresponding number from the saline soils was just short of 24%. The dominant taxon in the saline soils was *Nocardiopsis*, representatives of which formed around 50% of the isolates.

The actinomycete diversity in the soils from the semi-desert zone was greater than that for the samples collected from the steppe zone, as shown in Table 6. Isolates from the saline soils were assigned to 14 genera, the corresponding number from the adjacent soils was eight. The most common isolates were streptomycetes, they accounted for 33% and 64% of strains recovered from the saline and adjacent soils, respectively. Relatively large numbers of *Actinotalea* strains were found in the saline soils (18.2%) whereas in the adjacent soils the second highest number of isolates belonged to the genus *Rhodococcus* (14.8%). Representatives of just two the non-streptomycete genera *Rhodococcus* and *Sediminihabitans* were isolated from both environments.

## 5.2 Biotechnologically Important Genera of Nocardiopsis and Rhodococcus Isolated from the Saline and Adjacent Soils

It is especially interesting that taxonomically diverse *Nocardiopsis* and *Rhodococcus* strains were isolated from the steppe and semi-desert zones given the biotechnological importance of members of these taxa (Bennur et al. 2014, 2015, Ibrahim et al. 2018, Cappelletti et al. 2020). In general, *Nocardiopsis* strains isolated from alkaline and saline soils (He et al. 2015) are mainly haloalkaliphilic. These organisms are increasingly seen as a rich source of specialized metabolites, including new antibiotics and alkaline thermostable enzymes (Bennur et al. 2014, 2015). Rhodococci are also considered to be a source of new antibiotics (Kitagawa and Tamura 2008, Doroghazi and Metcalf 2013, Ceniceros et al. 2017), but are better known as agents of bioremediation and for the biocatalytic production of high value chemicals from low value substrates (Zhou et al. 2005, Kim et al. 2018). The application of phylogenomic methods showed that rhodococci could be assigned to several well-defined species-groups (Sangal et al. 2016, 2019) while the genera *Aldersonia* (Nouioui et al. 2018) and *Spelaeibacter* (Kim et al. 2022) were introduced for species previously classified in the genus *Rhodococcus*.

As for the Mongolian isolates, the eight *Nocardiopsis* strains recovered from the steppe saline soils and included in the 16S rRNA gene sequence analyses were most closely related to *Nocardiopsis exhalans* (5 strains, 99.93%) and *Nocardiopsis valliformis* (3 strains, 99.86%). These isolates were markedly different from the two *Nocardiopsis* isolates from the semi-desert soils which were most closely related to *Nocardiopsis dassonvillei* subsp. *dassonvillei* (100%) and *Nocardiopsis terrae* (99.66%). All the *Nocardiopsis* strains isolated from the hypersaline soils (pH 9.1–10.3) were considered to be moderate halophiles, apart from *N. dassonvillei* subsp. *dassonvillei* MN07-A0388, which was halotolerant (Ara et al. 2013). One of the moderate halophylic strains grew in the presence of 15% NaCl, three in the presence of 12% NaCl and six strains at a NaCl concentration of 10%. All the isolates grew from pH 7–9. Seven of the isolates were found to grow from 10°–37°C and the remaining ones from 10°–45°C (Ara et al. 2013). It seems likely that some of these isolates will prove to be novel given that the type-strain *N. valliformis* was found to be a new species, despite sharing the 16S rRNA gene sequence similarity of 99.93% with *N. exhalans* ES10.1ᵀ (Yang et al. 2008). Consequently, the Mongolian isolates should be studied further to determine whether they contain BGCs predicted to express for novel metabolites.

All but one of the 13 *Rhodococcus* strains isolated from the semi-desert soils were from the samples taken next to the saline soils, the exception was from a saline soil (Table 6). Three of these isolates were most closely related to *Rhodococcus cerastii* (99.5–100% similarity), another three to *Rhodococcus erythropolis* (96.0–99.4% similarity) and two each to *Rhodococcus fascians* (99.7–99.9% similarity), *Rhodococcus maanshanensis* (99.3% similarity) and *Rhodococcus yunnanensis* (99.7% similarity) and the remaining strain to *Rhodococcus kroppenstedtii* (99.5% similarity). It can be concluded that these rhodococci are taxonomically diverse as they fall into several of the species-groups defined by Sangal et al. (2016, 2019). It is also interesting that all of the rhodoccoci were isolated from the semi-desert soils, a result supporting the view that these organisms are resistant to drought and low nutrient levels (LeBlanc et al. 2008, Ceniceros et al. 2017). As for the *Nocardiopsis* strains, the putatively novel *Rhodococcus* isolates are worthy of additional study.

## 5.3 Antimicrobial Activities of Actinomycetes from the Saline and Adjacent Soils

Primary screening of dereplicated actinomycetes from extreme habitats against panels of microbial indicator strains is a key step in the discovery of new drugs needed to control the spread of multidrug resistant pathogens (Goodfellow and Fiedler 2010, Sayed et al. 2019). Considerable attention has been paid towards screening the gifted actinomycetes from extreme biomes (Jose and Jebakumar 2014, Sayed et al. 2019, Corral et al. 2020).

In the present study, 131 actinomycetes from the steppe saline and adjacent soils (16 and 48 strains, respectively) and from the corresponding semi-desert soils (17 and 50 strains, respectively) were examined for their ability to inhibit the growth of five test-organisms, namely *B. subtilis*, *S. aureus*, *E. coli*, *S. cerevisiae* and *A. awamori*, representing Gram-positive and Gram-negative bacteria, yeast and microscopic fungi, as shown in Fig. 2. One hundred of the isolates had been identified as streptomycetes and the others as non-streptomycete taxa, namely the genera *Actinoplanes* (3), *Isoptericola* (2), *Kitasatospora* (7), *Micromonospora* (3), *Nesterenkonia* (1), *Nocardia* (2), *Nocardiopsis* (10), *Nonomuraea* (2) and *Rhodococcus* (1). The results of these studies are shown in Fig. 2.

In total, 56% of the streptomycetes showed activity against at least one of the indicator organisms, the corresponding number for the non-streptomycete isolates was slightly higher at 58%. Further, 44% of the isolates inhibited the growth of the *B. subtilis* strain, the corresponding percentages for the *E. coli*, *S. aureus*, *S. cerevisiae* and *A. awamori* strains were 22%, 43%, 26% and 29%, respectively. Interestingly, the non-streptomycetes from the steppe saline soils only showed activity against the *A. awamori* and *S. cerevisiae* strains whereas those from the steppe adjacent soils

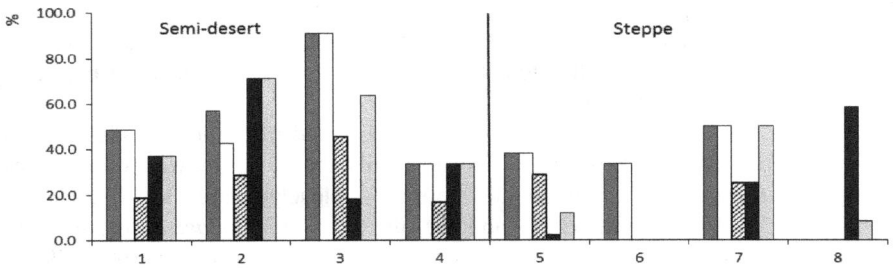

**Figure 2.** Antimicrobial activities of actinomycetes isolated from saline and adjacent soils of the steppe and semi-desert zones.

Percentage of strains inhibiting the growth of *B. subtilis* (dark grey column); *S. aureus* (clear column); *E. coli* (hatched column); *S. cerevisiae* (black column) and *A. awamori* (light grey column). 1-streptomycetes of the semi-desert adjacent soils; 2-non-streptomycetes from the semi-desert adjacent soils; 3-streptomycetes of the semi-desert saline soils; 4-non-streptomycetes from the semi-desert saline soils; 5-streptomycetes from the steppe adjacent soils; 6-non-streptomycetes from the steppe adjacent soils; 7-streptomycetes from the steppe saline soils; 8-non-streptomycetes from the steppe saline soils.

**Table 7.** Number of actinomycetes showing antimicrobial activity isolated from the saline and adjacent soils of the steppe and semi-desert zones.

| Zone | Steppe | | Semi-desert | |
|---|---|---|---|---|
| Location | Saline | Adjacent | Saline | Adjacent |
| Total studied | 16 | 48 | 17 | 50 |
| Active strains (%) | 62.5 | 43.8 | 76.5 | 60.0 |

were active only against the *B. subtilis* and *S. aureus* strains. It can be seen from Table 7, the saline soils were richer in actinomycetes with antimicrobial activity than the adjacent soils.

It is particularly interesting that all of the *Actinoplanes* strains, six out of the seven *Kitasatospora* strains and nine of the ten *Nocardiopsis* strains showed activity against one or more of the indicator organisms. The *Kitasatospora* strains shared high 16S rRNA gene sequence similarities (99.7–99.9%) with the type-strain of *K. albolonga* (formely *S. albolongus*). Strains bearing the latter name produce antibiotics, namely bafilomycins and proceomycin (Tsukiura et al. 1964, Yin et al. 2017). It is also interesting that some of the activity profiles were strain-specific even among isolates assigned to the same species.

It can be concluded from these initial studies that most of strains isolated from the steppe and semi-desert zones, notably from the saline soils, are bioactive, a result in good agreement with those from previous studies (Basilio et al. 2003, Jose and Jebakumar 2014, Corral et al. 2020). However, Sayed et al. (2019) reported that isolates from saline and hyper-saline soils had shown little bioactivity though this may be a function of the experimental procedures used. Indeed, Goodfellow and Fiedler (2010) noted, that activity reflected experimental conditions, such as the nutrient composition of production media.

# 6. Conclusions

It is evident from the original studies reported in this chapter that the diverse extreme biomes which are an integral part of the Mongolian countryside are a prolific source of gifted actinomycetes, notably taxonomically novel bioactive isolates of potential value for agricultural, industrial and medical biotechnology. The taxonomic diversity found in the various Mongolian biomes is in line with that reported for other extreme ecosystems, as exemplified by actinomycete diversity reported for Atacama Desert soils (Goodfellow et al. 2018). It is especially significant that extreme habitats in Mongolia are a prolific source of putatively novel strains belonging to rarely isolated taxa, including the genera *Actinotalea*, *Cellulomonas*, *Isoptericola* and *Microbacterium*, as well as to better known genera including *Micromonospora*, *Nocardiopsis*, *Rhodococcus* and *Streptomyces*. The strain library generated in the present study provides an invaluable resource for further studies, not least gene mining analyses designed to detect new BGCs of interest. In the broader context there is a need to discover links between taxonomic diversity, functional traits and the biotechnological potential of actinomycetes. Finally, it is important that the pristine extreme biomes of Mongolia are preserved so that future generations of microbiologists can build upon what has been achieved to date. Future studies should be extended to permafrost regions of Mongolia, especially as it is known that such habitats are a source of anticancer compounds (Silva et al. 2020).

# Acknowledgment

The author thanks the authorities of the Institute of Biology (IB-MAS), the Mongolian Academy of Sciences (MAS), Science and Technology Fund of Mongolia, and the authorities of the National Institute of Technology and Evaluation (NITE), Japan, for supporting much of the work reported in this chapter. I am very grateful to Dr. Katsuhiko Ando, the project leader from the Japanese side of the joint project, for his support. I am also indebted to members of the Laboratory of Microbiology (IB-MAS) and the Japanese members of the joint team for their contributions. I would like to thank Prof. Michael Goodfellow (Newcastle University, UK) who spent much time editing this chapter. Finally, I am grateful to Assoc. Prof. Ipek Kurtböke (University of the Sunshine Coast, Australia) for inviting me to prepare this chapter and for her encouragement and patience.

# References

Adrio, J.L. and Demain, A.L. (2014). Microbial enzymes: Tools for biotechnological processes. *Biomolecules*, 4: 117–139.

Al-shaibani, M.M., Mohamed, R.M.S., Sidik, N.M., Enshasy, H.A.E., Al-Gheethi, A., Noman, E. et al. (2021). Biodiversity of secondary metabolites compounds isolated from Phylum *Actinobacteria* and its therapeutic applications. *Molecules*, 26: 4504.

Andreyuk, E.I., Kogan, S.B. and Vladimirova, E.V. (1968). Biosynthesis of free amino acids by soil actinomycetes. *Agricultural Biology*, 2: 258–261 (in Russian).

Ara, I., Daram, D., Baljinova, T., Yamamura, H., Bakir, M.A., Suto, M. et al. (2013). Isolation, classification, phylogenetic analyses and scanning electron microscopy of halophilic, halotoletrant and alkaliphilic actinomycetes isolated from hypersaline soil. *African Journal of Microbiology Research*, 7(4): 298–308.

Ara, I., Tsetseg, B., Daram, D., Suto, M. and Ando, K. (2011a). *Pseudonocardia mongoliensis* sp. nov., and *Pseudonocardia khuvsgulensis* sp. nov., isolated from Mongolian soil. *International Journal of Systematic and Evolutionary Microbiology*, 61: 747–756.

Ara, I., Tsetseg, B., Daram, D., Suto, M. and Ando, K. (2011b). *Actinophytocola burenkhanensis* sp. nov., isolated from soil of Mongolia. *International Journal of Systematic and Evolutionary Microbiology*, 61: 1033–1038.

Ara, I., Tsetseg, B., Daram, D., Suto, M. and Ando, K. (2012a). *Herbidospora mongoliensis* sp. nov., isolated from soil, and reclassification of *Herbidospora osyris* and *Streptosporangoum claviforme* as synonyms of *Herbidospora cretacea*. *International Journal of Systematic and Evolutionary Microbiology*, 62: 2322–2329.

Ara, I., Tsetseg, B., Daram, D., Suto, M. and Ando, K. (2012b). *Cryptosporangium mongoliense* sp. nov., isolated from soil. *International Journal of Systematic and Evolutionary Microbiology*, 62: 2480–2484.

Ara, I., Yamamura, H., Tsetseg, B., Daram, D. and Ando, K. (2010a). *Luteipulveratus mongoliensis* gen. nov., sp. nov., an actinobacterial taxon in the family *Dermacoccaceae*. *International Journal of Systematic and Evolutionary Microbiology*, 60: 574–579.

Ara, I., Yamamura, H., Tsetseg, B., Daram, D. and Ando, K. (2010b). *Actinoplanes toevensis* sp. nov. and *Actinoplanes tereljensis* sp. nov., isolated from Mongolian soil. *International Journal of Systematic and Evolutionary Microbiology*, 60: 919–927.

Arasu, M.V., Esmail, G.A. and Al-Dhabi, N.A. (2016). Chapter 9. Hypersaline actinomycetes and their biological applications. pp. 229–245. *In*: Dharumadurai, D. and Jiang, Y. (eds.). *Actinobacteria - Basics and Biotechnological Applications*. Intech.

Aretz, W., Meiwes, J., Seibert, G., Vobis, G. and Wink, J. (2000). Friulimicins: Novel lipopeptide antibiotics with peptidoglycan synthesis inhibiting activity from *Actinoplanes friuliensis* sp. nov. I. Taxonomic studies of the producing microorganism and fermentation. *Journal of Antibiotics (Tokyo)*, 53: 807–815.

Baatar, B., Chiang, P.-W., Rogozin, D.Yu., Wu, Y.-T., Tseng, C.-H., Yang, C.-Y. et al. (2016). Bacterial communities in three saline meromictic lakes in Central Asia. *PLoS ONE*, 11(3): e0150847.

Baltz, R.H. (2017). Gifted microbes for genome mining and natural product discovery. *Journal of Industrial Microbiology and Biotechnology*, 44: 573–588.

Baltz, R.H. (2021). Genome mining for drug discovery: Progress at the front end. *Journal of Industrial Microbiology and Biotechnology*, 48: kuab044.

Barka, E.A., Vatsa, P., Sanchez, L., Gaveau-Vaillant, N., Jacquard, C., Klenk, H.-P. et al. (2016). Taxonomy, physiology and natural products of *Actinobacteria*. *Microbiology and Molecular Biology Reviews*, 80: 1–43.

Basilio, A., González, I., Vicente, M.F., Gorrochategui, J., Cabello, A., González, A. et al. (2003). Patterns of antimicrobial activities from soil actinomycetes isolated under different conditions of pH and salinity. *Journal of Applied Microbiology*, 95: 814–823.

Batima, P., Batnasan, N. and Lehner, B. (2004). Freshwater systems of the Great lakes basin, Mongolia: Opportunities and challenges in the face of climate change. *WWF Programme Office*, Ulaanbaatar, 95 pp.

Batjargal, Z. and Enkhbat, A. (eds.). (1998). Biological diversity in Mongolia. First national report. *Ministry for Nature and the Environment*. ADMON Printing House, Ulaanbaatar, 106 pp.

Bennur, T., Kumar, A.R., Zinjarde, S. and Javdekar, V. (2014). *Nocardiopsis* species as potential sources of diverse and novel extracellular enzymes. *Applied Microbiology and Biotechnology*, 98: 9173–9185.

Bennur, T., Kumar, A.R., Zinjarde, S. and Javdekar, V. (2015). *Nocardiopsis* species: A potential source of bioactive compounds. *Journal of Applied Microbiology*, 120: 1–16.

Bérdy, J. (2005). Bioactive microbial metabolites. *Journal of Antibiotics*, 58(1): 1–26.

Bespalov, N.D. (1951). Soils of the Mongolian People's Republic. *USSR's Academy of Sciences Publishing House*, Moscow, 318 pp (in Russian).

Boondaeng, A., Suriyachadkun, C., Ishida, Y., Tamura, T., Tokuyama, S. and Kitpreechavanich, V. (2011). *Herbidospora sakaeratensis* sp. nov., isolated from soil, and reclassification of *Streptosporangium claviforme* as a later synonym of *Herbidospora cretacea*. *International Journal of Systematic and Evolutionary Microbiology*, 61(Pt 4): 777–80.

Bouchek-Mechiche, K., Gardan, L., Normand, P. and Jouan, B. (2000). DNA relatedness among strains of *Streptomyces* pathogenic to potato in France: Description of three new species, *S. europaeiscabiei* sp. nov. and *S. stelliscabiei* sp. nov. associated with common scab, and *S. reticuliscabiei* sp. nov. associated with netted scab. *International Journal of Systematic and Evolutionary Microbiology*, 50: 91–99.

Brackman, G., Cos, P., Maes, L., Nelis, H.J. and Coenye, T. (2011). Quorum sensing inhibitors increase the susceptibility of bacterial biofilms to antibiotics *in vitro* and *in vivo*. *Antimicrobial Agents and Chemotherapy*, 55: 2655–2661.

Broome, J.D. (1981). L-Asparaginase: Discovery and development as a tumor-inhibitory agent. *Cancer Treatment Reviews*, 65: 111–114.

Bulgamaa, B. (2010). Mongolia distributes shares of mineral wealth to its citizens. *Mongolia Today*, 2(15): 4–7.

Bull, A.T. (2011). Actinobacteria of the extremobiosphere. pp. 1203–1240. *In*: Horikoshi, K. (ed.). *Extremophiles*, Handbook. Springer.

Bull, A.T., Ward, A.C. and Goodfellow, M. (2000). Search and discovery strategies for biotechnology: The paradigm shift. *Microbiology and Molecular Biology Reviews*, 64: 573–606.

Cappelletti, M., Presentato, A., Piacenza, E., Firrincieli, A., Turner, R.J. and Zannoni, D. (2020). Biotechnology of *Rhodococcus* for the production of valuable compounds. *Applied Microbiology and Biotechnology*, 104: 8567–8594.

Carro, L., Nouioui, I., Sangal, V., Meier-Kolthoff, J.P., Trujillo, M.E., Montero-Calasanz, M.C. et al. (2018). Genome-based classification of micromonosporae with a focus on their biotechnological and ecological potential. *Scientific Reports*, 8: 525.

Ceniceros, A., Dijkhuizen, L., Petrusma, M. and Medema, M.H. (2017). Genome-based exploration of the specialized metabolic capacities of the genus *Rhodococcus*. *BMC Genomics*, 18: 593.

Chakraborty, S. and Newton, A.C. (2011). Climate change, plant diseases and food security: An overview. *Plant Pathology*, 60: 2–14.

Chankhamhaengdecha, S., Hongvijit, S., Srichaisupakit, A., Charnchai, P. and Panbangred, W. (2013). Endophytic actinomycetes: A novel source of potential acyl homoserine lactone degrading enzymes. *BioMed Research International*, 782847.

Chen, P., Zhang, C., Ju, X., Xiong, Y., Xing, K. and Qin, S. (2019). Community composition and metabolic potential of endophytic actinobacteria from coastal salt marsh plants in Jiangsu, China. *Frontiers in Microbiology*, 10: 1063.

Chukwuneme, C.F., Babalola, O.O., Kutu, F.R. and Ojuederie, O.B. (2020). Characterization of actinomycetes isolates for plant growth promoting traits and their effects on drought tolerance in maize. *Journal of Plant Interactions*, 15(1): 93–105.

Corral, P., Amoozegar, M.A. and Ventosa. A. (2020). Halophiles and their biomolecules: Recent advances and future applications in biomedicine. *Marine Drugs*, 18: 33.

Costa, D.J.S., Kothe, E., Abate, C.M. and Amoroso, M.J. (2012). Unraveling the *Amycolatopsis tucumanensis* copper-resistome. *Biometals*, 25: 905–917.

Daram, D. and Tsetseg, B. (2001). Distribution of streptomycetes in soils of the Trans-Altai and Dzungarian Gobi. *Proceedings of the Institute of Biology, MAS*. Ulaanbaatar, Mongolia, 23: 181–184 (in Mongolian).

Daram, D. and Tsetseg, B. (2014). Antimicrobial activity and production of L-asparaginase by actinomycetes isolated from the forest-steppe soils. pp. 150–157. *In*: Puntsag, T. (ed.). *Microorganisms in Mongolia – 2014*. Proceedings of the Scientific Conference Dedicated to the 90th Anniversary of Academician and the 40th Anniversary of the Division of Microbiology, IB-MAS. December 24, 2014. Ulaanbaatar, Mongolia (in Mongolian).

DasSarma, S. (2006). Extreme halophiles are models for astrobiology. *Microbe*, 1: 120–126.

Davaadorj, B., Terekhova, L.P., Tsetseg, B., Laiko, A.V. and Puntsag, T. (1993). *Streptomyces phaeofaciens*-51, a producer of a new antibiotic belonging to the aureolic acid group. *Antibiotics and Chemotherapy. Moscow, Russia*, 38(6): 11–14 (in Russian).

Deepthi, M.K. and Devamma, M.N. (2012). L-Asparaginase activity of *Pseudonocardia* sp. isolated from mangrove soils along the east coast of Southern India. *Current Biotica*, 6(2): 189–197.

Demain, A. (2007). The business of biotechnology. *Industrial Biotechnology*, 3(3): 269–283.

De Kievit, T.R. and Iglewski, B.H. (2000). Bacterial quorum sensing in pathogenic relationships. *Infection and Immunity*, 68: 4839–4849.

Doroghazi, J.R. and Metcalf, W.W. (2013). Comparative genomics of actinomycetes with a focus on natural product biosynthetic genes. *BMC Genomics*, 14: 611.

Dugarjav, J. (1977). Chromatographic study of antibiotics produced by antagonistic actinomycetes 22, 78, 142 and 137. *Proceedings of the Institute of Natural Compounds*. Ulaanbaatar, Mongolia, 3: 110–118 (in Mongolian).

Egorov, A.N. (1993). Mongolian salt lakes: Some features of their geography, thermal patterns, chemistry and biology. *Hydrobiologia*, 267: 13–21.

El Baz, S., Baz, M., Barakate, M., Hassani, L., El Gharmali, A. and Imziln, B. (2015). Resistance to and accumulation of heavy metals by actinobacteria isolated from abandoned mining areas. *The Scientific World Journal* Article ID 761834, 14 pp.

Enkh-Amgalan, J., Kawasaki, H., Tsetseg, B. and Seki, T. (1998). Molecular diversity of actinomycetes from Mongolian soil. *Annual Report of ICBiotech. International Center for Biotechnology, Osaka University*, 21: 223–237.

Enkh-Amgalan, J., Kawasaki, H., Tsetseg, B. and Seki, T. (1999). Rare actinomycetes isolated from Mongolian Gobi Desert soils. pp. 89–94. *In*: Cheng-Lin Jiang and Li-Hua Xu (eds.). *Collected Papers. First International Conference on Biology of Actinomycetes under Extreme Environments*. August 2–3, 1999. Kunming, China.

Enkh-Amgalan, J., Komaki, H., Daram, D., Ando, K. and Tsetseg, B. (2012). Diversity of nonribosomal peptide synthetase and polyketide synthase genes in the genus *Actinoplanes* found in Mongolia. *Journal of Antibiotics*, 65: 103–108.

Fedorova, G.B., Katrukha, G.S., Arkhangel'skaya, N.M., Laiko, A.V., Sumarukova, I.G., Tokareva, N.A. et al. (1993). A novel antibiotic of the aureolic acid group: Isolation, identification, physicochemical and biological properties. *Antibiotics and Chemotherapy. Moscow, Russia*, 38(7): 7–10 (in Russian).

Fisch, K.M. (2013). Biosynthesis of natural products by microbial iterative hybrid PKS–NRPS. *RSC Advances*, 3: 18228–18247.

Genilloud, O. (2017). Actinomycetes: Still a source of novel antibiotics. *Natural Products Reports*, 34: 1203–1232.

Genilloud, O. (2018). Mining actinomycetes for novel antibiotics in the omics era: Are we ready to exploit this new paradigm? *Antibiotics*, 7: 85.

Golinska, P., Wypij, M., Agarkar, G., Rathod, D., Dahm, H. and Rai, M. (2015). Endophytic actinobacteria of medicinal plants: Diversity and bioactivity. *Antonie van Leeuwenhoek*, 108: 267–89.

Gong, C.S., Chen, L.F. and Tsao, G.T. (1980). Purification and properties of glucose isomerase of *Actinoplanes missouriensis*. *Biotechnology & Bioengineering*, 22: 833–845.

Goodfellow, M., Kämpfer, P., Busse, H-J., Trujillo, M.E., Suzuki, K-i., Ludwig, W. et al. (eds.). (2012). The *Actinobacteria*. Part A and B. *In*: *Bergey's Manual of Systematic Bacteriology*. 2nd edn. vol 5. Springer. New York.

Goodfellow, M. and Fiedler, H.P. (2010). A guide to successful bioprospecting: Informed by actinobacterial systematics. *Antonie van Leeuwenhoek*, 98: 119–142.

Goodfellow, M., Nouioui, I., Sanderson, R., Xie, F. and Bull, A.T. (2018). Rare taxa and dark microbial matter: Novel bioactive actinobacteria abound in Atacama Desert soils. *Antonie van Leeuwenhoek*, 111: 1315–1332.

Gunin, P.D., Vostokova, E.A., Bazha, S.N., Dugarzhav, Ch., Ulziikhutag, N. and Prischepa, A.V. (2007). Ecosystems of Mongolia: Diversity, present state and conservation. pp. 4–21. *In*: Gunin, P.D. and Drobyshev, Yu.I. (eds.). *Ecosystems of Inner Asia: Issues of Research and Conservation*. Moscow (in Russian).

Gupta, R.S. (2019). Commentary: Genome-based taxonomic classification of the phylum *Actinobacteria*. *Frontiers in Microbiology*, 10: 206.

Hamedi, J., Dehhaghi, M. and Mohammdipanah, F. (2015). Isolation of extremely heavy metal resistant strains of rare actinomycetes from high metal content soils in Iran. *The International Journal of Environmental Research*, 9(2): 475–480.

Hayakawa, M., Yamamura, H., Sakuraki, Y., Ishida, Y., Hamada, M., Otoguro, M. et al. (2010). Diversity analysis of actinomycetes assemblages isolated from soils in cool-temperate and subtropical areas of Japan. *Actinomycetologica*, 24: 1–11.

He, S.-T., Zhi, X.-Y., Jiang, H., Yang, L.-L., Wu, J.-Y., Zhang, Y.-G. et al. (2015). Biogeography of *Nocardiopsis* strains from hypersaline environments of Yunnan and Xinjiang Provinces, western China. *Scientific Reports*, 5: 13323.

Hop, D.V., Sakiyama, Y., Binh, C.T.T., Otoguro, M., Hang, D.T., Miyadoh, S. et al. (2011). Taxonomic and ecological studies of actinomycetes from Vietnam: Isolation and genus-level diversity. *Journal of Antibiotics*, 64: 599–606.

Hoskisson, P.A. and Seipke, R.F. (2020). Cryptic or silent? The known unknowns, unknown knowns, and unknown unknowns of secondary metabolism. *mBio*, 11: e02642-20.

Huang, Y., Wang, L., Lu, Zh., Hong, L., Liu, Zh., Tan, G.Y.A. et al. (2002). Proposal to combine the genera *Actinobispora* and *Pseudonocardia* in an emended genus *Pseudonocardia*, and description of *Pseudonocardia zijingensis* sp. nov. *International Journal of Systematic and Evolutionary Microbiology*, 52: 977–982.

Ibrahim, A.H., Desoukey, S.Y., Fouad, M.A., Kamel, M.S., Gulder, T.A.M. and Abdelmohsen, U.R. (2018). Natural product potential of the genus *Nocardiopsis*. *Marine Drugs*, 16: 147.

Igarashi, Y., Iida, T., Sasaki, T., Saito, N., Yoshida, R. and Furumai, T. (2002). Isolation of actinomycetes from live plants and evaluation of antiphytopathogenic activity of their metabolites. *Actinomycetologica*, 16: 9–13.

Itgel, O. (1995). *Control of Potato Diseases during Storage*. Ph.D. Dissertation Agricultural University. Ulaanbaatar, Mongolia (in Mongolian).

Ivanov, K., Stoimenova, A., Obreshkova, D. and Saso, L. (2013). Biotechnology in the production of pharmaceutical industry ingredients: Amino acids. *Biotechnology & Biotechnological Equipment*, 27: 2.

Janso, J.E. and Carter, G.T. (2010). Biosynthetic potential of phylogenetically unique endophytic actinomycetes from tropical plants. *Applied and Environmental Microbiology*, 76: 4377–4386.

Jensen, P.R., Williams, P.G., Oh, D.C., Zeigler, L. and Fenical, W. (2007). Species-specific secondary metabolite production in marine actinomycetes of the genus *Salinispora*. *Applied and Environmental Microbiology*, 73: 1146–1152.

Jiang, Y., Li, G., Li, Q., Zhang, K., Jiang, L., Chen, X. et al. (2019). Diversity and some bioactivities of soil actinomycetes from southwestern China. *BioRxiv* preprint. Doi; https://doi.org/10.1101/692814, July 4, 2019.

Jose, P.A. and Jebakumar, S.R.D. (2014). Unexplored hypersaline habitats are sources of novel actinomycetes. *Frontiers in Microbiology*, 5: Article 242.

Ju, X., Cao, X., Sun, Y., Wang, Z., Cao, C., Liu, J. et al. (2012). Purification and characterization of a fibrinolytic enzyme from *Streptomyces* sp. XZNUM 00004. *World Journal of Microbiology and Biotechnology*, 28: 2479–286.

Juboi, H., Basik, A.A., Shamsul, S.A.G., Arnold, Ph., Schmitt, E.K., Sanglier, J.-J. et al. (2015). *Luteipulveratus halotolerans* sp. nov., an actinobacterium (*Dermacoccaceae*) from forest soil. *International Journal of Systematic and Evolutionary Microbiology*, 65: 4113–4120.

Khamna, S., Yokota, A. and Lumyong, S. (2009). L-asparaginase production by actinomycetes isolated from some Thai medicinal plant rhizosphere soils. *International Journal of Integrative Biology (IJIB)*, 6(1): 22–26.

Kim, D., Choi, R.Y., Yoo, M., Zyletra, C. and Kim, C. (2018). Biotechnological potential of *Rhodococcus* biodegratative pathways. *Journal of Microbiology and Biotechnology*, 28(7): 1037–1051.

Kim, S.M., Lee, S.D., Koh, Y.S. and Kim, I.S. (2022). *Antrihabitans stalagmiti* sp. nov., isolated from a larva cave and a proposal to transfer *Rhodococcus cavernicola* Lee et al. 2020 to a new genus *Spelaeibacter* as *Spelaeibacter cavernicola* gen. nov. comb. nov. *Antonie van Leeuwenhoek*, 115: 521–532.

Kinoshita, S., Udaka, S. and Shimono, M. (1957). Amino acid fermentation: 1. Production of L-glutamic acid by various microorganisms. *The Journal of General and Applied Microbiology*, 3: 193–205.

Kitagawa, W. and Tamura, T. (2008). Three types of antibiotics produced from *Rhodococcus erythropolis* strains. *Microbes and Environments*, 23: 167–171.

Komaki, H., Ichikawa, N., Oguchi, A., Hamada, M., Tamura, T. and Fugita. N. (2015). Genome-based analysis of non-ribosomal peptide synthetase and type-I polyketide synthase gene clusters in all type strains of the genus *Herbidospora. BMC Research Notes*, 8: 548.

Krasilnikov, N.A. (1958). Microorganisms of soil and higher plants. *USSR Academy of Sciences Publishing House*, Moscow, 464 pp.

Kudo, T., Itoh, T., Miyadoh, S., Shomura, T. and Seino, A. (1993). *Herbidospora* gen. nov., a new genus of the family *Streptosporangiaceae* Goodfellow et al. *International Journal of Systematic and Evolutionary Microbiology*, 43(2): 319–28.

Kurapova, I., Zenova, G.M., Sudnitsyn, I.I., Kizilova, A.K., Manucharova, N.A., Norovsuren, Zh. et al. (2012). Thermotolerant and thermophilic actinomycetes from soils of Mongolia desert steppe zone. *Microbiology*, 81: 98–108.

Kurtböke, I. (2010). Biodiscovery from microbial resources: Actinomycetes leading the way. *Microbiology Australia*, 31(2): 53–56.

Lambert, D.H. and Loria, R. (1989a). *Streptomyces acidiscabies* sp. nov. *International Journal of Systematic and Evolutionary Microbiology*, 39: 393–396.

Lambert, D.H. and Loria, R. (1989b). *Streptomyces scabies* sp. nov., nom. rev. *International Journal of Systematic and Evolutionary Microbiology*, 39: 387–392.

Landau, N.S. and Egorov, N.S. (1971). Study on production of proteases with fibrinolytic activity by *Nocardia* sp. strain 1. *Mikrobiologiya*, 40(5): 829–832 (in Russian).

Landau, N.S. and Egorov, N.S. (1996). Accumulation and characterization of proteolytic enzymes from *Nocardia minima. Mikrobiologiya*, 65(1): 42–47 (in Russian).

Lazzarini, A., Cavaletti, L., Toppo, G. and Marinelli, F. (2000). Rare genera of actinomycetes as potential sources of new antibiotics. *Antonie van Leeuwenhoek*, 78: 399–405.

LeBlanc, J.C., Gonzalves, E.R. and Mohn, W.W. (2008). Global response to desiccation stress in the soil actinomycete *Rhodococcus jostii* RHA1. *Applied and Environmental Microbiology*, 74(9): 2627–2636.

Leuchtenberger, W., Huthmacher, K. and Drauz, K. (2005). Biotechnological production of amino acids and derivatives: current status and prospects. *Applied and Environmental Microbiology*, 69: 1–8.

Li, M., Xu, J., Yao, Z., Jiang, Y., Zhou, H., Jiang, W. et al. (2017). The anti-quorum sensing activity and bioactive substance of a marine derived *Streptomyces. Biotechnology & Biotechnological Equipment*, 31(5): 1007–1015.

Liu, D., Zhao, W. and Xiao, K. (2004). *Overview of Potato Scab in China*. University of Hebei, China, 182–189.

Liu, S., Wang, T., Lu, Q., Li, F., Wu, G., Jiang, Z. et al. (2021). Bioprospecting of soil-derived actinobacteria along the Alar-Hotan desert highway in the Taklamakan desert. *Frontiers in Microbiology*, 12: 604999.

Martin, D.K., Vicente, O., Beccari, T., Kellermayer, M., Koller, M., Lal, R. et al. (2021). A brief overview of global biotechnology. *Biotechnology & Biotechnological Equipment*, 35(sup1): 5–14.

Mathers, C.D. and Loncar, D. (2006). Projections of global mortality and burden of disease from 2002 to 2030. *PLoS Medicine*, 3(11): 442.

Matsumoto, A. and Takahashi, Y. (2017). Endophytic actinomycetes: Promising source of novel bioactive compounds. *Journal of Antibiotics*, 70: 514–519.

McGinnies, W.G., Goldman, B.J. and Paylore, P. (eds.). (1968). *Deserts of the World*. Tucson, Univ. Arizona, 188 pp.

Meier-Kolthoff, J.P., Göker, M., Spröer, C. and Klenk, H.P. (2013). When should a DDH experiment be mandatory in microbial taxonomy? *Archives of Microbiology*, 195: 413–418.

Meklat, A., Sabaou, N., Zitouni, A., Mathieu, F. and Lebrihi, A. (2011). Isolation, taxonomy, and antagonistic properties of halophilic actinomycetes in Saharan soils of Algeria. *Applied and Environmental Microbiology*, Sept., 6710–6714.

Miyajima, K., Tanaka, F., Takeuchi, T. and Kuninaga, S. (1998). *Streptomyces turgidiscabies* sp. nov. *International Journal of Systematic and Evolutionary Microbiology*, 48: 495–502.

Mohammadipanah, F. and Wink, J. (2015). Actinobacteria from arid and desert habitats: Diversity and biological activity. *Frontiers in Microbiology*, 6: 1541.

Molina, L., Constantinescu, F., Michel, L., Reimmann, C., Duffy, B. and Defago, G. (2003). Degradation of pathogen quorum-sensing molecules by soil bacteria: A preventive and curative biological control mechanism. *FEMS Microbiology Ecology*, 45: 71–81.

Mukhtar, S., Zaheer, A., Aiysha, D., Malik, K.A. and Mehnaz, S. (2017). Actinomycetes: A source of industrially important enzymes. *The Journal of Proteomics & Bioinformatics*, 10: 316–319.

Muramatsu, H., Shahab, N., Tsurumi, Y. and Hino, M. (2003). A comparative study of Malaysian and Japanese actinomycetes using a simple identification method based on partial 16S rDNA sequence. *Actinomycetologica*, 17: 33–43.

Murzaev, E.M. (1952). *Mongolian People's Republic*. Physico-geographical description. Second edition. State Publishing House of Geographic Literature, Moscow, 472 pp. (in Russian).

Näsholm, T., Huss-Danell, K. and Hugberg, P. (2000). Uptake of organic nitrogen in the field by four agriculturally important plant species. *Ecology*, 81: 1155–1161.

Nett, M., Ikeda, H. and Moore, B.S. (2009). Genomic basis for natural product biosynthetic diversity in the actinomycetes. *Natural Products Reports*, 26(11): 1362–1384.

Norovsuren, Zh. and Filippova, S.N. (2020). Antagonistic properties of endophytic streptomycetes isolated from barley (*Hordeum vulgare* L.). 2020. Reports of the TSHA: collection of articles. Moscow: Publishing House of the Russian State Agrarian University-MSHA named after K.A. Timiryazev. Iss. 292. Part. IV: 76–78 (in Russian).

Norovsuren, Zh., Zenova, G.M. and Mosina, L.V. (2007a). Actinomycetes in the rhizosphere of plants of the semi-desert soils of Mongolia. *Soil Science*, 4: 457–460 (in Russian).

Norovsuren, Zh., Oborotov, G.V., Zenova, G.M., Zvyagintsev, D.G. and Aliev, R.A. (2007b). Haloalkaliphilic actinomycetes in soils of Mongolian desert steppes. *Biology Bulletin*, 34(4): 417–422.

Nouioui, I., Carro, L., Garcнa-Lypez, M., Meier-Kolthoff, J.P., Woyke, T., Kyrpides, N.C. et al. (2018). Genome-based taxonomic classification of the phylum *Actinobacteria*. *Frontiers in Microbiology*, 9: 02007.

OECD (2009). *The Bioeconomy to 2030*. Designing a Policy Agenda, 19 p. ttps://doi.org/10.1787/9789264056886-en.

Oerke, E.C. (2006). Crop losses to pests. *The Journal of Agricultural Science*, 144: 31–43.

Okoro, C.K., Brown, R., Jones, A.L., Andrews, B.A., Asenjo, J., Goodfellow, M. et al. (2009). Diversity of culturable actinomycetes in hyper-arid soils of the Atacama Desert, Chile. *Antonie van Leeuwenhoek*, 95: 121–133.

Onkhor, G., Tsetseg, B. and Puntsag, T. (1983). Distribution of actinomycetes in soils of the desert zone and their antagonistic properties. *Proceedings of the Institute of General and Experimental Biology*. Ulaanbaatar, Mongolia, 18: 73–80 (in Mongolian).

Onkhor, G., Tsetseg, B. and Puntsag, T. (1984). Composition of actinomycete species in desert soils of Mongolia. *Proceedings of the Institute of General and Experimental Biology*. Ulaanbaatar, Mongolia, 19: 168–174 (in Mongolian).

Oren, A. and Garrity, G.M. (2021). Valid publication of the names of forty-two phyla of prokaryotes. *International Journal of Systematic and Evolutionary Microbiology*, 71: 005056.

Orozco-Mosqueda, M.d.C. Flores, A., Rojas-Sánchez, B., Urtis-Flores, C.A., Morales-Cedeño, L.R., Valencia-Marin, M.F. et al. (2021). Plant growth-promoting bacteria as bioinoculants: Attributes and challenges for sustainable crop improvement. *Agronomy*, 11: 1167.

Otto, M. (2004). Quorum-sensing control in staphylococci—A target for antimicrobial drug therapy? *FEMS Microbiology Ecology*, 241: 135–141.

Oyunbileg, N. and Tsetseg, B. (2002). Resistance of soil actinomycetes isolated from mining areas to heavy metals. *Transactions of the Science and Technology University*. Ulaanbaatar, Mongolia, 3(49): 186–196 (in Mongolian).

Oyunbileg, N., Davaapurev, B.-O., Tsetseg, B., Iizaka, Y., Fukumoto, A., Kato, F. et al. (2021a). Bioactive compounds and molecular diversity of endophytic actinobacteria isolated from desert plants. *IIOP Conference Series: Earth and Environmental Science*, 908: 012008.

Oyunbileg, N., Iizaka, Y., Hamada, M., Davaapurev, B.-O., Fukumoto, A., Tsetseg, B. et al. (2021b). *Actinocatenispora comari* sp. nov., an endophytic actinomycete isolated from aerial parts of *Comarum salesowianum*. *International Journal of Systematic and Evolutionary Microbiology*, 71: 004861.

Park, D.H., Kim, J.S., Kwon, S.W., Wilson, C., Yu, Y.M., Hur, J.H. et al. (2003). *Streptomyces luridiscabiei* sp. nov., *Streptomyces puniciscabiei* sp. nov. and *Streptomyces niveiscabiei* sp. nov., which cause potato common scab disease in Korea. *International Journal of Systematic and Evolutionary Microbiology*, 53: 2049–2054.

Pavlovicha, D.Ya. (1978). *Actinomycetes of Latvia*. Distribution and Biological Characteristics. Zinatne, Riga, 195 pp. (in Russian).

Peng, Y., Yang, X. and Zhang, Y. (2005). Microbial fibrinolytic enzymes: An overview of source, production, properties, and thrombolytic activity *in vivo*. *Applied Microbiology and Biotechnology*, 69: 126–132.

Polkade, A.V., Mantri, S.S., Patwekar, U.J. and Jangid, K. (2016). Quorum sensing: An under-explored phenomenon in the phylum *Actinobacteria*. *Frontiers in Microbiology*, 7: 131.

Puntsag, T., Onkhor, G. and Tsetseg, B. (1977). Taxonomic characterization of strain 225. *Proceedings of the Institute of Natural Compounds*. Ulaanbaatar, Mongolia, 3: 71–84 (in Mongolian).

Rachev, R., Sachanekova, M., Darakchieva Tzvetkova, R., Gesheva, R., Puntzag, T. et al. (1987). Orange pigments of *Streptomyces* sp. c-72. I. Purification and antimicrobial activity. *Acta Microbiologica Bulgarica. Sofia, Bulgaria*, 21: 57–61.

Raju, E.V.N. and Divakar, G. (2014). An overview on microbial fibrinolytic proteases. *International Journal of Pharmaceutical Sciences and Research (IJPSR)*, 5(3): 643–656.

Salam, N., Jiao, J.-Y., Zhang, X.-T. and Li, W.-J. (2020). Update on the classification of higher ranks in the phylum *Actinobacteria*. *International Journal of Systematic and Evolutionary Microbiology*, 70: 1331–1355.

Sangal, V., Goodfellow, M., Jones, A.L., Schwalbe, E.C., Blom, J., Hoskisson, P.A. et al. (2016). Next-generation systematics: An innovative approach to resolve the structure of complex prokaryotic taxa. *Scientific Reports*, 6: 38392.

Sangal, V., Goodfellow, M., Jones, A.L., Soviour, R.J. and Suteliffer, R.C. (2019). Systematics of the genus *Rhodococcus* based on whole genome analyses. *In*: Alvarez, H.M. (ed.). *Biology of Rhodococcus*. *Microbiology Monographs* 16. Springer.

Sayed, A.M., Hassan, M.H.A., Alhadrami, H.A., Hassan, H.M., Goodfellow, M. and Rateb, M.E. (2019). Extreme environments: Microbiology leading to specialized metabolites. *Journal Applied Microbiology*, 1–28.

Schlatter, D.C. and Kinkel, L.L. (2014). Global biogeography of *Streptomyces* antibiotic inhibition, resistance, and resource use. *FEMS Microbiology Ecology*, 88: 386–397.

Schwientek, P., Szczepanowski, R., Rückert, C., Kalinowski, J., Klein, A., Selber, K. et al. (2012). The complete genome sequence of the acarbose producer *Actinoplanes* sp. SE50/110. *BMC Genomics*, 13: 112.

Sevastiyanov, D.V., Egorov, A.N., Rasskazov, A.A. and Luvsandorzh, N. (2014). Hydrochemistry of Mongolian lakes. pp. 72–97. *In*: Dgebuadze, Yu. (ed.). *Limnology and Palaeolimnology of Mongolia*. Moscow (in Russian).

Shrivastava, A., Khan, A.A., Khurshid, M., Kalam, M.A., Jain, S. and Singhal, P.K. (2016). Recent developments in L-asparaginase discovery and its potential as anticancer agent. *Critical Reviews in Oncology/Hematology*, 100: 1–10.

Silva, G.M., Bezerra, R.P., Teixeira, J.A., Porto, T.S., Lima-Filho, J.L. and Porto, A.F. (2015). Fibrinolytic protease production by new *Streptomyces* sp. DPUA 1576 from Amazon lichens. *Electronic Journal of Biotechnology*, 18: 16–19.

Silva, L.J., Crevelin, E.J., Souza, D.T., Lacerda-Junior, G.V., de Oliveira, V.M., Ruiz, A.L.T.G. et al. (2020). Actinobacteria from Antarctica as a source for anticancer discovery. *Scientific Reports*, 10: 13870.

Singh, R. and Dubey, A.K. (2018). Diversity and applications of endophytic actinobacteria of plants in special and other ecological niches. *Frontiers in Microbiology*, 9: 1767.

Skalon, I.S. (1977). Species composition and peculiarities of microflora development of plant communities in zonal stations of the MPR. *In*: *Biological Resources and Natural Conditions of the Mongolian People's Republic*. Vol. II. Flora and fauna of Mongolia. Leningrad, Nauka, 157–192 (in Russian).

Somma, S., Gastaldo, L. and Corti, A. (1984). Teicoplanin, a new antibiotic from *Actinoplanes teichomyceticus* nov. sp. *Antimicrobial Agents and Chemotherapy*, 26: 917–923.

Sorokin, D.Y., van Pelt, S., Tourova, T.P. and Evtushenko, L.I. (2009). *Nitriliruptor alkaliphilus* gen. nov., sp. nov., a deep lineage haloalkaliphilic actinobacterium from soda lakes capable of growth on aliphatic nitriles, and proposal of *Nitriliruptoraceae* fam. nov. and *Nitriliruptorales* ord. nov. *International Journal of Systematic and Evolutionary Microbiology*, 59: 248–253.

Sorokin, D.Y., Khijniak, T.V., Zakharycheva, A.P., Elcheninov, A.G., Hahnke, R.L., Boueva, O.V. et al. (2021). *Natronoglycomyces albus* gen. nov., sp. nov., a haloalkaliphilic actinobacterium from a soda solonchak soil. *International Journal of Systematic and Evolutionary Microbiology*, 71: 004804.

Strange, R.N. and Scott, P.R. (2005). Plant disease: A threat to global food security. *Annual Review of Phytopathology*, 43: 83–116.

Subramani, R. and Aalbersberg, W.A. (2012). Marine actinomycetes: An ongoing source of novel bioactive metabolites. *Microbiology Research*, 167(10): 571–580.

Taechowisan, Th., Chanaphat, S., Ruensamran, W. and Phutdhawong, W.S. (2014). Antibacterial activity of new flavonoids from *Streptomyces* sp. BT01: An endophyte in *Boesenbergia rotunda* (L.) Mansf. *Journal of Applied Pharmaceutical Research*, 4(04): 008–013.

Taechowisan, Th., Chaisaeng, S. and Phutdhawong, W.S. (2017). Antibacterial, antioxidant and anticancer activities of biphenyls from *Streptomyces* sp. BO-07: An endophyte in *Boesenbergia rotunda* (L.) Mansf A. *Food and Agricultural Immunology*, 28(6)1: 330–1346.

Takahashi, Y. and Nakashima, T. (2018). Actinomycetes, an inexhaustible source of naturally occurring antibiotics. *Antibiotics*, 7: 45.

Tsegmid, Sh. (1962). Physico-geographical regionalization of Mongolian People's Republic. *Proceedings of the Academy of Sciences of the USSR, Series Geography*, 5: 34–41 (in Russian).

Tseng, M., Yang, S.F. and Yuan, G.F. (2010). *Herbidospora yilanensis* sp. nov. and *Herbidospora daliensis* sp. nov., from sediment. *nternational Journal of Systematic and Evolutionary Microbiology*, 60(Pt 5): 1168–72.

Tsetseg, B. (1973). Actinomycetes with antagonistic properties in the rhizosphere of some plants of the Gobi Desert. *Proceedings of the Institute of Biology*. Ulaanbaatar, Mongolia, 8: 73–78 (in Russian).

Tsetseg, B. (1974). Toxicity of culture fluids of some actinomycetes. *Science Transactions NUM. Biology.* Ulaanbaatar, Mongolia, 49: 75–77. (in Russian).

Tsetseg, B. (2003). Chapter 9. Unique microbiota of Mongolia. pp. 196–221. *In*: Badarch, D., Zilinskas, R.A. and Balint, P.J. (eds.). *Mongolia Today: Science, Culture, Environment and Development*. RoutledgeCurzon, London.

Tsetseg, B. (2008). Plant-microbe interactions: Actinomycetes in plant rhizosphere of Mongolian dry steppe ecosystem. pp. 145–149. *In*: Boldgiv, B. (ed.). *Fundamental and Applied Issues of Ecology and Evolutionary Biology. Proceedings of the International Conference*. Ulaanbaatar, Mongolia, 25 April 2008.

Tsetseg, B. (2010). Antimicrobial activity and taxonomic characteristics of *Streptomyces levoris*-605, a producer of polyene antibiotic. pp. 5–9. *In*: *Modern Problems of Microbiology of Central Asia*. Materials of the Conference with International Participation. Ulan-Ude, 27–28 May 2010. Ulan-Ude, Buryat State University Publishing House (in Russian).

Tsetseg, B. (2012). Production of extracellular amino acids by soil actinomycetes of different natural zones of Mongolia. *Sci. Transact. NUM. Biology*. Ulaanbaatar, Mongolia, 374: 73–82 (in Russian).

Tsetseg, B. (2018). Diversity and genetic resources of Mongolian microorganisms, their application in biotechnology and future prospects. pp. 40–51. *In*: Batjargal, B., Dugersuren, Zh., Oyunbileg, Yu. and Tsetseg, B. (eds.). *Development of Biotechnology and Application of Genetic Resources*. Collection of Papers of the Theoretical and Practical Conference. Ulaanbaatar. NUM Press Centre (in Mongolian).

Tsetseg, B. and Ando, K. (2013). Taxonomic and ecological studies of microorganisms in Mongolia and the utilization. *Report on the Joint Project*, 308 pp.

Tsetseg, B. and Badamgarav, T.-U. (2021). Phylogenetic diversity of actinobacteria of the genus *Kribbella* and their potential for production of extracellular enzymes. pp. 603–605. *In*: *Diversity of Soils and Biota of Northern and Central Asia*. The IV All-Russian Conference with International Participation. 15–18 June 2021. Ulan-Ude, Russia.

Tsetseg, B. and Nyamsuren, G. (2013). A new pathogenic streptomycete isolated from potato scab lesions in Mongolia. pp. 248–250. *In: Agricultural Science to Agricultural Production in Mongolia, Siberian Region of RF, Kazakhstan and Bulgaria.* XVI International Conference 29–30 May 2013. Agricultural University. UB, Mongolia. Part I.

Tsetseg, B., Ganbaatar, Ts. and Puntsag, T. (1999). Evaluation of actinomycetes isolated from the Gobi Desert soils as producers of extracellular enzymes. pp. 95–101. *In:* Jiang, C.L. and Xu, L.H. (eds.). *Biology of Actinomycetes under Extreme Environments.* Collected Papers of the First International Conference. 2–3 August 1999. Kunming, China.

Tsetseg, B., Daram, D., Onkhor, G. and Puntsag, T. (1994). Distribution of streptomycetes in Mongolian Gobi Desert soils and their biological activity. p. 229. *In: Theses of the 9th International Symposium on the Biology of Actinomycetes.* July 10–15, 1994. Moscow, Russia.

Tsetseg, B., Enkh-Amgalan, J. and Ando, K. (2016a). Taxonomic and ecological studies of microorganisms in Mongolia and utilization. *Report on the Joint Project*, 366 pp.

Tsetseg, B., Kudo, T., Daram, D. and Enkh-Amgalan, J. (1997). Preliminary results on isolation of actinomycetes of rare genera from Mongolian soils. *In: Abstracts of the 10th International Symposium on Biology of Actinomycetes.* May 27–30, 1997, Beijing: poster 7P24.

Tsetseg, B., Landau, N.S. and Egorov, N.S. (2016b). Isolation and characterization of proteases with fibrinolytic and thrombolytic activities produced by *Nocardia* sp. strain 2. p. 18. *In: Chemical Investigation and Utilization of Natural Resources.* Book of Abstracts of the Fourth International Conference. 8–10 July 2016. Ulaanbaatar, Mongolia.

Tsetseg, B., Nyamsuren, G. and Gantuya, T. (2008). Antagonistic and plant growth promotion activities of soil actinomycetes isolated from potato fields in Mongolia. p. 80. *In: Biological Control of Bacterial Plant Diseases.* The 2nd International Symposium. 4–7 November 2008, Regal Sun Resort, Lake Buena Vista (Orlando), Florida, USA.

Tsetseg, B., Suto, M., Marjangul, N., Yasuta, T., Ara, I., Yamamura, H. et al. (2011). Taxonomic diversity of bacteria and archaea isolated from Mongolia. *International Conference: Ecology and Geochemical Activity of Microorganisms in Extreme Environments.* Ulan-Ude-Ulaanbaatar. September 5–16, 2011: 180–183.

Tsukiura, H., Okanishi, M., Koshiyama, H., Ohmori, T., Miyaki, T. and Kawaguchi, H. (1964). *Proceomycin*, a new antibiotic. *Journal of Antibiotics (Tokyo)*, 17: 223–229.

Tumurbaatar, B. and Mijiddorj, B. (2006). Permafrost and permafrost thaw in Mongolia. pp. 41–48. *In:* Goulden, C.E., Sitnikova, T., Gelhaus, J. and Boldgiv, B. (eds.). *The Geology, Biodiversity and Ecology of Lake Huvsgul (Mongolia).* Backhuys Publishers, Leiden, The Netherlands.

Umarov, K.U. and Yakunin, G.N. (1974). Characterization of brown desert-steppe soils of Bulgan station. pp. 11–25. *In: Structure and Dynamics of the Steppe and Desert Ecosystems of Mongolian People's Republic.* Leningrad (in Russian).

UNDP. (2012). The future we want: Biodiversity and ecosystems—driving sustainable development. *United Nations Development Programme Biodiversity and Ecosystems Global Framework* 2012–2020. New York, 69 pp.

Urgamal, M., Gundegmaa, V., Baasanmunkh, Sh., Oyuntsetseg, B., Darikhand, D. and Munkh-Erdene, T. (2019). Additions to the vascular flora of Mongolia - IV. *Proceedings of the Mongolian Academy of Sciences*, 59(2): 41–53.

Varalakshmi, V. (2013). L-Asparaginase: An enzyme of medicinal value. *International Journal of Green and Herbal Chemistry (IJGHC)*, 2(3): 544–554.

Vasilchenko, L.G., Ariunaa, J., Tsetseg, B., Pristavka, A.A. and Rabinovich, M.L. (1996). Comparison of neutral and alkaline cellulase activity of selected fungi and streptomycetes. pp. 1526–1533. *In: Biomass for Energy and the Environment.* Proceedings of the 9th European Bioenergy Conference. V. 3. June 24–27, 1996. Copenhagen, Denmark.

Verma, N., Kumar, K., Kaur, G. and Anand, S. (2007). L-asparaginase: A promising chemotherapeutic agent. *Critical Reviews in Biotechnology*, 27: 45–62.

Vostokova, E.A., Gunin, P.D., Rachkovskaya, E.I. et al. (1995). Ecosystems of Mongolia: Distribution and present condition. Biological resources and natural conditions of Mongolia. *Proceedings of the Joint Russian-Mongolian Complex Biological Expedition*, vol. 39. "Nauka" Publishing House, Moscow, 223 pp (in Russian).

Wang, Y., Zhang, J.S., Ruan, J.S., Wang, Y.M. and Ali, S.M. (1999). Investigation of actinomycete diversity in the tropical rainforests of Singapore. *Journal of Industrial Microbiology and Biotechnology*, 23: 178–187.

Watanabe, K., Makino, H., Sasamoto, M., Kudo, Y., Fujimoto, J. and Demberel, Sh. (2009). *Bifidobacterium mongoliense* sp. nov., from airag, a traditional fermented mare's milk product from Mongolia. *nternational Journal of Systematic and Evolutionary Microbiology*, 59: 1535–1540.

WHO. (2015). *Global Action Plan on Antimicrobial Resistance*; World Health Organization: Geneva, Switzerland.

Widyastuti, Y. and Ando, K. (2010). Taxonomic and ecological studies of fungi and actinomycetes in Indonesia. *Report on the Joint Project*, 601–602.

Xie, F. and Pathom-aree, W. (2021). Actinobacteria from desert: Diversity and biotechnological applications. *Frontiers in Microbiology*, 12: 765531.

Yadav, N. and Yadav, A.N. (2019). Actinobacteria for sustainable agriculture. *Journal of Applied Biotechnology and Bioengineering*, 6(1): 38–41.

Yang, R., Zhang, L.P., Guo, L.G., Shi, N., Lu, Z. and Zhang, X. (2008). *Nocardiopsis valliformis* sp. nov., an alkaliphilic actinomycete isolated from alkali lake soil in China. *International Journal of Systematic and Evolutionary Microbiology*, 58: 1542–1546.

Yin, M., Li, G., Jiang, Y., Han, L., Huang, X., Lu, T. et al. (2017). The complete genome sequence of *Streptomyces albolongus* YIM 101047, the producer of novel bafilomycins and odoriferous sesquiterpenoids. *Journal of Biotechnology*, 262: 89–93.

Zavarzin, G.A. (1993). Epicontinental soda lakes as probable relict biotops or terrestrial biota formation. *Mikrobiologiya*, 62(5): 789–800.

Zhirnov, L.V., Gunin, P.D., Adiya, Ya. and Bazha, S.N. (2005). *Strategy of Conservation of Ungulates in Arid Zones of Mongolia*. Moscow, 323 pp. (in Russian).

Zhou, Z., Hashimoto, Y. and Kobayashi, M. (2005). Nitrile degradation by *Rhodococcus*: Useful microbial metabolism for industrial productions. *Actinomycetologica*, 19: 18–26.

Zhu, J., Miller, M.B., Vance, R.E., Dziejman, M., Bassler, B.L. and Mekalanos, J.J. (2002). Quorum-sensing regulators control virulence gene expression in *Vibrio cholerae*. *Proceedings of the National Academy of Sciences, USA*, 99: 3129–34.

# Index

# Biography

**D. İpek Kurtböke**

Dr. Kurtböke's experiences in the field of biodiscovery with actinomycetes date back to 1982 when she was first involved in the large-scale production of antibiotic gentamicin in Eczacıbaşı İlaç A.Ş. in İstanbul, Türkiye. Subsequently she was at the University of Milan in Italy (1983–86) for graduate research followed by a Ph.D. at the University of Liverpool, UK (1990). Her most significant contribution has been the development of a novel isolation technique that selectively cultures rare actinomycetes with industrial importance which was adopted and applied by leading pharmaceutical companies since the 1990s. Since taking up her first post-doctoral position at the University of Western Australia (1990), she has established bio-resource libraries for joint screening ventures with leading pharmaceutical companies in different settings in Australia. She was one of the key scientists involved in the establishment of AMRAD Discovery Technologies Pty. Ltd.'s Australia's largest bioresource library in Melbourne (1995–2000). Since 2001 she has been at the University of the Sunshine Coast (UniSC) in Queensland, Australia, currently Associate Professor and teaching and conducting research in the fields of applied, industrial and environmental microbiology.

Dr. Kurtböke's methodological strength in the field of actinomycetology played a key role in the detection of novel actinomycetes and contributed towards the establishment of yet another microbial library of bioactive actinomycetes at the UniSC. The library has been used for research and teaching activities at the UniSC as well as in collaborative partnership with regional, national, and international institutions for the discovery of new drugs, agro-biologicals, enzymes, and environmentally friendly biotechnological innovations. She has been an Executive Board Member of the World Federation of Culture Collections (WFCC) since 2000, currently serving her second term as the President of the Federation (2017–2020, 2021–2024). She is also one of the members of the International Committee on Taxonomy of Viruses (ICTV), Bacterial Viruses Subcommittee. She has editorial duties in different journals including *Marine Drugs, Diversity and Frontiers Marine Science/Marine Biotechnology.*

She is also the editor of the books titled: ***Bacteriophages*** (2012, InTech), ***Microbial Resources-From Functional Existence in Nature to Industrial Applications*** (2017, Academic Press, Elsevier) and ***Importance of Microbiology Teaching and Microbial Resource Management for Sustainable Futures*** (2022, Academic Press, Elsevier) and all of which brought experts in the fields of microbial ecology, taxonomy, culture collections and industrial microbiology together to highlight the importance of diverse microbial resources for global sustainability and biotechnological innovations.

For Product Safety Concerns and Information please contact our EU
representative GPSR@taylorandfrancis.com
Taylor & Francis Verlag GmbH, Kaufingerstraße 24, 80331 München, Germany

www.ingramcontent.com/pod-product-compliance
Lightning Source LLC
Chambersburg PA
CBHW060814220326
41598CB00022B/2614

*9 7 8 1 0 3 2 5 2 0 5 7 5*